T0180796

SPRINGER HANDBOOK OF AUDITORY RESEARCH

Series Editors: Richard R. Fay and Arthur N. Popper

SPRINGER HANDBOOK OF AUDITORY RESEARCH

Volume 1: The Mammalian Auditory Pathway: Neuroanatomy
Edited by Douglas B. Webster, Arthur N. Popper, and Richard R. Fay

Volume 2: The Mammalian Auditory Pathway: Neurophysiology
Edited by Arthur N. Popper and Richard R. Fay

Volume 3: Human Psychophysics
Edited by William Yost, Arthur N. Popper, and Richard R. Fay

Volume 4: Comparative Hearing: Mammals
Edited by Richard R. Fay and Arthur N. Popper

Volume 5: Hearing by Bats
Edited by Arthur N. Popper and Richard R. Fay

Volume 6: Auditory Computation
Edited by Harold L. Hawkins, Theresa A. McMullen, Arthur N. Popper, and Richard R. Fay

Volume 7: Clinical Aspects of Hearing
Edited by Thomas R. Van de Water, Arthur N. Popper, and Richard R. Fay

---

Forthcoming Volumes (partial list)

Development of the Auditory System
Edited by Edwin Rubel, Arthur N. Popper, and Richard R. Fay

The Cochlea
Edited by Peter Dallos, Arthur N. Popper, and Richard R. Fay

Plasticity in the Auditory System
Edited by Edwin Rubel, Arthur N. Popper, and Richard R. Fay

Douglas B. Webster
Arthur N. Popper
Richard R. Fay

Editors

The Mammalian Auditory Pathway: Neuroanatomy

With 110 Illustrations

Springer-Verlag

New York Berlin Heidelberg London Paris
Tokyo Hong Kong Barcelona Budapest

Douglas B. Webster
Kresge Hearing Research Laboratory
    of the South
Department of Otorhinolaryngology
Louisiana State University
New Orleans, LA 70112-2234

Arthur N. Popper
Department of Zoology
University of Maryland
College Park, MD 20742

Richard R. Fay
Parmly Hearing Institute and
Department of Psychology
Loyola University of Chicago
Chicago, IL 60626

Series Editors: Richard R. Fay and Arthur N. Popper

Cover illustration: Golgi-impregnated neurons from the medial geniculate body of a cat. The form and distribution can be related to the neurochemical circuitry of the auditory thalamus. This is a detail from Figure 6.4A, p. 239.

Library of Congress Cataloging-in-Publication Data
The mammalian auditory pathway: Neuroanatomy / Douglas B. Webster, Arthur N. Popper, Richard R. Fay, editors.
    p.   cm. — (Springer handbook of auditory research; v. 1)
    Includes bibliographical references and index.
    1. Auditory pathways—Anatomy. 2. Neuroanatomy. I. Webster, Douglas B., 1934– . II. Popper, Arthur N. III. Fay, Richard R. IV. Series.
    [DNLM: 1. Anatomy, Comparative. 2. Auditory Pathways—anatomy & histology. WV 272 N4932]
    QM451.N48 1991
    599'.048—dc20
    DNLM/DLC   91-5016
    for Library of Congress   CIP

Printed on acid-free paper.

Production managed by Terry Kornak; manufacturing supervised by Jacqui Ashri

Typeset by Impressions, Madison, Wisconsin, a division of Edwards Brothers, Inc.
Printed and bound by Edwards Brothers, Inc., Ann Arbor, Michigan
Printed in the United States of America.

9 8 7 6 5 4 3 2

ISBN 0-387-97678-7 Springer-Verlag New York Berlin Heidelberg (hardcover)
ISBN 3-540-97678-7 Springer-Verlag Berlin Heidelberg New York (hardcover)
ISBN 0-387-97800-3 Springer-Verlag New York Berlin Heidelberg (softcover)
ISBN 3-540-97800-3 Springer-Verlag Berlin Heidelberg New York (softcover)

Series Preface

The *Springer Handbook of Auditory Research* presents a series of comprehensive and synthetic reviews of the fundamental topics in modern auditory research. It is aimed at all individuals with interests in hearing research including advanced graduate students, postdoctoral researchers, and clinical investigators. The volumes will introduce new investigators to important aspects of hearing science and will help established investigators to better understand the fundamental theories and data in fields of hearing that they may not normally follow closely.

Each volume is intended to present a particular topic comprehensively, and each chapter will serve as a synthetic overview and guide to the literature. As such, the chapters present neither exhaustive data reviews nor original research that has not yet appeared in peer-reviewed journals. The series focusses on topics that have developed a solid data and conceptual foundation rather than on those for which a literature is only beginning to develop. New research areas will be covered on a timely basis in the series as they begin to mature.

Each volume in the series consists of five to eight substantial chapters on a particular topic. In some cases, the topics will be ones of traditional interest for which there is a solid body of data and theory, such as auditory neuroanatomy (Vol. 1) and neurophysiology (Vol. 2). Other volumes in the series will deal with topics which have begun to mature more recently, such as development, plasticity, and computational models of neural processing. In many cases, the series editors will be joined by a co-editor having special expertise in the topic of the volume.

Richard R. Fay
Arthur N. Popper

Preface

When students or colleagues have asked us where they could learn about the organization of the mammalian central auditory system, we have had no good place to send them. Neuroanatomy texts treat the subject superficially, and often incorrectly. Review articles assume a previous sophisticated knowledge. The original literature is too specialized for the novice. We saw a need for a different source of information.

This volume is meant to answer that need. It is intended for basic researchers, graduate and postdoctoral students, and for our colleagues from other fields who wish to learn about auditory neuroanatomy—including basic scientists as well as audiologists, otologists, and speech-language pathologists.

Chapter 1 presents an overview of the entire neural auditory system, from the periphery through the auditory cortex, with an emphasis on the human system. The following six chapters treat each region of the auditory pathway in detail. Each chapter is authored by an expert who not only understands the specialty but can write about it clearly for the non-specialist.

This volume is the first in the series; it presents the structural basis for what will follow. The companion Volume 2 has parallel chapters presenting the neurophysiology of the mammalian auditory system. Succeeding volumes will deal specifically with the cochlea, human and animal psychoacoustics, development, and the auditory neurophysiology and neuroanatomy of nonmammalian vertebrates.

As editors we are indebted to the chapter authors for their time and effort, and to the staff of Springer-Verlag for their enthusiastic cooperation.

Douglas B. Webster
Arthur N. Popper
Richard R. Fay

Contents

Contributors

Nell Beatty Cant
Department of Neurobiology, Duke University Medical Center, Durham, NC 27710, USA

Michael F. Huerta
Department of Anatomy, The University of Connecticut Health Center, Farmington, CT 06032, USA

Douglas L. Oliver
Department of Anatomy, The University of Connecticut Health Center, Farmington, CT 06032, USA

David K. Ryugo
Center for Hearing Sciences, Departments of Otolaryngology–Head and Neck Surgery and Neuroscience, Johns Hopkins University School of Medicine, Baltimore, MD 21205, USA

Ilsa R. Schwartz
Department of Surgery/Otolaryngology, Yale University School of Medicine, New Haven, CT 06510, USA

W. Bruce Warr
Boys Town National Research Hospital, Omaha, NE 68131, USA

Douglas B. Webster
Kresge Hearing Research Laboratory of the South, Department of Otorhinolaryngology, Louisiana State University, New Orleans, LA 70112-2234, USA

Jeffery A. Winer
Division of Neurobiology, Department of Molecular and Cell Biology, University of California, Berkeley, CA 94720-2097, USA

1

An Overview of Mammalian Auditory Pathways with an Emphasis on Humans

Douglas B. Webster

1. Introduction

This chapter provides a succinct description of the entire mammalian central auditory system—an overview, before we embark on the specific, highly detailed chapters which follow. It also emphasizes what is known of the structure of the human central auditory pathways, including how they are similar and how dissimilar to those of other mammals.

Biomedical research depends heavily upon the study of nonhuman animals, and this is nowhere more true than in the study of the human nervous system. Today we know more about the nervous system than could have been dreamed possible even a generation ago. These advances are due to powerful modern research methods which rely on invasive techniques such as implanting electrodes within the brain to record nerve impulses, and injecting chemical "markers" or specific antibodies into the brain.

Indeed, few techniques are available for the direct study of the nervous system in humans. These include: (1) noninvasive techniques such as psychoacoustics and surface recordings of physiological activity (auditory brainstem responses, middle latencies, late potentials); (2) anatomical studies of human autopsy material; and (3) radiological techniques such as magnetic resonance imaging and positron emission tomography.

However, comparative studies have demonstrated great similarities among mammals in the structure and function of many parts of the brain, including the central auditory pathways. Furthermore, where function and neural connections are dissimilar, neural structures are comparably dissimilar. Thus it is considered valid to infer human neural connections and functions from data obtained from nonhuman mammals when neural structures are closely comparable. The neural structures of most of the human auditory system have been described, but the connections and details of neural function have been obtained primarily from nonhuman mammals—particularly cats, rodents, and (to a lesser extent) nonhuman primates.

In broadest terms, the neural parts of the mammalian auditory system are:

Inner and outer hair cells of the cochlea,
Spiral ganglion and cochlear portion of the vestibulocochlear nerve,
Cochlear nuclear complex,
Trapezoid body,
Superior olivary complex,
Lateral lemniscus, including its nuclei,
Inferior colliculus,
Brachium of the inferior colliculus,
Medial geniculate body,
Thalamocortical auditory radiation,
Auditory portions of the cerebral cortex.

2. Innervation of the Hair Cells of the Organ of Corti

Mammalian cochleas usually have a single row of inner hair cells and three rows of outer hair cells, arranged parallel to the length of the cochlear duct and extending from base to apex along the basilar membrane. Occasionally there are four or even five rows of outer hair cells for short distances (Smith and Takasaka 1971). At the basal ends of the hair cells are synapses with vestibulocochlear nerve fibers. The apical ends, at the endolymphatic surface, bear stereocilia and a cuticular plate. In these features, humans have a typical mammalian organ of Corti.

2.1 Afferent Innervation

The most studied pattern of afferent innervation is that of cats (Spoendlin 1972): 90–95% of the primary afferent neurons synapse only with inner hair cells, with each inner hair cell receiving synapses from about 20 primary neurons. Usually these primary "radial fibers" do not branch, and each primary neuron makes a single synapse with an inner hair cell.

The situation is somewhat different in humans. Many of the afferent radial primary auditory fibers branch shortly before the inner hair cells and synapse on two or three inner hair cells (Nadol 1983a) (Fig. 1.1).

In cats, the remaining 5–10% of primary afferent fibers turn basalward, branch repeatedly, and make synapses, smaller than those on inner hair cells, on as many as 15 to 20 outer hair cells (Spoendlin 1972). Each outer hair cell receives 6–10 afferent endings.

In humans, each outer hair cell receives 4–8 afferent synaptic endings (Nadol 1983b).

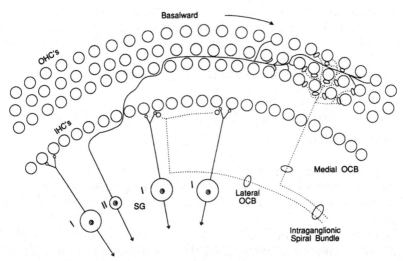

FIGURE 1.1. Schematic diagram of the innervation of the human cochlea based on the data of Nadol and his colleagues (Nadol, 1990; Nadol, Burgess, and Reisser 1990). The three rows of outer hair cells (OHCs) and one row of inner hair cells (IHCs) are shown for a short segment of one cochlear turn. Solid fibers are of afferent neurons and dashed fibers are of efferent neurons. Three Type I spiral ganglion neurons (I) and one Type II spiral ganglion neuron (II) are shown. Type I cells comprise 88% and type II 12% of the afferent neurons. A single medial olivocochlear bundle (OCB) fiber to outer hair cells and a single lateral olivocochlear bundle fiber to type I spiral ganglion fibers are shown. SG, spiral ganglion.

2.2 Efferent Innervation

[See Chapter 7 (Warr) for full coverage of this subject.]

The mammalian cochlea also has efferent synapses—which are the axonal endings of the olivocochlear bundle. Their neuronal cell bodies lie in the superior olivary complex. Their axons, partially crossed and partially uncrossed, travel out of the brain with the vestibular portion of the vestibulocochlear nerve; at the base of the modiolus they split off from the vestibular portion of the nerve and travel as the anastomosis of Oort to join the cochlear portion of the vestibulocochlear nerve. After traveling through Rosenthal's canal as the intraganglionic spiral bundle, the olivocochlear axons leave the spiral ganglion and travel out through the osseous spiral lamina. They exit through small holes, called the habenula perforata, and enter the organ of Corti.

In the organ of Corti, the smaller axons of the olivocochlear bundle make vesiculated synapses with those peripheral processes of spiral ganglion neurons that form afferent synapses with the inner hair cells (Fig. 1.1). These synapses are numerous in all mammals studied, including humans (Spoendlin 1972; Nadol 1990).

The larger axons of the olivocochlear bundle, after passing through the osseous spiral lamina, course across the tunnel of Corti as tunnel radial fibers and make large, vesiculated synaptic endings with outer hair cells (Fig. 1.1). These efferent endings on outer hair cells are more numerous in cats and guinea pigs than in humans (Nadol 1990).

3. Spiral Ganglion and Cochlear Division of the Vestibulocochlear Nerve

[See Chapter 2 (Ryugo) for full coverage of this subject.]

The cell bodies of the afferent cochlear fibers comprise the spiral ganglion, which is located in Rosenthal's canal within the modiolus of the cochlea.

Spiral ganglion neurons that innervate the inner hair cells are called type I spiral ganglion neurons. The peripheral processes of these bipolar neurons pass through the osseous spiral lamina, exit through the habenula perforata, and innervate inner hair cells. In cats and most other mammals these neurons (including their cell bodies) are myelinated except for the portions of the peripheral processes between the habenula perforata and their terminations (Spoendlin 1972). In humans most of the cell bodies are not myelinated (Nadol, Burgess, and Reisser 1990). The central fibers of these neurons extend down the modiolus through the internal auditory meatus and enter the brain at the cerebellopontine angle. In cats and other nonhuman mammals studied, type I neurons comprise 90–95% of the spiral ganglion cells. In humans, 88% of spiral ganglion neurons are large; presumably they are type I neurons and innervate inner hair cells (Nadol, Burgess, and Reisser 1990).

The other 5–10% (in cats) of spiral ganglion neurons innervate the outer hair cells and are called type II spiral ganglion neurons. They are typically smaller than type I cells, and in most mammals they are unmyelinated. In cats, many type II neurons are unipolar rather than bipolar; however, the nerve fiber bifurcates after leaving the cell body, with a peripheral branch going to the organ of Corti and a central branch to the brain (Kiang et al. 1982).

Type II peripheral fibers cross the tunnel of Corti just above the bases of the pillar cells. They then turn basalward and travel between Deiters' cells for a considerable distance before finally branching and terminating in small bouton synapses on several outer hair cells. Type II central fibers course with type I central fibers through the modiolus and internal auditory meatus to enter the brain at the cerebellopontine angle as part of the cochlear division of the vestibulocochlear nerve.

In humans, 12% of spiral ganglion neurons are small and presumably type II. A few of these type II neurons in humans (Nadol, 1981, 1984; Nadol, Burgess, and Reisser 1990) and chimpanzees form reciprocal synapses with outer hair cells (Nadol 1990). The morphology of reciprocal synapses suggests that they may be both afferent and efferent.

Physiological studies which have recorded from individual nerve fibers of nonhuman mammals have demonstrated that type I fibers have very sharp best frequencies—that is, an individual nerve fiber is excited by an extremely narrow band of frequencies. The role of type II spiral ganglion cells is unclear: no one has yet been able to record nerve impulses from type II spiral ganglion cells, and it is not even certain that they conduct action potentials.

4. Cochlear Nuclear Complex

[See Chapter 3 (Cant) for full coverage of this subject.]

The cochlear nuclear complex spans the border between the pons and the medulla at the cerebellopontine angle. All primary auditory fibers—that is, the central fibers of the spiral ganglion neurons—terminate here (Sando 1965; Webster 1971), after passing down the modiolus and leaving the temporal bone through the internal auditory meatus. Each primary auditory fiber bifurcates as it enters the cochlear nuclear complex (Lorente de Nó 1933). An ascending branch goes to the anterior ventral cochlear nucleus (AVCN); a descending branch goes first to the posterior ventral cochlear nucleus (PVCN) and then to the dorsal cochlear nucleus (DCN) (Fig. 1.2) (Lorente de Nó 1933). As they travel in the cochlear nuclei, each fiber makes synaptic connections with cells of AVCN, PVCN, and DCN.

Fibers whose cell bodies lie in the apex of the cochlea, and which therefore carry low-frequency information, bifurcate as soon as they enter the cochlear nuclear complex. The ascending branch stays in the ventrolateral portion of AVCN; the descending branch stays in the ventrolateral portions of PVCN and DCN. On the other hand, fibers whose cell bodies lie in the basal, high-frequency portion of the cochlea go to the dorsomedial portion of the ventral cochlear nucleus (between AVCN and PVCN) before bifurcating. The ascending branches pass through the dorsomedial portion of AVCN; the descending branches go to dorsomedial portions of PVCN and DCN (Fig. 1.2). Thus there is a cochleotopic—and therefore tonotopic—organization in each of the three divisions of the cochlear nuclei, with low-frequency information represented ventrolaterally and high-frequency information dorsomedially in each division (Sando 1965; Webster 1971). This cochleotopic organization is maintained throughout the central auditory pathways and is found in all mammals studied; presumably it is also the case in humans.

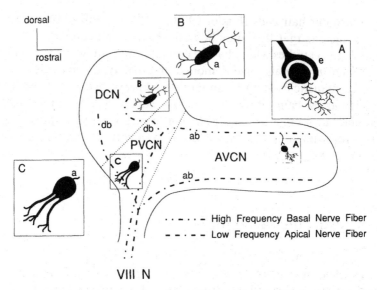

FIGURE I.2. Schematic diagram of a dorsolateral view of the human cochlear nuclear complex with its three divisions, anterior ventral cochlear nucleus (AVCN), posterior ventral cochlear nucleus (PVCN), and dorsal cochlear nucleus (DCN). Two spiral ganglion nerve fibers (VIIIN) are shown bifurcating into ascending branches (ab) and descending branches (db). Note that high-frequency (basal) fibers course in the dorsomedial portion and low-frequency (apical) fibers course in the ventrolateral portion of the cochlear nuclear complex. A single spherical bushy cell of AVCN is shown in A with its axon (a) and an endbulb of Held (e) synapsing on it. A single stellate cell of DCN is shown in B with its axon (a). A single octopus cell of PVCN is shown in C with its axon (a).

4.1 Anterior Ventral Cochlear Nucleus

Although AVCN contains several neuronal types, bushy cells predominate. Bushy cells are characterized by having only one or two very thick dendrites, with each dendrite branching profusely so that it resembles a bush (Brawer, Morest, and Kane 1974; Webster and Trune 1982). In the anterior part of AVCN, bushy cells have spherical cell bodies and are called spherical bushy cells (Fig. 1.2). In the posterior part of AVCN, their cell bodies are more oval and they are called globular bushy cells.

Ascending branches of type I spiral ganglion neurons synapse with the cell bodies of bushy cells, forming very large calyceal endings called the endbulbs of Held (Ryugo and Fekete 1982). Because of these very large synapses, a large amount of neurotransmitter passes from the presynaptic, ascending branches to the postsynaptic bushy cells. Each action potential carried by an ascending branch of a cochlear nerve fiber can therefore release enough neurotransmitter to depolarize the postsynaptic cell and cause an action potential in the postsynaptic bushy cell. Thus the primary-type responses are retained in the bushy cells.

Bushy cell axons are large and myelinated. In the AVCN they form the ventral acoustic stria, which courses through the ventral part of the inferior portion of the pontine medulla, and, as it approaches the midline, becomes the trapezoid body.

4.2 Posterior Ventral Cochlear Nucleus

Like AVCN, PVCN also contains several cell types; octopus cells are the most distinctive (Osen 1969; Brawer, Morest, and Kane 1974). Each octopus cell has several large, scantily branched dendrites extending from one side of its large cell body (Fig. 1.2) (like the tentacles of an octopus). These dendrites course at right angles to the descending branches of type I and type II spiral ganglion neurons. Many axons of the descending branches make small bouton synaptic endings onto the dendrites or cell bodies of the octopus cells. Because of this, the octopus cells are more sensitive to bands of frequencies than to pure frequencies.

The axons of the octopus cells leave PVCN and curve dorsal to the restiform body, forming the intermediate acoustic stria (stria of Held), which then enters the pontine tegmentum.

4.3 Multipolar Cells of AVCN and PVCN

Scattered throughout both AVCN and PVCN are multipolar cells (Osen 1969; Brawer, Morest, and Kane 1974; Cant 1982; Smith and Rhode 1989). They vary in size and in the details of their morphology, but they characteristically have irregularly shaped cell bodies from which extend several dendrites, each with several branches. These dendritic branches may be aligned in parallel or extend in a star-like fashion. Multipolar cells receive bouton synaptic endings, mainly on their dendrites, from spiral ganglion cell axons.

The axons of the multipolar cells of the AVCN exit the cochlear nuclei via the ventral acoustic stria and extend into the pontine tegmentum, becoming part of the trapezoid body. The axons of the multipolar cells of the PVCN exit the cochlear nuclei via the intermediate acoustic stria along with octopus cell axons.

The ventral cochlear nucleus of humans is made up of neuronal types and organizations similar to other mammals described above (Bácsik and Strominger 1973; Dublin 1974; Moore and Osen 1979; Richter 1983; Adams 1986; Moore 1987). One must presume that synaptic connections and functions are similar.

4.4 Dorsal Cochlear Nucleus

In most mammals the DCN is a laminated structure. Its outer layer, called the molecular layer, has few neurons. Deep to this is a fusiform layer, which usually contains a single layer of fusiform cells. Deep to the fusiform layer is the central region, containing many and varied stellate (multipolar) neurons (Osen 1969).

The synaptic morphology of the DCN is complex: some of these cells receive input from the descending branches of spiral ganglion neurons, and, at least in nonhuman mammals, individual neurons of the DCN respond to tonal and complex stimuli in complex ways.

In addition to receiving synapses from descending branches of spiral ganglion neurons, DCN neurons also receive abundant synaptic input from other portions of the brain. Axons of the stellate cells leave the DCN as the dorsal acoustic stria (stria of Monakow), which courses dorsal to the restiform body and then enters the pontine tegmentum.

The human DCN, although large, appears degenerate and disorganized compared to that of most mammals (Moore and Osen 1979; Adams 1986). It lacks the usual laminar structure and many of the cell types. Stellate cells are the most numerous (Fig. 1.2); they seem to resemble the cells in the central region of other mammals. These multipolar neurons have irregular cell bodies with dendrites extending mostly parallel to the free surface of the nucleus (Adams 1986).

5. Superior Olivary Complex

[See Chapter 4 (Schwartz) for full coverage of this subject.]

The superior olivary nuclei lie in the ventral portion of the tegmentum of the medulla in many mammals. In humans, because of the much larger pons, the superior olivary nuclei are in the inferior portion of the pontine tegmentum.

The superior olivary nuclei receive most of the synaptic input from the AVCN via the ventral acoustic stria and trapezoid body. There are three major nuclei in the superior olivary complex: the lateral superior olivary nucleus (LSO), the medial superior olivary nucleus (MSO), and the medial nucleus of the trapezoid body (MNTB) (Fig. 1.3) (Irving and Harrison 1967). These three major nuclei are surrounded by more diffuse groups of nuclei, collectively called the periolivary nuclei.

The size of the superior olivary complex varies greatly among mammalian groups, as does the relative size of each nucleus. In humans, MSO is the largest; LSO is quite small (Strominger and Hurwitz 1976); MNTB is sometimes thought to be nonexistent, although it has been at least tentatively identified (Richter et al. 1983).

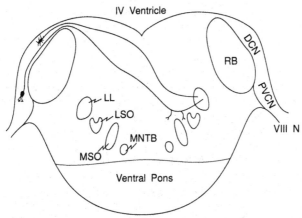

FIGURE 1.3. Schematic diagram of a transverse section through the human brain-stem at the pontomedullary junction where the cochlear portion of the eighth cranial nerve (VIIIN) enters into the cochlear nuclear complex (PVCN = posterior ventral cochlear nucleus; DCN = dorsal cochlear nucleus). A single stellate cell of DCN is shown with its axon coursing over the restiform body (RB), crossing the midline below the IV ventricle, and forming part of the ascending lateral lemniscus (LL). A single octopus cell of PVCN is shown with its axon coursing over the RB, crossing the midline, sending collaterals to the periolivary nuclei, and forming part of the ascending LL. LSO, lateral superior olivary nucleus; MNTB, medial nucleus of the trapezoid body; MSO, medial superior olivary nucleus.

5.1 Medial Superior Olivary Nucleus

While MSO is the largest of the three in humans, it is extremely small in several other mammals, most of which, like mice and rats, hear poorly at low frequencies (Irving and Harrison 1967).

The principal neurons of MSO are multipolar. They receive bilateral synaptic input from the axons of spherical bushy cells of both left and right AVCN (Fig. 1.4). Thus it is in MSO that the first binaural processing of auditory information takes place.

It is also here that coding for horizontal sound localization first occurs, particularly for low-frequency sounds (Masterton et al. 1975). Since low-frequency sounds have long wavelengths, when they strike an animal's head they are diffracted, not reflected. Therefore, there is essentially no intensity difference between low-frequency sounds received in right and left ears, even if the sound comes from a very lateral source. However, there is a time difference, measured in microseconds, between when the sound reaches the right ear and when it reaches the left ear. Because of this time difference, there is also a phase difference. These differences between the right and left ear input determine how MSO neurons will respond and carry the code for horizontal auditory localization.

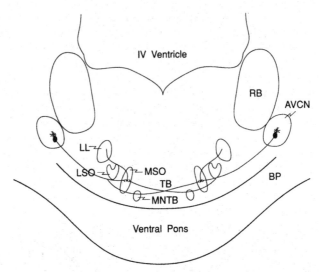

FIGURE 1.4. Schematic diagram of a transverse section of the human brainstem just rostral to that of Figure 1.3. This section through the anterior ventral cochlear nucleus (AVCN) shows a single spherical bushy cell from each side sending its axon to both the ipsilateral and contralateral medial superior olivary nuclei (MSO). MSO cells send their axons into the ipsilateral lateral lemnisus (LL). BP, brachium pontis; LSO, lateral superior olivary nucleus; MNTB, medial nucleus of the trapezoid body; RB, restiform body; TB, trapezoid body.

Axons of MSO neurons contribute to the ipsilateral lateral lemniscus (Fig. 1.4).

5.2 Medial Nucleus of the Trapezoid Body (MNTB)

The MNTB is made up of large, multipolar principal cells which receive their synaptic input via large calyceal endings from the contralateral globular bushy cells of AVCN (Fig. 1.5) (Jean-Baptiste and Morest 1975). There is more high- than low-frequency representation in MNTB. The axons of its principal cells terminate on the ipsilateral principal neurons of LSO (Fig. 1.5).

5.3 Lateral Superior Olivary Nucleus (LSO)

This nucleus is made up of relatively small multipolar principal cells, which receive bouton-type synaptic endings from the ipsilateral spherical bushy cell axons of AVCN (Fig. 1.5), and the ipsilateral principal cell axons of MNTB (Fig. 1.5). Like MNTB neurons, those of LSO also receive more high- than low-frequency information. They code for auditory localization in the horizontal plane, using high frequency cues (Erulkar

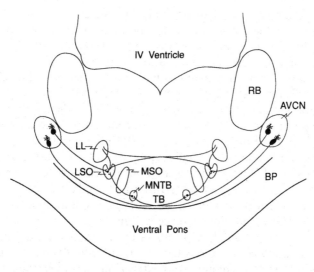

FIGURE 1.5. Schematic diagram of a transverse section identical to Figure 1.4 but showing neural connections to the medial nucleus of the trapezoid body (MNTB) and lateral superior olivary nucleus (LSO). A single globular bushy cell in each anterior ventral cochlear nucleus (AVCN) is shown sending its axon via the trapezoid body (TB) to the contralateral MNTB. MNTB cells send their axons to the ipsilateral LSO. A single spherical bushy cell of each AVCN is shown sending its axons to the ipsilateral LSO. LSO cells send their axons bilaterally to enter both lateral lemnisci (LL). BP, brachium pontis; MSO, medial superior olivary nucleus; RB, restiform body.

1972). Since high frequencies have short wavelengths, sound is reflected from the head rather than diffracted by it. Therefore there are intensity differences between the sounds reaching the right and left ear. The activity of LSO neurons depends on these intensity differences (Boudreau and Tsuchitani 1970).

The axons of LSO neurons contribute bilaterally to the lateral lemnisci (Fig. 1.5).

5.4 Periolivary Nuclei and the Olivocochlear Bundle

[See Chapter 7 (Warr) for full coverage of this subject.]

The multipolar cell bodies of olivocochlear bundle neurons form two separate groups of neurons, which are parts of the periolivary nuclei (Warr and Guinan 1979).

The medial group is composed of multipolar neurons which lie in the medial portion of the periolivary nuclei. Most, but not all, of their axons extend contralaterally and leave the brain as part of the vestibular division of the vestibulocochlear nerve. These large axons leave the vestibular division at the base of the cochlea and join the cochlear division as the anastomosis of Oort, or vestibulocochlear anastomosis. They then spiral within Rosenthal's canal as the intraganglionic spiral bundle before passing out through the osseous spiral lamina and habenula perforata to the organ of Corti. There they travel radially across the tunnel of Corti and terminate as large vesiculated boutons on outer hair cells (Fig. 1.1).

The lateral group of olivocochlear neurons lies in the lateral portion of the periolivary nuclei and within LSO. It is composed of small, multipolar neurons which send their axons mostly, but not entirely, ipsilaterally, to follow the same course as the medial group of olivocochlear axons through the anastomosis of Oort and intraganglionic spiral bundle. However, once in the organ of Corti these axons terminate as small, vesiculated terminal boutons on the peripheral processes of type I spiral ganglion neurons, immediately below the inner hair cells (Fig. 1.1).

Thus the olivocochlear bundle fibers of the medial group are mostly contralateral and terminate exclusively on outer hair cells as axosomatic endings, whereas the smaller-sized lateral olivocochlear fibers are primarily ipsilateral and terminate as axoaxonic terminals on the peripheral fibers of type I spiral ganglion cells under the inner hair cells. Although details of the human olivocochlear bundle have not been worked out, the terminal endings in the organ of Corti are similar to those in cats (Nadol 1990).

5.5 Stapedius Reflex

The superior olivary complex also plays an essential role in the stapedius reflex. Some axons from both LSO and MSO synapse with facial nucleus neurons that innervate the stapedius muscle. Thus they play an interneuron role in the stapedius reflex. The simplest stapedius reflex arc, involving the fewest possible neurons, would consist of (1) type I spiral ganglion neurons, (2) bushy cells of AVCN, (3) principal neurons of MSO (and/or LSO), and (4) facial nucleus neurons innervating the stapedius muscle.

6. The Lateral Lemniscus

[See Chapter 4 (Schwartz) for full coverage of this subject.]

The lateral lemniscus is a major auditory tract extending from the lower pontine tegmentum to the inferior colliculus. It forms just lateral to the superior olivary complex and is composed of axons from several locations (Brunso-Bechtold, Thompson, and Masterton 1981):

Contralateral multipolar neurons of the ventral cochlear nucleus, via the ventral acoustic stria and trapezoid body;

Contralateral stellate cells of DCN, via the dorsal acoustic stria (Fig. 1.3);
Ipsilateral principal cells of MSO (Fig. 1.4);
Both ipsilateral and contralateral principal neurons of LSO (Fig. 1.5); and
Contralateral octopus cells of PVCN, via the intermediate acoustic stria
  (Fig. 1.3).

All these axon groups except the last travel in the lateral lemniscus to
the inferior colliculus. Axons of the contralateral octopus cells of PVCN
end in the ventral nucleus of the lateral lemniscus (VNLL). The axons
of neurons from VNLL run as part of the lateral lemniscus to the inferior
colliculus.

Thus the lateral lemniscus contains second order neurons from the
ventral and dorsal cochlear nuclei; third order neurons from the MSO
and VNLL nuclei; and fourth order neurons from LSO. In addition, axons
from the intermediate and dorsal nuclei of the lateral lemniscus contribute
to the upper part of the lateral lemniscus.

It is precisely because the lateral lemniscus contains second, third, and
fourth order axons that there is not a one-to-one correspondence between
structures of the ascending auditory pathways and the individual waves
of auditory brainstem responses (ABRs). Since there is a delay at each
synapse within any neural system, the timing of synchronous firings
within the lateral lemniscus reflects these different synaptic delays from
second to third and fourth order auditory neurons. Because of these different delays in different groups of axons within the lateral lemniscus,
the lateral lemniscus must contribute to at least three of the waves of the
auditory brainstem response.

Most axons of the lateral lemniscus terminate on neurons of the ipsilateral central nucleus of the inferior colliculus. However, the lateral lemniscus also contains interstitial nuclei. These are the ventral nucleus, the
intermediate nucleus, and the dorsal nucleus of the lateral lemniscus.
Many, if not all, octopus cells project from PVCN to the contralateral
ventral nucleus of the lateral lemniscus, often with large calyceal endings.
The lateral lemniscal nuclei all receive some synaptic input from lateral
lemniscal fibers, either as terminal endings or collateral endings. Most of
the axons leaving these interstitial nuclei join the other axons of the lateral
lemniscus and project to the ipsilateral central nucleus of the inferior
colliculus; but some neurons of the dorsal nucleus of the lateral lemniscus
send axons across the midline as the commissure of Probst, to synapse
on neurons of the contralateral central nucleus of the inferior colliculus
(Fig. 1.6).

The human lateral lemniscus and its nuclei have been described from
autopsy material (Olszewski and Baxter 1954; Ferraro and Minkler 1977)
and are apparently similar to those of the cat.

7. The Inferior Colliculus

[See Chapter 5 (Oliver and Huerta) for full coverage of this subject.]

The inferior colliculus is the major midbrain auditory center and the largest auditory structure of the brainstem. Its largest nucleus is the central nucleus (Fig. 1.6). Surrounding it are smaller auditory nuclei, which, structurally and functionally, are more parts of the descending than ascending auditory pathways (Oliver and Morest 1984).

The central nucleus of the inferior colliculus (CNIC) is composed primarily of small, multipolar fusiform cells with tufted dendrites. These dendrites are oriented obliquely across the central nucleus of the inferior colliculus and have a cochleotopic (and therefore also a tonotopic) organization. These principal cells of CNIC receive most of their input from the lateral lemniscus, and some from the commissure of Probst.

The commissure of the inferior colliculus crosses the midline, and interconnects neurons of right and left CNIC (Fig. 1.6). A minority of these commissural fibers bypass the contralateral CNIC and contribute their axons to the contralateral brachium of the inferior colliculus. Most axons of principal cells of CNIC, however, remain ipsilateral and form

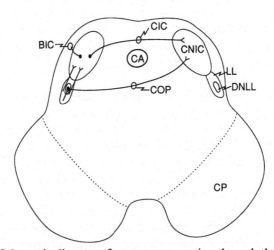

FIGURE 1.6. Schematic diagram of a transverse section through the human brainstem at the level of the inferior colliculus. Lateral lemniscal (LL) fibers terminate in the central nucleus of the inferior colliculus (CNIC) and dorsal nucleus of the lateral lemniscus (DNLL). Most CNIC cells send their axons via the brachium of the inferior colliculus (BIC) to the medial geniculate nucleus of the thalamus (not shown). A few CNIC cells send their axons to the contralateral CNIC via the commissure of the inferior colliculus (CIC). DNLL cells send axons via the commissure of Probst (COP) to the contralateral CNIC. CA, cerebral aqueduct; CP, cerebral peduncle.

the large brachium of the inferior colliculus. This tract lies on the dorsolateral surface of the midbrain and leaves the midbrain to enter the posterior portion of the thalamus. There the fibers end massively in the medial geniculate body, which is the principal auditory nucleus of the thalamus.

The human inferior colliculus has an organization very similar to that of cats (Geniec and Morest 1971).

8. The Medial Geniculate Body

[See Chapter 6 (Winer) for full coverage of this subject.]

The medial geniculate body (MGB) is the major auditory nucleus of the thalamus; it is readily divisible into ventral, dorsal, and medial (or magnocellular) divisions (Winer 1985). The overall organization of MGB is similar in cats and humans; the major difference is that the neuropil between cell bodies is more extensive in humans (Winer 1984).

8.1 The Ventral Division of the Medial Geniculate Body

This division is composed primarily of bitufted neurons in arcing, parallel, tonotopically organized laminae. Most fibers of the brachium of the inferior colliculus synapse on cells of the ventral division of MGB. The bitufted neurons of the ventral division project their axons via the sublenticular portion of the internal capsule to primary auditory cortex (Fig. 1.7). The primary auditory cortex, and to a lesser extent auditory association cortices, project many axons back to the ventral division of MGB. This division appears to be the portion of MGB that transmits specific auditory discriminative information to the cerebral cortex (Winer 1984, 1985).

8.2 The Dorsal Division of the Medial Geniculate Body

The organization of the dorsal division of MGB is extremely complex. As many as 10 subdivisions have been described, containing at least eight neuronal types (Winer 1984, 1985). The dorsal division of MGB receives diffuse axonal terminals from CNIC, the peripheral portion of inferior colliculus, brainstem reticular neurons, the ventral division of MGB, and other thalamic nuclei. In turn, neurons of the dorsal division of MGB project axons primarily to association auditory cortex. It has been proposed that the dorsal division of MGB functions in maintaining and directing auditory attention (Winer 1984).

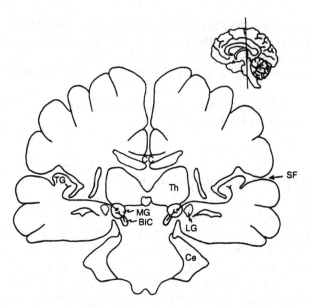

FIGURE 1.7. Schematic diagram of a coronal section through the human brain at the level of the medial geniculate body (MG). Brachium of the inferior colliculus (BIC) axons terminate in MG and MG cells send their axons via the sublenticular portion of the internal capsule to the transverse gyri of Heschl (TG). Insert shows the plane of section. CC, corpus callosum; Ce, cerebellum; LG, lateral geniculate body; SF, Sylvian fissure; Th, thalamus.

8.3 The Medial Division of the Medial Geniculate Body

The medial, or magnocellular, portion of MGB is composed of a variety of relatively large, multipolar neurons. It receives axons from vestibular nuclei, spinal cord nuclei, and superior colliculus; and a small amount of auditory information from the vicinity of the superior olivary complex and lateral lemniscus. It projects axons, albeit sparsely, to all auditory cortices, some nonauditory cortices, the putamen, and the amygdala. All auditory cortical regions and some nonauditory cortical regions project diffusely back to the medial division via the internal capsule. It has been proposed that the medial division of MGB functions as a multisensory arousal system (Winer 1984, 1985).

9. Transverse Gyri of Heschl

[See Chapter 6 (Winer) for full coverage of this subject.]

In all mammals, the primary auditory cortex lies in the temporal lobe; in humans it lies more specifically in the transverse gyri of Heschl (Celesia

1976). These gyri lie on the superior surface of the temporal lobe, continuous with, but not part of, the superior temporal gyrus (Fig. 1.8). There may be one to three—rarely four—transverse gyri of Heschl in each hemisphere (Campain and Minckler 1976). The smoother portion of the superior surface of the temporal lobe caudal to the transverse gyri is named the planum temporale (Fig. 1.8) which was larger on the left than on the right in 65 of 100 human brains sampled (Geschwind and Levitsky 1968).

There is confusion and inconsistency in the literature concerning the definition of the human primary auditory cortex and its precise location within the temporal lobe. In most human neuroanatomy textbooks, what is called primary auditory cortex is really part of the superior temporal gyrus rather than the gyri of Heschl, which are not visible on the surface of the brain until one opens the lateral fissure.

The human primary auditory cortex, or koniocortex, is variably called Area 41, A1, TC, or KAm and KAt, depending on the authority. Cytoarchitectonic studies attempting to identify human koniocortex have described it entirely, or almost entirely, within the first transverse gyrus of Heschl (Galaburda and Sanides 1980).

The parakoniocortex (area 42, TB, or PA, depending on the authority) has a quite different cytoarchitecture. It surrounds the koniocortex and extends a variable distance beyond the transverse gyri of Heschl onto the planum temporale (Galaburda, Sanides, and Geschwind 1978; Seldon 1981a,b). Some authorities limit human primary auditory cortex to the koniocortex; others include both koniocortex and parakoniocortex. This

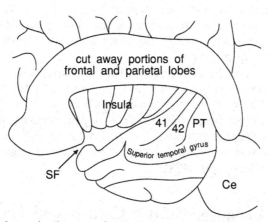

FIGURE 1.8. Schematic diagram of a dissection of the left human cerebrum. Portions of the frontal and parietal lobes have been cut away above the Sylvian fissure (SF), exposing the insula and superior surface of the temporal lobe. Primary auditory cortex (41) is shown on the first transverse gyrus of Heschl and parakoniocortex (42) is shown on the second transverse gyrus of Heschl. Ce, cerebellum; PT, planum temporale.

latter definition fits the scant human physiological data. There are apparently no consistent right/left asymmetries in the size of either koniocortex or parakoniocortex in the human.

In cats and in nonhuman primates there is a precise tonotopic organization of koniocortex. One assumes this would also be true in humans. In cats, at least, the long isofrequency bands of auditory cortex contain alternating sub-bands. One sub-band responds to tonal stimuli by excitatory responses from both right and left ears; alternate sub-bands respond by excitatory responses from the contralateral ear and inhibitory responses from the ipsilateral ear. In cats and nonhuman primates there are multiple tonotopic maps in the parakoniocortices, some incomplete, others complete. The same is likely true in humans. In the few human cases of bilateral loss of the transverse gyri with retention of area 22, there is a substantial loss of auditory sensitivity but retention of frequency discrimination and language comprehension.

10. Planum Temporale and Superior Temporal Gyrus

The planum temporale, behind the parakoniocortex, and the posterior part of the superior temporal gyrus, roughly behind the vein of Labbé, comprise area 22 (Tpt of some authors). Cytoarchitecture indicates that area 22 also extends somewhat onto the parietal operculum and inferior parietal lobule in humans (Galaburda, Sanides, and Geschwind 1978).

In over 90% of humans, the left area 22 is "Wernicke's area"—the speech receptive area. Some authors also include the entire inferior parietal lobule (angular gyrus, 39, and supramarginal gyrus, 40) in Wernicke's area; but most authors consider these to be higher order association areas which also receive visual and somesthetic information for multimodel integration of sensory information.

Measurements of the surface area of the planum temporale in human brains have shown that the left side is larger in 65% and the right side is larger in 11%, with no difference in 24% (Geschwind and Levitsky 1968). Furthermore, when area 22 (Tpt), as determined cytoarchitectonically, was specifically measured in four brains, the left side was found to be larger in all four, varying from 1.1 to 7.3 times larger (Galaburda, Sanides, and Geschwind 1978).

11. Other Language-Related Cortices and Their Connections

The major auditory long association pathway, the arcuate fasciculus, interconnects area 22, the inferior parietal lobule, and the area triangularis of the inferior frontal gyrus. The supramarginal gyrus (area 40) and angular gyrus (area 39) make up the inferior parietal lobule. This is a complex association area which integrates visual, auditory, and somesthetic information and is therefore of great importance in the visual and somesthetic aspects of language such as reading and writing.

The area triangularis on the inferior frontal gyrus, areas 44 and 45, is better known as Broca's area, which is important in expressive speech and language, particularly in the left hemisphere of humans (Geschwind 1970).

12. Summary and Synthesis

The tonotopic organization established in the cochlea is sustained throughout the mammalian auditory system, and, in fact, provides its only consistent organizational feature.

The primary type I spiral ganglion neurons are sharply tuned and therefore bring frequency-specific information to the neurons of the cochlear nuclear complex. After branching, each primary fiber synapses with each of the principal cell types of the cochlear nuclear complex: bushy cells of AVCN, octopus cells of PVCN, multipolar cells through VCN, and stellate cells of DCN. Thus, information from the cochlea goes to four distinct neuronal groups in the cochlear nuclear complex. Each of these groups codes the information differently and initiates a distinct pathway from the cochlear nuclei to the inferior colliculus.

Bushy cells of AVCN receive endbulbs of Held from primary neurons and pass this frequency-specific information to the superior olivary complex via the ventral acoustic stria and trapezoid body. Superior olivary neurons receive information from both right and left cochlear nuclei; they code this binaural information for horizontal auditory localization and send their axons to the inferior colliculi via the lateral lemniscus.

Each octopus cell of PVCN receives bouton-type synapses from many primary neurons, each tuned to a slightly different frequency. Therefore, octopus cells have broadly tuned best frequencies. They respond to brief stimuli with a burst of nerve impulses at the beginning of each stimulus. Such "on" type responses carry information about transients such as those found in speech. Octopus cell axons travel in the intermediate acoustic stria, providing collaterals to periolivary neurons, and terminate in the contralateral ventral nucleus of the lateral lemniscus (VNLL). VNLL neurons project their axons to the inferior colliculi via the lateral lemniscus.

Multipolar cells throughout VCN receive many bouton synapses from primary neurons. These multipolar neurons respond to brief tonal stimuli with repetitive bursts of action potentials and are called "chopper" units. Their axons exit the cochlear nuclei via either the intermediate or ventral acoustic stria; they make some synaptic endings in periolivary neurons but principally project to the contralateral inferior colliculus via the lateral lemniscus.

Stellate cells of DCN receive bouton synapses from primary neurons. Their axons form the dorsal acoustic stria (of Monakow), cross the midline, and ascend in the opposite lateral lemniscus to the inferior colliculus.

These multiple parallel auditory pathways from the cochlear nuclear complex to the inferior colliculus terminate primarily in the central nucleus of the inferior colliculus. This convergence of auditory information onto a single midbrain nucleus poses a major problem in understanding the functional anatomy of the central auditory system. For although the central nucleus of the inferior colliculus can be subdivided, there is no clear evidence that the pathways leading to it terminate in separate subdivisions. For now, we remain challenged by the significance of the apparent fact that separate parallel ascending auditory pathways established at the level of the cochlear nuclear complex converge onto a single neuropil in the central nucleus of the inferior colliculus.

The major ascending auditory pathway from the central nucleus of the inferior colliculus is via the brachium of the inferior colliculus to the ventral division of medial geniculate body, and from there via the internal capsule to the primary auditory complex in the transverse gyri of Heschl.

Much work is needed to clarify the connections between the dorsal and medial divisions of MGB and its ventral division; and even more, perhaps, to reveal their functional interactions. Similarly, many details of the organization of primary and secondary auditory cortices remain to be determined, as well as how they interact with other parts of the cerebral cortex.

In short, we have a fair understanding of the functional anatomy of the central auditory system below the level of the inferior colliculus. From the midbrain through the cerebral cortex, the basic functional anatomy still needs clarification and synthesis.

Acknowledgments.   I greatly appreciate extensive editing by Molly Webster. While writing this chapter I was supported by NIH Grants DC-00206 and DC-00379.

References

Adams JC (1986) Neuronal morphology in the human cochlear nucleus. Arch Otolaryngol—Head & Neck Surgery 112:1253–1261.

Bácsik RD, Strominger NL (1973) The cytoarchitecture of the human anteroventral cochlear nucleus. J Comp Neur 147:281–290.

Boudreau JC, Tsuchitani C (1970) Cat superior olive S-segment cell discharge to tonal stimulation. In: Neff WD (ed) Contributions to Sensory Physiology, Vol. 4. New York: Academic Press, pp. 143–213.

Brawer JR, Morest DK, Kane EC (1974) The neuronal architecture of the cochlear nucleus of the cat. J Comp Neur 155:251–300.

Brunso-Bechtold JK, Thompson GC, Masterton RB (1981) HRP study of the organization of auditory afferents ascending to central nucleus of inferior colliculus in cat. J Comp Neur 197:705–722.

Campain R, Minckler J (1976) A note on the gross configuration of the human auditory cortex. Brain Lang 3:318–323.

Cant NB (1982) The fine structure of two types of stellate cells in the anterior division of the anteroventral cochlear nucleus of the cat. Neuroscience 6:2643–2655.

Celesia GG (1976) Organization of auditory cortical areas in man. Brain 99:403–413.

Dublin WB (1974) Cytoarchitecture of the cochlear nuclei. Arch Otolaryngol 100:355–359.

Erulkar SD (1972) Comparative aspects of spatial localization of sound. Physiol Rev 52:236–360.

Ferraro JA, Minckler J (1977) The human lateral lemniscus and its nuclei. The human auditory pathways: A quantitative study. Brain Lang 4:277–294.

Galaburda A, Sanides F (1980) Cytoarchitectonic organization of the human auditory cortex. J Comp Neur 190:597–610.

Galaburda A, Sanides F, Geschwind N (1978) Human brain: Cytoarchitectonic left-right asymmetries in the temporal speech region. Arch Neur 35:812–817.

Geniec P, Morest DK (1971) The neuronal architecture of the human posterior colliculus: A study with the Golgi method. Acta OtoLaryngol Suppl 295:1–33.

Geschwind N (1970) The organization of language and the brain. Science 170:940–944.

Geschwind N, Levitsky W (1968) Human brain: Left-right asymmetries in temporal speech region. Science 161:186–187.

Irving R, Harrisom JM (1967) The superior olivary complex and audition: A comparative study. J Comp Neur 130:77–86.

Jean-Baptiste M, Morest DK (1975) Transneuronal changes of synaptic endings and nuclear chromatin in the trapezoid body following ablations in cats. J Comp Neur 162:111–133.

Kiang NYS, Rho JM, Northrop CC, Liberman MC, Ryugo DK (1982) Hair-cell innervation by spiral ganglion cells in adult cats. Science 217:175–177.

Lorente de Nó R (1933) Anatomy of the eighth nerve. III. General plan of structure of the primary cochlear nuclei. Laryngoscope 43:327–350.

Masterton B, Thompson GC, Bechtold JK, Robards MJ (1975) Neuroanatomical basis of binaural phase-difference analysis for sound localization: A comparative study. J Comp Physiol Psychol 89:379–386.

Moore JK (1987) The human auditory brain stem: A comparative view. Hear Res 29:1–32.

Moore JK, Osen KK (1979) The cochlear nuclei in man. Am J Anat 154:393–417.

Nadol Jr JB (1981) Reciprocal synapses at the base of outer hair cells in the organ of Corti of man. Ann Otol Rhinol Laryngol 90:12–17.

Nadol Jr JB (1983a) Serial section reconstruction of the neural poles of hair cells in the human organ of Corti. I. Inner hair cells. Laryngoscope 93:599–614.

Nadol Jr JB (1983b) Serial section reconstruction of the neural poles of hair cells in the human organ of Corti. II. Outer hair cells. Laryngoscope 93:780–791.

Nadol Jr JB (1984) Incidence of reciprocal synapses on outer hair cells of the human organ of Corti. Ann Otol Rhinol Laryngol 93:247–250.

Nadol Jr JB (1990) Synaptic morphology of inner and outer hair cells of the human organ of Corti. J Elect Micr Tech 15:187–196.

Nadol Jr JB, Burgess BJ, Reisser C (1990) Morphometric analysis of normal human spiral ganglion cells. Ann Otol Rhinol Laryngol 99:340–348.

Oliver DL, Morest DK (1984) The central nucleus of the inferior colliculus in the cat. J Comp Neur 222:237–264.

Olszewski J, Baxter D (1954) Cytoarchitecture of the Human Brain Stem. Philadelphia: JB Lippincott Co., pp. 116–117.

Osen KK (1969) Cytoarchitecture of the cochlear nuclei in the cat. J Comp Neur 136:453–484.

Richter E (1983) The ventral cochlear nucleus in human brains. Adv Oto-Rhino-Laryngol 31:59–71.

Richter EA, Norris BE, Fullerton BC, Levine RA, Kiang NYS (1983) Is there a medial nucleus of the trapezoid body in humans? Am J Anat 168:157–166.

Ryugo DK, Fekete DM (1982) Morphology of primary axosomatic endings in the anteroventral cochlear nucleus of the cat: A study of the endbulbs of Held. J Comp Neur 210:239–257.

Sando I (1965) The anatomical interrelationships of the cochlear nerve fibers. Acta Otolaryngol 59:417–436.

Seldon HL (1981a) Structure of the human auditory cortex. I. Cytoarchitectonics and dendritic distributions. Brain Res 229:277–294.

Seldon HL (1981b) Structure of the human auditory cortex. II. Axon distributions and morphological correlates of speech perception. Brain Res 229:295–310.

Smith CA, Takasaka T (1971) Auditory receptor organs of reptiles, birds, and mammals. In: Neff WD (ed) Contributions to Sensory Physiology, Vol. V. New York: Academic Press, pp. 129–178.

Smith PS, Rhode WS (1989) Structural and functional properties distinguish two types of multipolar cells in the ventral cochlear nucleus. J Comp Neur 282:595–616.

Spoendlin H (1972) Innervation densities of the cochlea. Acta Otolaryngol 73:235–248.

Strominger NL, Hurwitz JL (1976) Anatomical aspects of the superior olivary complex. J Comp Neur 170:485–497.

Warr WB, Guinan Jr JJ (1979) Efferent innervation of the organ of Corti: Two separate systems. Brain Res 173:152–155.

Webster DB (1971) Projection of the cochlea to cochlear nuclei in Merriam's kangaroo rat. J Comp Neur 143:323–340.

Webster DB, Trune DR (1982) Cochlear nuclear complex of mice. Am J Anat 163:103–130.

Winer JA (1984) The human medial geniculate body. Hear Res 15:225–247.

Winer JA (1985) The medial geniculate body of the cat. Adv Anat Embryol Cell Biol 86:1–98.

2

The Auditory Nerve: Peripheral Innervation, Cell Body Morphology, and Central Projections

DAVID K. RYUGO

1. Introduction

In mammals, all known auditory information enters the brain by way of the cochlear division of the vestibulocochlear nerve, hereafter referred to as the auditory nerve. Primary neurons, whose cell bodies reside in the spiral ganglion of the cochlea, send peripheral processes out to the organ of Corti to contact the acoustic receptor cells; the central processes or axons bundle together to form the auditory nerve. The terminus of the auditory nerve is the cochlear nucleus. In this way, primary neurons convey the output of the receptors to neurons of the cochlear nucleus. There are two types of receptors, inner hair cells and outer hair cells (Retzius 1884; Ramón y Cajal 1909), two populations of primary neurons (Munzer 1931; Spoendlin 1973), and many neuron classes in the cochlear nucleus (Lorente de Nó 1933; Osen 1969; Brawer, Morest, and Kane 1974). In turn, the cells of the cochlear nucleus give rise to all central auditory pathways. In a general way, the role of the cochlear nucleus is to receive incoming auditory nerve discharges, to preserve or transform the signals, and to distribute outgoing activity to higher brain centers. In order to understand the earliest stages of stimulus coding in the auditory system, we need to know (1) the nature of the signals conveyed by auditory nerve fibers, (2) their source in the periphery, and (3) their destination in the brain. This report shall review the progress that has been made along these lines of investigation.

2. Peripheral Innervation

In mammals, there are four rows of hair cells arrayed longitudinally along the basilar membrane; one row of inner hair cells (IHCs) and three rows of outer hair cells (OHCs). In the cat, there are roughly 3000 IHCs, 9000 OHCs, and 50,000 myelinated axons in the auditory nerve (Retzius 1884; Gacek and Rasmussen 1961); in the mouse, there are 765 IHCs, 2526

OHCs, and 12,350 fibers passing to the organ of Corti (Ehret 1983); and in the rat, there are 960 IHCs, 3470 OHCs, and 15,800 spiral ganglion neurons (Keithley and Feldman 1979, 1982; Berglund and Ryugo 1991). One problem with these quantitative data is that different methods have been used which affect the results. For example, in the cat, the unmyelinated axons were not counted and so they are inferred to number around 2500 based on counts of small ganglion cell bodies (Spoendlin 1971; Kiang et al. 1984). In the mouse, the counts at the habenular openings include both myelinated and unmyelinated primary fibers, but also include the efferent fibers. In humans, there are between 2800–4400 IHCs, 11,200–16,000 OHCs, 31,400 cochlear nerve fibers, and 25,000–30,000 cell bodies in the spiral ganglion (Retzius 1884; Rasmussen 1940; Bredberg 1968). In addition to methodological differences in data collection, there are actual individual variations within a species as well as fundamental variations across species. Quantitative data are nevertheless important because comparative studies in neuronal structure and function can be correlated with life styles of various species. Comparative studies can often provide valuable insight into the behavioral correlates of central nervous system variations (e.g., Irving and Harrison 1967; Masterton et al. 1975).

2.1 Fine Structure of Cochlear Innervation

The nature in which hair cells are connected to the brain will influence how we consider the function of the separate hair cell populations and how we model information passing through the auditory nerve. For example, models of afferent input to the brain will obviously be different if all hair cells receive an identical pattern of innervation or if IHCs have a different innervation pattern than do OHCs. Synapses are located at the base of IHCs and OHCs in the organ of Corti where two types of endings are present (Engström and Wersäll 1958; Kimura 1975; Nadol 1983a,b). One kind of ending contains many round vesicles and is part of the efferent system whose cell bodies reside in the brain stem. The other kind contains few vesicles, exhibits a thickening of its membrane where it apposes the hair cell, and is part of the afferent system whose cell bodies reside in the spiral ganglion.

2.1.1 Inner Hair Cells

There are roughly from 10–30 afferent terminal endings on each IHC in cats (Spoendlin 1969; Liberman 1980a) and a variable number of endings under IHCs of other species (e.g., Nadol 1988a). Moreover, this number appears dependent upon the region of the cochlea sampled (Keithley and Schreiber 1987; Liberman, Dodds, and Pierce 1990). Regardless, the endings of the IHCs are almost exclusively afferent, a conclusion based on their nonvesiculated appearance and their abutment against synaptic bodies of the hair cell (Kimura 1975). Each small afferent ending typically forms a single synapse with the hair cell (Spoendlin 1973; Liberman 1980a). At the synapse, the plasma membranes of the afferent terminal and hair cell are separated by a narrow gap (100–150 Å) and the afferent terminal membrane bears a dense, fuzzy coat (Fig. 2.1). The region of membrane bearing this coat as reconstructed from serial sections would appear as an elliptical plate (dimensions roughly 0.7 × 0.5 μm). Within the hair cell cytoplasm, positioned along a groove in the presynaptic membrane, is found the synaptic body around which is clustered round vesicles, approximately 35 nm in diameter. The synaptic body is an electron-dense, elongated structure oriented parallel to the long axis of the synaptic plate. Synaptic bodies range in length from 0.2–0.6 μm and 0.06–0.12 μm in diameter. The endings originate from radial fibers of the spiral ganglion which pass through the habenular openings to reach the base of the IHC (Smith 1961; Spoendlin 1973; Liberman 1980a). Radial fibers

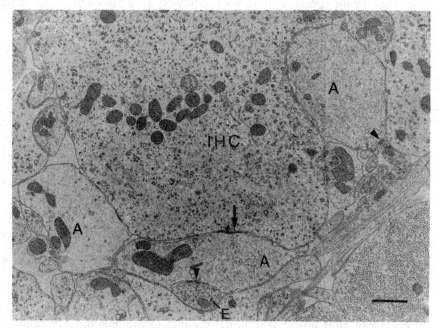

FIGURE 2.1. Afferent nerve endings (A) near the base of an IHC in a squirrel monkey. One afferent ending receives a synapse (arrow) characterized by a marked thickening of the postsynaptic membrane and a presynaptic body surrounded by a halo of round vesicles. An efferent ending (E), filled with synaptic vesicles, forms a synapse (arrowhead) on the afferent ending. Efferent endings do not usually make synaptic contact directly with the IHC. Scale bar equals 1 μm. (This micrograph is courtesy of Dr. R.S. Kimura.)

in the mature cat rarely if ever branch (Spoendlin 1973; Liberman 1980a, 1982a,b), but if they do, they are restricted to the region apical to the 1 kHz point (Liberman, Dodds, and Pierce 1990). Consequently, individual type I spiral ganglion neurons tend to receive all of their input from a single IHC. A single IHC, however, is divergent in that it sends information to the brain by way of many type I neurons, and each type I neuron makes synaptic contact with many different cell types in the cochlear nucleus.

Vesiculated endings do not usually make direct contact with IHCs. Instead, they establish what appears to be a typical axodendritic synapse on the afferent process postsynaptic to the hair cell (Smith 1961; Smith and Rasmussen 1963; Liberman 1980b). In the basal half of the cochlea, each radial fiber received approximately 30 efferent synapses, whereas in the apical half, there were from 35–40 efferent synapses per radial fiber (Liberman, Dodds, and Pierce 1990). These efferents belong to what is commonly called the lateral olivocochlear system (Warr and Guinan 1979; Guinan, Warr, and Norris 1983; White and Warr 1983). The cell bodies of these efferents are relatively small and fusiform in shape, stain positively for the presence of acetylcholinesterase, and lie within the superior olivary complex, lateral to the medial superior olive (Warr 1975). The axons of the lateral olivocochlear efferents are thin and unmyelinated (Brown et al. 1988b).

2.1.2 Outer Hair Cells

There are two types of endings under OHCs. Small, nonvesiculated endings are considered to be afferent and large, vesiculated endings are considered to be efferent. In the cat, there are between 4 and 10 afferent terminal endings under each OHC in the cochlear base (20–50 kHz region) and this number gradually increases to about 10–20 afferent endings under each OHC in the cochlear apex (below 3 kHz region; Simmons and Liberman 1988; Liberman, Dodds, and Pierce 1990). At the synapse, the width of the hair cell membrane tends to be slightly thicker than that of the afferent terminal, separated by a gap of approximately 150 Å (Fig. 2.2). In 30–40% of the cases, a synaptic body is present but variable in size and shape and surrounded by a few vesicles; in the remaining cases, no synaptic body is observed (Dunn 1975; Nadol 1983b). These afferent endings are typically surrounded by vesiculated endings and are sometimes postsynaptic to them (Kimura 1975). Because individual outer spiral fibers can emit up to 60 terminal endings (Simmons and Liberman 1988), a single type II spiral ganglion neuron receives converging inputs from many OHCs, and a single OHC sends its information to the brain by way of a small but variable number of type II neurons. In contrast to the innervation density observed under IHCs, the number of afferent terminals per OHC tends to fall from apex to base (Liberman, Dodds, and Pierce 1990).

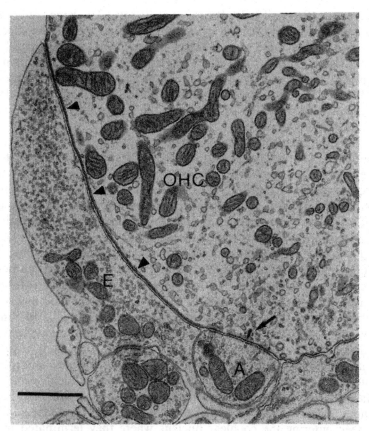

FIGURE 2.2. Nerve endings at the base of an OHC of a rhesus monkey. The afferent nerve ending (A) is relatively small and the synaptic contact with the hair cell (arrow) is characterized by a membrane thickening, a presynaptic body, and a few synaptic vesicles. The efferent ending (E) is typically large, filled with synaptic vesicles, and makes direct contact with the outer hair cell. In addition to the synaptic vesicles, the efferent ending opposes a long subsynaptic cystern (arrowheads) inside the hair cell. Scale bar equals 1 μm. (This micrograph is courtesy of Dr. R.S. Kimura.)

Efferent endings are typically larger (3 μm in diameter) than the afferent endings (1 μm) and they generally make direct contact with the hair cell (Smith and Sjöstrand 1961; Spoendlin 1969; Kimura 1975; Nadol 1983b). The membrane appositions between efferent terminals and OHC are smooth with patches of electron-dense material appearing along the pre-synaptic membrane. The membranes on both sides of the synaptic gap are equal in thickness and the synaptic gap (approximately 190 Å) is slightly larger than those of the afferent endings. These efferent endings are from the so-called medial olivocochlear system and have thick myelinated axons (Brown 1987b; Brown et al. 1988b). The larger cell bodies of the superior olivary complex, located medial to the medial superior olive, give rise to these thick efferent axons (Warr and Guinan 1979; Guinan, Warr, and Norris 1983; White and Warr 1983). Efferent innervation of OHCs becomes less frequent in the apical regions of the cochlea (Smith 1961; Ishii and Balough 1968; Guinan, Warr, and Norris 1983; Liberman, Dodds, and Pierce 1990) and the size of the efferent boutons becomes smaller (Brown 1987b).

Medial olivocochlear efferent fibers (but not lateral olivocochlear efferent fibers) in mouse, gerbil, and cat have been shown to send collaterals to the granule cell regions of the cochlear nucleus ipsilateral to their cochlear target (Brown et al. 1988b). These observations are based on the horseradish peroxidase (HRP) method that permits the continuous tracing of individually stained axons. The time-consuming and technically demanding reconstructions required of this method, however, make it difficult to have a large sample size. Nevertheless, these single fiber data are consistent with retrograde labeling studies, where labeled cell bodies of efferent neurons were never (Covey, Jones, and Casseday 1984; Spangler et al. 1987) or rarely (Adams 1983) found in regions known to be the origins of the lateral efferent system, whereas many labeled somata were found associated with the medial efferent system following tracer injections in the cochlear nucleus. Unfortunately, one study using an alternative population method in the gerbil reached a somewhat different conclusion (Ryan et al. 1987). Following cochlear incubation with D[³H]aspartic acid and subsequent autoradiographic processing, lateral (but not medial) somata were labeled in the brain stem and labeled fibers were seen entering the cochlear nucleus where they terminated in the central part of the ventral cochlear nucleus, a region where no HRP-labeled olivocochlear efferents were observed. Ryan et al. (1987) concluded that fibers of the lateral efferent system branched to innervate the cochlear nucleus. This conflict is resolvable if there exists a subpopulation of efferents from the lateral system that are not revealed using HRP methods.

The differential distribution of efferent endings along the cochlear spiral implies that the system will have less "feedback" influence on afferent information originating from more apical regions, where low frequencies are encoded. Electrophysiological data are consistent with this idea (Wiederhold 1970). When electrical shocks are applied to the olivocochlear bundle, sound-evoked responses of IHCs (Brown and Nuttall 1984) and type I afferents (Wiederhold and Kiang 1970) are reduced, presumably due to activation of the medial olivocochlear efferents (Gifford and Guinan 1987). The idea is that the thick, myelinated medial efferents have much lower thresholds and greater ability to follow the high-frequency shock rates than do the thin, unmyelinated lateral efferents. Because the medial efferents project to the OHCs and the affected neural responses are from type I afferents innervating IHCs, efferent modification of the OHCs is hypothesized to be mechanically coupled to the IHCs, thereby altering IHC output (Brown and Nuttall 1984; Kiang et al. 1986). This altered receptor output is then reflected in the responses of the type I auditory nerve fibers.

2.1.3 Human Cochlear Anatomy

The fundamental structural plan of the mammalian cochlea is remarkably similar across species. Although there are some differences when comparing humans to other mammals, the functional significance of these differences remains to be determined. The afferent innervation density for both IHCs and OHCs is somewhat lower in humans than what has been found in other mammals, such as guinea pig and cat (Nadol 1988a,b). There are roughly 10 afferent endings per IHC throughout the cochlear length, but for OHCs, the number is four per OHC in the base and rises to eight per OHC in the apex. Furthermore, it is common for a single radial fiber to branch and innervate up to three adjacent IHCs, and for an IHC to form multiple synapses with a single afferent terminal (Nadol 1983a). This situation is different from the unbranched, single synapse afferents in the cat (Spoendlin 1973; Liberman 1980a), but the functional significance of such differences is not known. In addition, there is evidence for reciprocal synapses under OHCs in humans (Nadol 1981, 1983b). That is, some afferent terminals demonstrate synaptic specializations consistent with transmission from neuron to hair cell. Since a single outer spiral fiber contacts many OHCs along a segment of cochlea, reciprocal synapses may be a mechanism whereby one OHC could affect the output of nearby OHCs in surrounding frequency regions. The result might produce a kind of "lateral inhibition" for sharpening frequency resolution.

2.2 Innervation Density

In order to understand the possible substrate for hearing sensitivity, the relationship between sensory receptor density, receptor innervation density, and frequency sensitivity has been investigated. Variations are present in the number of IHCs and OHCs per mm along the organ of Corti (Ehret and Frankenreiter 1977; Ramprashad et al. 1978; Bruns and Schmieszek 1980; Bohne, Kenworthy, and Carr 1982; Liberman 1982b; Burda 1984) and in hair cell innervation density (Spoendlin 1972; Morrison, Schindler, and Wersäll 1975; Nadol 1983a,b; Keithley and Schreiber 1987). An average number of type I afferent terminals per IHC can be calculated based on figures published. The number is 8–11 for human (Held 1926; Guild et al. 1931; Nadol 1988a), 10 for guinea pig (Firbas 1972), 13 for mouse (Ehret and Frankenreiter 1977; Ehret 1979; Keithley and Feldman 1983), 10–26 for cat (Gacek and Rasmussen 1961; Spoendlin 1969, 1973; Liberman 1980a; Keithley and Schreiber 1987; Liberman, Dodds, and Pierce 1990), 8–24 for horseshoe bat (Bruns and Schmieszek 1980), and 70 for the little brown bat (Ramprashad et al. 1978). Innervation density, however, is not homogeneous along the cochlear length. In the cat, for example, innervation density rises progressively from 10 per IHC in the apex to approximately 30 per IHC in the base (Keithley and Schreiber 1987; Liberman, Dodds, and Pierce 1990).

Significantly less is known about afferent innervation density for OHCs, an issue complicated by the amount of outer spiral fiber branching (Smith 1975). Published counts of innervation density per single OHC yield 7–10 for man (Held 1926), 5–15 for guinea pig (Smith and Sjöstrand 1961), 6–10 for cat (Spoendlin 1969), and 3 for horseshoe bat (Bruns and Schmieszek 1980). The variation in the number of afferent terminals per OHC also differs with respect to the row in which the OHC resides (Liberman, Dodds, and Pierce 1990). With the exception of the little brown bat, fluctuations of innervation density across the above-mentioned mammals seems relatively minor and may not have any significant effect on auditory function. Within a species, however, absolute threshold sensitivity tends to correspond with frequency regions having the highest inner and outer hair cell density and the highest nerve fiber density (Ramprashad et al. 1979; Ehret 1983). It is a working hypothesis that by considering structural variations in the context of acoustic requirements and behavioral adaptations of individual species, we may gain insight into those structural features that contribute to specific features of auditory function.

Since the audiogram for the population of single auditory nerve fibers is similar to the behavioral audiogram in cats (Liberman and Kiang 1978), behavioral thresholds might possibly be explainable at the level of single primary neurons, since an animal should not be more sensitive than its basic units. A different role for hair cell and/or innervation density could be proposed whereby frequency discrimination is the result of variations in receptor density per octave. One part to this argument is derived from the observation that in the cochlea of the greater horseshoe bat, there is a widely expanded frequency representation between 83 and 83.2 kHz creating an "acoustic fovea" (Bruns and Schmieszek 1980). This behaviorally important frequency range is the constant frequency segment of the bat's signal used to echolocate moving prey. The other part of the argument arises by analyzing the number of IHCs per octave and comparing discrimination sensitivity. Between the 1 and 2 kHz region of the cochlea, humans have roughly 360 IHCs, whereas cats have 270 IHCs; within this range of hearing, humans are better at making frequency discriminations (Schuknecht 1960; Liberman 1982b; Nadol 1983a; Fay 1988).

2.3 Ganglion Cells and Hair Cell Innervation

The studies described so far have limited their analyses primarily to the organ of Corti. The examination of restricted regions is obviously easier than that of large regions. This research "strategy" is highly practical when studying primary auditory neurons because the cell bodies and peripheral processes are encased in the bony capsule of the cochlea and the central processes terminate several millimeters away in the brain. This situation leads to nontrivial technical problems in the study of cochlear neurons because bone must be removed or decalcified in order to access the cochlear tissue. The histological preparation of bone using calcium chelating agents (e.g., dilute hydrochloric acid or ethylenediamine tetraacetic acid) can compromise cochlear tissue if adequate care is not taken to preserve the tissue while simultaneously decalcifying the bone. Another technical obstacle in studying primary neurons (and in fact all neurons) is that no single staining procedure reveals all aspects of a neuron. For example, Nissl stains reveal ribosomes and chromatin material, and so provide cytological details of the cell body and nucleus but not of the processes (ribosomes and chromatin are not present in dendrites or axons). Protargol, a proteinaceous silver compound, presumably condenses neurofilaments into neurofibrils, thereby revealing the skeletal form of the cell body and its processes (Gray and Guillery 1966). The Golgi method randomly stains individual neurons in their entirety (except for the myelinated axons) by way of a silver or mercuric chromate precipitate, but obscures all internal features (Ramón-Moliner 1970; Valverde 1970). Thus, an image of primary auditory neurons has historically been formed by combining observations from different staining procedures, sometimes from different species, and often from animals of different ages.

The classic descriptions of primary auditory neurons were based on the Golgi method applied to neonatal mice and cats (Ramón y Cajal 1909; Lorente de Nó 1937). The Golgi method stains approximately 1–5% of the cell population, revealing individual neurons or parts of neurons as opaque entities but in a mostly uncontrolled fashion (Valverde 1970). Axons and dendrites from different sources are typically intermixed and only in rare instances can a process be followed for any distance before it becomes entangled with other processes. It was within the limits of these methods that three categories of primary neurons were originally described: those neurons that innervated exclusively IHCs, those that innervated exclusively OHCs, and those that contacted both IHCs and OHCs (Retzius 1892; von Ebner 1903; Lorente de Nó 1937; Polyak, McHugh, and Judd 1946).

On the basis of somatic staining characteristics and size criteria, two populations of spiral ganglion cells have been reported (Munzer 1931;

Suzuki, Watanabe, and Osada 1963; Kellerhals, Engström, and Ades 1967; Spoendlin 1973). The problem was how to relate the two populations of ganglion cells to the three patterns of peripheral innervation. Electron microscopic observations confirmed the presence of two major populations of primary neurons: large myelinated type I ganglion cells and small unmyelinated type II ganglion cells (Spoendlin 1973). The advantage of the electron microscope is the great magnification and resolution of structural details; the disadvantage is its restricted view. Another consideration is that the tissue sections are very thin (only 70 nm thick), so for all practical purposes, it is impossible to use the electron microscope to trace processes from the cell bodies to their innervation sites in the organ of Corti. The distance involved in such a task would require thousands of sections.

Spoendlin (1973) performed some important studies that provided strong but indirect evidence for linking the two types of afferent fibers in the organ of Corti with the two types of ganglion cell bodies. He examined the nonvesiculated endings postsynaptic to the hair cells in normal cats and in cats whose auditory nerve had been severed. Nonvesiculated endings in contact with IHCs degenerated following auditory nerve damage, whereas those in contact with OHCs remained intact. By correlating the disappearance of IHC endings with the disappearance of type I ganglion cell bodies, Spoendlin postulated that type I cells must give rise to the radial fibers contacting IHCs. Since the afferent endings under OHCs and type II cell bodies remained, Spoendlin concluded that type II neurons must give rise to outer spiral fibers innervating OHCs. Spoendlin (1975) also described a so-called type III ganglion cell in his pathological cases, suggesting that they innervated OHCs, but later stated (1979) that they innervated IHCs. Type III ganglion cells are not present in normal cochleas, and their function in pathological cochleas is unknown. In the basal region of the cochleas of neonatal cats and rats, giant fibers were described in Golgi material which projected to a group of IHCs; such giant fibers, however, were not observed in animals older than a few days and their cell bodies of origin were not defined (Perkins and Morest 1975). In short, the correspondence between the separate populations of ganglion cells and the different innervation patterns in the organ of Corti depends on the age of the animal under study and the condition of its cochlea.

2.4 Current Status

Spoendlin's hypothesized innervation pattern for normal adult cats finally received direct confirmation by studies that were able to trace individual afferent fibers continuously from their cell bodies in the spiral ganglion to their terminations in the organ of Corti (Kiang et al. 1982; Ginzburg and Morest 1983; Berglund and Ryugo 1987; Brown 1987a). Cochlear innervation in adolescent and mature animals revealed a segregated pattern of hair cell innervation. The population of large cell bodies gave rise to peripheral processes, called radial fibers, that contacted one or rarely two IHCs, whereas the population of small cell bodies gave rise to peripheral processes, called outer spiral fibers, that contacted many OHCs (Fig. 2.3).

The various descriptions in cochlear innervation may be accounted for by ignoring observations made in pathological cases and by postulating that the innervation of both IHCs and OHCs by a single fiber normally represents an immature and transient pattern. The notion that differences

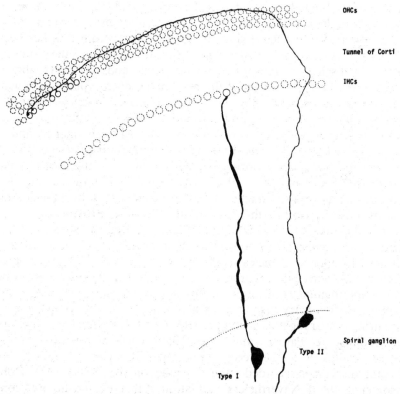

FIGURE 2.3. Composite drawing tube reconstruction of HRP-labeled type I and type II neurons in the mouse. These neurons are representative of the two populations of ganglion neurons in mammals. The type I neuron gives rise to a relatively thick process, projects radially towards the organ of Corti, and terminates on a single IHC. The type II neuron sends a continuously thin process that crosses the tunnel, spirals basally, and then terminates on several OHCs. Scale bar equals 10 μm. (From Berglund and Ryugo 1987, Wiley-Liss Publishers.)

in afferent innervation exist between cochleas of immature and mature animals is consistent with current ideas that exuberant projections retract and the innervation field becomes more restricted as the animal ages (e.g., Innocenti, Fiori, and Caminiti 1977; Jackson and Parks 1982; Stanfield and O'Leary 1985). It is now generally accepted that there is a complete segregation of pathways from the two classes of hair cell receptors by way of the two populations of primary neurons.

2.5 Structure–Function Correlates

The physiological responses of individual spiral ganglion neurons can be monitored as they are conveyed along the axon using microelectrodes inserted into the auditory nerve (e.g., Kiang et al. 1965; Evans 1972). All single unit activity recorded from the auditory nerve, however, is derived from the myelinated axons of type I neurons and virtually nothing is known about the response properties of type II neurons (Liberman 1982a; Robertson 1984). For these kinds of single unit recordings, it has been suggested that under a limited range of stimulus conditions the temporal pattern of discharges transmitted in any single auditory nerve fiber is qualitatively similar throughout the myelinated fiber population. Quantitatively, however, single fiber responses differ in frequency selectivity, spontaneous discharge rate, and threshold. Frequency selectivity refers to a fiber's tendency to be most sensitive to a single frequency, and is best illustrated by a "threshold tuning curve." A tuning curve describes the level and frequency coordinates of a neuron's response area to tonal sitmuli (see Fig. 3 of Evans 1972). The tip of this curve indicates the neuron's characteristic frequency (CF, that frequency to which the neuron is most sensitive) and its threshold in dB SPL to that frequency.

Since the time of von Békésy (1960), it has been accepted that the mammalian cochlea acts as a mechanical frequency analyzer, maximally sensitive to high frequencies in the basal region and sensitive to progressively lower frequencies in regions located progressively closer to the apex. Experimental verification of this notion has been derived by the mapping of frequency responses as a function of cochlear position by measuring basilar membrane motion (Rhode 1971; Khanna and Leonard 1982; Johnstone, Patuzzi, and Yates 1986), cochlear microphonic responses (Dallos 1971), IHC receptor potentials (Russell and Sellick 1978), or spike discharges recorded from the ganglion (Kohllöffel 1975; Robertson et al. 1980). Nevertheless, validation of this relationship was considered incomplete because direct access to the organ of Corti and spiral ganglion was not possible for all cochlear regions. "Place" maps were also generated by plotting the position of hair cell lesions as they related to behavioral threshold changes (Schuknecht 1953) or to alterations in the discharge patterns of auditory nerve fibers (Liberman and Kiang 1978). These latter studies suffer from some ambiguity because of the difficulty in standardizing criteria for identifying histological damage. That is, one wonders how much histological damage must be evident before function is affected, and whether hair cell appearance at the light microscopic level is sufficient criterion for accessing normal hair cell function. A definitive frequency map was finally generated for the cat by intracellular labeling of single auditory neurons of known CFs and identifying the specific IHC upon which the primary neuron terminated (Liberman 1982b). From such data we know that CF indicates the longitudinal location of the inner hair cell innervated by that fiber (Fig. 2.4). The sensitivity and range in frequency for the population of fibers reflect the hearing capabilities of the animal in question (Fay 1988).

For any particular frequency range, spontaneous discharge rate (SR) can vary from near zero to >100 spikes/sec. Spike discharges in auditory nerve fibers occur in the absence of acoustic stimulation. Fibers having different SRs display systematic differences in their threshold at CF, dynamic range, and discharge rate characteristics in response to sound (Sachs and Abbas 1974; Liberman 1978, 1990; Kim and Molnar 1979; Evans and Palmer 1980; Schalk and Sachs 1980). In the population of auditory nerve fibers, there is a striking bimodal distribution of SR where fibers tend to have rates <10 spikes/sec or >30 spikes/sec (Liberman 1978; Evans and Palmer 1980). On average, 30–40% of the fibers have SRs <20 spikes/sec and are placed in the low SR group, whereas the remaining fibers have SRs >20 spikes/sec and are placed in the high SR group. An electron microscopic study of serial sections through two IHCs at the 2.2 kHz region and two IHCs at the 2.6 kHz region revealed that each radial fiber was unbranched and formed a single synapse with one IHC (Liberman 1980a). Each IHC was innervated by 22, 26, 26, and 30 radial fibers, roughly equivalent to what has been reported for cats by

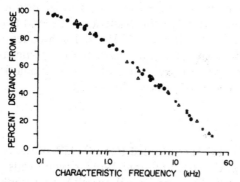

FIGURE 2.4. Relationship between CF of auditory nerve fibers and the relative location of their terminations in the cochlea. Data were determined by intracellular recording and staining methods in cats, and the different symbols are not important for our purposes. (From Liberman 1982b, AIP Publishers.)

Spoendlin (1972, 1973). Forty percent of the 104 radial fibers contacting the side of hair cell facing the modiolus were small in diameter and relatively poor in mitochondria (Liberman 1980a). In contrast, the remaining fibers were larger in diameter, richer in mitochondria, and tended to contact the side of the hair cell facing the pillar cells, but could be found anywhere along the side of the hair cell. The natural conclusion was that high SR fibers were the thicker fibers contacting the pillar side of the hair cell and that low SR fibers were the thinner fibers contacting the modiolar side of the hair cell.

A direct test of this notion was performed by labeling single auditory nerve fibers using intracellular recording and staining techniques, and these relationships were generally confirmed (Liberman 1982a). That is, high SR fibers were on average thicker in diameter and their afferent endings tended to be located on the pillar sector of the IHC. In contrast, low SR fibers were thinner and their endings were typically found in the modiolar sector of the IHC (Fig. 2.5). These differences in the anatomical and physiological properties between SR groups, present across the entire range of frequencies, suggest that the separate SR groups may play fundamentally different roles in the process of auditory perception.

3. Spiral Ganglion

In the bony core of the cochlea, a channel called Rosenthal's canal coils from base to apex in parallel with the cochlear spiral. In this channel reside the cell bodies of primary neurons, collectively called the spiral ganglion. Each cell body exhibits two processes, one extending toward the organ of Corti and the other projecting into the auditory nerve. von Ebner (1903) credits Corti as having first described the cell bodies of primary auditory neurons, circa 1850, where spiral ganglion neurons in the pig were characterized as being bipolar in shape and rather uniform in size. Only much later was the existence of at least two populations of ganglion cells reported (Munzer 1931) and subsequently confirmed on the basis of differences in size and staining properties as seen with the electron microscope (Suzuki, Watanabe, and Osada 1963; Kellerhals, Engström, and Ades 1967; Spoendlin 1973).

3.1 Cell Types

Spoendlin (1971, 1973) concluded that 90–95% of the ganglion cell population in the cat consisted of large, bipolar type I cells and 5–10% were smaller type II cells. The type I ganglion cell has a cytoplasm with a great number of ribosomes and cytoplasmic organelles, whereas the type II cell has a more filamentous cytoplasm with fewer ribosomes (Fig. 2.6). The cell bodies of type II neurons tend to lie in the periphery of Rosenthal's

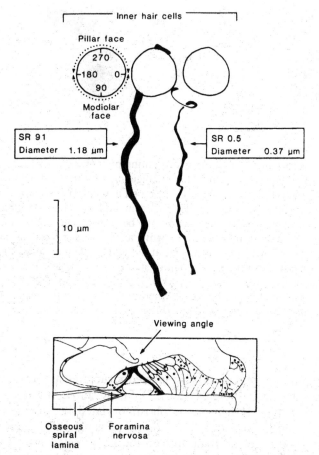

FIGURE 2.5. Schematic drawing of two radial fibers and their terminations upon a single IHC when viewed from above. OHCs would be toward the top of the figure. The high SR fiber is thicker and terminates on the pillar side of the IHC. In contrast, the low SR fiber is thinner and terminates on the modiolar side of the IHC. (From Liberman 1982a, AAAS Publishers.)

canal, towards the osseous spiral lamina (Robertson 1984; Berglund and Ryugo 1987). The histological affinity of type II cell bodies for neurofilament stains (Kiang et al. 1984; Berglund and Ryugo 1986) is consistent with the specific labelling by protargol of proteins that form neurofilaments (Gambetti, Antilio-Gambetti, and Papasozomenos 1982).

Monoclonal antibodies directed against the 200 kD neurofilament protein have unambiguously labeled the cell bodies of type II neurons of the mammalian spiral ganglion (Berglund and Ryugo 1986, 1991; Romand, Hafidi, and Despres 1987). Because tissue stained with basic dyes or protargol still has the cell types intermixed, one principal advantage of

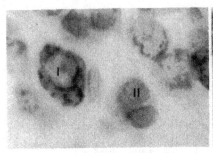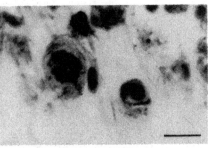

FIGURE 2.6. Left panel: Toluidine blue staining of cochlear spiral ganglion neurons in rat. The type II neuron (II) is distinguished by its small cell body, pale cytoplasm, and dark staining nucleus. The type I neuron (I) has a larger cell body, displays prominent Nissl bodies in its cytoplasm, and has a pale staining nucleus. A satellite cell nucleus is evident against the bottom of the type II neuron and against the side of the type I neuron. Right Panel: The same cells as shown on the left but stained with protargol (and at a slightly different focal plane). The type II cytoplasm and nucleus are darkly stained. In contrast, the nucleus but not the cytoplasm of the type I stains darkly. The implication is that the cytoplasm of type I neurons is rich in ribosomes, whereas that of type II neurons is rich in neurofilaments. Scale bar equals 10 μm. (From Berglund and Ryugo 1986, Elsevier Publishers.)

immunostaining is that positively labeled structures are obviously distinct from unlabeled structures (Fig. 2.7). Furthermore, it was shown that the epitope in type II neurons is phosphorylated because removal of phosphate groups resulted in no staining of ganglion cell bodies, and immunostaining cochlear tissue with antibodies directed against the non-phosphorylated 200 kD neurofilament protein labeled both types of ganglion cells. Monoclonal antibodies directed against the 68 kD or 160 kD neurofilament proteins also labeled the somata of both types of ganglion cells. The antibody method is consistent with ultrastructural descriptions of neurofilament distribution and is able to label selectively the population of type II neurons in a variety of mammals including humans (Berglund and Ryugo 1991). Under normal circumstances, this method can greatly facilitate studies involving counts of type II neurons, plots of their distribution within the ganglion, or morphometry. The application of immunocytochemical techniques may also prove useful for grouping ganglion cell subpopulations by identifying molecules associated with particular structural proteins, metabolic enzymes, or neurotransmitter candidates.

The most unambiguous distinction between the two types of ganglion cells is their segregated innervation of inner versus outer hair cells. Other somatic characteristics of ganglion cells are not universal across species. For example, the presence of a myelin sheath around type I neurons is

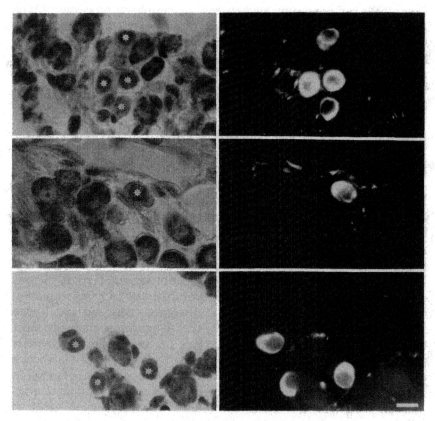

FIGURE 2.7. Immunostaining of type II neurons in rat spiral ganglion. The left panel shows tissue stained with toluidine blue and the right panel shows the same tissue stained with the antibody RT-97. Note how the type II neurons (stars) are preferentially labeled with this antibody which is directed against the phosphorylated 200-kD neurofilament protein. (From Berglund and Ryugo 1986, Elsevier Publishers.)

frequent but not constant; there are as many myelinated as unmyelinated small cells in the human (Ota and Kimura 1980). In the cat, the nucleus of type II neurons is eccentric and lobulated, but such is not the case in humans and rodents (Spoendlin 1981, 1982; Spoendlin and Schrott 1988). Somatic shape is also distinctive in cats where type I neurons tend to be bipolar and type II neurons tend to be pseudomonopolar (Kiang et al. 1982), but in rodents, nearly all ganglion cells appear bipolar (Berglund and Ryugo 1987, 1991; Brown et al. 1988a). Structural idiosyncracies have been found in the cochleas of the macaque monkey where two new ganglion cell classes have been described in addition to the usual type I and type II categories. Dendrodendritic synapses were observed within

Rosenthal's canal, presumably between the peripheral processes of type II ganglion cells, and synapses of efferent and sympathetic origin were found on the perikarya and processes of the unmyelinated neurons (Kimura 1986, 1987). Finally, the distinctions in cell body size between the two cell types diminish as the size of the animal decreases (Brown et al. 1988a). Clearly, more studies of a comparative nature are required in order to fully appreciate the functional significance of these variations.

3.2 Process Characteristics

In the vicinity of the cell body, the processes of the two types of ganglion cells differ in appearance (Kiang et al. 1982; Berglund and Ryugo 1987; Brown 1987a; Brown et al. 1988a). In the cat, for example, the larger type I neurons exhibit a peripheral process whose average diameter is much smaller (0.56 μm) than that of the central process (1.73 μm). In contrast, the type II neurons exhibit processes whose mean diameters are virtually identical (central = 1.25 μm; peripheral = 1.23 μm; Kiang et al. 1982). For the smaller rodents (mice and gerbils), the processes of type II neurons must extend some 100–200 μm away from the cell body before unambiguous diameters differences can be discerned (Berglund and Ryugo 1987; Brown et al. 1988a) In rodents, type II processes gradually taper to less than 1 μm in diameter, whereas type I processes remain relatively constant at 1–2.5 μm in diameter). As the body weight of the species decreases, so does the distinction between the two populations of ganglion cells using cell body silhouette area (Fig. 2.8). Irrespective of these morphologic variations, it is accepted that there are two principal ganglion cell classes.

Type I neurons appear to have a spatial organization within the ganglion. The cell body silhouette area is largest (on average 350 μm^2) for low frequency neurons (<0.2 kHz) and becomes progressively smaller until it levels off (275 μm^2) at about 4 kHz; cell body area does not appear to be related to SR (Liberman and Oliver 1984). The position of intracellularly labeled ganglion cell bodies was mapped along the length of Rosenthal's canal as a function of CF in cats (Keithley and Schreiber 1987). The frequency of organization of the ganglion is similar to that of the organ of Corti: low frequencies are located apically and progressively higher frequencies are located at progressively more basal regions. There is also a spatial relationship between SR and cell body location within Rosenthal's canal (Kawase and Liberman 1991). That is, auditory nerve fibers tend to maintain the regional segregation established at the IHC in the spiral ganglion. Fibers of low SR are much more common in the half of the ganglion nearest scala vestibuli, whereas fibers of high SR fibers are more common in the region of the ganglion nearest scala tympani. This observation is consistent with fibers leaving the organ of Corti and not crossing each other. The modiolar side of the IHC projects

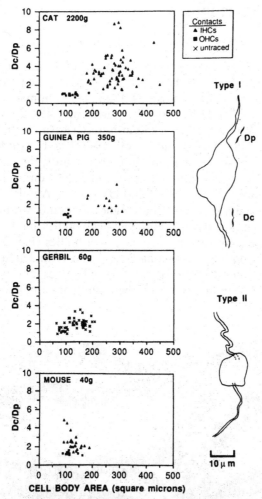

FIGURE 2.8. Morphometry of HRP-labeled spiral ganglion neurons from four species. The legend indicates those neurons whose processes were traced to IHCs, OHCs, or could not be traced to their endings. The ratios of the diameters of the central (Dc) and peripheral (Dp) processes are plotted against cell body silhouette area. Right: drawing tube reconstructions of labeled ganglion cells showing the segments over which Dc and Dp were measured. The drawings are from a type I neuron traced to an IHC and a type II neuron traced to OHCs in a guinea pig. (From Brown et al. 1988a, Wiley-Liss Publishers.)

through the scala vestibuli pole of the ganglion and the pillar side of the IHC projects through the scala tympani pole of the ganglion. These data directly confirm a previously proposed feature of spatial organization within the ganglion (Leake and Snyder 1989).

4. Central Projections

4.1 Auditory Nerve

Fibers of the auditory nerve travel in the central core of the cochlea, called the modiolus, toward the cochlear nucleus in the brain. There are approximately 50,000 fibers in cat, 31,400 fibers in human, 31,250 fibers in rhesus monkey, 31,240 fibers in squirrel monkey, 24,000 fibers in guinea pig, 15,800 in rat, and 12,250 in mouse (Rasmussen 1946; Gacek and Rasmussen 1961; Alving and Cowan 1971; Ehret 1979; Keithley and Feldman 1979). Irrespective of absolute fiber counts, the nerve contains roughly 90–95% thick, myelinated fibers and 5–10% unmyelinated fibers (Alving and Cowan 1971; Arnesen and Osen 1978; Anniko and Arnesen 1988). Recent evidence in cats unambiguously demonstrated that thick fibers are myelinated and arise from type I cell bodies and that thin fibers are unmyelinated and arise from type II cell bodies (Fig. 2.9; Ryugo et al. 1991). The proportion of myelinated and unmyelinated fibers matches the proportion of the two types of ganglion cell bodies for other mammals (Kiang et al. 1984), thereby making the argument extremely strong that

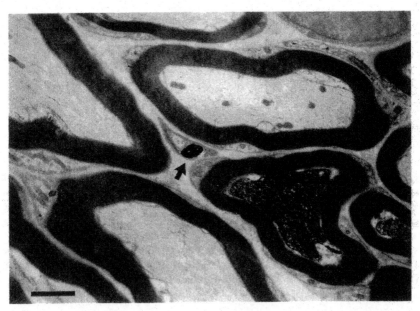

FIGURE 2.9. Electronmicrograph of an HPR-labeled thick and thin fiber in the auditory nerve. The thick fiber is myelinated and arises from a type I spiral ganglion neuron. The thin fiber (arrow) is unmyelinated and arises from a type II spiral ganglion neuron. Scale bar equals 0.5 μm. (From Ryugo et al. 1991, Wiley-Liss Publishers.)

the two types of fibers in the auditory nerve arise respectively from the two types of ganglion cells.

The diameters of the myelinated fibers appear, on average, to be thicker for apical fibers and thinner for basal fibers in cat, squirrel monkey, and mouse (Alving and Cowan 1971; Arnesen and Osen 1978; Liberman and Oliver 1984; Anniko and Arnesen 1988). The diameter difference across these particular species is roughly 0.5 μm. For all frequencies, the axons of high SR fibers are thicker on average than those of low SR fibers (Liberman and Oliver 1984; Kawase and Liberman 1991). It also appears that axons get progressively thicker with increased distance from the cell body. At 1–2 mm from the ganglion, diameters tend to stabilize following as much as a 50% increase.

Fibers originating from the apex of the cochlea occupy the middle of the nerve trunk with more basal fibers spiralling and coursing peripheral to the more apical ones (Arnesen and Osen 1978). This spatial arrangement of cochlear fibers is orderly within the nerve trunk and continues into the cochlear nucleus (Sando 1965; Arnesen, Osen, and Mugnaini 1978), and is reflected by the tonotopic organization of the nerve (Liberman and Kiang 1978). It seems that both fiber types adhere to this cochleotopic relationship (Brown et al. 1988a).

4.2 Cochlear Nucleus

4.2.1 Type I Fibers

The myelinated auditory nerve fibers are qualitatively similar to one another. Figure 2.10 illustrates several features common to most fibers. The "root branch" crosses the Schwann-glia border, ascends for a distance into the nucleus, and then bifurcates. This characteristic bifurcation gives rise to an ascending branch and a descending branch. The ascending branch is directed anteriorly through the anteroventral cochlear nucleus (AVCN) and the descending branch is directed posteriorly through the posteroventral cochlear nucleus (PVCN), toward and usually into the dorsal cochlear nucleus (DCN). The ascending, descending, and root branches generally maintain straight trajectories, exhibit the largest diameters, and are commonly referred to as "parent" branches.

4.2.2 Root Branch

The portion of the fiber extending between the spiral ganglion and the bifurcation was classically called the radicular or root branch (Ramón y Cajal 1909). The total length of this root branch, whether originating from apical or basal regions, is approximately the same, namely 7 to 8 mm from their entrance into the spiral lamina to the tips of their ascending branches (Arnesen and Osen 1978; Fekete et al. 1984). Although the cell bodies of basal fibers are "closer" to the Schwann-glia border

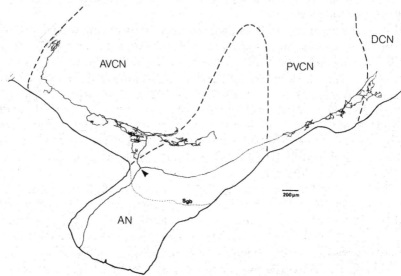

FIGURE 2.10. Drawing tube reconstruction of single auditory nerve fiber (CF = 0.45 kHz, SR = 1.2s/s) in a left cochlear nucleus. Anterior is toward the left and dorsomedial is toward the top. Upon entering the nucleus, the root branch bifurcates (arrowhead) to form the ascending and descending branches. This fiber forms a relatively planar arborization within the nucleus whose spatial position is related to the location of the bifurcation. Abbreviations: AN, auditory nerve; AVCN, anteroventral cochlear nucleus; DCN, dorsal cochlear nucleus; PVCN, posteroventral cochlear nucleus; Sgb, Schwann-glial border. (From Ryugo and Rouiller 1988, Wiley-Liss Publishers.)

than are those of apical fibers, the axons of basal fibers must penetrate further into the nucleus before bifurcating and arborizing. That is, fibers arising from the apex (the low-frequency region) ramify just after crossing the Schwann-glia border to distribute endings in the ventral portion of the cochlear nucleus. Fibers arising from progressively more basal regions of the cochlea (higher-frequency regions) bifurcate and arborize in progressively more dorsal regions of the cochlear nucleus (Fig. 2.11). This orderly arrangement in the distribution of auditory nerve fibers represents what is termed a "cochleotopic" projection, observed originally in Golgi stained material (Ramón y Cajal 1909; Lorente de Nó 1933).

This cochleotopic projection has been verified by several different methods. One method was a population approach where small lesions were placed in different locations along the cochlear spiral, and the resulting distribution of silver grains attracted by degenerating fibers and terminals was plotted (Sando 1965; Osen 1970; Webster 1971; Moskowitz and Liu 1972; Noda and Pirsig 1974). Other population studies made

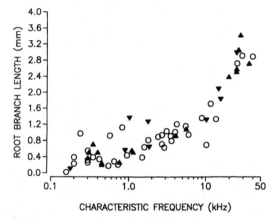

FIGURE 2.11. Relationship of fiber CF to root branch length (measured from Schwann-glia border to bifurcation). These data reveal that the location of the bifurcation is systematically related to fiber CF (r = 0.95, p < 0.001) but not fiber SR. Symbols: Open circles, SR > 18 s/s; filled triangles, SR < 18 s/s.

small HRP deposits in restricted sectors of the spiral ganglion and mapped the distribution of anterogradely labeled central axons (Leake and Snyder 1989; Brown and Ledwith 1990). From both types of anterograde tracing studies, projection "bands" of stained fibers and terminals were revealed whose regional locations corresponded systematically to the lesion or injection site in the cochlea. The other method used the intracellular technique where the task was to record the characteristic frequency of a single fiber, inject the fiber with HRP, and then reconstruct the labeled axonal arborization within the cochlear nucleus (Fig. 2.12). These results of auditory nerve projections (Fekete et al. 1984; Rouiller and Ryugo 1988; Ryugo and Rouiller 1988; Wright et al. 1991) are consistent with the electrophysiological data reporting the tonotopic organization for the cochlear nucleus (Rose, Galambos, and Hughes 1959; Bourk, Mielcarz, and Norris 1981; Spirou, May, and Ryugo 1989).

4.2.3 Ascending Branch

For every auditory nerve fiber, the ascending branch arises at the bifurcation of the root branch, exhibits a relatively straight trajectory into the anterior division of the AVCN, and terminates as a large, axosomatic ending called the endbulb of Held (Ramón y Cajal 1909). The mean (± SD) length of the ascending branch is relatively uniform (2.33 ± 0.53 mm in cat) and unrelated to fiber CF or SR. Each parent branch gives rise to an average of nine primary collaterals, a number also unrelated to fiber CF or SR (Fekete et al. 1984). The collateral ramifications of most ascending branches, especially those of high SR fibers, are short and

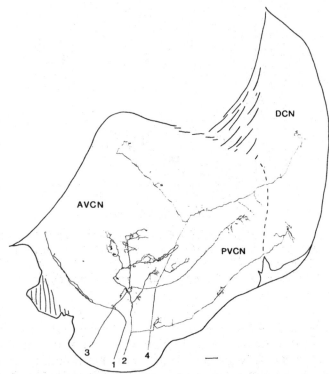

FIGURE 2.12. Drawing tube reconstruction of four intracellularly labeled auditory nerve fibers in a left cochlear nucleus. There is a systematic relationship between fiber CF and location of arborization in the nucleus, but its manifestation can be distorted when viewed in only two dimensions. In the AVCN, for example, the ascending branches of fiber 2 and fiber 3 project out of the plane of the paper towards the viewer. Fiber 1 has no descending branch, CF = 0.2 kHz, SR = 112 s/s; Fiber 2, CF = 1.2 kHz, SR = 1.1 s/s; Fiber 3 has a descending branch that does not enter the DCN, CF = 4.5 kHz, SR = 49.6 s/s; Fiber 4, CF = 20 kHz, SR = 61 s/s. AVCN, anteroventral cochlear nucleus; PVCN, posteroventral cochlear nucleus; DCN, dorsal cochlear nucleus. (Scale bar equal 100 μm.)

relatively simple. The relative thickness of terminal collaterals roughly corresponds to the size of the terminal swellings. The overall complexity of the ascending branch arborization, however, is strongly correlated to fiber SR (Fekete et al. 1984; Ryugo and Rouiller 1988). In cat, the number of branch points for low SR fibers is on average, more than twice than that for high SR fibers. The number of branch points plus one equals the number of terminal swellings, and the number of terminal swellings is a rough estimate of the number of neurons synaptically contacted by that fiber. In addition, the mean collateral length for low SR fibers is 55% greater than that of high SR fibers, indicating that synaptic information for low SR fibers is distributed to a wider region.

One interpretation for these observations is that low SR fibers convey information to more neurons distributed over a wider region of the AVCN than do high SR fibers. The data may be related to an idea proposed by Stevens and Davis (1938) in which the perception of loudness was hypothesized to be proportional to the number of active neurons. The excitation of high threshold, low SR fibers by loud sounds would not only increase the pool of active auditory nerve fibers but also could produce a spread of activity to many additional neurons in the AVCN. Such recruitment might be important because a saturation of discharge rate for high SR fibers is already produced by only moderate levels of sound.

There is also circumstantial evidence implicating low SR fibers as part of the neural circuit mediating the middle ear muscle reflex. This reflex occurs in the presence of a loud noise whereby muscles of the middle ear contract and exert feedback control on the cochlear response (Borg 1972). The sensory limb of this reflex passes through the auditory nerve, and the unmyelinated fibers can be eliminated from consideration because their conduction time will be too long (>10 msec) to permit their participation in the reflex (Kiang, Keithley, and Liberman 1983). Neurons that reside in the AVCN apparently contribute to this reflex because transection of the output pathway of the AVCN abolishes the reflex; there is no change in the reflex following transection of the output pathways of the DCN or the PVCN (Borg 1973). The low SR fibers have high thresholds, and therefore are still within their operating range in the presence of loud sounds (85–95 dB) that evoke the middle ear muscle reflex (Liberman and Kiang 1984). Since it is in the AVCN where low SR fibers exhibit their arbor specializations, the current task is to discover which neurons might preferentially receive inputs from the low SR fibers and project in turn to the motor nuclei of cranial nerves V and VII.

4.2.4 Descending Branch

The descending branch is directed posteriorly through the PVCN and usually (85% of the cases), but not always, enters the DCN (Fekete et al. 1984). In cats, each parent branch is 3.38 ± 0.64 mm in length and gives rise to an average of 11 primary collaterals. The morphology of the descending branch arborization appears to be independent of fiber SR, but there are some features of the fiber that are related to fiber CF. Auditory nerve fibers having CFs below 1–2 kHz do not enter the central region of the PVCN, known as the octopus cell region; not surprisingly, low-frequency units are poorly represented in this region (Ritz and Brownell 1982).

On the other hand, fibers with CFs greater than 4 kHz emit collaterals which stream ventrally in long parallel arrays (Fig. 2.12). The orientation of this streaming is perpendicular to the descending parent but roughly parallel to the primary dendrites of octopus cells. Octopus cells represent a physiological class of "on" units, which respond primarily at the onset of a tone burst (Godfrey et al. 1975; Ritz and Brownell 1982). This assignment has been directly confirmed in which "on" units were intracellularly labeled with HRP (Rhode et al. 1983; Rouiller and Ryugo 1984). In general, "on" units are among the most broadly tuned cells in the PVCN when measured 20 dB above threshold, yet when measured at 10 dB above threshold they are as sharply tuned as other unit types in the PVCN (Godfrey et al. 1975). The parallel alignment of dendrites and collateral ramifications may provide an anatomical basis for both the sharpness of frequency selectivity near threshold and the broad tuning at higher intensities. Certain spatial relationships might permit a maximal number of synaptic contacts to be made between a few fibers of similar CFs and the dendrites of a particular octopus cell. Such input would be expected to dominate the cell's response at low stimulus levels, thereby accounting for the sharp tuning near threshold. Primary fibers having higher or lower CFs would make relatively fewer contacts with that octopus cell because of spatial restrictions; consequently, at low stimulus levels, their inputs would not be sufficient to activate the cell. At higher stimulus levels, more primary fibers representing a wider range of frequencies would be activated and their summed activity converging onto the octopus cell might account for the broader frequency response (compared to other cochlear nucleus units or auditory nerve fibers).

Upon entering the DCN, individual fibers ramify and terminate within a relatively narrow band of the deep polymorphic layer, also called layer III (Fig. 2.12). Rarely do fibers reach as far as the fusiform or pyramidal cell layer (also called layer II); no terminals were ever observed in the molecular layer although fibers occasionally passed through this layer on their way elsewhere. The arborization in DCN is sparse, producing only a few *en passant* and terminal swellings. These swellings are relatively small and are found in the neuropil. The topographic banding of terminal arbors in the DCN (Wright et al. 1991) is consistent with the tonotopic arrangement of this nucleus (Rose, Galambos, and Hughes 1959; Spirou, May, and Ryugo 1989).

4.2.5 Type II Axons

There is evidence that the unmyelinated fibers maintain a cochleotopic projection into the nucleus. When the central axons of type II spiral ganglion neurons are traced into the cochlear nucleus, their bifurcations are typically clustered tightly around the bifurcations of the myelinated axons. Likewise, most type II axons send branches toward the AVCN and PVCN following the labeled type I axons from the same region of the spiral ganglion (Brown 1987a; Brown et al. 1988a). Individual type II axons were completely reconstructed along with those of type I axons in gerbil and mouse from basal regions of the cochlea, and both fiber types were overlapping in their trajectories (Fig. 2.13). In the mouse, type II fibers from various cochlear regions exhibit a cochleotopic projection (Berglund and Brown 1989).

Type II axons differ markedly from type I fibers in the complexity of their arborizations (Brown et al. 1988a; Ryugo et al. 1991). In gerbil, type I fibers give rise to an average (± SD) of 29 ± 9 primary collaterals and 80 ± 27 terminal swellings. In contrast, type II fibers exhibit 4 ± 2 pri-

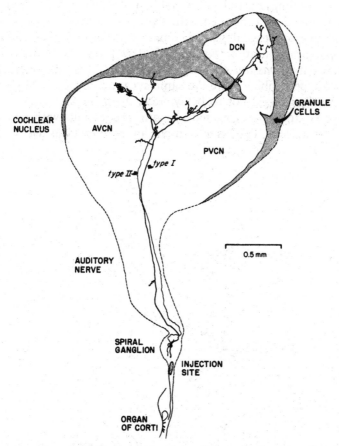

FIGURE 2.13. Drawing tube reconstruction of the central axons of an identified type I neuron (thick line) and an identified II neuron (thin line) through the auditory nerve and into the cochlear nucleus of a gerbil. Although the trajectories of these fibers are similar, the terminations of type II neurons are consistently separated from those of type I neurons because they project into granule cell regions. This drawing is a two-dimensional projection of 21 sections, each 80 μm in thickness. AVCN, anteroventral cochlear nucleus; PVCN, posteroventral cochlear nucleus; DCN, dorsal cohlear nucleus. (Modified from Brown et al. 1988a, Wiley-Liss Publishers.)

mary collaterals and 7 ± 3 terminal swellings; these terminal swellings are all bouton endings. Although the axon trajectories coincide with those of type I fibers, their final destination is in regions containing high densities of granule cells. That is, the terminal swellings of type II projections are distributed in granule cell regions of the dorsal and lateral VCN, the granule cell lamina separating the DCN from the VCN, and the granule cell layer of the DCN. Furthermore, there is a general size difference between the terminal swellings of the two fiber types (Fig. 2.14). For completely reconstructed type II fibers, there was an average of 128 ± 62 *en passant* swellings central to the cell body (Brown et al. 1988a). Many *en passant* swellings examined with the electron microscope, however, did not form synapses and instead, appeared as empty varicosities typical of a pathological condition (Ryugo et al. 1991). Because it is not known to what extent *en passant* swellings are synaptic, it is unknown whether type I and type II fibers share the same synaptic targets. If, however, *en passant* swellings tend not to be synaptic, then the two types of fibers may have completely segregated synaptic connections with neurons in the cochlear nucleus. This issue is important because ideas regarding how

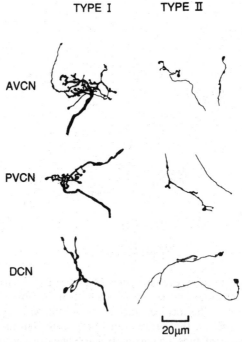

TYPE I   TYPE II

AVCN

PVCN

DCN

20μm

FIGURE 2.14. Drawing tube tracings of terminal swellings typical for type I and type II neurons in the mouse. Type II neurons exhibit thinner fibers and generally smaller terminals than do type I neurons. (From Brown et al. 1988a, Wiley-Liss Publishers.)

the central nervous system processes information from the hair cell receptors will depend substantially on whether the two primary fiber types have separate or overlapping connections.

4.2.6 Type I Fiber Ending Categories

Terminal boutons are the most common category of ending and the endbulbs of Held are the most conspicuous for type I auditory nerve fibers (Fig. 2.15). Held (1893) originally suggested on the basis of analysis of Golgi material in kittens that boutons and endbulbs were the two types of endings emitted by auditory nerve fibers. This conclusion has generally been supported by subsequent investigations in other animals (Ramón y Cajal 1909; Lorente de Nó 1933; Harrison and Irving 1966; Feldman and Harrison 1969; Brawer and Morest 1975). In the mature cat, the majority of endings fall into one of the these two groups although some endings, called modified or small endbulbs, have distinctly intermediate characteristics (Rouiller et al. 1986).

All ending types were found on the ascending branch and all but endbulbs were found on the descending branch. As a result, there is a considerable range in ending size within the VCN. In contrast, the DCN contains a fairly homogeneous population of small primary endings. Despite the variations in morphological appearance, HRP-labeled endings in adult cats could be placed into one of three descriptive categories on the basis of size and shape. Each category showed specific characteristics with respect to spatial distribution in the cochlear nucleus, yet the proportion of the different ending categories appears independent of the fiber CF or SR.

Small endings (approximately 5–7 μm^2) were primarily terminal boutons, represented 94% of all endings, and were distributed throughout the neuropil of the cochlear nucleus. Terminal boutons of low SR fibers are on average 1.5 μm^2 smaller but more numerous compared to those of high SR fibers, a difference especially pronounced for the ascending branch ($p < 0.01$). There is no light microscopic appearance of a relationship between bouton morphology and fiber CF. *En passant* swellings are small (approximately 4 μm^2), constant in size for both SR groups of fibers, and distributed in neuropil. Because they were not always found to form synapses, they are not strictly considered endings (Rouiller et al. 1986).

Intermediate endings (20–40 μm^2) consisted of modified endbulbs which mostly formed axosomatic contacts. This category composed 4% of the ending population and was found in close proximity to the perikarya of primarily globular cells, but also of octopus and spherical cells. These endings from low SR fibers were on average smaller than those of high SR fibers, but there was no difference in numbers. Furthermore, these endings did not vary in their parent branch distribution with respect

FIGURE 2.15. Selected collaterals and endings of a type I fiber. (A) Terminal boutons and *en passant* swellings in the deep layer of the DCN; (B) Long, ventrally directed collaterals terminate as boutons of varying size in the central region of the PVCN; (C, D) Small "modified" endbulbs terminating against the cell bodies of globular cells in the region of the auditory nerve root; (E) Terminal boutons and *en passant* swellings in the AVCN; (F) Endbulb of Held in the anterior division of the AVCN. (From Fekete et al. 1984, Wiley-Liss Publishers.)

to fiber SR, nor did they exhibit morphological features related to fiber CF.

The third category of endings contained exclusively endbulbs of Held (Fig. 2.16). These large endings (200–350 μm^2) represent 2% of the ending population and make axosomatic contact with spherical cells in the anterior division of AVCN and occasionally with globular cells in the posterior division of AVCN. Fibers having CFs below 4 kHz gave rise to the largest endbulbs. There were no systematic variation in endbulb size or branch distribution that correlated with fiber SR.

The similarity in average endbulb size across the SR groups was unexpected, especially in light of their differences in appearance. That is, endbulbs of high SR fibers tend to have larger but fewer swellings and lobules compared to endbulbs of low SR fibers (Fig. 2.17). There is a subjective impression that low SR fibers gave rise to lacy, delicate structures, yielding endbulbs that appear to have greater complexity in form. The ratio, silhouette area divided by silhouette perimeter (with the units dropped), was used to provide an objective value for representing each endbulb. This value is referred to as a "form factor" and it separated endbulbs into two almost nonoverlapping populations according to fiber SR, irrespective of fiber CF or endbulb size (Sento and Ryugo, 1989). Endbulbs from low SR fibers had form factors <0.52, whereas those from high SR fibers had form factors >0.52. These endbulb observations, coupled with root branch length data (Fig. 2.11), suggest that two fundamental properties of auditory nerve fibers, CF and SR group, can be

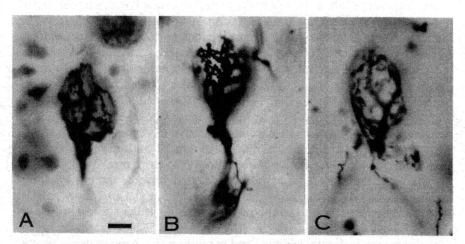

FIGURE 2.16. Photomicrographs of endbulbs of Held in the anterior division of the AVCN from adult cats. Endbulb morphology is revealed by (A) HRP methods, (B) Golgi-Kopsch methods, and (C) Golgi-Cox methods. Scale bar equals 10 μm. (From Ryugo and Fekete 1982, Wiley-Liss Publishers.)

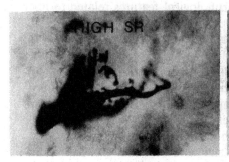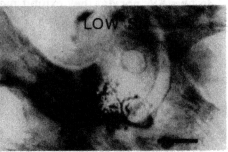

FIGURE 2.17. Photomicrographs of endbulbs revealing SR-related differences in morphology. The endbulb from a high SR fiber (CF = 1.1 kHz, SR = 56 s/s) is compared to that from a low SR fiber (CF = 1.0 kHz; SR = 0.01 s/s). These endbulbs are from the same cat but opposite cochlear nuclei. Note that the endbulb from the low SR fiber has more but smaller lobulations and swellings compared to that of the endbulb from the high SR fiber. Scale bar equals 10 μm. (From Ryugo and Sento 1991, Wiley-Liss Publishers.)

roughly estimated without having to perform the difficult intracellular electrophysiological experiments.

4.2.7 Primary Synapses

Electron microscopic studies have revealed at least four types of terminals that may be distinguished on the basis of differences in synaptic vesicle size and shape, all of which make synaptic contact with second order neurons of the cochlear nucleus (Lenn and Reese 1966; Cohen 1972; Gentschev and Sotelo 1973; Ibata and Pappas 1976; Cant and Morest 1979; Tolbert and Morest 1982). For the purpose of this review, it is significant that all terminals having large (approximately 50–60 nm in diameter), clear, round vesicles disappear after cochlear ablation (Cohen 1972; Cant and Morest 1979; Tolbert and Morest 1982). Such terminals have the same morphological features as those labeled with HRP following extracellular deposits in the auditory nerve (Ryugo and Fekete 1982) or intracellular injections of individual type I auditory nerve fibers (Rouiller et al. 1986; Ryugo and Sento 1991). In typical HRP-labeled endings, mitochondria, numerous clear, round vesicles, and postsynaptic densities are visible (Fig. 2.18). The vesicles, 54.5 ± 4.4 μm (mean ± SD) in diameter, accumulate on the presynaptic side of the synapse, in close proximity to a thickening of the cell membrane. The extracellular space in this region is slightly widened and contains a thin band of dense, fuzzy material. A segment of the postsynaptic cell membrane exhibits a prominent density that is coextensive with the thickened presynaptic membrane, giving the synaptic contact an asymmetrical appearance. The

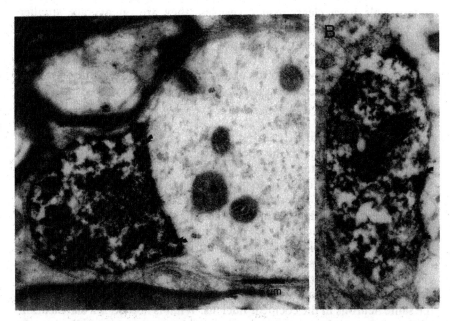

FIGURE 2.18. Electron micrographs comparing the morphology of the central synapses from terminals of type II (A) and type I (B) spiral ganglion neurons in cat. The terminal of the type II neuron contains small round vesicles and a membrane apposition that is largely synaptic (flanked by arrows). In contrast, the terminal of the type I neuron contains large round vesicles and a punctate postsynaptic density (arrows). (From Ryugo et al. 1991, Wiley-Liss Publishers.)

synapse is relatively small (Fig. 2.18, arrows) compared to the total membrane apposition. In the case of the endbulbs, the postsynaptic density often exhibits a slight convexity, whereas for boutons, the postsynaptic density is typically flat. Twenty HRP-labeled terminals in different regions of the cochlear nucleus from intracellularly characterized type I auditory nerve fibers have been examined, and qualitatively, they all have similar morphology (Rouiller et al. 1986).

Synapses from type I auditory nerve fibers are distributed throughout the cochlear nucleus, primarily against cell bodies and dendrites. These terminations do not enter the superficial granule cell regions of the dorsal and lateral parts of the VCN, the lamina of granule cells separating the DCN from the VCN, or the granule cell and molecular layers of the DCN (Osen 1970; Fekete et al. 1984; Brown et al. 1988a). The implication is that type I primary fibers connect with neurons which project axons to higher auditory centers. This situation is in marked contrast to that of type II fibers, which most likely connect with local circuit neurons of the granule cell regions (Fig. 2.18A). The synaptic region of type II fibers is quantitatively different from that of type I fibers. That is, although type

II terminals have clear, round vesicles, they are on average smaller (46.3 ± 4.6 μm in diameter) than those of type I fibers, and much of their membrane apposition is synaptic (Fig. 2.18A, arrows). Although the sample size is small, there is a distinct difference in the size of the post-synaptic plaque compared to the extent of membrane apposition for synapses of the two types of auditory nerve fibers: for synapses of type I fibers, this ratio is 6–26%, whereas for synapses of type II fibers the ratio is 42–77% (Ryugo et al. 1991).

There seems to be little question that type I fibers transmit information from IHCs to neurons of the cochlear nucleus. Type II fibers also have a morphology consistent with the capability to convey "sensory" information from OHCs directly to neurons in the cochlear nucleus. Such communication may not necessarily be directly related to the acoustic environment but rather pertain to the functional status of the OHCs, perhaps providing feedback to the brain analogous to muscle spindle afferents. This idea, however, would not preclude OHCs from participating in mechanisms that determined the responses of IHCs, and influencing the information carried by type I fibers. Nor would it preclude the possibility that both sets of afferents might synaptically converge on the same neurons in the cochlear nucleus. What is needed is a clarification of the response characteristics and central connections of the type II ganglion cells.

4.2.8   Central Projections in Humans

We have very few details concerning the central projections of the auditory nerve in humans. Information has been limited due to the obvious ethical and philosophical considerations, coupled with technological constraints in studying postmortem material. The thickness of the auditory nerve ranges from 1.15–2.62 mm with a mean (± S.D.) of 1.76 ± 0.53 (Natout et al. 1987). Cochlear nerve fibers range in diameter from 3–11 μm with more than half in the 3–5 μm range; no mention has been made of the unmyelinated fibers (Natout et al. 1987). Presently, however, there are methods to study the central projections of single fibers in postmortem tissue. In formalin-fixed tissue, HRP (Kageyama and Meyer 1987; Haber 1988) and carbocyanine dyes (e.g., Godement et al. 1987) have been successfully applied to study neurons and fiber pathways. The fluorescent carbocyanine dyes can be stabilized using a photoconversion procedure (McConnell, Ghosh, and Shatz 1989). These methods open up whole new vistas for neuroanatomical research in the human nervous system.

5. Summary

No description of the auditory nerve can be considered complete until we have much greater knowledge of the details of synaptic connections between primary afferents, peripheral receptors, and neuron classes in the brain. Considerable information is waiting to be gathered. Such a description is necessary if we are to understand the structural basis by which neural activity from the auditory nerve is "processsed" by the cochlear nucleus, because the coding process is hypothesized to be influenced by definable features of the relationship between pre- and post-synaptic cells. We still need to know the specific locations (i.e., cell body, proximal or distal dendrite, axon hillock) that certain terminal types (from type I or type II fibers) make on identifiable cells.

Nevertheless, at this time it still seems safe to propose that the connections between the cochlea and the cochlear nucleus are represented by two separate systems, both of which have the apparatus for chemical transmission in the central nervous system (Fig. 2.19). There is one system for the rapid conduction of auditory information which arises from a punctate region of the basilar membrane and is conveyed by IHCs and the myelinated axons of type I neurons. A parallel system arises from a length of basilar membrane and utilizes OHCs and the unmyelinated axons of type II neurons. Afferent information from both types of hair cell receptors can apparently influence widely separate regions within the cochlear nucleus, but the effective stimulus, the latency to activation, and the neuronal targets appear to be profoundly different.

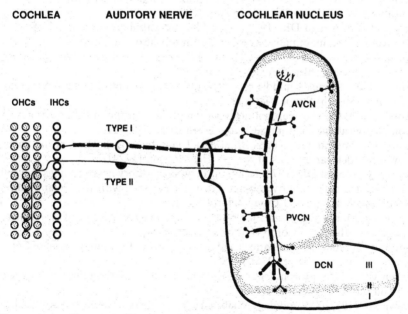

FIGURE 2.19. Summary diagram of the connections between the cochlea and the cochlear nucleus. Two separate but parallel systems convey information from the acoustic receptors to the brain.

Acknowledgements. The author was supported by NIH grant R01 DC00232 during the preparation of this chapter. Special thanks to Tan Pongstaporn and Debora D. Wright for technical assistance, and to Dr. R.S. Kimura for providing micrographs used in Figures 2.1 and 2.2.

References

Adams JC (1983) Cytology of periolivary cells and the organization of their projections. J Comp Neurol 215:275–289.

Alving BM, Cowan WM (1971) Some quantitative observations on the cochlear division of the eighth nerve in the squirrel monkey (*Saimiri sciureus*). Brain Res 25:229–239.

Anniko M, Arnesen AR (1988) Cochlear nerve topography and fiber spectrum in the pigmented mouse. Arch Otorhinolaryngol 245:155–159.

Arnesen AE, Osen KK (1978) The cochlear nerve in the cat: Topography, cochleotopy, and fiber spectrum. J Comp Neurol 178:661–678.

Arnesen AE, Osen KK, Mugnaini E (1978) Temporal and spatial sequence of anterograde degeneration in the cochlear nerve fibers of the cat. A light microscopic study. J Comp Neurol 178:679–696.

Berglund AM, Brown MC (1989) Axonal trajectories of type-II spiral ganglion cells from various cochlear regions in mice. Soc Neurosci Abstr 15:742.

Berglund AM, Ryugo DK (1986) A monoclonal antibody labels type II cells of the spiral ganglion. Brain Res 383:327–332.

Berglund AM, Ryugo DK (1987) Hair cell innervation by spiral ganglion neurons in the mouse. J Comp Neurol 255:560–570.

Berglund AM, Ryugo DK (1991) Neurofilament antibodies and spiral ganglion neurons of the mammalian cochlea. J Comp Neurol (in press).

Bohne BA, Kenworthy A, Carr CD (1982) Density of myelinated nerve fibers in the chinchilla cochlea. J Acoust Soc Am 72:102–107.

Borg E (1972) Acoustic middle ear reflexes: A sensory-control system. Acta Otolaryngol (Stockh) Suppl 304:1–34.

Borg E (1973) On the neuronal organization of the acoustic middle ear reflex. A physiological and anatomical study. Brain Res 49:101–123.

Bourk TR, Mielcarz JP, Norris BE (1981) Tonotopic organization of the anteroventral cochlear nucleus of the cat. Hear Res 4:215–241.

Brawer JR, Morest DK (1975) Relations between auditory nerve endings and cell types in the cat's anteroventral cochlear nucleus seen with the Golgi method and Nomarski optics. J Comp Neurol 160:491–506.

Brawer JR, Morest DK, Kane EC (1974) The neuronal architecture of the cochlear nucleus of the cat. J Comp Neurol 155:251–300.

Bredberg G (1968) Cellular pattern and nerve supply of the human organ of Corti. Acta Otolaryngol (Suppl 236) 1–135.

Brown MC (1987a) Morphology of labeled afferent fibers in the guinea pig cochlea. J Comp Neurol 260:591–604.

Brown MC (1987b) Morphology of labeled efferent fibers in the guinea pig cochlea. J Comp Neurol 260:591–604.

Brown MC, Berglund, AM, Kiang NYS, Ryugo DK (1988a) Central trajectories of type II spiral ganglion neurons. J Comp Neurol 278:581–590.

Brown MC, Ledwith JV (1990) Projections of thin (type-II) and thick (type-I) auditory-nerve fibers into the cochlear nucleus of the mouse. Hear Res 49:105–118.

Brown MC, Liberman MC, Benson TE, Ryugo DK (1988b) Brainstem branches from olivocochlear axons in cats and rodents. J Comp Neurol 278:591–603.

Brown MC, Nuttall AL (1984) Efferent control of cochlear inner hair cell responses in the guinea-pig. J Physiol (Lond) 354:625–646.

Bruns V, Schmieszek E (1980) Cochlear innervation in the greater horseshoe bat: Demonstration of an acoustic fovea. Hear Res 3:27–43.

Burda H (1984) Guinea pig cochlear hair cell density; Its relation to frequency discrimination. Brain Res 14:315–317.

Cant NB, Morest DK (1979) The bushy cells in the anteroventral cochlear nucleus of the cat. A study with the electron microscope. Neuroscience 4:1925–1945.

Cohen ES (1972) Synaptic Organization of the Caudal Cochlear Nucleus of the Cat. Doctoral Thesis, Harvard University, Cambridge, MA.

Covey E, Jones DR, Casseday JH (1984) Projections from the superior olivary complex to the cochlear nucleus in the tree shrew. J Comp Neurol 226:289–305.

Dallos P (1971) On the limitations of cochlear microphonic measurements. J Acoust Soc Am 49:1141–1154.

Dunn RA (1975) A comparison of Golgi-impregnated innervation patterns and fine structural synaptic morphology in the cochlea of the cat. Doctoral Thesis, Harvard University, Cambridge, MA.

Ehret G (1979) Quantitative analysis of nerve fibre densities in the cochlea of the house mouse (*Mus musculus*). J Comp Neurol 193:73–88.

Ehret G (1983) Peripheral anatomy and physiology II. In: Willott JF (ed) The Auditory Psychobiology of the Mouse. Springfield, IL: Charles C Thomas, pp. 169–200.

Ehret G, Frankenreiter M (1977) Quantitative analysis of cochlear structures in the house mouse in relation to mechanisms of acoustic information processing. J Comp Physiol 122:65–85.

Engström H, Wersäll J (1958) Structure and innervation of the inner ear sensory epithelia. Int Rev Cytol 7:535–585.

Evans EF (1972) The frequency response and other properties of single fibres in the guinea-pig cochlear nerve. J Physiol 226:263–287.

Evans EF, Palmer AR (1980) Relationship between the dynamic range of cochlear nerve fibers and their spontaneous activity. Exp Brain Res 40:115–118.

Fay RR (1988) Hearing in Vertebrates. Winnetka, IL: Hill-Fay Associates.

Fekete DM, Rouiller EM, Liberman MC, Ryugo DK (1984) The central projections of intracellularly labeled auditory nerve fibers in cats. J Comp Neurol 229:432–450.

Feldman ML, Harrison JM (1969) The projection of the acoustic nerve to the ventral cochlear nucleus of the rat. A Golgi study. J Comp Neurol 137:267–294.

Firbas W (1972) Uber anatomische Anpassungen des Hörogans an die Aufnahme hoher Frequenzen. Mschr Ohr hk Laryngol-Rhinol (Vienna) 106:105–156.

Gacek RR, Rasmussen GL (1961) Fiber analysis of the statoacoustic nerve of guinea pig, cat and monkey. Anat Rec 139:455–463.

Gambetti P, Antilio-Gambetti L, Papasozomenos S (1982) Bodian's silver method stains neurofilament polypeptides. Science 213:1521–1522.

Gentschev T, Sotelo C (1973) Degenerative patterns in the ventral cochlear nucleus of the rat after primary deafferentation. An ultrastructural study. Brain Res 62:37–60.

Gifford ML, Guinan JJ (1987) Effects of electrical stimulation of medial olivocochlear neurons on ipsilateral and contralateral cochlear responses. Hear Res 29:179–194.

Ginzberg RD, Morest DK (1983) A study of cochlear innervation in the young cat with the Golgi method. Hearing Res 10:227–246.

Godement P, Vanselow J, Thanos S, Bonhoeffer F (1987) A study in developing visual systems with a new method of staining neurones and their processes in fixed tissue. Development 101:697–713.

Godfrey, DA, Kiang NYS, Norris BE (1975) Single unit activity in the posteroventral cochlear nucleus of the cat. J Comp Neurol 162:247–268.

Gray EG, Guillery RW (1966) Synaptic morphology in the normal and degenerating nervous system. Int Rev Cytol 19:111–182.

Guild SR, Crowe SJ, Bunch CC, Polvogt LM (1931) Correlations of differences in the density of innervation of the organ of Corti with differences in the acuity of hearing, including evidence as to the location in the human cochlea of the receptors for certain tones. Acta Otolaryngol (Stockh) 15:269–308.

Guinan JJ, Warr WB, Norris BE (1983) Differential olivocochlear projections from lateral vs medial zones of the superior olivary complex. J Comp Neurol 221:358–370.

Haber S (1988) Tracing intrinsic fiber connections in postmortem human brain with WGA-HRP. J Neurosci Methods 23:15–22.

Harrison JM, Irving R (1966) Ascending connections of the anterior ventral cochlear nucleus in the rat. J Comp Neurol 126:51–64.

Held H (1926) Die Cochlea der Säuger und der Vögel, ihre Entwicklung und ihr Bau. In: Bethe A, v Bergman G, Ellinger A (ed) Handbuch der Normalen und Pathologischen Physiologie, Vol. XI. Berlin: J Springer, pp. 467–534.

Ibata Y, Pappas GD (1976) The fine structure of synapses in relation to the large spherical neurons in the anterior ventral cochlear (sic) of the cat. J Neurocytol 5:395–406.

Innocenti GM, Fiori L, Caminiti R (1977) Exuberant projection into the corpus callosum from the visual cortex of newborn cats. Neurosci Lett 4:237–242.

Irving R, Harrison JM (1967) The superior olivary complex and audition: A comparative study. J Comp Neurol 130:77–86.

Ishii D, Balough Jr. K (1968) Distribution of efferent nerve endings in the organ of Corti. Their graphic reconstruction in cochleae by localization of acetylcholinasterase activity. Acta Otolaryngol 66:282–288.

Jackson H, Parks TN (1982) Functional synapse elimination in the developing avian cochlear nucleus with simultaneous reduction in cochlear nerve axon branching. J Neurosci 2:1736–1743.

Johnstone BM, Pattuzzi R, Yates GK (1986) Basilar membrane measurements and the traveling wave. Hear Res 22:147–153.

Kageyama GH, Meyer RL (1987) Dense HRP filling in pre-fixed brain tissue for light and electron microscopy. J Histochem Cytochem 35:1127–1136.

Kawase K, Liberman MC (1991) Spatial organization of the spiral ganglion according to spontaneous discharge rate. Assn Res Otolaryngol Abst p. 17.

Keithley EM, Feldman ML (1979) Spiral ganglion cell counts in an age-graded series of rat cochleas. J Comp Neurol 188:429–442.

Keithley EM, Feldman ML (1982) Hair cell counts in an age-graded series of rat cochleas. Hear Res 8:249–262.

Keithley EM, Feldman ML (1983) The spiral ganglion and hair cells of the Bronx waltzer mice. Hear Res 12:381–391.

Keithley EM, Schreiber RC (1987) Frequency map of the spiral ganglion in the cat. J Acoust Soc Am 81:1036–1042.

Kellerhals B, Engström H, Ades HW (1967) Die Morphologie des Ganglion spirale Cochleae. Acta Otolaryngol Supp 226:6–33.

Khanna SM, Leonard DGB (1982) Basilar membrane tuning in the cat cochlea. Science 215:305–306.

Kiang NYS, Watanabe T, Thomas LC, Clark LF (1965) Discharge Patterns of Single Fibers in the Cats Auditory Nerve. Cambridge: MIT Press.

Kiang NYS, Rho JM, Northup CC, Liberman MC, Ryugo DK (1982) Hair-cell innervation by spiral ganglion cells in adult cats. Science 217:175–177.

Kiang NYS, Keithley EM, Liberman MC (1983) The impact of auditory nerve experiments on cochlear implant design. Ann NY Acad Sci 405:114–121.

Kiang NYS, Liberman MC, Gage JS, Northrup CC, Dodds LW, Oliver ME (1984) Afferent innervation of the mammalian cochlea. In: Bolis L, Keynes RD, Maddrell HP (eds) Comparative Physiology of Sensory Systems. Cambridge: Cambridge University Press, pp. 143–161.

Kiang NYS, Liberman MC, Sewell WF, Guinan JJ (1986) Single unit clues to cochlear mechanisms. Hear Res 22:171–182.

Kim DO, Molnar CE (1979) A population study of cochlear nerve fibres: Comparison of spatial distributions of average rate and phase-locking measures of responses to single tones. J Neurophysiol 42:16–30.

Kimura RS (1975) The ultrastructure of the organ of Corti. Int Rev Cytol 42:173–222.

Kimura RS (1986) An electron microscopic study of cochlear nerve fibers followed serially from spiral ganglion to organ of Corti. Ear Res Jpn 17:4–7.

Kimura RS, Bongiorno CL, Iverson NA (1987) Synapses and ephapses in the spiral ganglion. Acta Otolaryngol Suppl 438:3–18.

Kohllöffel LUE (1975) A study of neurone activity in the spiral ganglion of the cat's basal turn. Arch Oto Rhino Laryngol 209:179–202.

Leake PA, Snyder RL (1989) Topographic organization of the central projections of the spiral ganglion in cats. J Comp Neurol 281:612–629.

Lenn NY, Reese TS (1966) The fine structure of nerve endings in the nucleus of the trapezoid body and the ventral cochlear nucleus. Am J Anat 118:375–389.

Liberman MC (1978) Auditory-nerve response from cats raised in a low-noise chamber. J Acoust Soc Am 53:442–455.

Liberman MC (1980a) Morphological differences among radial afferent fibers in the cat cochlea: An electron microscopic study of serial sections. Hear Res 3:45–63.

Liberman MC (1980b) Efferent synapses in the inner hair cell area of the cat cochlea: An electron microscopic study of serial sections. Hear Res 3:189–204.

Liberman MC (1982a) Single-neuron labeling in the cat auditory nerve. Science 216:1239–1241.

Liberman MC (1982b) The cochlear frequency map for the cat: Labelling auditory-nerve fibers of known characteristic frequency. J Acoust Soc Am 72:1441–1449.

Liberman MC (1990) Effects of chronic cochlear de-efferentation on auditory-nerve response. Hear Res 49:209–224.

Liberman MC, Kiang NYS (1978) Acoustic trauma in cats: Cochlear pathology and auditory-nerve activity. Acta Otolaryngol Suppl 358:1–63.

Liberman MC, Kiang NYS (1984) Single-neuron labeling and chronic cochlear pathology. Stereocilia damage and alterations in rate- and phase-level functions. Hear Res 16:75–90.

Liberman MC, Oliver ME (1984) Morphometry of intracellular labeled neurons of the auditory nerve: Correlations with functional properties. J Comp Neurol 223:163–176.

Liberman MC, Dodds LW, Pierce S (1990) Afferent and efferent innervation of the cat cochlea: Quantitative analysis with light and electron microscopy. J Comp Neurol 301:443–460.

Lorente de Nó R (1933) Anatomy of the eighth nerve. III. General plan of structure of the primary cochlear nuclei. Laryngoscope 43:327–350.

Lorente de Nó R (1937) The sensory endings in the cochlea. Laryngoscope (St. Louis) 47:373–377.

Masterton RB, Thompson GC, Bechtold JK, RoBards MJ (1975) Neuroanatomical basis of binaural phase-difference analysis for sound localization: A comparative study. J Comp Physiol Psych 89:379–386.

McConnell SK, Ghosh A, Shatz CJ (1989) Subplate neurons pioneer the first axon pathway for the cerebral cortex. Science 245:978–982.

Morrison D, Schindler RA, Wersäll J (1975) A quantitative analysis of the afferent innervation of the organ of Corti in guinea pig. Acta Otolaryngol 79:11–23.

Moskowitz N, Liu JC (1972) Central projections of the spiral ganglion of the squirrel monkey. J Comp Neurol 144:335–344.

Munzer FT (1931) Uber markhaltige Ganglienzellen. Z Mikrosk Anat Forsch 24:286–361.

Nadol JB (1981) Reciprocal synapses at the base of outer hair cells in the organ of Corti of man. Ann Oto Rhinol Laryngol 90:12–17.

Nadol JB (1983a) Serial section reconstruction of the neural poles of hair cells in the human organ of Corti. I. Inner hair cells. Laryngoscope 93:599–614.

Nadol JB (1983b) Serial section reconstruction of the neural poles of hair cells in the human organ of Corti. II. Outer hair cells. Laryngoscope 93:780–791.

Nadol JB (1988a) Comparative anatomy of the cochlea and auditory nerve in mammals. Hear Res 34:253–266.

Nadol JB (1988b) Innervation densities of inner and outer hair cells of the human organ of Corti. ORL 50:363–370.

Natout MAY, Terr LI, Linthicum Jr FH, House WF (1987) Topography of vestibulocochlear nerve fibers in the posterior cranial fossa. Laryngoscope 97:954–958.

Noda Y, Pirsig W (1974) Anatomical projection of the cochlea to the cochlear nuclei of the guinea pig. Arch Otorhinolaryngol 208:107–120.

Osen KK (1969) Cytoarchitecture of the cochlear nuclei in the cat. J Comp Neurol 136:453–484.

Osen KK (1970) Course and termination of the primary afferents in the cochlear nuclei of the cat. Arch Ital Biol 108:21–51.

Ota CY, Kimura RS (1980) Ultrastructural study of the human spiral ganglion. Acta Otolaryngol 89:53–62.

Perkins RE, Morest DK (1975) A study of cochlear innervation patterns in cats and rats with the Golgi method and Nomarski optics. J Comp Neurol 63:129–158.

Polyak SL, McHugh G, Judd DK (1946) The Human Ear in Anatomical Transparencies. Elmsford, NY: Sonotone.

Ramón-Moliner E (1970) The Golgi-Cox technique. In: Nauta WJH, Ebbesson SOE (eds) Contemporary Research Methods in Neuroanatomy. New York, NY: Springer-Verlag, pp. 32–55.

Ramón y Cajal S (1909) Histologie du Système nerveux de l'Homme et des Vertébrés, Vol 1. Madrid: Instituto Ramón y Cajal, pp. 774–838.

Ramprashad F, Money KE, Landolt JP, Laufer J (1978) A neuroanatomical study of the cochlea of the little brown bat (*Myotis lucifugus*). J Comp Neurol 178:347–363.

Ramprashad F, Landolt JP, Money KE, Clark D, Laufer J (1979) A morphometric study of the cochlea of the little brown bat (Myotis lucifugus) J Morphol 160:345–368.

Rasmussen GL (1940) Studies of the VIIIth cranial nerve of man. Laryngoscope 50:67–83.

Rasmussen GL (1946) The olivary peduncle and other fiber connections of the superior olivary complex. J Comp Neurol 84:141–219.

Retzius G (1884) Das Gehörorgan der Wirbeltiere. II. Das Gehörorgan der Reptilien, derr Vögel und der Säugetiere. Stockholm: Samson and Wallin.

Retzius G (1892) Die Endigungsweise des Gehörnerven. Biolog Untersuchungen, Neue Folfe, III. Leipzig: Vogel.

Rhode WS (1971) Observations of the vibration of the basilar membrane in squirrel monkeys using the Mössbauer technique. J Acoust Soc Am 49:1218–1231.

Rhode WS, Oertel D, Smith PH (1983) Physiological response properties of cells labeled intracellularly with horseradish peroxidase in cat ventral cochlear nucleus. J Comp Neurol 213:448–463.

Ritz LA, Brownell WE (1982) Single unit analysis of the posteroventral cochlear nucleus of the decerebrate cat. Neuroscience 7:1995–2010.

Robertson D (1984) Horseradish peroxidase injection of physiologically characterized afferent and efferent neurons in the guinea pig spiral ganglion. Hear Res 15:113–121.

Robertson D, Cody AR, Bredberg G, Johnstone BM (1980) Response properties of spiral ganglion neurons in cochleas damaged by direct mechanical trauma. J Acoust Soc Am 67:1295–1303.

Romand R, Hafidi A, Despres G (1987) Immunocytochemical localization of neurofilament protein subunits in the spiral ganglion of the adult rat. Brain Res 462:167–173.

Rose JE, Galambos R, Hughes JR (1959) Microelectrode studies of the cochlear nuclei of the cat. Bull Johns Hopkins Hospital 104:211–251.

Rouiller EM, Cronin-Schreiber R, Fekete DM, Ryugo DK (1986) The central projections of intracellularly labeled auditory nerve fibers in cats: An analysis of terminal morphology. J Comp Neurol 249:261–278.

Rouiller EM, Ryugo DK (1984) Intracellular marking of physiologically characterized neurons in the ventral cochlear nucleus of the cat. J Comp Neurol 225:167–186.

Russel IJ, Sellick PM (1978) Intracellular studies of hair cells in the mammalian cochlea. J Physiol 284:261–290.

Ryan AF, Schwartz IR, Helfert RH, Keithley EM, Wang ZX (1987) Selective retrograde labeling of lateral olivocochlear neurons in the brainstem based on preferential uptake of ³H-D-aspartic acid in the cochlea. J Comp Neurol 255:606–616.

Ryugo DK, Fekete DM (1982) Morphology of primary axosomatic endings in the anteroventral cochlear nucleus of the cat: A study of the endbulbs of Held. J Comp Neurol 210:239–257.

Ryugo DK, Rouiller EM (1988) The central projections of intracellularly labeled auditory nerve fibers in cats: Morphometric correlations with physiological properties. J Comp Neurol 271:130–142.

Ryugo DK, Sento S (1991) Synaptic connections of the auditory nerve in cats: Relationship between endbulbs of Held and spherical bushy cells. J Comp Neurol 305:35–48.

Ryugo DK, Dodds LW, Benson TE, Kiang NYS (1991) Unmyelinated axons of the auditory nerve in cats. J Comp Neurol (in press).

Sachs MB, Abbas PJ (1974) Rate versus level functions for auditory nerve fibers in cats: Tone-burst stimulation. J Acoust Soc Am 56:1835–1847.

Sando I (1965) The anatomical interrelationships of the cochlear nerve fibers. Acta Otolaryngol 59:417–436.

Schalk TB, Sachs MB (1980) Nonlinearities in auditory-nerve fiber responses to bandlimited noise. J Acoust Soc Am 67:903–913.

Schuknecht HF (1953) Techniques for study of cochlear function and pathology in experimental animals. Arch Oto-Laryngol 58:377–397.

Schuknecht HF (1960) Neuroanatomical correlates of auditory sensitivity and pitch discrimination in the cat. In: Rasmussen GL, Windle WF (eds) Neural Mechanisms of the Auditory and Vestibular Systems. Springfield, IL: Charles C Thomas, pp. 76–90.

Sento S, Ryugo DK (1989) Endbulbs of Held and spherical bushy cells in cats: Morphological correlates with physiological properties. J Comp Neurol 280:553–562.

Simmons DD, Liberman MC (1988) Afferent innervation of outer hair cells in adult cats: I. Light microscopic analysis of fibers labeled with horseradish peroxidase. J Comp Neurol 270:132–144.

Smith CA (1961) Innervation pattern of the cochlea. The internal hair cell. Ann Otol Rhinol Laryngol 70:1–24.

Smith CA (1975) Innervation of the cochlea of the guinea pig by use of the Golgi stain. Ann Otol Rhinol Laryngol 84:443–458.

Smith CA, Rasmussen GL (1963) Recent observation on the olivocochlear bundle. Ann Otol Rhinol Laryngol 72:489–497.

Smith CA, Sjöstrand FS (1961) Structure of the nerve endings on the external hair cells of the guinea pig cochlea as studied by serial section. J Ultrastruct Res 5:523–556.

Spangler KM, Cant NB, Henkel CK, Farley GR, Warr WB (1987) Descending projections from the superior olivary complex to the cochlear nucleus of the cat. J Comp Neurol 259:452–465.

Spirou GA, May BJ, Ryugo DK (1989) 3-Dimensional frequency mapping in the cat dorsal cochlear nucleus. Soc Neurosci Abst 15:744.

Spoendlin H (1969) Innervation patterns in the organ of Corti of the cat. Acta Otolaryngol (Stockh) 67:239–254.

Spoendlin H (1971) Degeneration behavior of the cochlear nerve. Arch Klin Exp Ohr- Nas- Kehlk Heilk 200:275–291.

Spoendlin H (1972) Innervation densities of the cochlea. Acta Otolaryngol 73:235–248.

Spoendlin H (1973) The innervation of the cochlea receptor. In: Moller AR (ed) Mechanisms in Hearing. New York: Academic Press, pp. 185–229.

Spoendlin H (1975) Retrograde degeneration of the cochlear nerve. Acta Otolaryngol 79:266–275.

Spoendlin H (1979) Neural connections of the outer hair cell system. Acta Otolaryngol 87:381–387.

Spoendlin H (1981) Differentiation of cochlear afferent neurons. Acta Otolaryngol 91:451–456.

Spoendlin H (1982) The innervation of the outer hair cell system. Am J Otol 3:274–278.

Spoendlin H, Schrott A (1988) The spiral ganglion and the innervation of the human organ of Corti. Acta Otolaryngol (Stockh) 105:403–410.

Stanfield BB, O'Leary DDM (1985) The transient corticospinal projections from the occipital cortex during the postnatal development of the rat. J Comp Neurol 238:236–248.

Suzuki Y, Watanabe A, Osada M (1963) Cytological and electron microscopic studies on the spiral ganglion cells of adult guinea pigs and rabbits. Arch Histol Jap 24:9–33.

Tolbert LP, Morest DK (1982) The neuronal architecture of the anteroventral cochlear nucleus of the cat in the region of the cochlear nerve root: Electron microscopy. Neuroscience 7:3053–3067.

Valverde F (1970) The Golgi method: A tool for comparative structural analyses. In: Nauta WJH, Ebbesson SOE (eds) Contemporary Research Methods in Neuroanatomy. New York: Springer-Verlag, pp. 12–31.

von Ebner B (1903) Die Endigung des Schneckennerven im Cortischen Organe. Kölliker's Handbuch der Gewebelehre des Menschen, III. Leipzig: Engelmann, pp. 944–960.

Warr WB (1975) Olivocochlear and vestibular efferent neurons of the feline brainstem: Their location, morphology, and number determined by retrograde axonal transport and acetylcholinesterase histochemistry. J Comp Neurol 161:159–182.

Warr WB, Guinan JJ (1979) Efferent innervation of the organ of Corti: Two separate systems. Brain Res 173:152–155.

Webster DB (1971) Projection of the cochlea to cochlear nuclei in Merriam's kangaroo rat. J Comp Neurol 143:323–340.

White JS, Warr WB (1983) The dual origins of the olivocochlear bundle in the albino rat. J Comp Neurol 219:203–214.

Wiederhold ML (1970) Variations in the effects of electrical stimulation of the crossed olivocochlear bundle on cat single auditory-nerve-fiber responses to tone bursts. J Acoust Soc Am 48:966–977.

Wiederhold ML, Kiang NYS (1970) Effects of electric stimulation of the crossed olivocochlear bundle on single auditory-nerve fibers in the cat. J Acoust Soc Am 48:950–965.

Wright DD, Spirou GA, May BJ, Ryugo DK (1991) Frequency representation in the dorsal cochlear nucleus of cats. Assn Res Otolaryngol Abst p. 140.

3

The Cochlear Nucleus: Neuronal Types and Their Synaptic Organization

NELL BEATTY CANT

1. Introduction

The auditory system, like other sensory systems, comprises multiple neuronal pathways, each conveying information from the periphery to the forebrain via distinct sets of neurons and each with its own role to play in the function of the system. In the cochlear nuclear complex, the part of the brain that receives inputs from the cochlea via the eighth nerve, the neurons that participate in the separate pathways are, to a large extent, distinguishable from one another morphologically and many are also segregated spatially, making them especially accessible for detailed study. The results of detailed anatomical studies of the different cell types provide a framework for relating and interpreting the results of studies of the physiology, connectivity and transmitter neurochemistry of the different pathways.

Three major divisions of the cochlear nuclear complex were recognized by the earliest investigators (e.g., Held 1891; Kölliker 1896; Ramón y Cajal 1909; Fuse 1913; reviewed by Harrison and Irving 1966a; Brawer, Morest, and Kane 1974). The dorsal cochlear nucleus is, in most mammals, a layered structure perched on the dorsolateral aspect of the brain stem. The ventral cochlear nucleus is divided into an anteroventral and a posteroventral part by the entering cochlear nerve fibers. Later investigators demonstrated that it is possible to define many more anatomical subdivisions of the ventral cochlear nucleus (Harrison and Irving 1965, 1966b; Osen 1969; Brawer, Morest, and Kane 1974). Using cytoarchitectonic criteria, including cell size and packing density, and information about differences in neuronal shape and dendritic patterns, Brawer, Morest, and Kane (1974) presented the most detailed parcellation of the ventral cochlear nuclei, which when combined with information about the synaptic organization of the different neuronal types in the nucleus, provides a way of distinguishing areas that are likely to have different functional roles in audition.

Underlying the cytoarchitectonic subdivisions are differences in the constituent neuronal populations. Although the early students of the auditory system recognized that the parts of the cochlear nuclei were not organized simply and that there were distinct morphological differences among the neurons, the first explicit classifications of the multiple cell types, based on their appearance in Nissl and silver stains, were presented by Harrison and his colleagues in rat (Harrison and Warr 1962; Harrison and Irving 1965, 1966b) and by Osen in cat (Osen 1969). Brawer, Morest, and Kane (1974) further distinguished cell types based on their appearance in Golgi preparations. These investigators demonstrated that some parts of the ventral cochlear nucleus contain distinguishable neuronal types that are restricted in location to that part. Other neuronal types are more widely distributed throughout several subdivisions and some do not seem to respect the boundaries of the subdivisions at all. In the dorsal cochlear nucleus, in most mammals although not in humans (Section 6.1), different neuronal types reside in different layers. The morphological cell types in both the dorsal and ventral cochlear nuclei have subsequently been demonstrated to differ on the basis of physiological and connectional data and therefore presumably, they have different functions.[1] While it would be misleading to suggest that any of these types represent entirely homogeneous populations of cells, the classifications provide a rational starting point for considering the connections of the cochlear nucleus.

The main emphasis of this chapter is a description of the synaptic organization of the different neuronal types in the cochlear nucleus. The next three sections are devoted to development of the necessary background for such a description. Section 2 contains a review of the major subdivisions of the cochlear nucleus. Section 3 introduces the light microscopic criteria that distinguish the individual neuronal types. Section 4 is devoted to a brief account of the sources of input to the cochlear nucleus. In Section 5, with the information from the preceding sections as a framework, the fine structure of each neuronal type is presented. Since the most detailed studies of the neurons of the cochlear nuclei have been done in the cat, all of the descriptions in the next four sections of this chapter are of the cells of that species unless otherwise noted. The organization of the cochlear nucleus of other mammals, including humans, is discussed in Section 6. The last section, Section 7, contains a discussion of some of the implications of current knowledge and questions that remain for future studies.

---

[1] It seems most likely that the cell types in the cochlear nucleus can be considered "natural types" as defined by Rodiek and Brening (1983) since, as will be shown, they can be distinguished on the basis of a number of characteristics that covary. However, explicit cluster analysis to demonstrate cell types in the cochlear nucleus has not yet been done.

2. Subdivisions of the Cochlear Nucleus

For the purposes of this chapter, we can consider the cochlear nuclear complex to be composed of seven distinct parts, the dorsal cochlear nucleus and six subdivisions in the ventral cochlear nucleus. This grouping serves to separate the major areas in which the synaptic organization of the different neuronal populations has been studied in some detail. Although most of these seven areas can be further subdivided, differences in the synaptic organization of the cell types populating further subdivisions have not yet been completely characterized.

Figure 3.1 illustrates the seven areas of the cochlear nucleus on photographs of horizontal sections. In the ventral cochlear nucleus, the entry zone of the eighth nerve fibers forms a boundary between the anteroventral and posteroventral cochlear nuclei (AVCN and PVCN, respectively). Both of these can be further divided into anterior and posterior divisions. These four parts of the ventral cochlear nucleus are surrounded by a fifth subdivision made up of areas of small cells, the peripheral cap of small cells of Osen (1969) and the marginal layer of Lorente de Nó (1981), and by a sixth subdivision made up of granule cell domains (Mugnaini, Warr, and Osen 1980).

2.1 AVCN: Anterior and Posterior Divisions

The AVCN is divided into two major divisions (Brawer, Morest, and Kane 1974). Both divisions contain large neurons relative to the neurons in the areas around the margins of the nucleus, which are considered to form separate subdivisions (see below). A boundary is easily discerned between the two divisions, since the cell bodies in the anterior part of the AVCN are smaller and are packed more closely together than those in the posterior division (Fig. 3.1). More subtle differences allow subdivision of these larger divisions (discussed in later sections). These two major divisions of the AVCN will be designated AVCN-A and AVCN-P.

2.2 PVCN: Anterior and Posterior Divisions

Brawer, Morest, and Kane (1974) subdivided the PVCN into seven parts using cytoarchitectonic criteria. Although they described differences in the neuronal populations of these parts, the synaptic organization of most of the neuronal types in the PVCN have not been studied in sufficient detail to discuss each subdivision in these terms. A simpler classification divides the PVCN into an anterior part, PVCN-A, which comprises the anterodorsal, anterior, ventral, dorsal, lateral, and posterior subdivisions of Brawer, Morest, and Kane (1974) and a posterior part, PVCN-P, corresponding to their central subdivision, which is the same as the octopus cell area of Osen (1969). The boundary separating these two parts of the

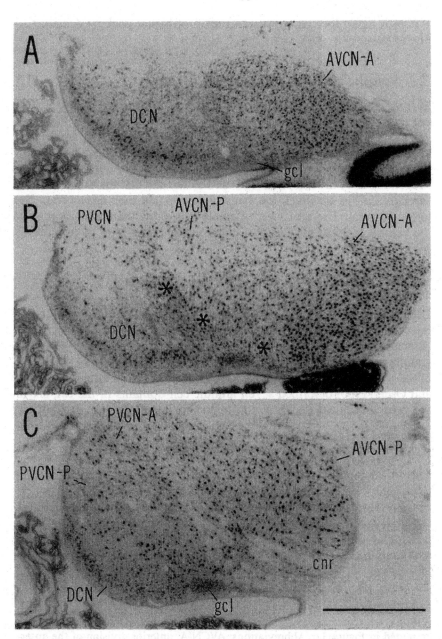

FIGURE 3.1. Horizontal sections at three levels through the cochlear nuclear complex of a cat. The 20 μm thick sections were stained with cresyl violet. The section in panel A is the most dorsal. That in panel C is the most ventral and that in panel B lies approximately halfway in between the other two. Medial is toward the top of the figure; anterior is to the right. The boundaries of the subdivisions

PVCN is easily located between a region of higher packing density and one of lower (Fig. 3.1).

2.3 Small Cell Cap and Marginal Layer

Surrounding the four divisions of large neurons in the AVCN and PVCN are areas of small neurons (Osen 1969; Lorente de Nó 1981; Fig. 3.1). On the lateral surface of the ventral cochlear nucleus, the small cells cluster into a region called the peripheral cap of small cells (Osen 1969). This area is distinguished from the areas of the nucleus containing larger neurons both by the size of the cells and by the fact that the axons running through the cap are much less heavily myelinated than are those in the large cell areas. The thin layer of small cells along the medial border of the ventral cochlear nucleus was called the marginal layer by Lorente de Nó (1981).

2.4 Granule Cell Domains

Large numbers of granule cells cover the superficial surface of much of the cochlear nucleus and form clusters of cells in several other locations (Brawer, Morest, and Kane 1974; Mugnaini, Warr, and Osen 1980). The granule cell domains are easily recognized because of the very small size of the cells and their very high packing density (Fig. 3.1). Mugnaini, Warr, and Osen (1980) described seven granule cell domains in the cat, six in the ventral cochlear nucleus and one in the dorsal cochlear nucleus.

2.5 DCN: Layers

The dorsal cochlear nucleus (DCN) of the cat is usually subdivided into three layers, which are clearly visible in Nissl-stained material (Fig. 3.1; Brawer, Morest, and Kane 1974). The molecular (or superficial) layer contains small, loosely packed cells. The fusiform cell (or granular) layer is made up of a single row of fusiform cells intermingled with small cells and granule cells. The dorsal cochlear nucleus is separated from the anteroventral cochlear nucleus by one of the granule cell domains. The

---

(FIGURE 3.1. *Continued*) are indicated on the drawings of these same sections presented in Figure 3.6. Abbreviations: AVCN-A, anterior division of the anteroventral cochlear nucleus; AVCN-P, posterior division of the anteroventral cochlear nucleus; PVCN-A, anterior division of the posteroventral cochlear nucleus; PVCN-P, posterior division of the posteroventral cochlear nucleus; DCN, dorsal cochlear nucleus; gcl, granule cell layer; cnr, cochlear nerve root. The asterisks in the middle panel indicate the location of the peripheral cap of small cells. Scale bar = 1 mm for all panels.

granule cells from this domain appear to form a band of cells continuous with the layer of granule cells that lie in the fusiform layer of the dorsal cochlear nucleus (Fig. 3.1). The polymorphic (or deep) layer of the DCN contains very large cells and scattered small cells.

2.6 Concept of Isofrequency Laminae

The auditory nerve enters the ventral cochlear nucleus at its ventral surface and each axon divides into an ascending and descending branch (Fig. 3.2). The ascending branch passes through the posterior and then the anterior division of the AVCN while the descending branch passes through the anterior and then the posterior division of the PVCN. Most descending branches ultimately enter the deep layer of the DCN. The small cell cap and probably also the marginal layer of small cells receive inputs from the cochlea (Osen 1970). In rodents, the granule cell areas receive inputs from type II auditory fibers (Brown et al. 1988a; see Section 4.1 and Chapter 2); these axons have not been followed to their termination zones in the cochlear nucleus of cats. Thus, each auditory nerve fiber passes through most of the major divisions of the cochlear nucleus. Each part, consequently, contains neurons responsive to the entire frequency range represented in the auditory nerve (Figs. 3.3, 3.4; Rose, Galambos, and Hughes 1959; Rose 1960; Bourk, Mielcarz, and Norris 1981). These axons, and therefore the neurons that they contact, are organized systematically according to their frequency specificity, with axons tuned to high frequencies located dorsally and axons tuned to low frequencies located ventrally. The AVCN has been modelled as a series of "isofrequency" planes or layers, each layer containing neurons maximally responsive to a small frequency range (Bourk, Mielcarz, and Norris 1981; Fig. 3.4). The PVCN and DCN presumably also contain such planes so that the entire cochlear nuclear complex could be conceived of as a series of layers of neurons with each subdivision represented in each layer. How the small cell areas fit into a scheme of isofrequency planes is not clear since their tonotopic sequence seems to be reversed compared to that of the adjacent large cell areas (Rose, Galambos, and Hughes 1959; Rose 1960; Osen 1969; Bourk, Mielcarz, and Norris 1981).

3. The Neuronal Classes and Their Locations

3.1 Neurons in the Ventral Cochlear Nucleus

Many of the neuronal types that Osen distinguished in Nissl preparations of the cochlear nucleus of the cat were related to the types of neurons defined in Golgi preparations by Morest and his students (Kane 1973, 1974a; Brawer, Morest, and Kane 1974; Cant and Morest 1979a; Cant

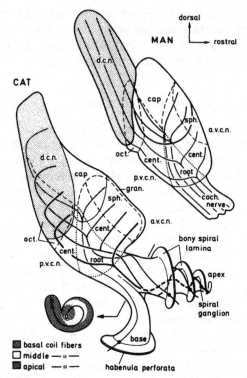

FIGURE 3.2. Lateral views of the cochlear nuclear complex of cat and man that illustrate the course of the cochlear nerve fibers from the spiral ganglion to their terminations in the complex. Axons enter the cochlear nuclear complex at its ventral aspect and bifurcate to form ascending and descending branches. The ascending branches pass through the length of the AVCN. The descending branches pass first through the PVCN and then enter the DCN. Abbreviations: DCN, dorsal cochlear nucleus; a.v.c.n., anteroventral cochlear nucleus; p.v.c.n., posteroventral cochlear nucleus; sph., spherical cell area, which occupies the anterior division of the AVCN (AVCN-A, cf. Fig. 3.1); cent., central part of the ventral cochlear nucleus, which comprises AVCN-P and PVCN-A (cf. Fig. 3.1); oct., octopus cell area, which occupies PVCN-P (cf. Fig. 3.1); cap, peripheral cap of small cells. The small diagram at the bottom left illustrates the arrangement of fibers from the base (b) to the apex (a) in the cochlea of the cat. These branch at different levels in the cochlear nucleus, giving rise to its tonotopic organization (Figs. 3.3 and 3.4). From A. Brodal, Neurological Anatomy, Oxford University Press, 1981. Used with permission. (Original figures appeared in Moore and Osen 1979, and Arnesen and Osen 1978.)

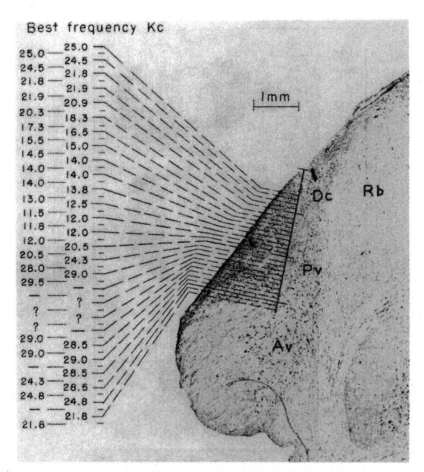

FIGURE 3.3. A section through the cochlear nucleus lying between the transverse and horizontal planes. The dorsal aspect at the top of the section lies more posterior than the ventral aspect at the bottom. Medial is to the right. The approximately vertical line drawn on the section represents the path of a recording electrode through the nucleus. The numbers on the left indicate the best frequencies of small clusters of units encountered at 50 μm steps. There is an orderly sequence from high to low best frequencies in the DCN (Dc) and a jump to higher frequencies as the electrode enters the PVCN (Pv). Abbreviations: Av, anteroventral cochlear nucleus; Rb, restiform body. [From Rose (1960); used with permission.]

1981; Kane, Puglisi, and Gordon 1981; Tolbert and Morest 1982a; reviewed by Cant and Morest 1984). Major neuronal classes are illustrated in Figure 3.5. There are two types of cells called bushy cells because their short primary dendrites give rise to a profusion of thin, lumpy appendages which resemble a shrub (Brawer, Morest, and Kane 1974; Fig. 3.5A). The

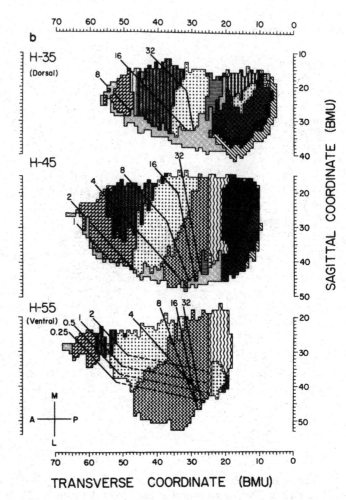

FIGURE 3.4. Drawings of horizontal sections through the cochlear nucleus of a cat taken at approximately the same levels as those shown in Figures 3.1 and 3.6. The top section (H-35) is the most dorsal. The bottom section (H-55) is the most ventral. Medial is to the top; anterior is to the left. Each subdivision of the cochlear nuclear complex is indicated in a different pattern and can be compared to those in Figure 3.6. The lines across the sections indicate locations of single unit recordings in which the characteristic frequency (CF) of the units was the same. As would be expected from the cochlear innervation pattern (Fig. 3.2), the most dorsal section contains units of high CF, whereas the most ventral sections also contain units of lower CF. [From Bourk, Mielcarz, and Norris (1981); used with permission.]

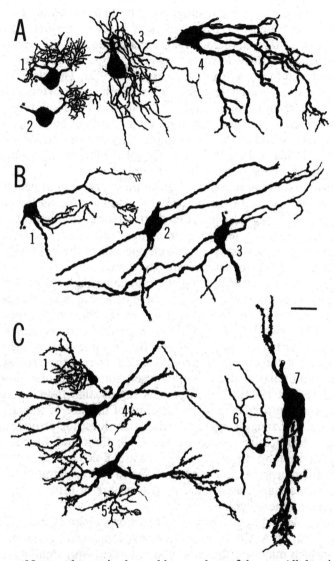

FIGURE 3.5. Neuronal types in the cochlear nucleus of the cat. All drawings were made of neurons impregnated by the Golgi method. (A) Cells typical of specific subdivisions in the ventral cochlear nucleus. Cells 1 and 2: Spherical bushy cells located in AVCN-A; Cell 3: Globular bushy cell located in AVCN-P; Cell 4: Octopus cell found in PVCN-P. (B) Multipolar cells typical of those found throughout the ventral cochlear nucleus. Cell 1: Medium-sized multipolar cell found in AVCN-A. Cells 2 and 3: Large multipolar cells found in AVCN-P and PVCN-A. (C) Neuronal types located in the dorsal cochlear nucleus. Cell 1: Cartwheel cell; Cells 2 and 3: Fusiform cells; Cells 4 and 5: Granule cells; Cell 6: Stellate cell; Cell 7: Giant cell. Scale bar = 10 μm for all panels. [Modified from Cant and Morest, 1984; used with permission.]

two types of bushy cells are distinguished by their appearance in Nissl-stained material (Osen 1969). The *spherical bushy cells,* when stained for Nissl substance, have a characteristic organization of the rough endoplasmic reticulum, which forms a "necklace" of patches around the perikaryon and also forms a "cap" next to the nucleus. The *globular bushy cells* have no distinguishable Nissl bodies. Rather, the cytoplasm has a homogeneous, granular appearance. The nucleus of the globular bushy cell is often located eccentrically. Both major types of bushy cells might be usefully subdivided further on anatomical grounds (Osen 1969; Tolbert and Morest 1982a; Cant 1991), but few physiological or connectional implications of such subclasses are yet apparent.

Another odd-looking neuronal type found in the ventral cochlear nucleus is the *octopus cell* (Fig. 3.5A). The octopus cell has long, sturdy primary dendrites that branch sparsely, extend for a considerable distance from the cell body, and terminate in a small tuft of appendages. All of the primary dendrites usually arise from one side of the cell body, giving the cell its distinctive appearance and its name.

In contrast to the unusual dendritic branching patterns of the bushy and octopus cells, the *multipolar cells* have dendritic branching patterns similar to neurons found throughout the brainstem, with primary dendrites extending away from all sides of the neuron and dividing to form secondary and tertiary branches (Fig. 3.5B). (In some parts of the ventral cochlear nucleus, these neurons have also been referred to as stellate cells; the terms multipolar and stellate are used somewhat interchangeably [Cant and Morest 1984].) Measurements suggest that there are *at least* four groups of multipolar neurons in terms of the size of the cell body, although the size ranges overlap so that size does not provide an unambiguous classification system (Osen 1969; Cant and Morest 1984). Until other criteria to distinguish the cells are discovered and correlated with size differences, the number of different types of multipolar cells will remain uncertain. Nissl-staining patterns are of little use in distinguishing among these cells.

The multipolar cells vary widely in the morphology of their dendritic trees (Fig. 3.5B). In AVCN-A, they often have long, relatively unbranched dendrites which may be either smooth or studded with small spines and short digitiform appendages (Cant and Morest 1979a). Some of the spiny dendrites terminate in a little spray of appendages far from the cell body. Small multipolar cells are quite variable in their morphology but usually have stellate or star-shaped dendritic fields. In AVCN-P, Tolbert and Morest (1982a) characterized the multipolar cells as stellate or elongate. The dendrites of the stellate cells radiate in several directions whereas those of elongate cells emerge from opposite poles of the cell body and extend away from the soma for long distances. In the PVCN, the multipolar cells appear to form a diverse group (Brawer, Morest, and Kane 1974). Smith and Rhode (1989) classified cells in PVCN-A into two groups based on

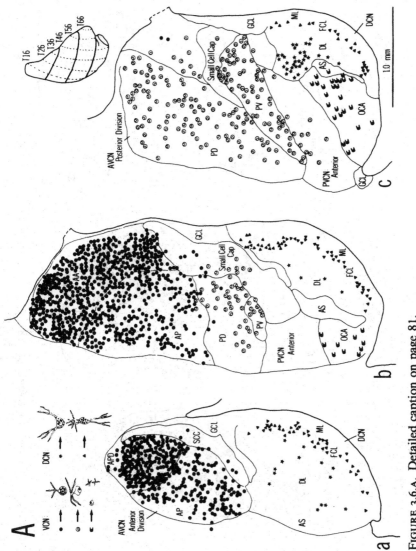

FIGURE 3.6.A. Detailed caption on page 81.

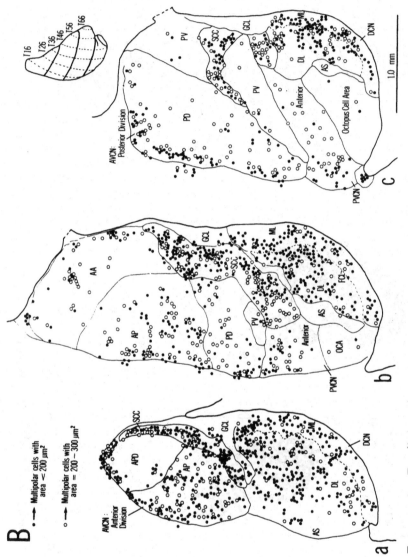

FIGURE 3.6.B. Detailed caption on page 81.

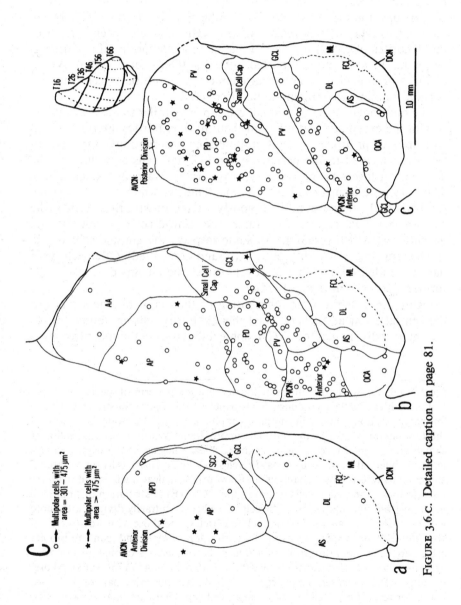

FIGURE 3.6.c. Detailed caption on page 81.

their physiological response properties. After establishing the class to which an individual cell belonged, they injected horseradish peroxidase into the cell to reveal its dendritic morphology. The dendritic trees of one group appeared to be stellate while those of the other tend to be elongated, similar to those described in the AVCN-P. In the PVCN of the mouse, two types of multipolar (stellate) cells have also been described; the dendrites of one group end in distinct and elaborate tufts of appendages that have not been reported in the cat (Oertel et al. 1990). Which of the morphological details that characterize multipolar cells are important functionally remains to be discovered.

The multipolar cells in the small cell cap are similar to those in AVCN-A with the exception that many of them have dendrites almost covered with spinous protrusions (Cant, unpublished observations). One type of small cell located in the peripheral cap of small cells as well as in the granule cell layers is the *Golgi cell* (Mugnaini, Warr, and Osen 1980). This very peculiar neuron has elaborate dendrites which branch near the cell body and curve around it sinuously. Mixed in with these cells is the smallest neuronal type that populates the ventral cochlear nucleus, the *granule cell,* which resembles in some respects the granule cells of the cerebellum (Fig. 3.4C; Mugnaini, Warr, and Osen 1980). Other cell types have been identified but have not yet been studied in any detail (see, e.g., Brawer, Morest, and Kane 1974).

Figure 3.6 illustrates the distribution of the major neuronal types in the ventral cochlear nucleus. The spherical bushy cells are found only in the anterior division of the AVCN. Some differences in the spherical bushy

---

FIGURE 3.6. Drawings that illustrate the distribution patterns of specific cell types in the cochlear nuclear complex. Each panel in this figure illustrates the three horizontal sections shown in the photographs in Figure 3.1. Anterior is toward the top; lateral is toward the left. The approximate location of each section is shown on the small inset drawing of the cochlear nucleus. (A) Distribution patterns of spherical bushy cells (closed circles); globular bushy cells (open circles) and octopus cells (octopus symbols) in the ventral cochlear nucleus and fusiform cells (triangles) and giant cells (stars) in the DCN. (B) Distribution of small and medium-sized multipolar cells. The cells have been arbitrarily divided into those with cross-sectional area $< 200~\mu m^2$ (closed circles) and those with cross-sectional area 200–300 μm^2 (open circles). (C) Distribution of large and giant multipolar cells. Large cells are arbitrarily defined as cells with cross-sectional area of 301–475 μm^2 (open circles); giant cells are those with area $> 475~\mu m^2$ (stars). Abbreviations: APD, posterodorsal part of AVCN-A; AP, posterior part of AVCN-A; AA, anterior part of AVCN-A; SCC, small cell cap; GCL, granule cell layer; PD, dorsal part of AVCN-P; PV, ventral part of AVCN-P; ML, molecular layer of DCN; FCL, fusiform cell layer of DCN; DL, deep layers of DCN; AS, acoustic striae; OCA, octopus cell area (PVCN-P). (Modified slightly from Cant and Morest (1984); used with permission.)

cells in the anterior and posterior parts of the anterior division have been noted (Cant and Casseday 1986; Cant 1991), but differences in their synaptic organization have not been described. In the most anterior part of AVCN-A, the spherical bushy cells are almost completely segregated from other cell types. The globular bushy cells are located mainly in AVCN-P and to a lesser extent in PVCN-A. The octopus cells are found only in PVCN-P, where they are almost completely segregated from other neuronal types. Thus, from rostral to caudal through an isofrequency lamina, three different types of cells predominate and their distributions do not overlap significantly.

The multipolar or stellate cells are less obviously confined to particular subdivisions. From one subdivision to another, however, there is substantial variation in the average size of the multipolar neurons (Osen 1969; Cant and Morest 1984). Those in AVCN-P and PVCN-A are large. Medium-sized multipolar cells are distributed mainly in the posterior part of AVCN-A, and in the small cell cap. Small multipolar cells are distributed mainly around the margins of the VCN and are especially common in the DCN. A small group of very large multipolar cells is scattered throughout the central core of the VCN, being found mostly in AVCN-P and PVCN-A.

3.2 Neuronal Types in the DCN

Like the ventral cochlear nucleus, the dorsal cochlear nucleus contains a variety of cell types, which are defined both by their morphology and by their location in different layers (Ramón y Cajal 1909; Lorente de Nó 1933, 1981; Brawer, Morest, and Kane 1974; Fig. 3.5C). The row of *fusiform cell* bodies defines the middle layer of the DCN. The extensive dendrites of the fusiform cells extend into the molecular layer where they are highly branched and studded with spines (the apical dendrites) and into the deep layer where they are less highly branched and are generally smooth (the basal dendrites). Both the basal and apical dendritic arbors are flattened in the plane perpendicular to the path of the parallel fibers (granule cell axons) and parallel to the plane of trajectory of the cochlear fibers (Blackstad, Osen, and Mugnaini 1984; Schweitzer and Cant 1984b [hamster]). The other large cell type in the DCN is the *giant cell,* which is located in the deep layer. Based on the morphology of their dendrites, these cells fall into five classes (Kane, Puglisi, and Gordon 1981). These are the elongate bipolar, elongate multipolar, globular, radiate, and oriented multipolar giant cells.

Several classes of small cells that can be identified on the basis of dendritic morphology are located in specific layers. The cell bodies of the *cartwheel cells* lie in the fusiform cell layer. They send highly branched, spine-studded dendrites into the molecular layer, where they have a characteristic curvy appearance. The *multipolar* or *stellate cells* are located in all three layers of the DCN; the *elongate cells* (including the corn cells of Lorente de Nó 1933) are located in the fusiform cell layer and throughout the deep layer. *Granule cells* in the DCN are found mainly in the fusiform cell layer.

4. Sources of the Inputs to the Cochlear Nucleus

As a preface to a description of the fine structure of the major neuronal types in the cochlear nucleus, this section contains a brief discussion of the sources of inputs to the different cell areas in the cochlear nucleus. Coverage of this topic is also found in Chapters 1, 2, and 7. In some but not all cases, the types of synaptic terminals (Section 5) contacting a particular neuronal type can be associated with a specific source of input.

4.1 Input from the Cochlea

All fibers entering the brain from the cochlea terminate in the cochlear nuclear complex (Ramón y Cajal 1909; Lorente de Nó 1933). Auditory nerve fibers can be classified on the basis of anatomical and functional considerations into at least three types: type I axons, which are further classified as having high or low levels of spontaneous rates (SR) of activity, and type II axons (Spoendlin 1971; Liberman 1978; Evans and Palmer 1980; Fekete et al. 1981; discussed in detail in Chapter 2). The trajectories of type II fibers in the cochlear nucleus have not yet been described in the cat so that the references to those fibers rely on data available from gerbils and mice (Brown et al. 1988a). All three types of primary auditory axons appear to bifurcate after entering the nucleus and to terminate in each of the main subdivisions of the ventral cochlear nucleus (Fig. 3.2). Most also enter the DCN, where they terminate in the deep and fusiform cell layers, although a few type I axons may not enter the DCN (Fekete et al. 1982). Unlike the type I fibers, the type II fibers appear to send branches into the granule cell domains as well as into the main body of the nucleus.

Type I primary afferent axons with different SR values differ morphologically. Both give rise to the large synaptic terminals known as the endbulbs of Held in the anterior AVCN and to smaller endings throughout their extent. However, the low (and medium) SR fibers give rise to more endings than do the high SR fibers (Rouiller et al. 1986) and these are, on average, smaller (Ryugo and Rouiller 1988). Lower frequency fibers of both SR types tend to have the largest endbulbs (Rouiller et al. 1986; Sento and Ryugo 1989). Sento and Ryugo (1989) showed that endbulbs on low to medium SR fibers differ in shape from those on the high SR fibers and that the spherical bushy cells receiving the low SR inputs are on average smaller than those receiving high SR inputs. It is not known whether, within an isofrequency lamina, the different types defined on the basis of their level of spontaneous activity are distributed differently. Sento and Ryugo (1989) provide evidence that the endbulbs of the two groups of type I fibers do not converge upon individual spherical bushy cells so that within an isofrequency lamina, different cells of the same variety are probably receiving different cochlear inputs.

4.2 Input from the Brain Stem

In addition to their cochlear input, the subdivisions of the cochlear nuclei are rich with the arborizations of inputs from other sources, both extrinsic and intrinsic to the complex. Each subdivision has a different character-istic pattern of terminations from these inputs (e.g., Cant and Morest 1978). Extrinsic sources of input to the neurons of the cochlear nucleus lie in the contralateral cochlear nucleus, the superior olivary complex, the nuclei of the lateral lemniscus and the inferior colliculus (Rasmussen 1960, 1967; Osen and Roth 1969; van Noort 1969; Adams and Warr 1976; Kane 1976, 1977a,b; Elverland 1977; Cant and Gaston 1982; Conlee and Kane 1982; Adams 1983; Osen et al. 1984; Spangler et al. 1987).

The inputs from the contralateral cochlear nucleus arise from a small number of multipolar neurons scattered throughout the dorsal and ven-tral nuclei (Adams and Warr 1976; Cant and Gaston 1982). Some of these neurons are among the largest multipolar neurons in the ventral cochlear nucleus (Cant and Gaston 1982). In guinea pigs, they probably use glycine as a neurotransmitter (Wenthold 1987; Benson and Potashner 1990). The major sites of termination of the crossed cochlear nuclear connections are the anterior division of the AVCN, the PVCN, and the superficial layers of the DCN, but other parts of the cochlear nucleus receive sparse terminations (Cant and Gaston 1982).

Almost all of the periolivary nuclei on both sides of the brain contain at least a few neurons that project to the cochlear nucleus. In the cat the main sources of the superior olivary input are the ipsilateral lateral nu-cleus of the trapezoid body and both the ipsilateral and contralateral ventral nuclei of the trapezoid body (Elverland 1977; Spangler et al. 1987). The inputs from the superior olivary complex are distributed throughout both the dorsal and ventral cochlear nuclei, although the input is not equally dense in all parts. Whether there are differences in the distribution of the three major sources is not clear, although all of the periolivary nuclei appear to project widely throughout the cochlear nucleus (Spangler et al. 1987). In both the cat and guinea pig, the periolivary nuclei appear to use as transmitters glycine or GABA or both (Adams and Wenthold 1987; Benson and Potashner 1990; Ostapoff, Morest, and Potashner 1990). Collaterals from the axons of the medial olivocochlear bundle project into the cochlear nucleus, particularly into the granule cell areas

(Osen and Roth 1969; Osen et al. 1984; Brown et al. 1988b). These latter inputs appear to use acetylcholine as their transmitter.

Projections to the cochlear nuclei from the nuclei of the lateral lemniscus arise in both the ventral and dorsal nuclei (VNLL and DNLL, respectively). Neurons in the ventrocaudal part of the VNLL appear to project to both the ipsilateral and contralateral DCN or PVCN (Adams and Warr 1976). The DNLL also sends projections to the PVCN and DCN of both sides (Kane and Finn 1977; Kane and Conlee 1979; Conlee and Kane 1982). The PVCN receives a relatively sparse input (Kane and Conlee 1979). The projection from the DNLL to the dorsal cochlear nucleus appears to be confined to the fusiform cell layer and deep layers (Conlee and Kane 1982). Projections to the AVCN from the nuclei of the lateral lemniscus have not been reported.

Neurons in the inferior colliculus project bilaterally to the dorsal cochlear nucleus, where they appear to terminate in the fusiform and deep layers (Lorente de Nó 1933; Rasmussen 1960; van Noort 1969; Conlee and Kane 1982). A sparse bilateral projection to the PVCN also arises in the inferior colliculus, but a projection to the AVCN has not been described.

4.3 Intrinsic Connections

Lorente de Nó (1933, 1976) described a rich intrinsic circuitry in the cochlear nuclear complex, including projections from the DCN to the VCN and from the VCN to the DCN. These connections have been studied in both Golgi and experimental material (Lorente de Nó 1933; Rasmussen 1967; Warr 1969; Kane and Finn 1977; Cant and Morest 1978; Oliver 1984; Snyder and Leake 1988). The cells in the VCN giving rise to the projection to the DCN include multipolar cells and granule cells (Kane and Finn 1977; Mugnaini, Warr, and Osen 1980; Adams 1983; Oertel et al. 1990). In the tree shrew, the projection from the VCN terminates mainly in the molecular layer of the DCN (Jones and Casseday 1979). It has been reported that spherical cells, globular cells and octopus cells also project to the DCN (Snyder and Leake 1988), although this does not seem consistent with the numerous reports that the axons of these neuronal types do not give rise to intrinsic collateral branches (e.g., Brawer, Morest, and Kane 1974; Tolbert, Morest, and Yurgelun-Todd 1982; Rhode, Oertel, and Smith 1983; Rouiller and Ryugo 1984; Wu and Oertel 1984; Smith and Rhode 1987; Oertel et al. 1990). Projections from the DCN to the AVCN arise in small cells in the deeper layers of the DCN (Lorente de Nó 1933, 1976; Adams and Wenthold 1987; Wenthold 1987). Many of the neurons that give rise to the projection appear to be glycinergic; a few may use GABA as a transmitter (Adams and Wenthold 1987). The projection from small cells in the deep DCN is also present in mouse (Wickesberg and Oertel 1988) where the axons appear to use glycine as an inhibitory transmitter and to contact both bushy and stellate [multipolar] cells (Wickesberg and Oertel 1990).

There are also intrinsic connections among neurons *within* subdivisions of the cochlear nucleus. The DCN contains an extensive intrinsic circuitry (Kane 1974a; Mugnaini, Warr, and Osen 1980; Mugnaini et al. 1980; Osen and Mugnaini 1981; Wouterlood and Mugnaini 1984; Wouterlood et al. 1984). There appear to be three types of inhibitory interneurons in the DCN: the stellate cells, the cartwheel cells and the Golgi cells. All of these types contain GAD and therefore presumably use the inhibitory neurotransmitter GABA. Many of the cell bodies that contain GAD are also immunoreactive to antibodies made to glycine (Osen, Ottersen, and Storm-Mathisen 1990), suggesting that some cells may release both GABA and glycine as neurotransmitters. Granule cells throughout the cochlear nucleus project to the molecular layer of the DCN (Mugnaini, Warr, and Osen 1980), where they form the parallel fibers that are in a position to terminate on the apical dendrites of the fusiform cells and on the dendrites of the stellate and cartwheel cells. The granule cells appear to use an excitatory amino acid as their transmitter (Godfrey et al. 1977, 1978; Oliver et al. 1983; Potashner 1983) and are therefore presumably excitatory. Numerous GAD-immunopositive boutons also contact the somatic surface of all of these neuronal types. Many of these are presumably from the intrinsic inhibitory interneurons. The Golgi cells appear to participate in a synaptic glomerulus similar to those in the cerebellum (Mugnaini, Warr, and Osen 1980; Mugnaini et al. 1980; Osen and Mugnaini 1981). That is, the Golgi cell dendrites are contacted by mossy fibers that also contact granule cell dendrites. The Golgi cells may give rise to inhibitory boutons that also enter the glomeruli (Osen and Mugnaini 1981). The intrinsic circuitry of the DCN is discussed further in Section 5.4.

In the ventral cochlear nucleus, the intrinsic circuitry is not as well understood as in the DCN, but there do appear to be extensive interconnections within the confines of the subdivisions (e.g., Snyder and Leake 1988). For example, axons of multipolar cells give rise to collaterals that terminate in the general vicinity of their own cell bodies (Rhode, Oertel, and Smith 1983; Wu and Oertel 1984; Oertel et al. 1990), but the specific types of neurons receiving the contacts are not known.

5. Fine Structure of the Different Neuronal Classes

The fine structure of most of the cell types described in Section 3 has been examined. With the electron microscope, it is possible to appreciate that the different cell types have different patterns of synaptic organization, providing further justification for the concept that the cochlear nucleus is composed of classes of cells that are functionally different. In this section, the patterns of synaptic input to the cell bodies and, insofar as it is known, to the dendrites of each cell type will be described. In some cases it is possible to indicate the sources of origin of some of the synapses on particular cell types. Much remains to be learned about the neurons of the cochlear nucleus at a fine structural level, but present knowledge provides a framework into which new facts can be incorporated as they become available.

5.1 General Features of Fine Structure of the Cochlear Nuclei

Just as the neurons in the cochlear nucleus have been classified into types, the synaptic terminals in the nucleus have been classified. The most useful criteria for distinguishing synaptic types in aldehyde-fixed sections through the cochlear nucleus have proven to be the size and shape of the synaptic vesicles. The types so distinguished often correlate with the specific sources of the input or with the presence of specific transmitter-related molecules. Four basic types of synaptic terminals are found in the cochlear nucleus (Fig. 3.7). These are terminals with (1) large, round synaptic vesicles; (2) small, round vesicles; (3) oval or pleomorphic vesicles; and (4) distinctly flattened vesicles. Within these basic categories, various subsets have been recognized in different parts of the cochlear nucleus. Some authors have also used the size of the terminal as a criterion for distinguishing types, but this does not seem so useful in distinguishing functional groups, since a single axon is known to give rise to many different sizes of terminals.

5.2 Fine Structure of Each Cell Type with a Restricted Location in the Ventral Cochlear Nucleus

In the ventral cochlear nucleus, the two types of bushy cells and the octopus cells are morphologically distinct and are segregated spatially into different zones. The spherical and octopus cells are, in some places, almost completely segregated from other neuronal types. This fact has made it relatively easy to analyze their synaptic organization using the electron microscope so that they are the most thoroughly characterized cell types in the ventral cochlear nucleus. The spherical and globular cells have prominent connections with the main nuclei of the superior olivary complex. They are therefore presumed to be important in carrying information used in the encoding of binaural differences in the time of arrival and intensity of an acoustic stimulus. Some of the fine structural specializations described below will be discussed in reference to this presumed function.

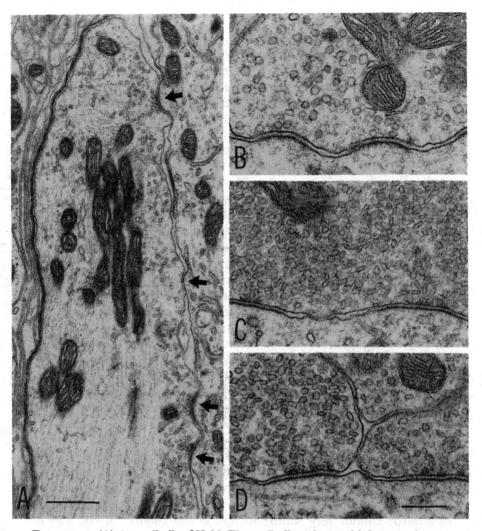

FIGURE 3.7. (A) An endbulb of Held. The endbulb makes multiple synaptic contacts (arrows) with the somatic surface of a spherical bushy cell. The synaptic vesicles, which cluster at the active zones, are large and round. Scale bar = 1 μm. (B-D) The most common types of synaptic terminals in the cochlear nuclear complex classified on the basis of the size and shape of their synaptic vesicles. (B) Terminal with large, round vesicles. The terminals arising from the cochlear nerve, including the endbulbs, contain this type of vesicle. (C) Terminal with small, flattened synaptic vesicles. Such terminals are often immunoreactive to antibodies to glycine. (D) Terminal with small, predominately oval synaptic vesicles. The vesicles are often referred to as "pleomorphic," since they range in shape from round to flat. Such terminals are often immunoreactive to antibodies to GAD or GABA. In some cases, they appear to contain both GABA and glycine. Scale bar for panels B-D = 0.25 μm.

5.2.1   Spherical Bushy Cells

Most authors who have examined the cochlear nucleus in the electron microscope have described the fine structure of the spherical bushy cells, enticed no doubt by the fact that these are the cells that receive input via the endbulbs of Held (*cat:* Lenn and Reese 1966; McDonald and Rasmussen 1971; Ibata and Pappas 1976; Cant and Morest 1979b; Adams and Mugnaini 1987; Saint Marie, Morest, and Brandon 1989; *rat:* Gentshev and Sotelo 1973; Sotelo, Gentschev, and Zamora 1976; *guinea pig:* Schwartz and Gulley 1978; Altschuler et al. 1986; Oberdorfer et al. 1988; Wenthold et al. 1988). Most of the studies appear to have been made of the *large* spherical cells, although the exact localization of the cells examined was not always specified.

A large spherical cell and the synaptic inputs impinging upon it are illustrated in Figure 3.8A. Both the cell bodies and proximal dendrites of spherical cells are contacted by numerous synaptic terminals, which are of several types. The most striking input is that from the endbulbs of Held. In the electron microscope, these appear as long finger-like terminals that make multiple synaptic contacts with the cell body or with the most proximal part of the dendrite(s). The endbulbs contain large, round synaptic vesicles and form multiple, short synaptic contacts that are slightly asymmetric (Figs. 3.7, 3.8A). Smaller terminal boutons that arise from the cochlear nerve have a similar fine structure. About 70% of the somatic surface of the spherical cell is apposed by synaptic terminals, about half of which contain large, round synaptic vesicles (Cant 1981). All of the terminals with this fine structure appear to arise from the cochlea (e.g. Cant and Morest 1979b).

The cochlear inputs to the spherical cells are excitatory (Pfeiffer 1966; Bourk 1976), but the neurotransmitter has not been identified conclusively. The multiple contacts made by one endbulb would appear to ensure secure synaptic transmission from the primary afferent fiber to the bushy cell. As expected, the spherical bushy cell has a response pattern to acoustic stimuli very similar to that of the auditory nerve, and it follows very high rates of activity in the nerve fibers (Pfeiffer 1966; Bourk 1976). As the spherical bushy cells appear to play a major role in the circuits necessary for sound localization (e.g., Cant and Casseday 1986), it is perhaps not surprising that this synapse is specialized for faithful transmission of the inputs from the auditory nerve. Unlike the cell soma, the bushy dendritic appendages of the spherical bushy cells appear to be contacted by relatively few terminals (Cant and Morest 1979) and their function is obscure.

The synaptic contacts with the spherical bushy cells that survive cochlear ablation are smaller than the endbulb and contain smaller synaptic vesicles. The large majority of these contain either oval or distinctly flattened vesicles. Many of the noncochlear terminals in the cat and

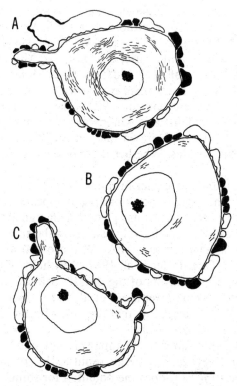

FIGURE 3.8. Drawings of a spherical bushy cell (A), a globular bushy cell (B) and an octopus cell (C) to illustrate some features of their fine structure. In this figure and in Figures 3.9 and 3.10, the only cell organelles that are indicated are the nucleus and nucleolus and stacks of granular endoplasmic reticulum (thin lines). Synaptic terminals that make contact with the cell surface are drawn in outline to indicate terminals that contain large, round synaptic vesicles or are filled in to indicate terminals that contain either flattened or oval synaptic vesicles. Similar types of synaptic terminals contact all three types of neurons, but only the spherical bushy cells are contacted by the endbulbs of Held (the very large terminal in A). Scale bar is equal to roughly 10 μm.

guinea pig appear to be GABAergic (*cat:* Adams and Mugnaini 1987; Saint Marie, Morest, and Brandon, 1989; *guinea pig:* Oberdorfer et al. 1988). In addition, in the guinea pig, many other noncochlear terminals appear to be glycinergic (Wenthold et al. 1987, 1988). Presumably these glycinergic terminals also exist in the cat, but they have not yet been demonstrated in that species. Possible sources of the glycinergic inputs to the spherical cells include the contralateral cochlear nucleus (Wenthold 1987), the superior olivary complex (Adams and Wenthold 1987; Wenthold et al. 1987; Benson and Potashner 1990 [guinea pig]); and the dorsal cochlear nucleus (Adams and Wenthold 1987; Saint Marie et al. 1991).

Possible sources of the GABAergic terminals also include the dorsal cochlear nucleus (Mugnaini 1985; Wenthold et al. 1986; Adams and Mugnaini 1987; Adams and Wenthold 1987; Saint Marie et al. 1991) and the superior olivary complex (Saint Marie, Ostapoff, and Morest 1986; Adams and Wenthold 1987; Ostapoff et al. 1988; Ostapoff, Morest, and Potashner 1990 [guinea pig]). All of these areas are known to project into the spherical cell area (e.g., Cant and Gaston 1982; Adams and Wenthold 1987; Spangler et al. 1987). Adams and Mugnaini (1987) also describe a population of GAD-positive neurons in the peripheral cap of small cells in the ventral cochlear nucleus. Whether these project to the spherical cells is unknown.

In some cases, neurotransmitter immunohistochemistry may correlate with terminal morphology. In the guinea pig, it appears that the terminals with distinctly flattened vesicles use glycine as a transmitter (Wenthold et al. 1988). The terminals with oval vesicles can be subdivided into two groups, those that appear to use GABA and those that may contain both GABA and glycine (Oberdorfer et al. 1987; Wenthold et al. 1988). Small neurons in the superficial layers of the dorsal cochlear nucleus also colocalize GABA and glycine (Wenthold et al. 1987; Osen, Ottersen, and Storm-Mathisen 1990), suggesting that they may be a source of this particular input to the spherical cells.

The function of the noncochlear inputs is not known, although the possibility of a contralateral glycinergic input to the spherical bushy cells is intriguing, suggesting as it does the potential for some sort of binaural processing before information is passed to the superior olivary complex. Another source of glycinergic inhibition to bushy cells was demonstrated in the mouse by Wu and Oertel (1986), although the cells they describe could have been globular bushy cells (see Section 5.2.2).

5.2.2   Globular Bushy Cells

The fine structure of the globular bushy cells in the cat has been studied in the most detail by Tolbert and Morest (1982b), Smith and Rhode (1987) and Saint Marie, Morest, and Brandon (1989). Like the somatic surface of the spherical cells, that of the globular cells is almost completely covered by synaptic terminals (Fig. 3.8B). The fine structure of the terminals is very similar to that described for spherical bushy cells, although, as would be expected from the light microscopic descriptions of the auditory nerve, the cochlear terminals are not as large as the endbulbs that contact those cells. The terminals arising in the cochlea contain large, round vesicles and make up a little under half of the terminal population. Terminals with oval vesicles make up approximately 40% of the population and those with flattened vesicles account for about 15% (Smith and Rhode 1987). The input to the proximal dendrites is like that on the soma, but the distribution of terminal types is quite different on the more distal dendrites, where most (over 90%) of the terminals contain flattened or oval vesicles (Ostapoff and Morest 1984; Smith and Rhode 1987). Like the peripheral dendrites of the spherical bushy cells, those of the globular bushy cells are relatively sparsely innervated.

Many of the noncochlear terminals appear to be GABAergic (Adams and Mugnaini 1987; Saint Marie, Morest, and Brandon 1989; Osen et al 1991). The cell bodies of globular cells are also contacted by glycinergic terminals (Osen et al. 1991). The sources of GABAergic and glycinergic input are potentially the same as for the spherical bushy cells, although this has not been demonstrated conclusively.

The globular bushy cells have been thought to correspond to the physiological response type known as "primary-like with notch" (Bourk 1976). Smith and Rhode (1987) demonstrated this relationship directly. Convergence of multiple primary inputs onto the cell body is likely to be responsible for the highly faithful first spike after the onset of an acoustic stimulus. The function of the noncochlear inputs is not known, although in the mouse, Wu and Oertel (1986) showed that bushy cells receive late inhibitory input when the auditory nerve is stimulated electrically in a slice preparation. This inhibition appears to be mediated by glycine.

5.2.3 Octopus Cells

Kane first described the synaptic organization of the octopus cells (Kane 1973, 1974b, 1977a; Schwartz and Kane 1977). A section through the soma of an octopus cell is illustrated in Figure 3.8C. The terminals that contact the soma and proximal dendrites of the octopus cells are of three main types and are similar to those in the AVCN. The most numerous type has large, round synaptic vesicles and makes multiple short synaptic contacts with the neuronal surface. These terminals cover 50% of the somatic surface and 70% of the proximal dendritic surface. Like the similar terminals in the AVCN, the terminals with large, round vesicles degenerate after cochlear ablation, although at 4 days after surgery not all terminals of this type have disappeared (Kane 1973). A second type of terminal has small, round-to-oval, or pleomorphic vesicles and contacts approximately 15% of the somatic and proximal dendritic surface. Kane (1973) presents evidence that some of these terminals may also arise in the cochlea, so that, unlike the cells so far studied in the AVCN, the octopus cells may receive two morphologically different types of inputs from the cochlea. Many of the primary inputs form contacts with the numerous somatic spines characteristic of this cell type (Kane 1973).

The axosomatic terminals on octopus cells which survive cochlear ablation contain small, flattened vesicles and are relatively sparse. Another type of terminal, found only on the distal dendrites of the octopus cells (Kane 1977a) contains small spherical vesicles. After very large lesions are made in the superior olivary complex, degeneration in the ipsilateral octopus cell area is confined mainly to the terminals on the distal dendrites, whereas in the contralateral octopus cell area, the somatic terminals with small, flattened vesicles appear to undergo degenerative changes (Kane 1977a). The exact sources of the terminals are not known, but some areas that appear to have a major input to the octopus cell area are the contralateral cochlear nucleus (Cant and Gaston 1982) and the ipsilateral posterior periolivary nucleus (Spangler et al. 1987).

Octopus cells are contacted by small, GAD-positive terminals containing small oval synaptic vesicles. These terminals are distributed in moderate density on both the cell bodies and dendrites (Adams and Mugnaini 1987; Saint Marie, Morest, and Brandon 1989). In the guinea pig, octopus cells contain glycine receptors postsynaptic to terminals with flat or oval synaptic vesicles (Wenthold et al. 1988). Oberdorfer et al. (1987) report that many puncta colabelled for both GABA and glycine are concentrated in the PVCN, although they do not mention which part. Potential sources of these presumably inhibitory inputs are much the same as those discussed above for the spherical and globular bushy cells. To what extent these three cell types share the same sources of inhibition is not known.

The octopus cells have a physiological response pattern known as an onset pattern; that is, they respond only at the onset of a tonal stimulus (Godfrey, Kiang, and Norris 1975; Rhode, Oertel, and Smith 1983). Whether the relatively few inhibitory terminals contacting the somatic surface contribute to this pattern is not known; the neurons appear to remain in a depolarized state throughout the duration of an auditory stimulus, although the depolarization is subthreshold (Rhode, Oertel, and Smith 1983; Romand 1978).

5.3 Fine Structure of Multipolar Cells in the Ventral Cochlear Nucleus

Multipolar cells, which are found throughout the ventral cochlear nucleus, have not been studied systematically with the electron microscope, but those in several specific areas have been described (Cant 1981; Tolbert, Morest, and Yurgelun-Todd 1982; Smith and Rhode 1989). Many of these neurons project directly to the inferior colliculus (e.g., Adams 1979) and so participate in one of the shortest pathways from the auditory periphery to the forebrain. The function of these neurons and their projection patterns are not yet understood. Physiologically, most of them appear to respond to tonal stimuli with a "chopper" or "onset" response pattern (Pfeiffer 1966; Bourk 1976; Rhode, Oertel, and Smith 1983; Rouiller and Ryugo 1984; Smith and Rhode 1989).

5.3.1 AVCN

Multipolar cells in both AVCN-A and AVCN-P have been studied with the electron microscope (Fig. 3.9). Cant (1981) concluded that the large multipolar (stellate) cells in AVCN-A can be classified into two groups based on their appearance in the electron microscope. The cells of the first group, the type I stellate cells, receive very few synaptic contacts on their somatic surface and thus must receive most of their inputs on their dendrites. The few contacts on the somatic surface and the many contacts on the proximal dendrites are similar to those on the bushy cells, including cochlear endings with their characteristic large, spherical vesicles

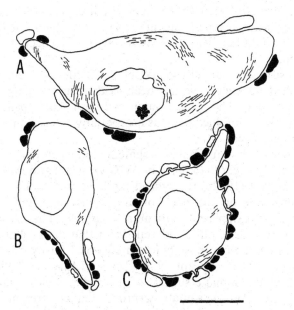

FIGURE 3.9. Drawings of multipolar cells in the ventral cochlear nucleus to illustrate some features of their fine structure. (A) Large multipolar cell from AVCN-P. Few synaptic terminals make contact with the somatic surface. Those that do may be of either cochlear or noncochlear origin. (B) Medium-sized multipolar cell typical of AVCN-A and the peripheral cap of small cells. Like the larger multipolar cells in more caudal parts of the nucleus, many smaller multipolar cells receive only a few synaptic contacts on the surface of the cell body. (C) Medium-sized multipolar cell found in AVCN-A. Unlike the cell shown in (B), this cell receives numerous synaptic contacts on its soma as well as on its dendrites. None of the multipolar cells receive contacts from endbulbs of Held, although all receive inputs from the cochlea via small synaptic terminals. The cells illustrated are based on cells located in the AVCN; similar types of multipolar cells are found in the PVCN (Smith and Rhode 1989). Scale bar is equal to roughly 10 μm for all drawings.

as well as inputs from other sources. The cochlear endings are never as large as endbulbs, so they probably correspond to the boutons that arise from cochlear nerve fibers. Whether the noncochlear sources of input to the type I stellate cells are the same as for the bushy cells remains to be determined, but the fine structure of the noncochlear terminals is similar.

The neurons in the second group of stellate cells, the type II stellate cells, are apposed by endings covering at least 70% of the cell surface. Dense clusters of terminals also form synapses with the dendrites as far as they can be followed from the soma. Thus, activity in these neurons reflects the interplay of the inputs to both the dendrites and the soma. Like the other cell types in AVCN-A, these neurons receive both cochlear and noncochlear inputs, although the fine structure of one of the noncochlear inputs is different from that of any of those on the other cell types, as they contain large, disk-shaped synaptic vesicles (Cant 1981). It does not appear that type II stellate cells receive their cochlear input from endbulbs, although definitive studies with serial sections have not been reported.

The smallest multipolar cells in AVCN-A, like those in the rest of the ventral cochlear nucleus, have not been studied in any detail with the electron microscope. The typical small cell with scant, pale perikaryoplasm and a highly infolded nucleus is especially prominent around the margins of this region.

At least some of the multipolar (stellate and elongate) cells in AVCN-P are considerably larger than those in AVCN-A (Cant and Morest 1984). Tolbert and Morest (1982b) recognized the large multipolar cells in the electron microscope on the basis of their very large and prominent stacks of granular endoplasmic reticulum (Fig. 3.9). The synaptic organization of these cells is quite different from that of the globular cells, the differences paralleling those between spherical bushy cells and type I stellate cells in AVCN-A (Cant 1981; Cant and Morest 1979b). Specifically, <15% of the somatic surface of the stellate cells is apposed by synaptic terminals. The axosomatic terminals that are present, as well as the more abundant terminals forming synapses with the proximal dendrites, are of the same types as those found contacting the globular bushy cells. The terminals with large, round vesicles arise in the cochlea. The sources of the noncochlear terminals are not known; possible sources were discussed in earlier sections.

5.3.2 PVCN

Smith and Rhode (1989) recorded physiologically from multipolar cells in the posteroventral cochlear nucleus and then labelled them and studied them in the electron microscope. They found two classes of cells. The first had the physiological response property known as a "chopper" pattern. The dendritic field of these neurons was stellate in shape. In the electron microscope, the soma of the chopper neurons was sparsely innervated. There was substantial input to the dendrites made up of both cochlear and noncochlear terminals as defined in other studies of the cochlear nucleus (above). The second cell type had a response pattern known as an "onset" pattern. Both the soma and proximal dendrites of these cells were heavily innervated by both cochlear and noncochlear terminals. The distal dendritic processes of both cell types appeared to be relatively sparsely innervated.

These two types of multipolar cells resemble in many respects those described in AVCN-A (Cant 1981), although in the light microscope the dendritic processes of the cells in the PVCN seem to be somewhat more elaborate.

5.3.3 Small Cell Cap

A published description of the fine structure of the neurons in the peripheral cap of small cells is not available. My unpublished observations indicate that the fine structure of the small cell cap resembles the granule cell areas as described by Mugnaini et al. (1980) more closely than it does the large cell areas described above. However, there are some important differences. The neurons of the cap region are fairly similar to one another in their fine structure. Most of them resemble the type I stellate cells of AVCN-A in that the cell somas receive few, if any, synaptic contacts. The synaptic input to the neurons appears, rather, to be mainly to the dendrites of the neurons. Most of the synaptic terminals in the small cell cap are one of two types. The predominant type is a terminal containing small, round, densely packed synaptic vesicles. These terminals form most of their synaptic contacts with spinous processes arising from the dendrites of the neurons in the small cell cap. These spinous contacts are the most characteristic elements in the neuropil of the small cell cap and are much more numerous than mossy fiber glomeruli, the predominant synaptic structure in the granule cell layers (Section 5.5), although glomeruli are also present in the small cell cap. Since all sizes of neurons in the cap may have spiny dendrites, it is not possible to be certain which cell types are receiving these synapses. The second type of terminal found in the small cell cap is one containing oval vesicles. These terminals almost always form synaptic contacts with the dendritic shafts of the neurons rather than with the spines. It is relatively rare to find terminals with large, round vesicles like those known to arise from the cochlear nerve in the rest of the nucleus.

Terminals that label with antibodies to GAD are present in the small cell cap, which also contains immunoreactive neuronal cell bodies (Adams and Mugnaini 1987). Presumably, the GAD reactivity is localized to the terminals with oval vesicles in keeping with the patterns in other parts of the nucleus. By analogy with the cerebellum and the dorsal cochlear nucleus (see below), it is possible that the terminals with small, round vesicles contain glutamate or a similar excitatory amino acid. Glutamate levels in the region of the small cell cap are among the highest in any part of the cochlear nuclear complex (Godfrey et al. 1977). The source of these terminals is not known.

Another input to the area of the peripheral cap of small cells is the medial olivocochlear system (Brown et al. 1988b). An electron microscopic study of the olivocochlear axons in the mouse led to the conclusion that the targets of these axons were dendrites of multipolar cells (Benson and Brown 1990). The mouse does not have an obvious cap of small cells, but the multipolar cells receiving these contacts did lie just below the superficial granule cell layer of the ventral cochlear nucleus. The highest choline acetyltransferase activity in the cochlear nucleus is present in the granule cell layers, but the regions adjacent to the granule cell layers (the area of the small cell cap) also contain a relatively high activity (Godfrey, Williams, and Matschinsky 1977). It is likely that some of the mossy fiber synapses present in the small cell cap, like those in the granule cell areas (Section 5.5), may use acetylcholine as a transmitter.

5.4 Fine Structure of the Dorsal Cochlear Nucleus

The fine structure of most of the neuronal types in the DCN has been described. The most thorough descriptions of the synaptic organization of the fusiform and giant cells in the cat were provided by Kane (1974a,b, 1977b; Kane, Puglisi, and Gordon 1981). Further specific details about the fusiform cells were added by Smith and Rhode (1985). Extensive studies of the several types of interneurons in the rat's DCN have been reported by Mugnaini and his associates (Mugnaini 1985; Wouterlood and Mugnaini 1984; Wouterlood et al. 1984). In the absence of comparable descriptions in the cat, their findings in the rat will be summarized below. Several authors have discussed the physiological response properties of the neurons in the DCN in relation to its known synaptic organization (e.g., Rhode, Smith, and Oertel 1983; Smith and Rhode 1985; Young et al. 1988).

5.4.1 Fusiform Cells

The fusiform cells can be distinguished easily in the electron microscope because of their large size and their confinement to a single layer. A section through one of these cells is illustrated in Figure 3.10A. Synaptic terminals cover the somatic surface and the proximal parts of both the apical and basal dendrites. Kane (1977b) distinguished eight types of synaptic terminals in the fusiform cell layer, using the size and location of the terminals as well as the size and shape of the synaptic vesicles as her criteria. For the most part, the terminal types defined in the DCN are very similar

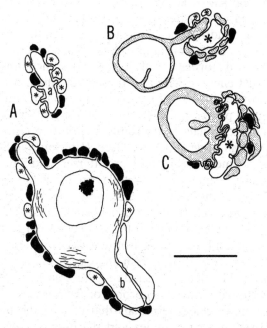

FIGURE 3.10. (A) Drawing to illustrate some features of the fine structure of the fusiform cells of the DCN. The cell body and both the apical (a) and basal (b) dendrites of the cells are covered with synaptic contacts. The terminals that contain round vesicles (unfilled outlines) may contain large, round vesicles as in other parts of the nuclear complex, but many of them contain round vesicles that are smaller than those in the cochlear nerve terminals. Terminals that contain small, round vesicles are marked with asterisks. They are especially numerous on the apical dendrites of these cells. (B) Drawing to illustrate some of the features of the fine structure of granule cells. The granule cell dendrites (stippled) participate in a glomerular structure whose center is occupied by a large mossy fiber that contains very small, round vesicles (large asterisk). The granule cell dendrites are also contacted by terminals containing flat or pleomorphic vesicles (filled profiles). (C) Drawing to illustrate features of the fine structure of a Golgi cell (Mugnaini et al. 1980). The highly infolded surface of this cell and its dendrites (stippled) form glomerular complexes with mossy fibers (large asterisk). As in the granule cell glomeruli, terminals with pleomorphic or flattened vesicles (filled profiles) also participate in the glomeruli. Scale bar is equal roughly to 10 μm for all drawings.

to those in the ventral cochlear nucleus, a fact that is not surprising since many of the sources of input may be the same.

The cochlear input to the fusiform cells is morphologically similar to that in the rest of the cochlear nuclear complex. The terminals are characterized by large, round synaptic vesicles and form multiple synaptic complexes with the soma and proximal dendrites, especially the basal dendrites. Smith and Rhode (1985) found only one cochlearlike terminal contacting the soma of a fusiform cell. All other cochlearlike terminals contacted the basal dendrites, being most common in the distal branches (38% of all terminals on the distal branches). Noncochlear inputs to the soma and proximal dendrites are of several types, including those with pleomorphic or flattened vesicles and those with small, round vesicles. Most of the terminals on the cell soma (and also on the initial segment of the axon) contain pleomorphic or flattened vesicles (Smith and Rhode 1985). Many of the terminals on the soma are immunopositive for GAD or GABA (Adams and Mugnaini 1987; Osen, Ottersen, and Storm-Mathisen 1990) or for glycine or for both (Osen, Ottersen, and Storm-Mathisen 1990). Terminals with small, round vesicles are especially numerous on the apical dendrites of the fusiform cells; these appear to arise predominately from the granule cell axons (parallel fibers; Kane 1974a; Mugnaini, Warr, and Osen 1980; Mugnaini et al. 1980). There is evidence that these inputs are excitatory and that the transmitter used is glutamate (Godfrey et al. 1977, 1978; Schwartz 1981; Hirsch and Oertel 1988 [mouse]; Manis 1989 [guinea pig]). The dendritic shafts of the apical dendrites receive contacts from terminals with pleomorphic or flat vesicles.

Sources of the noncochlear terminals that contact the fusiform cells include the superior olivary complex and the inferior colliculus (Kane 1977b). Lesions in the superior olivary complex lead to the degeneration of terminals with pleomorphic or flat vesicles apposed to the soma and primary dendrites of fusiform cells. Lesions of the inferior colliculus lead to degeneration of terminals with flattened vesicles apposed to fusiform cell dendrites. Other terminal types in the DCN also degenerate after these lesions but whether they contact fusiform cells is not known. Rapid Golgi studies of the small cells in the DCN led to the conclusion that all three types form contacts with the fusiform cells (Lorente de Nó 1933; Kane 1974a), a suggestion corroborated by the studies of the fine structure of these cells in rats (Mugnaini et al. 1980; Wouterlood and Mugnaini 1984; Wouterlood et al. 1984; Section 5.4.3). Many of these interneurons appear to use GABA or glycine as a neurotransmitter (e.g., Osen, Ottersen, and Storm-Mathisen 1990) so they could provide at least one source of the GAD- or glycine-positive terminals that contact the fusiform cell surface. Other potential sources of inputs to the apical dendrites of the fusiform cells are the contralateral cochlear nucleus (Cant and Gaston 1982) and the ipsilateral AVCN (Jones and Casseday 1979 [tree shrew]). Other potential sources of input to the basal dendrites include the nuclei of the lateral lemniscus (Conlee and Kane 1977). Recurrent collaterals of fusiform cell axons may terminate on any part of the fusiform cell surface (Smith and Rhode 1985). Their terminals contain small round vesicles which are, on average, smaller than those in the granule cell axons.

5.4.2 Giant Cells

One study of the synaptic organization of the giant cells of the deep layer of the DCN has been published (Kane, Puglisi, and Gordon 1981). Five types of giant cells can be distinguished on the basis of their dendritic arborization patterns and the arrangement of synaptic inputs on their somatic and dendritic surfaces. The primary dendrites of all five kinds of giant cell are contacted by synaptic inputs. The elongate multipolar cells and radiate cells receive many somatic inputs as well. Although the sources of the inputs were not established, many of the axodendritic and, where they were present, axosomatic terminals resemble cochlear inputs to the fusiform cells. This type, with large, round vesicles, forms contacts with all types of giant cells, but the pattern of distribution of inputs may be different on each type (Kane, Puglisi, and Gordon 1981). Collaterals of fusiform cells contact the dendrites of at least some giant cells (Smith and Rhode 1985). Other potential inputs to these neurons are the same as those that may contact the basal dendrites of the fusiform cells (Section 5.4.1). Differences in the physiology or projection patterns of the different types of giant cells have not been reported so it is not possible to speculate on the functional significance of the different synaptic arrangements.

5.4.3 Small Cells

The fine structure of the cartwheel and stellate cells located in the superficial layers of the DCN have been studied in the most detail by Mugnaini, Wouterlood and their colleagues (Wouterlood and Mugnaini 1984; Wouterlood et al. 1984; Mugnaini 1985) in the rat. Although it is not known whether all of the details discovered by them also apply to the cat, the main results of their studies will be summarized here. The fine structure of the cartwheel cell was studied by Wouterlood and Mugnaini (1984). Its perikaryon and dendritic shafts are covered by synaptic terminals, all of which contain pleomorphic vesicles and which may be of several types. The profuse dendritic spines characteristic of this cell type make synaptic contact with terminals with round vesicles, many of which are axons of granule cells (Mugnaini 1985). The sources of the other terminals are not known; potential sources are other small cells and any source of projections to the superficial layers.

The stellate cells described by Wouterlood et al. (1984) were found only in the outer two layers of the DCN. Small cells in the deeper layers have not been described in electron microscopic studies. There are few axosomatic contacts on the stellate cells. Terminals with round vesicles, many arising from granule cell axons (Mugnaini 1985), contact the soma, dendritic shaft and spines. Terminals with pleomorphic vesicles contact the soma and dendritic shafts. In the rat, dendrosomatic or dendrodendritic appositions between stellate cells are characterized by gap junctions (Wouterlood et al. 1984; Mugnaini 1985). Gap junctions linking neurons have also been observed in the AVCN of the rat (Sotelo, Gentschev, and Za-mora 1976), but have not been found in the cat.

Golgi cells are infrequent in the DCN (Wouterlood and Mugnaini 1984). Since they were first described in reference to the granule cell domains found throughout the cochlear nuclear complex (Mugnaini, Warr, and Osen 1980; Mugnaini et al. 1980), they will be considered in Section 5.5 with the granule cells.

5.5 Fine Structure of Granule Cell Areas

The granule cell areas were studied in the electron microscope by Mug-naini et al. (1980). A restricted part of the domain, that in the dorsal cochlear nucleus, was also studied by Kane (1974a,b, 1977b). The small somas of the granule cells are largely apposed by astroglial processes or by neighboring granule cells. Occasionally a synaptic terminal containing either round or pleomorphic vesicles makes contact with the soma.

Granule cell dendrites participate in complex synaptic nests which re-semble cerebellar glomeruli (Fig. 3.10B). A large mossy fiber terminal lies in the center of the glomerulus and is surrounded by thin granule cell dendrites. On their outer surface the dendrites are contacted by boutons containing pleomorphic vesicles. The entire complex is surrounded by astrocytic processes. Mugnaini et al. (1980) suggest, based on its mor-phology and analogy with the cerebellum, that the mossy fiber synapse is excitatory. The physiological response properties of granule cells are unknown. Sources of mossy endings appear to be diverse. In the DCN, possible sources include the collaterals of fusiform cells (Smith and Rhode 1985), primary afferent terminals (Kane 1974a,b), and the inferior col-liculus (Conlee and Kane 1977). Sources of input to the granule cell layers include collaterals of the medial olivocochlear system, which Brown et al. (1988b) suggest may be a source of mossy fibers. Consistent with this is the observation that some mossy fibers are positive for acetylcholin-esterase (AChE) (McDonald and Rasmussen 1971) and that they originate in the superior olivary complex (Osen et al. 1984). It is not likely, how-ever, that all mossy fibers are AChE-positive (Osen et al. 1984).

Small cells in the granule cell domains were called Golgi cells by Mug-naini, Warr, and Osen (1980); and Mugnaini et al. (1980). The Golgi cell body often exhibits numerous "hairy" appendages which engage in syn-aptic complexes with large mossy-like endings (Fig. 3.10C). Golgi cell dendrites are associated with a glomerular complex different from that entered into by the granule cell dendrites (Fig. 3.10B). There is a large centrally located mossy fiber, but it is surrounded by large Golgi cell dendrites. Two other kinds of boutons contact the Golgi cell, both on its soma and in the glomerular complex. These are small boutons containing small round vesicles and those containing pleomorphic vesicles. At least some of the terminals with small, round vesicles may arise from granule cell axons since they resemble the known terminals of those axons in the molecular layer of the DCN (Mugnaini et al. 1980).

Wenthold et al. (1988) report glycine receptor immunoreactivity in the neuropil of the granule cell layer of guinea pig. Nerve terminals immunoreactive for GAD are also present in the granule cell domain (Adams and Mugnaini 1987). By analogy with other areas of the cochlear nucleus, it is likely that these substances are found in the small terminals with small, pleomorphic vesicles, but this has not been demonstrated directly.

Another source of input to the granule cell areas is the type II primary afferent axons (Section 4.1). The target of these terminals is not known. Brown et al. (1988b) emphasize the interesting possibility that the medial olivocochlear fibers, which project to the outer hair cells in the cochlea and the type II afferent fibers, which innervate the outer hair cells, could interact in some way in the granule cell areas of the cochlear nucleus.

6. The Neuronal Types in Other Species

The neurons in the cochlear nucleus have been described in a number of other species. In what follows, only the major differences from the cat will be emphasized. Except for studies of the guinea pig and rat, few electron microscopic studies of these other species have been undertaken so that little can be said about similarities or differences in the fine structure of the cells.

6.1 Human and Monkey

In many respects, the human ventral cochlear nucleus resembles that of other mammals (Moore and Osen 1979a). The cochlear nerve fibers enter the nucleus ventrally and bifurcate to form ascending and descending branches (Fig. 3.2). Endbulbs are found on the anterior branch in the anterior parts of the nucleus (Moore and Osen 1979a; Adams 1986). The pattern of innervation indicates that the tonotopic organization of the human cochlear nucleus is similar to that of the cat in the ventral cochlear nucleus but along a different axis in the DCN (Fig. 3.2; Moore and Osen 1979a). Covering the lateral surface of the ventral cochlear nucleus is a large marginal zone which seems to correspond to the peripheral cap of small cells of the cat. In the human, this area is relatively greatly enlarged. The dorsal cochlear nucleus is large, but the lamination typical of other mammals is not seen in humans or in some other species of primates (Moore and Osen 1979a; Moore 1980; see below). Areas of granule cells are absent in many species of primates as are plexuses of AChE-positive fibers and terminals typically seen in the granule cell layer of other animals (Moore and Osen 1979a).

Many of the neuronal types first described in the cat can also be recognized in the human and other primates (Strominger and Strominger 1971; Bacsik and Strominger 1973; Konigsmark 1973; Moore and Osen 1979a,b; Moore 1980; Adams 1986). The ventral cochlear nucleus contains spherical, globular and octopus cells located approximately in the same relative locations in which they are found in cats (Moore and Osen 1979a,b). Large and small spherical cells have not been distinguished. The largest subdivisions in the human appear to be a multipolar cell area located in the vicinity of the nerve root and a laterodorsal cap of small cells. The AChE staining of the neuropil in the area of the cap is relatively strong, as it is in cats and rodents.

In Golgi preparations (Adams 1986), similarities between cell types in the human and in other animals are obvious. Bushy cells are found throughout the human ventral cochlear nucleus, being most abundant rostrally. Multipolar cells are common. The dendritic trees of many of these cells are aligned with respect to the entering auditory nerve fibers, as has also been described for the cat (Tolbert, Morest, and Yurgelun-Todd 1982). Even in the rostral pole of the nucleus, multipolar cells are intermingled with the bushy cells so that no region containing only bushy cells is apparent. Octopus cells lie in the caudal portion of the nucleus. As Moore and Osen (1979b) point out, the cell types that project to the inferior colliculus (in the cat) appear to have increased in relative numbers in primates. Those cell types that have the superior olivary complex as their main target appear relatively less numerous.

The dorsal cochlear nucleus of the human is relatively large, but it is not clearly organized into layers. Moore (1980) studied a variety of prosimian and anthropoid primates and has shown that there are differences in the location, depth and extent of the granular layer of the DCN. In prosimians, the DCN is similar to that of other mammals with the exception that there is in addition to the other layers, a superficial or external layer of granule cells. There are species of anthropoid primates in which only the external layer of granule cells is present and others in which there is no granular layer at all. In these animals, the fusiform cells lose their orientation and become located in the central region of the nucleus.

In the human, pyramidal (fusiform) cells can be identified, but they lack a clear radial orientation and are scattered among other neuronal types in an area of "cochlear fiber neuropil"(Moore and Osen 1979a). Adams (1986) argues that these neurons are not equivalent to the fusiform cells of the cat. The issue remains to be settled. The other neuronal types include small elongate cells and relatively rare giant cells situated in the deeper region of the nucleus. Cartwheel cells do not appear to be present in humans (Adams 1986). Moore and Osen suggest that the entire dorsal cochlear nucleus of the human is equivalent to the fusiform and deep layers in other mammals (see also Adams 1986). This fits with the observation that primates have very few if any granule cell areas, since the molecular layer in other animals is made up largely of the parallel fibers that are the axons of the granule cells and their postsynaptic targets. Unfortunately, our lack of knowledge of function of the DCN renders speculation about the significance of the differences between primates and other animals meaningless (cf. Moore and Osen 1979b).

The similarities between the human and other mammals in terms of cell types in the cochlear nucleus, especially the ventral division, allows us to form hypotheses about the human auditory system based on data from experimental animals. We can begin to address such hypotheses using immunocytochemistry and tract tracing methods that rely on compounds that can be used in dead, fixed tissue.

6.2 Rodents

The first species in which neuronal types in the cochlear nucleus were described in detail was the rat (Harrison and Warr 1962; Harrison and Irving 1965, 1966; Harrison and Feldman 1970). A few studies of the rat, mainly of the DCN, have been undertaken since then (Kromer and Moore 1976; Mugnaini, Warr, and Osen 1980; Mugnaini et al. 1980; Wouterlood and Mugnaini 1984; Wouterlood et al. 1984). Several immunocytochemical studies are also available (Mugnaini 1985; Shiraishi et al. 1985; Moore and Moore 1987).

Species of rodents that have received considerable attention are the mouse (Mugnaini, Warr, and Osen 1980; Mugnaini et al. 1980; Martin 1981a,b; Browner and Baruch 1982; Trune 1982a,b; Webster and Trune 1982; Willard and Ryugo 1983; Wu and Oertel 1984; Ryugo and Willard 1985; Oertel and Wu 1989; Oertel et al. 1990) and the guinea pig (Pirsig 1968; Schwartz and Gulley 1978; Treck and Pirsig 1979; Altschuler et al. 1986; Hackney and Pick 1986; Moore 1986; Wenthold 1987; Wenthold et al. 1988). A recent study of the cochlear nucleus of the chinchilla provides a very detailed description and contains a discussion of the similarities to and differences from the cat (Morest, Hutson, and Kwok 1990). Reports describing some of the neuronal types in hamster (Schweitzer and Cant 1984), gerbil (Nordeen, Killackey, and Kitzes 1983; Roberts and Ribak 1987), kangaroo rat (Webster 1971), and mountain beaver (Merzenich, Kitzes, and Aitkin 1973) have also appeared.

In all of these species, neuronal classes similar to those in the cat have been identified. Webster and Trune (1982) made a thorough study of the relationship between the types identified in Nissl stains and those identified in Golgi impregnations and concluded that all of the types described in the cat are also present in the mouse. A major difference between rodents and cats is the presence in rodents of a group of cells known as the acoustic nerve nucleus or cochlear root nucleus, which was first described in the rat (Harrison and Warr 1962; Harrison, Warr, and Irving 1962; Harrison and Irving 1966b; Harrison and Feldman 1970; Merchan et al. 1988). This nucleus is also present in mouse (Webster and Trune 1982), chinchilla (Morest, Hutson, and Kwok 1990) and gerbil (Chamberlain 1977). The nucleus is made up of neurons that occupy the central part of the nerve, extending from the border between glial and Schwann cells to the point of entry into the cochlear nucleus. The neurons receive inputs from the auditory nerve (Harrison, Warr, and Irving 1962; Merchan et al. 1988); no other source of endings has been described. The very large neurons in this nucleus send large axons into the trapezoid body, but their targets are not known (Harrison, Warr, and Irving 1962). Webster and Trune (1982) concluded that these neurons represent displaced globular cells, but Merchan et al. (1988) and Osen et al. (1991) provide several lines of evidence that they represent a separate population of cells. For example, these authors demonstrated that, unlike globular cells, the neurons of the cochlear root nucleus do not receive GABAergic or glycinergic inputs on their cell bodies. GABAergic terminals do contact these neurons but only on their dendrites.

Another apparent difference from cats in many species of rodents is the relative decrease in the size of the large spherical cell area. The size of the large spherical cell area appears to correlate with the frequency range to which a species is sensitive, however, since those rodents that hear low frequency sounds appear to have a large spherical cell area closer in relative size to that of the cat (eg., the kangaroo rat [Webster 1971] and the gerbil [Cant, unpublished observations]). Finally, some rodent species have a greatly enlarged granule cell domain. This is true in the mountain beaver (Merzenich, Kitzes, and Aitkin 1973) and has been noted in another burrowing rodent, the porcupine (Fuse 1920, quoted by Mugnaini, Warr, and Osen 1980).

6.3 Other Mammals

Other mammals in which the neuronal types of the cochlear nucleus have been described in some detail include rabbits (Disterhoff, Perkins, and Evans 1980; Perry and Webster 1981), moles (*Monera,* Kudo et al. 1990), porpoises (*Phocaena phocaena;* Osen and Jansen 1965) and bats (see below). The cochlear nucleus of the rabbit is very similar to that of the cat and rodents. The mole resembles other burrowing mammals in having relatively enlarged granule cell layers. The bats and porpoises show some major differences from the cat.

The two species of bats that have received attention are the mustache bat (*Pteronotus parnellii;* Zook and Casseday 1982a,b; Zook and Leake 1989) and the horseshoe bat (*Rhinolophus rouxi;* Schweizer 1981; Feng and Vater 1985). Zook and Casseday (1982a) emphasize that although there are several marked differences between the AVCN of the mustache bat and other mammals, these differences "stand out against a setting of auditory nuclei that appears very similar to that of other mammals."

This species has no evident large spherical cells in its AVCN. This is similar to but more extreme than the case in other animals lacking low-frequency hearing. Very small round cells are plentiful in the rostral AVCN and in Golgi material bushy cells can be identified (Pollak and Casseday 1989) so that it seems likely that small spherical cells are present. However, these cells appear to project to the inferior colliculus (Zook and Casseday 1982b). It is unclear whether any of the small spherical cells of the cat project to the colliculus, but such a projection has not been definitely shown. A second difference in the AVCN of the mustache bat is the presence of a marginal subdivision that contains only large multipolar cells. Such segregation of large multipolar cells does not occur in other mammals. These multipolar cells project to the inferior colliculus and may process high-frequency information (Zook and Casseday 1982b). The PVCN of the mustache bat is similar to that of the cat, and its DCN is relatively small. The fusiform cell layer contains few granule cells and lamination is not distinct. The cochlear nucleus of the horseshoe bat is similar to that of the mustache bat, but the DCN in the former species is considered to have two parts: a laminated part devoted to lower frequencies and a nonlaminated part devoted to the higher frequencies. There is also a large concentration of large multipolar cells in a segregated region, but in the horseshoe bat, this is considered part of the PVCN.

The cochlear nucleus of the common porpoise appears more similar to that of primates than to that of any other species so far described (Osen and Jansen 1965). The DCN is small and lacks lamination. There is an extensive area in the dorsal VCN that appears to be equivalent to the small cell cap of the cat (cf. Osen 1969). There is no granule cell layer. The lack of a granule cell layer in those species with a greatly enlarged small cell cap is interesting in light of marked similarities in structure in the small cell cap and the granule cell layer in the cat (Section 5.3.3).

7. Summary and Discussion

Both neurophysiological and neuroanatomical studies have contributed to the current view that the large brainstem structure known as the cochlear nucleus actually comprises a group of nuclei, each containing distinct populations of neurons. The different types of cells form populations that differ in their morphology, physiology, neurotransmitter chemistry and connectivity. Experiments in which single cells were injected with horse-radish peroxidase after their physiological response properties were recorded show that there is very good agreement between particular morphological categories and physiological ones (Rhode, Oertel, and Smith 1983; Rhode, Smith, and Oertel 1983; Rouiller and Ryugo 1984; Smith and Rhode 1987, 1989). The recognition of the different neuronal populations in the cochlear nucleus leads to the view that the auditory system is made up of multiple neuronal pathways (e.g., Warr 1982). Presumably the different pathways have different functions in auditory signal processing. In some cases, we have insight into what that role is. For example, the spherical and globular bushy cells project into the superior olivary complex and appear to play a central role in binaural processing. The multipolar cells are components of pathways that for the most part bypass the superior olivary complex and terminate in the inferior colliculus. They thus participate in the shortest pathways from the auditory periphery to the forebrain. The functional significance of their pattern of projection is as yet obscure.

Studies of the neuronal types in the cochlear nucleus are important for several reasons. First, a clear understanding of the distribution of the cell types, which in many cases does not follow subdivision boundaries, is essential for the proper interpretation of experiments on connections, both efferent and afferent. Second, a solid anatomical framework is provided for the interpretation of physiological and immunocytochemical data. Further, recognition of neuronal classes provides a starting point from which to explore the working hypothesis that these different neuronal populations give rise to functionally different ascending systems. The synaptic organization and topography of these systems can be compared and contrasted and related to the presumed function(s) of the pathways. Experiments both at the light and electron microscopic level can be designed to take advantage of the spatial segregation of particular types. Although current understanding of the neuronal types is useful for framing hypotheses and experiments, there is continuing need for a refinement of our understanding of the neuronal classes. It would not be correct to assert that any of the categories of cells described in this chapter are completely homogeneous. For example, there may be systematic differences in the sizes of cells in a population or in the extent of their dendritic fields or even in their patterns of synaptic organization (see below). In addition, there could be subtypes within the currently recognized categories (and in some cases, there almost certainly are) but, if so, the differences are subtle and of limited use in designing and interpreting experiments at the present time.

A number of interesting problems remain to be solved. While we know that the different neuronal types are organized with respect to the entering nerve fibers, thus preserving frequency specificity, it is not known whether the cells of a given type *within* an isofrequency plane have the same or different patterns of synaptic organization. The organization within the isofrequency planes is only poorly understood. Is the organization of each large spherical cell, to take one example, similar or are different large spherical cells within an isofrequency region organized differently? Leake and Snyder (1989) provide evidence that there is an orderly topographic arrangement of the vertical dimension of the spiral ganglion that maps into the cochlear nucleus along its medial to lateral dimension (orthogonal to the frequency representation). This could indicate that there are differences in synaptic organization of the neurons in a given isofrequency lamina.

There are a number of issues related to the distribution of the inputs, both cochlear and noncochlear, to the neurons of the cochlear nucleus that remain to be clarified. It is not known, for example, whether the two classes of type I cochlear inputs differ in their distribution on the cell body and dendrites of the cells of a given type. The fine structure of type II terminals has not been examined in any detail; it is not known how much of the cochlear input to different cell types is via these neurons (cf. Brown et al. 1988a). There is also a need for a better understanding of the distribution of the inputs from descending pathways with respect to cell types. Almost all cells in the cochlear nucleus receive primary input from the cochlea. If one assumes that the same information reaches all terminal branches of the cochlear nerve fibers (an assumption not yet tested), it must be the numbers and arrangements of the cochlear inputs along with their relationships to noncochlear inputs that determine, in large part, the different physiological response properties of the different cell types.

The recognition by Harrison and his colleagues and by Osen of the distinct cell types in the cochlear nucleus and of their differential distribution was a seminal contribution toward the effort to develop an understanding of the organization of the auditory pathways. The neuronal types have distinctive patterns of projections to the superior olivary complex, lateral lemniscus and inferior colliculus. It is reasonable to suggest that the function of each of these pathways is different. The ability to recognize the different cellular components at the level of the cochlear nucleus allows us to study each functional pathway with the various neuroanatomical, neurophysiological, neuropharmacological and immunocytochemical techniques currently at our disposal. Knowledge of the specific interconnections of the neuronal types will allow us to explore in detail the principles of synaptic organization and connectivity that underlie their function.

Acknowledgements   Preparation of this chapter was supported in part by a grant from the National Institute on Deafness and Other Communication Disorders, DC00135. Unpublished studies on the synaptic organization of the peripheral cap of small cells were funded in part by a grant from the Deafness Research Foundation.

References

Adams JC (1983) Multipolar cells in the ventral cochlear nucleus project to the dorsal cochlear nucleus and the inferior colliculus. Neurosci Lett 37:205–208.

Adams JC (1986) Neuronal morphology in the human cochlear nucleus. Arch Otolaryngol Head Neck Surg 112:1253–1261.

Adams JC, Mugnaini E (1987) Patterns of glutamate decarboxylase immuno-staining in the feline cochlear nuclear complex studied with silver enhancement and electron microscopy. J Comp Neurol 262:375–401.

Adams JC, Warr WB (1976) Origins of axons in the cat's acoustic striae determined by injection of horseradish peroxidase into severed tracts. J Comp Neurol 182:519–538.

Adams JC, Wenthold RJ (1987) Immunostaining of GABA-ergic and glycinergic inputs to the anteroventral cochlear nucleus. Soc Neurosci Abstr 13:1259.

Altschuler RA, Betz H, Parakkal MH, Reeks KA, Wenthold RJ (1986) Identification of glycinergic synapses in the cochlear nucleus through immunocytochemical localization of the postsynaptic receptor. Brain Res 369:316–320.

Arnesen AR, Osen KK (1978) The cochlear nerve in the cat: Topography, cochleoptopy and fiber spectrum. J Comp Neurol 178:661–678.

Bacsik RD, Strominger NL (1973) The cytoarchitecture of the human anteroventral cochlear nucleus. J Comp Neurol 147:281–290.

Benson CG, Potashner SJ (1990) Retrograde transport of [^{3}H] glycine from the cochlear nucleus to the superior olive in the guinea pig. J Comp Neurol 296:415–426.

Benson TE, Brown MC (1990) Synapses formed by olivocochlear axon branches in the mouse cochlear nucleus. J Comp Neurol 295:52–70.

Blackstad TW, Osen KK, Mugnaini E (1984) Pyramidal neurones of the dorsal cochlear nucleus: A Golgi and computer reconstruction study in cat. Neuroscience 13:827–854.

Bourk TR (1976) Electrical response of neural units in the anteroventral cochlear nucleus of the cat. Massachusetts Institute of Technology, Ph.D. Dissertation, Cambridge, MA.

Bourk TR, Mielcarz JP, Norris BE (1981) Tonotopic organization of the anteroventral cochlear nucleus of the cat. Hear Res 4:215–241.

Brawer, JR, Morest DK, Kane EC (1974) The neuronal architecture of the cochlear nucleus of the cat. J Comp Neurol 160:491–506.

Brown, MC, Berglund AM, Kiang NYS, Ryugo DK (1988a) Central trajectories of type II spiral ganglion neurons. J Comp Neurol 278:581–590.

Brown, MC, Liberman MC, Benson TE, Ryugo DK (1988b) Brainstem branches from olivocochlear axons in cats and rodents. J Comp Neurol 278:591–603.

Browner RH, Baruch A (1982) The cytoarchitecture of the dorsal cochlear nucleus in the 3-month- and 26-month-old C57BL/6 mouse: A Golgi impregnation study. J Comp Neurol 211:115–138.

Cant NB (1981) The fine structure of two types of stellate cells in the anterior division of the anteroventral cochlear nucleus of the cat. Neuroscience 6:2643–2655.

Cant NB (1982) Identification of cell types in the anteroventral cochlear nucleus that project to the inferior colliculus. Neurosci Lett 32:241–246.

Cant NB (1991) Projections to the lateral and medial superior olivary nuclei from the spherical and globular bushy cells of the anteroventral cochlear nucleus. In: Altschuler RA, Hoffman DW, Bobbin RB, Clopton B (eds) Neurobiology of Hearing, Vol. 2, The Central Auditory System. New York: Raven Press, pp. 99–119.

Cant NB, Casseday JH (1986) Projections from the anteroventral cochlear nucleus to the lateral and medial superior olivary nuclei. J Comp Neurol 247:457–476.

Cant NB, Gaston KC (1982) Pathways connecting the right and left cochlear nuclei. J Comp Neurol 212:313–326.

Cant NB, Morest DK (1978) Axons from non-cochlear sources in the anteroventral cochlear nucleus of the cat. A study with the rapid Golgi method. Neuroscience 3:1003–1029.

Cant NB, Morest DK (1979a) Organization of the neurons in the anterior division of the anteroventral cochlear nucleus of the cat. Light-microscopic observations. Neuroscience 4:1909–1923.

Cant NB, Morest DK (1979b) The bushy cells in the anteroventral cochlear nucleus of the cat. A study with the electron microscope. Neuroscience 4:1925–1945.

Cant NB, Morest DK (1984) The structural basis for stimulus coding in the cochlear nucleus of the cat. In: Berlin CI (ed) Hearing Science: Recent Advances. San Diego: College-Hill Press, pp. 371–421.

Chamberlain SC (1977) Neuroanatomical aspects of the gerbil inner ear: Light microscopic observations. J Comp Neurol 171:193–204.

Conlee JW, Kane ES (1982) Descending projections from the inferior colliculus to the dorsal cochlear nucleus in the cat: an autoradiographic study. Neuroscience 7:161–178.

Disterhoff JF, Perkins RE, Evans S (1980) Neuronal morphology of the rabbit cochlear nucleus. J Comp Neurol 192:687–702.

Elverland HH (1977) Descending connections between the superior olivary and cochlear nuclear complexes in the cat studied by autoradiographic and horseradish peroxidase methods. Exper Brain Res 27:397–412.

Evans EF, Palmer AR (1980) Relationship between the dynamic range of cochlear nerve fibres and their spontaneous activity. Exp Brain Res 40:115–118.

Fekete DM, Rouiller EM, Liberman MC, Ryugo DK (1982) The central projections of intracellularly labeled auditory nerve fibers in cats. J Comp Neurol 229:432–450.

Feng A, Vater M (1985) Functional organization of the cochlear nucleus of rufous horseshoe bats (*Rhinolophus rouxi*): Frequencies and internal connections are arranged in slabs. J Comp Neurol 235:529–553.

Fuse G (1913) Das Ganglion ventrale und das Tuberculum acusticum beim einigen Sauglingen und beim Menschen. Arbeiten aus dem Hirnanatomischen Institut in Zuerich, 7:1–210.

Gentschev T, Sotelo C (1973) Degenerative patterns in the ventral cochlear nucleus of the rat after primary deafferentation. An ultrastructural study. Brain Res 62:37–60.

Godfrey DA, Kiang NYS, Norris BE (1975) Single unit activity in the posteroventral cochlear nucleus of the cat. J Comp Neurol 162:247–268.

Godfrey DA, Carter JA, Berger SJ, Lowry OH, Matschinsky FM (1977) Quantitative histochemical mapping of candidate transmitter amino acids in cat cochlear nucleus. J Histochem Cytochem 25:417–431.

Godfrey DA, Williams AD, Matchinsky FM (1977) Quantitative histochemical mapping of enzymes of the cholinergic system in cat cochlear nucleus. J Histochem Cytochem 25:397–416.

Godfrey DA, Carter JA, Lowry OH, Matschinsky FM (1978) Distribution of gamma-aminobutyric acid, glycine, glutamate and aspartate in the cochlear nucleus of the rat. J Histochem Cytochem 26:118–126.

Hackney CM, Pick GF (1986) The distribution of spherical cells in the anteroventral cochlear nucleus of the guinea pig. Br J Audiology 20:215–220.

Harrison JM, Feldman ML (1970) Anatomical aspects of the cochlear nucleus and superior olivary complex. Contrib Sensory Physiol 4:95–142.

Harrison JM, Irving R (1965) The anterior ventral cochlear nucleus. J Comp Neurol 124:15–42.

Harrison JM, Irving R (1966a) Ascending connections of the anterior ventral cochlear nucleus in the rat. J Comp Neurol 126:51–64.

Harrison JM, Irving R (1966b) The organization of the posterior ventral cochlear nucleus in the rat. J Comp Neurol 126:391–402.

Harrison JM, Warr WB (1962) A study of the cochlear nuclei and ascending auditory pathways of the medulla. J Comp Neurol 119:341–379.

Harrison JM, Warr WB, Irving R (1962) Second order neurons in the acoustic nerve. Science 138:893–895.

Held H (1891) Die centralen Bahnen des Nervus acusticus bei der Katze. Arch Anat Physiol Anat Abtil 15:271–291.

Hirsch JA, Oertel D (1988) Synaptic connections in the dorsal cochlear nucleus of mice, in vitro. J Physiol 396:549–562.

Ibata Y, Pappas GD (1976) The fine structure of synapses in relation to the large spherical neurons in the anterior ventral cochlear [nucleus] of the cat. J Neurocytol 5:395–406.

Jones DR, Casseday JH (1979) Projections to laminae in dorsal cochlear nucleus in the tree shrew, *Tupaia glis*. Brain Res 160:131–133.

Kane ESC (1973) Octopus cells in the cochlear nucleus of the cat: Heterotypic synapses upon homeotypic neurons. Int J Neurosci 5:251–279.

Kane ES (1974a) Synaptic organization in the dorsal cochlear nucleus of the cat: A light and electron microscopic study. J Comp Neurol 155:301–330.

Kane ES (1974b) Patterns of degeneration in the caudal cochlear nucleus of the cat after cochlear ablation. Anat Rec 179:67–92.

Kane ES (1976) Descending projections to specific regions of cat cochlear nucleus: A light microscopic study. Exper Neurol 52:372–388.

Kane ES (1977a) Descending inputs to the octopus cell area of the cat cochlear nucleus: an electron microscopic study. J Comp Neurol 173:337–354.

Kane ES (1977b) Descending inputs to the cat dorsal cochlear nucleus: An electron microscopic study. J Neurocytol 6:583–605.

Kane ES, Conlee JW (1979) Descending inputs to the caudal cochlear nucleus of the cat: Degeneration and autoradiographic studies. J Comp Neurol 4:759–783.

Kane ES, Finn RC (1977) Descending and intrinsic inputs to dorsal cochlear nucleus of cats: A horseradish peroxidase study. Neuroscience 2:897–912.

Kane ES, Puglisi SG, Gordon BS (1981) Neuronal types in the deep dorsal cochlear nucleus of the cat. I. Giant neurons. J Comp Neurol 198:483–513.

Kölliker A (1896) Handbuch der Gewebelehre des Menschem Bd. 2. Leipzig: Wilhelm Engelman.

Konigsmark BW (1973) Cellular organization of the cochlear nuclei in man. J Neuropath Exp Neurol 32:153–154.

Kromer LF, Moore RY (1976) Cochlear nucleus innervation by central norepinephrine neurons in the rat. Brain Res 118:531–537.

Kudo M, Nakamura Y, Tokuno H, Kitao Y (1990) Auditory brain stem in the mole (*Mogera*): Nuclear configurations and the projections to the inferior colliculus. J Comp Neurol 298:400–412.

Leake PA, Snyder RL (1989) Topographic organization of the central projections of the spiral ganglion in cats. J Comp Neurol 281:612–629.

Lenn NJ, Reese TS (1966) The fine structure of nerve endings in the nucleus of the trapezoid body and the ventral cochlear nucleus. Am J Anat 118:375–390.

Liberman, MC (1978) Auditory-nerve response from cats raised in a low-noise environment. J Acoust Soc Am 63:442–455.

Lorente de Nó R (1933) Anatomy of the eighth nerve—III. General plans of structure of the primary cochlear nuclei. Laryngoscope 43:327–350.

Lorente de Nó R (1976) Some unresolved problems concerning the cochlear nerve. Ann Otol Rhinol Lar Suppl 34, Vol 85:1–28.

Lorente de Nó R (1981) The Primary Acoustic Nuclei. New York: Raven Press.

McDonald DM, Rasmussen GL (1971) Ultrastructural characteristics of synaptic endings in the cochlear nucleus having acetylcholinesterase activity. Brain Res 28:1–18.

Manis PB (1989) Responses to parallel fiber stimulation in the guinea pig dorsal cochlear nucleus in vitro. J Neurophysiol 61:149–161.

Martin MR (1981) Morphology of the cochlear nucleus of the normal and reeler mutant mouse. J Comp Neurol 197:141–152.

Martin MR (1981b) Acetylcholinesterase-positive fibers and cell bodies in the cochlear nuclei of normal and reeler mutant mice. J Comp Neurol 197:153–167.

Merchan MA, Collia F, Lopez DE, Saldaña E (1988) Morphology of coclear root neurons in the rat. J. Neurocytol 17:711–725.

Merzenich MM, Kitzes L, Aitkin L (1973) Anatomical and physiological evidence for auditory specialization in the mountain beaver (Aplodontia rufa). Brain Res 58:331–344.

Moore JK (1980) The primate cochlear nuclei: loss of lamination as a phylogenetic process. J Comp Neurol 193:609–629.

Moore JK (1986) Cochlear nuclei: Relationship to the auditory nerve. In: Altschuler RA, Hoffman DW, Bobbin RP (eds) Neurobiology of Hearing: The Cochlea. New York: Raven Press, pp. 283–301.

Moore JK, Moore RY (1987) Glutamic acid decarboxylase-like immunoreactivity in brainstem auditory nuclei of the rat. J Comp Neurol 260:157–174.

Moore JK, Osen KK (1979a) The cochlear nuclei in man. J Comp Neurol 154:393–418.

Moore JK, Osen KK (1979b) The human cochlear nuclei. Exp Brain Res Suppl II:36–44.

Morest DK, Hutson KA, Kwok S (1990) Cytoarchitectonic atlas of the cochlear nucleus of the chinchilla, Chinchilla laniger. J Comp Neurol 300:230–248.

Mugnaini E (1985) GABA neurons in the superficial layers of the rat dorsal cochlear nucleus: Light and electron microscopic immunocytochemistry. J Comp Neurol 235:61–81.

Mugnaini E, Warr WB, Osen KK (1980) Distribution and light microscopic features of granule cells in the cochlear nuclei of cat, rat and mouse. J Comp Neurol 191:581–606.

Mugnaini E, Osen KK, Dahl A-L, Friedrich VL, Korte G (1980) Fine structure of granule cells and related interneurons (termed Golgi cells) in the cochlear nuclear complex of cat, rat and mouse. J Neurocytol 9:537–570.

Nordeen KW, Killackey HP, Kitzes LM (1983) Ascending auditory projections to the inferior colliculus in the adult gerbil, *Meriones unguiculatus*. J Comp Neurol 214:131–143.

Oberdorfer MD, Parakkal MH, Altschuler RA, Wenthold RJ (1987) Colocalization of glycine and GABA in the cochlear nucleus. Neurosci Abstr 13:544.

Oberdorfer MD, Parakkal MH, Altschuler RA, Wenthold RJ (1988) Ultrastructural localization of GABA-immunoreactive terminals in the anteroventral cochlear nucleus of the guinea pig. Hear Res 33:229–238.

Oertel D, Wu SH (1989) Morphology and physiology of cells in slice preparations of the dorsal cochlear nucleus of mice. J Comp Neurol 283:228–247.

Oertel D, Wu SH, Garb MW, Dizak C (1990) Morphology and physiology of cells in slice preparations of the posteroventral cochlear nucleus of mice. J Comp Neurol 295:136–154.

Oliver DL (1984) Dorsal cochlear nucleus projections to the inferior colliculus in the cat: A light and electron microscopic study. J Comp Neurol 224:155–172.

Oliver DL, Potashner SJ, Jones DR, Morest DK (1983) Selective labelling of spiral ganglion and granule cells with D-aspartate in the auditory system of cat and guinea pig. J Neurosci 3:455–472.

Osen KK (1969) Cytoarchitecture of the cochlear nuclei in the cat. J Comp Neurol 136:453–484.

Osen KK (1970) Course and termination of the primary afferents in the cochlear nuclei of the cat. Arch Ital Biol 108:21–51.

Osen KK, Jansen J (1965) The cochlear nuclei in the common porpoise, *Phocaena phocaena*. J Comp Neurol 125:223–258.

Osen KK, Mugnaini E (1981) Neuronal circuits in the dorsal cochlear nucleus. In: Syka J, Aitkin L (eds) Neuronal Mechanisms of Hearing, New York: Plenum, pp. 119–125.

Osen KK, Roth K (1969) Histochemical localization of cholinesterases in the cochlear nuclei of the cat with notes on the origin of acetylcholinesterase-positive afferents and the superior olive. Brain Res 16:165–185.

Osen KK, Mugnaini E, Dahl A-L, Christiansen AH (1984) Histochemical localization of acetylcholinesterase in the cochlear and superior olivary nuclei. A reappraisal with emphasis on the cochlear granule cell system. Arch Ital Biol 122:169–212.

Osen KK, Ottersen OP, Storm-Mathisen J (1990) Colocalization of glycine-like and GABA-like immunoreactivities: A semiquantitative study of individual neurons in the dorsal cochlear nucleus of cat. In: Ottersen OP, Storm-Mathisen J (eds) Glycine Neurotransmission. New York: John Wiley & Sons, pp. 417–451.

Osen KK, Lopez DE, Slyngstad TA, Ottersen OP, Storm-Mathisen J (1991) GABA-like and glycine-like immunoreactivities of the cochlear root nucleus in rat. J. Neurocytol 20:17–25.

Ostapoff E-M, Morest DK (1984) Analysis of synapses to bushy cells in the posterior anteroventral cochlear nucleus (AVCN-P) of the cat. Neurosci Abstracts 10:842.

Ostapoff E-M, Morest SK, Potashner SJ (1990) Uptake and retrograde transport of [^{3}H] GABA from the cochlear nucleus to the superior olive in the guinea pig. J Chem Neuroanatomy 3:285–295.

Ostapoff E-M, Staatz-Benson C, Morest DK, Potashner SJ, Saint Marie RL (1988) GABA and glycine immunoreactivity of descending and commissural inputs to the cochlear nucleus in guinea pig. Soc Neurosci Abstr 14:489.

Perry DR, Webster WR (1981) Neuronal organization of the rabbit cochlear nucleus: some anatomical and electrophysiological observations. J Comp Neurol 197:623–638.

Pfeiffer RR (1966) Classification of response patterns of spike discharges for units in the cochlear nucleus: Tone burst stimulation. Exp Brain Res 1:220–235.

Pirsig W (1968) Regionen, zelltypen und synapsen im ventralen nucleus cochlearis des meerschwinschens. Arch Klin Exp Ohr-Nas-Kehlkopfheilk 192:333–350.

Pollak GD, Casseday JH (1989) The Neural Basis of Echolocation in Bats. Berlin: Springer-Verlag.

Potashner SJ (1983) Uptake and release of D-aspartate in the guinea pig cochlear nucleus. J Neurochem 41:1094–1101.

Ramón y Cajal S (1909) Histologie du systéme nerveaux de l'homme et des vertèbrès (1952 reprint). Madrid: Instituto Ramón y Cajal.

Rasmussen GL (1960) Efferent fibers of the cochlear nerve and cochlear nucleus. In: Rasmussen GL, Windle W (eds) Neural Mechanisms of the Auditory and Vestibular system. Springfield, IL: Charles C Thomas.

Rasmussen GL (1967) Efferent connections of the cochlear nucleus. In: Graham AB (ed) Sensorineural Hearing Processes and Disorders. Boston: Little, Brown.

Rhode WS, Oertel D, Smith PH (1983) Physiological response properties of cells labeled intracellularly with horseradish peroxidase in cat ventral cochlear nucleus. J Comp Neurol 213:448–463.

Rhode WS, Smith PH, Oertel D (1983) Physiological response properties of cells labeled intracellularly with horseradish peroxidase in cat dorsal cochlear nucleus. J Comp Neurol 213:426–447.

Roberts RC, Ribak CE (1987) GABAergic neurons and axon terminals in the brainstem auditory nuclei of the gerbil. J Comp Neurol 258:267–280.

Rodieck RW, Brening RK (1983) Retinal ganglion cells: properties, types, genera, pathways and trans-species comparisons. Brain Behav Evol 23:121–164.

Romand R (1978) Survey of intracellular recording in the cochlear nucleus of the cat. Brain Res 148:43–65.

Rose JE (1960) Organization of frequency sensitive neurons in the cochlear nuclear complex of the cat. In: Rasmussen GL, Windle W (eds) Neural Mechanisms of the Auditory and Vestibular Systems. Springfield, IL: Charles C Thomas, pp. 116–136.

Rose JE, Galambos R, Hughes JR (1959) Microelectrode studies of the cochlear nuclei of the cat. Bull Johns Hopkins Hosp 104:211–251.

Rouiller EM, Ryugo DK (1984) Intracellular marking of physiologically characterized cells in the ventral cochlear nucleus of the cat. J Comp Neurol 225:167–186.

Rouiller EM, Cronin-Schreiber R, Fekete DM, Ryugo DK (1986) The central projections of intracellularly labeled auditory nerve fibers in cats: An analysis of terminal morphology. J Comp Neurol 249:261–278.

Ryugo DK, Rouiller EM (1988) Central projections of intracellularly labeled auditory nerve fibers in cats: Morphometric correlations with physiological properties. J Comp Neurol 271:130–142.

Ryugo DK, Willard FH (1985) The dorsal cochlear nucleus of the mouse: A light microscopic analysis of neurons that project to the inferior colliculus. J Comp Neurol 242:381–396.

Saint Marie RL, Benson CG, Ostapoff E-M, Morest DK (1991) Glycine immunoreactive projections from the dorsal to the anteroventral coclear nucleus. Hearing Res 51:11–28.

Saint Marie RL, Morest DK, Brandon CJ (1989) The form and distribution of GABAergic synapses on the principal cell types of the ventral cochlear nucleus of the cat. Hear Res 42:97–112.

Saint Marie RL, Ostapoff E-M, Morest DK (1986) Co-localization of ³H-GABA and GABA-like immunoreactivity in superior olivary neurons retrogradely labeled from guinea pig cochlear nucleus. Soc Neurosci Abstr 12:1269.

Schwartz AM, Gulley RL (1978) Non-primary afferents to the principal cells of the rostral anteroventral cochlear nucleus of the guinea pig. Am J Anat 153:489–508.

Schwartz AM, Kane EC (1977) Development of the octopus cell area in the cat ventral cochlear nucleus. Am J Anat 148:1–14.

Schwartz IR (1981) The differential distribution of label following uptake of ³H-labeled amino acids in the dorsal cochlear nucleus of the cat. Exper Neurol 73:601–617.

Schweizer H (1981) The connections of the inferior colliculus and the organization of the brainstem auditory system in the greater horseshoe bat (*Rhinolophus ferrumequinum*). J Comp Neurol 201:25–49.

Schweitzer LF, Cant NB (1985a) Differentiation of the giant and fusiform cells in the dorsal cochlear nucleus of the hamster. Developmental Brain Res 20:69–82.

Schweitzer LF, Cant NB (1985b) Development of oriented dendritic fields in the dorsal cochlear nucleus of the hamster. Neuroscience 16:969–978.

Sento S, Ryugo DK (1989) Endbulbs of Held and spherical bushy cells in cats: Morphological correlates with physiological properties. J Comp Neurol 280:5520–562.

Shiraishi T, Senba E, Tohyama M, Wu J-Y, Kubo T, Matsunaga T (1985) Distribution and fine structure of neuronal elements containing glutamate decarboxylase in the rat cochlear nucleus. Brain Res 347:183–187.

Smith PH, Rhode WS (1985) Electron microscopic features of physiologically characterized, HRP-labeled fusiform cells in the cat dorsal cochlear nucleus. J Comp Neurol 237:127–143.

Smith PH, Rhode WS (1987) Characterization of HRP-labeled globular bushy cells in the cat anteroventral cochlear nucleus. J Comp Neurol 266:360–375.

Smith PH, Rhode WS (1989) Structural and functional properties distinguish two types of multipolar cells in the ventral cochlear nucleus. J Comp Neurol 282:595–616.

Snyder RL, Leake PA (1988) Intrinsic connections within and between cochlear nucleus subdivisions in cat. J Comp Neurol 278:209–225.

Sotelo C, Gentschev T, Zamora AJ (1976) Gap junctions in ventral cochlear nucleus of the rat. A possible new example of electronic junctions in the mammalian central nervous system. Neuroscience 1:5–7.

Spangler KM, Cant NB, Henkel CK, Farley GR, Warr WB (1987) Descending projections from the superior olivary complex to the cochlear nucleus of the cat. J Comp Neurol 259:452–465.

Spoendlin HH (1971) Degeneration behavior of the cochlear nerve. Arch Ohr Nas Kehlk Heilk 200:275–291.

Strominger NL, Strominger AI (1971) Ascending brain stem projections of the anteroventral cochlear nucleus in the rhesus monkey. J Comp Neurol 143:217–242.

Tolbert LP, Morest DK (1982a) The neuronal architecture of the anteroventral cochlear nucleus of the cat in the region of the cochlear nerve root: Golgi and Nissl methods. Neuroscience 7:3013–3030.

Tolbert LP, Moret DK (1982b) The neuronal architecture of the anteroventral cochlear nucleus of the cat in the region of the cochlear nerve root: Electron microscopy. Neuroscience 7:3053–3068.

Tolbert LP, Morest DK, Yurgelun-Todd DA (1982) The neuronal architecture of the anteroventral cochlear nucleus of the cat in the region of the cochlear nerve root: Horseradish peroxidase labelling of identified cell types. Neuroscience 7:3031–3052.

Treeck HH, Pirsig W (1979) Differentiation of nerve endings in the cochlear nucleus on morphological and experimental basis. Acta Otolaryngol 87:47–60.

Trune DR (1982a) Influence of neonatal cochlear removal on the development of mouse cochlear nucleus. I. Number, size, and density of its neurons. J Comp Neurol 209:409–424.

Trune DR (1982b) Influence of neonatal cochlear removal on the development of mouse cochlear nucleus: II. Dendritic morphometry of its neurons. J Comp Neurol 209:425–434.

van Noort J (1969) The Structure and Connections of the Inferior Colliculus. An Investigation of the Lower Auditory System. Van Gorcum, Assen.

Warr WB (1969) Fiber degeneration following lesions in the posteroventral cochlear nucleus of the cat. Exper Neurol 23:140–155.

Warr WB (1982) Parallel ascending pathways from the cochlear nucleus: Neuroanatomical evidence of functional specialization. In: Neff WD (ed) Contributions to Sensory Physiology, Vol 7. New York: Academic Press, pp. 1–38.

Webster DB (1971) Projection of the cochlea to cochlear nuclei in Meriam's kangaroo rats. J Comp Neurol 143:323–340.

Webster DB, Trune DR (1982) Cochlear nuclear complex of mice. Am J Anat 163:103–130.

Wenthold RJ (1987) Evidence for a glycinergic pathway connecting the two cochlear nuclei: an immunocytochemical and retrograde transport study. Brain Res 415:183–187.

Wenthold RJ, Zempel JM, Parakkal MH, Reeks KA, Altschuler RA (1986) Immunocytochemical localization of GABA in the cochlear nucleus of the guinea pig. Brain Res 380:7–18.

Wenthold RJ, Huie D, Altschuler RA, Reeks KA (1987) Glycine immunoreactivity localized in the cochlear nucleus and superior olivary complex. Neuroscience 3:897–912.

Wenthold RJ, Parakkal MH, Oberdorfer MD, Altschuler RA (1988) Glycine receptor immunoreactivity in the ventral cochlear nucleus of the guinea pig. J Comp Neurol 276:423–435.

Wickesberg RE, Oertel D (1988) Tonotopic projection from the dorsal to the anteroventral cochlear nucleus of mice. J Comp Neurol 268:389–399.

Wickesberg RE, Oertel D (1990) Delayed, frequency-specific inhibition in the cochlear nucleus of mice: A mechanism for monaural echo suppression. J Neurosci 10:1762–1768.

Willard FH, Ryugo DK (1983) Anatomy of the central auditory system. In: Willott JF (ed) The Auditory Psychobiology of the Mouse. Springfield, IL: Charles C Thomas, pp. 201–304.

Wouterlood FG, Mugnaini E (1984) Cartwheel neurons of the dorsal cochlear nucleus: A Golgi-electron microscope study in rat. J Comp Neurol 227:136–157.

Wouterlood FG, Mugnaini E, Osen KK, Dahl A-L (1984) Stellate neurons in rat dorsal cochlear nucleus with combined Golgi impregnation and electron microscopy: synaptic junctions and mutual coupling by gap junctions. J Neurocytol 131:639–664.

Wu SH, Oertel D (1984) Intracellular injection with horseradish peroxidase of physiologically characterized stellate and bushy cells in slices of mouse anteroventral cochlear nucleus. J Neurosci 4:1577–1588.

Wu SH, Oertel D (1986) Inhibitory circuitry in the ventral cochlear nucleus is probably mediated by glycine. J Neurosci 6:2691–2706.

Young ED, Shofner WP, White JA, Robert J-M, Voigt HF (1988) Response properties of cochlear nucleus neurons in relationship to physiological mechanisms. In: Edelman GM, Gall WE, Cowan WM (eds) Auditory Function. Neurobiological Bases of Hearing. New York: John Wiley and Sons, pp. 277–312.

Zook JM, Casseday JH (1982a) Cytoarchitecture of auditory system in lower brainstem of the mustache bat, *Pteronotus parnellii*. J Comp Neurol 207:1–13.

Zook JM, Casseday JH (1982b) Origin of ascending projections to inferior colliculus in the mustache bat, *Pteronotus parnellii*. J Comp Neurol 207:14–28.

Zook JM, Leake PA (1989) Connections and frequency representation in the auditory brainstem of the mustache bat, *Pteronotus parnelli*. J Comp Neurol 290:243–261.

4

The Superior Olivary Complex and Lateral Lemniscal Nuclei

ILSA R. SCHWARTZ

1. Introducing the Superior Olivary Complex (SOC) and Lateral Lemniscus (LL)

The superior olivary complex (SOC) is a group of interrelated nuclei located on each side of the brainstem at the level of the cochlear nuclei (CN). The SOC lies just rostral to the facial nucleus and between the roots of the facial and abducens nerves. The SOC is the first level of the auditory system at which there is integration of information from the two ears onto individual neurons. The integrated information is transferred to other cells within the SOC for further processing and to higher centers, the nuclei of the lateral lemniscus (LL) and inferior colliculus (IC), as well as back down to the CN and cochlea. The lateral lemnisci are bands of fibers running from the lateral side and rostral end of each SOC to the ipsilateral IC. Groups of neurons, the LL nuclei, are located within the LL. Axons from the CN and trapezoid body also contribute to the LL. The LL nuclei integrate information directly from the CN, from the SOC, and from the contralateral LL before transmitting it to the IC. The LL neurons may also interact with reticular formation pathways.

Several neuronal types are recognized in each of the nuclei of the CN, SOC, and LL. Neuronal types vary not only in size, shape, and location, but also in the specific neuronal populations from which they receive synapses and in the neurons to which they send terminals. The projections from different neuronal types in the various CN subdivisions to specific neuronal types in the various SOC and LL nuclei form the basis for major differences in functional capabilities of these nuclei.

2. The Superior Olivary Complex (SOC)

The SOC is composed of three well-defined cell groups [usually the medial and lateral superior olivary nuclei (MSO and LSO) and the medial nucleus of the trapezoid body (MNTB)][1] surrounded by more loosely organized

---

[1] In the older literature the LSO is also referred to as the main nucleus or the S segment while the MSO is also called the accessory nucleus. See Ollo and Schwartz (1979) for a discussion of nomenclature.

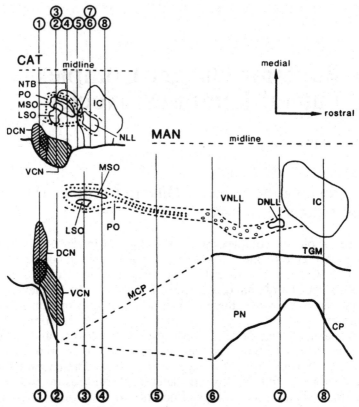

FIGURE 4.1. Location of the SOC and LL in cat and human (Fig. 3 in Moore, 1987a, with permission). Comparable levels through the cat and human brain stem indicated in this figure are illustrated in Figure 4.2. Correspondence drawings show the relative size of brainstem structures in cat and human.

periolivary cell groups. Although the SOC has been extensively studied in a number of mammalian and nonmammalian species (see Table 4.1), in this chapter I will focus on its organization as defined in the cat, some rodent species, and humans.

In the cat there are several well-defined periolivary cell groups distributed around the well-defined MSO, LSO, and MNTB (see Figures 4.1 to 4.4). These include a ventral nucleus of the trapezoid body (VNTB) located ventral and medial to the MSO, a lateral nucleus of the trapezoid body (LNTB) ventral and lateral to the LSO, a ventrolateral periolivary group (VLPO) between the LSO and MSO ventrally, a ventromedial periolivary group (VMPO) between the MSO and MNTB ventrally, a dorsomedial periolivary group (DMPO) (the superior paraolivary nucleus, or SPN, in rodents) medial to the LSO and MSO dorsally, a dorsal periolivary group (DPO) dorsal to the LSO, a dorsolateral periolivary group

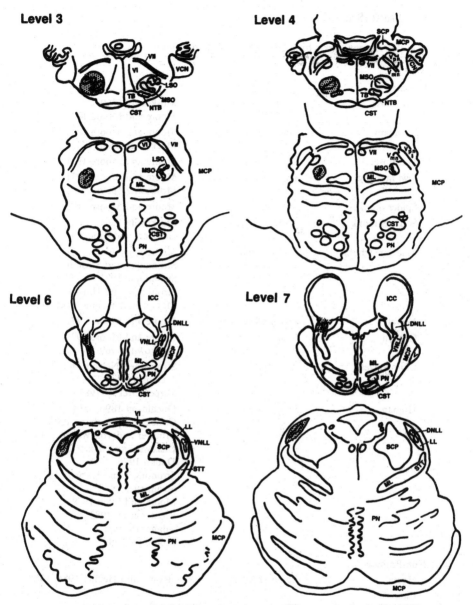

FIGURE 4.2. Location of the SOC and LL in cat and human redrawn from Figures 8, 10, 14, and 16 in Moore, 1987a, with permission. Levels 3, 4, 6, and 7 from Figure 4.1. Level 3 (Figure 8 from Moore 1987a). Cross section through the cat (upper) and human (lower) brain stems at the level of the caudal superior olivary complex. The olivary nuclei are labeled on the right side of the brain and cross-hatched on the left side. Level 4 (Figure 10 in Moore 1987a). Cross sections through the cat and human brain stems at the middle level of the superior olivary complex. Level 6 (Figure 14 from Moore 1987a). Cross sections through the cat and human brain stems at the level of the caudal portion of the lateral lemniscus. Level 7 (Figure 16 from Moore 1987a). Cross sections through the cat and human brain stems at the level of the dorsal nucleus of the lateral lemniscus.

TABLE 4.1. Animals in which the SOC has been studied.

Mammals

  Primates

Slow loris	*Nycticebus coucang*	Moskowitz 1969 Moore and Moore 1971
Bushbaby	*Galago senegalensis*	Moore and Moore 1971 Moskowitz 1966, 1969
Marmoset	*Saguinus oedipus*	Moore and Moore 1971
Owl monkey	*Aotus trivirgatus*	Moore and Moore 1971 Moskowitz 1969
Spider monkey	*Ateles sp.*	Moore and Moore 1971
Squirrel monkey	*Saimiri sciureus*	Moskowitz 1965 Goldberg and Moore 1967
Macaque (rhesus)	*Macaca mulatta*	Moore and Moore 1971 Strominger and Strominger 1971 Strominger 1973 Strominger and Hurwitz 1976 Moore, Karpas, and Moore 1977
Vervet	*Cercopithecus aethiops*	Moore and Moore 1971
Gibbon	*Hylobates lar* *Holobates hooklock*	Moore and Moore 1971 Moskowitz 1969
Chimpanzee	*Pan paniscus*	Strominger, Nelson, and Dougherty 1977 Carpenter et al. 1987
Human	*Homo sapiens*	Donaldson 1890, 1891 Castex and Marchand 1906 Brouwer and von Walree 1914 Olszewski and Baxter 1954 Dublin 1976 Strominger and Hurwitz 1976 Strominger 1978 Moore and Osen 1979 Richter et al. 1983 Adams 1986a,b Moore 1987a,b

  Non-Primate

Horse	*Equus caballus*	Heffner and Heffner 1984, 1986
Cats		
Domestic cat	*Felis catus*	Ramón y Cajal 1899 Morest 1968a,b Scheibel and Scheibel 1974 Schwartz 1977 Tsuchitani 1978 Glendenning et al. 1981 Cant 1984
Northern native cat	*Dasyurus hallucatus*	Aitkin, Byers, and Nelson 1986
Dog	*Canis canis*	Goldberg and Brown 1968, 1969

TABLE 4.1. *Continued*

Rabbit	*Oryctolagus cuniculus*	Ramón y Cajal 1899
		Borg 1973a,b
Hedgehog	*Hemiechinus auritus*	Moore and Moore 1971
	Erinaceus europaeus	Brown and Howlett 1972
Chinchilla	*Chinchilla laniger*	Perkins 1973
		Finlayson and Caspary 1989
Rat		
Laboratory rat	*Rattus norvegicus*	Harrison and Warr 1962
		Harrison and Feldman 1970
		Casey and Feldman 1985
		—ᵃ
Kangaroo rat	*Dipodomys merriami*	Webster, Ackerman, and Longa 1968
		Browner and Webster 1975
Guinea pig	*Cavia poncellus*	Brown 1985
		Winter, Robertson, and Cole 1989
Mole		
European mole	*Talpa europaea*	Aitkin, Horseman, and Bush 1982
	Mogera robusta	Kudo et al. 1988
Weasel		
Least weasel	*Mustela nivalis*	Heffner and Heffner 1987
Ferret	*Mustela putorius*	Brunso-Bechtold, Henkel, and Linville 1990
		Henkel and Brunso-Bechtold 1990
Marsupials		
Brush-tailed possum	*Tichosurus vulpecula*	Aitkin, Bush, and Gates 1978
		Aitkin and Kenyon 1981
North American opossum	*Didelphus virginianus*	Willard and Martin 1984
Mouse	*Mus musculus*	Ramón y Cajal 1899
		Ross 1962
		Sidman, Angevine, and Taber-Pierce 1971
		—ᵃ
C57BL6		Ollo and Schwartz 1979
Deer mouse	*Peromyscus bairdi*	Ross 1969
Gerbil	*Meriones unguiculatus*	Nordeen, Killackey, and Kitzes 1983
		Schwartz and Yu 1986
		Helfert and Schwartz 1987
		Helfert, Schwartz, and Ryan 1988
		Ryan, Schwartz, and Keithley 1989
		—ᵃ
Tree Shrew	*Tupaia glis*	Covey, Jones, and Casseday 1984
Elephant shrew	*Elephantus sp.*	Moskowitz 1965

TABLE 4.1. *Continued*

Bat	Species unknown	Babmindra and Zharskaya 1980, 1982
Mustache bat	*Pteronotus parnellii*	Zook and Casseday 1982a,b, 1985
		Feng and Vater 1985
		Bishop and Henson 1987
		Zook and Casseday 1987
Horseshoe bat	*Rhinolophus rouxi*	Schweizer 1981
		Casseday, Covey, and Vater 1988
		Ross, Pollak, and Zook 1988
		Zook and DiCaprio 1988
		Casseday et al. 1989
		Vater and Feng 1990
Big brown bat	*Eptesicus fuscus*	Covey and Casseday 1986
A Japanese bat	*Miniopterus schreibersi fuliginosus*	Nakajima 1971
A Japanese bat	*Vespertilio superans*	Nakajima 1971
Dolphin		Zvorykin 1964
Birds		
Chicken	*Galus galus*	Smith and Rubel 1979
		Rubel and Durham 1985
		Conlee and Parks 1986
		Nudo and Masterton 1986
Amphibians		
Frog		
Bull frog	*Rana catesbiana*	Fuller and Ebbesson 1973
		Rubinson and Skiles 1975
		Wilczynski 1981
		Gorodetskaia and Bibikov 1985
Leopard frog	*Rana pipiens*	Feng 1986a,b
Reptiles		
Lizard		
Iguana	*Iguana iguana*	Foster and Hall 1978
Turtle		
Painted turtle	*Pseudemys scripta*	Belekhova et al. 1985
		Kunzle 1986

[a] Many more studies not listed.

(DLPO) dorsal and lateral to the LSO, as well as an anterior periolivary group (APO) rostral to the LSO and MSO, and a posterior periolivary group (PPO) caudal to the LSO and MSO. In humans separate periolivary groups are not distinguished; rather, periolivary cells form a discontinuous sphere around the MSO and LSO. Groups of cells in this continuum are similar in appearance to cells found in various cat periolivary nuclei. Cells similar to cat MNTB cells are found in the human, but do not coalesce to form a distinct nucleus (Richter et al. 1983; Moore 1987a).

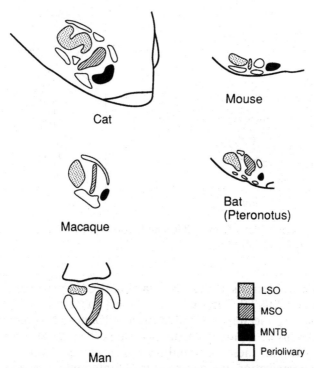

FIGURE 4.3. Comparison of the size and location of olivary nuclear groups in the cat, mouse, monkey (Macaque), bat (Pteronotus) and human. Cat, mouse, monkey and human redrawn after Moore and Moore (1971) reprinted by permission of Wiley-Liss, a division of John Wiley and Sons, Inc. Bat figure drawn after information in Zook and Casseday 1982a.

Periolivary cell groups vary somewhat between species in their size, orientation, shape, and cellular composition, but the morphology, chemistry, input to and projections of, some of their component neurons show considerable similarity across a number of mammalian species studied (e.g., Moore 1987a).

It is helpful to think about the different neural classes as belonging to systems of neurons performing various functions. Among the major functions performed by the SOC and LL nuclei are: (1) sound detection, (2) orientation and localization of sounds in space, (3) complex signal processing, and (4) feedback controls of cochlear mechanisms.

There are several parallel projections ascending from each CN to higher brain centers, including the SOC and LL, which may serve these various functions. Projections from spherical cells in both anterior ventral CNs (AVCN) converge on neurons in the MSO, which project in turn largely to the ipsilateral IC. The LSO receives information directly from the spherical bushy cells of the ipsilateral AVCN, and indirectly from the

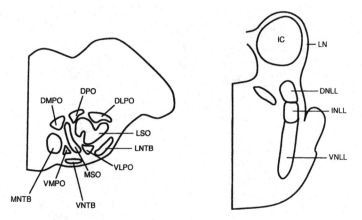

FIGURE 4.4. Nomenclature of the SOC and LL on hypothetical transverse sections through the cat brain stem.

globular bushy neurons of the contralateral AVCN by way of neurons in the ipsilateral MNTB. LSO neurons project in turn to both ICs. The differences in connections to the LSO and MSO mean that different types of comparisons of inputs from the two ears are made at different sites. The physiology of these LSO and MSO neurons is complex, involving tonotopic organization and comparisons of interaural time and intensity differences. A consideration of the detailed physiology is beyond the scope of this chapter, but see Aitkin, Irvine, and Webster (1984) for a brief summary, also Chapter 4 in Volume 2 of this series. The integration of binaural input and the parallel lines insure that information critical to coding locations in space reach higher centers. As the visual system maps the visual world seen by the retina onto the visual cortex, so the SOC and LL provide information to the IC permitting the construction of a map of auditory space (Knudsen and Konishi 1978; Semple et al. 1983; Moore et al. 1984; Poon et al. 1990). As a consequence of the convergence of binaural input to neurons of the MSO and LSO, unilateral lesions of the auditory system above the level of the CN cannot produce a unilateral deafness.

Several cell groups within the SOC, largely in periolivary areas, have been identified by a variety of methods as the cells of origin of the olivocochlear efferents which exit the brain and innervate the CN and the cochlea. These olivocochlear (OC) cells mediate the modulating effects of the brain on the reception and transmission of information from the cochlea. They are recognized to form at least two groups: the medial and lateral systems. These groups differ in their location within the SOC, their chemistry, the inputs they receive from other neurons, and in the places in the CN and cochlea to which they project. They are discussed in greater detail in Chapter 7.

2.1 Medial Superior Olive (MSO)

2.1.1 Location and General Features

In most mammals that have been studied the MSO is a well-defined sheet of cells located medial to a prominent LSO.[2] The presence and size of the MSO shows variability related to the frequency range in which the animal hears and its ability to localize sounds using binaural time cues (Masterton et al. 1975). Usually, the MSO is well-developed in mammals with sensitive low-frequency hearing (Harrison and Irving 1966a; Irving and Harrison 1967; Moore and Moore 1971). The MSO generally is larger in animals with large heads (Moore and Moore 1971). The MSO is very large in echolocating bats with a constant-frequency component in their call (Neuweiler, Bruns, and Schuller 1980; Schnitzler and Henson 1980) and is virtually absent in some bats which rely on their visual systems and do not have sonar systems. The hedgehog has no obvious MSO and is unable to use binaural phase information (Masterton et al. 1975). In transverse sections the MSO appears crescent shaped in cat (Figs. 4.4 and 4.5), but is straighter, shorter and more ventral in gerbil, mouse and rat. In the human the superior olivary nuclei are poorly defined, but a column of cells comprising an MSO can be detected in transverse sections at the level of the VIth nerve rootlet and the tract of the VIIth nerve (Strominger and Hurwitz 1976; Strominger 1978). The frequency of sound stimulation to which MSO neurons respond best varies systematically across the sheet of cells, being highest at the ventrolateral pole and lowest at the dorsomedial pole (see Fig. 4.6) (Guinan, Norris, and Guinan 1972; Tsuchitani 1977).

2.1.2 Histology and Cytoarchitectonic Features

Cell stains show the MSO to consist of a rather tightly packed layer of fusiform cell bodies. In transverse plastic sections of osmicated cat MSO, distinct dendrite-rich zones are clearly seen on either side of the central cell band (see Fig. 3 in Schwartz 1977). A few marginal cells are located at the outer edges of the dendrite-rich zone, at the ends of the central cell dendrites. The marginal cells are oriented at right angles to the medially and laterally directed dendrites of the central cells (Fig. 4.5B). There is a rather sharp boundary between the dendritic zone and orthogonally oriented encircling myelinated axons.

2.1.3 Morphology of Individual MSO Neuronal Types

In cat, mouse, and gerbil the MSO contains four cell types: principal bipolar, principal multipolar, marginal, and rostrocaudally elongated neurons (Fig. 4.5B). Both types of principal neurons have their long axis

---

[2] For an extensive discussion of the homologies between bat cell groups and the MSO see Vater and Feng (1990).

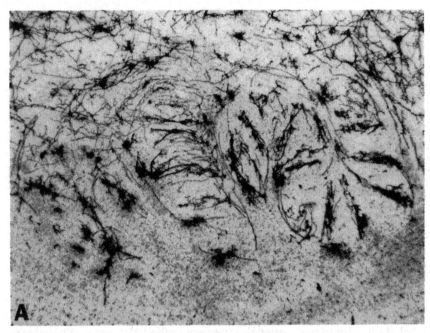

FIGURE 4.5A. A transverse section through the MSO and LSO of a young cat. Bipolar principal neurons are seen in the MSO, while the rostrocaudally flattened LSO bipolar principal cells and dendritic arbors appear as slabs orthogonal to the S-shaped curvature of the LSO. Golgi–Cox preparation.

oriented perpendicular to the sheet of cells. The majority of MSO neurons are bipolar.

Principal neurons which appear bipolar in transverse sections are actually disc-shaped and have dendritic trees flattened in the rostrocaudal direction (Fig. 4.5B, cell A). In contrast, the dendritic spread of multipolar neurons extends in all planes (Fig. 4.5B, cell B). Most dendrites of marginal cells extend over the surface of the MSO at the ends of the dendrites of principal neurons which span the cellular sheet (Fig. 4.5B, cell C). An occasional marginal cell dendrite extends into the body of the MSO, but not more than halfway across the sheet of cells (LaVilla 1898; Ramón y Cajal 1899; Scheibel and Scheibel 1974; Schwartz 1977). All MSO cell classes are largely covered by synaptic terminals over their somata and proximal dendrites (Perkins 1973; Lindsey 1975; Schwartz 1977). Synaptic terminals on principal bipolar and multipolar neurons are similar. Marginal cell somata receive a much higher proportion of endings containing small vesicles and endings which show high affinity uptake of γ-aminobutyric acid (GABA) and/or glycine (GLY) (Schwartz 1978, 1980, 1982, 1984a,b). With immunocytochemical staining no somatic labeling with GLY antibodies is observed in guinea pig MSO (Helfert et al. 1989),

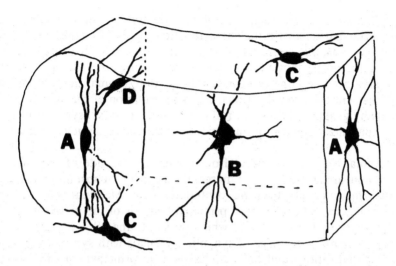

FIGURE 4.5B. Schematic representation of MSO cell types as seen in Golgi preparations in young cat: (A) bipolar principal cells seen in transverse section at the left, with the disc shape of their flattened dendritic arborization seen in the sagittal plane on the right; (B) a multipolar principal neuron with its dendrites extending in all planes; (C) marginal cells at the lateral and medial surfaces of the MSO (top and bottom respectively); (D) a rostrocaudally elongated neuron in the central cell band.

although the vast majority of central cell band perikarya, medial dendrites and marginal neurons are contacted by a moderate number of terminals immunoreactive for glycine [GLY(+)]. Adams and Mugnaini (1990) have shown a heavy distribution of calbindin immunostained [CaBP(+)] terminals around cat MSO neuronal somata and proximal dendrites that arise from collaterals of MNTB axons passing through the MSO. Their data suggest that these CaBP(+) terminals are also GLY(+). The distribution of GABA-immunoreactive [GABA(+)] terminals is sparse throughout the MSO. However a very small percentage of MSO neurons in the gerbil are intensely GABA(+) (Schwartz and Yu 1986). They are similar in size, shape and orientation to unlabeled neurons and receive a similar number of GLY(+) terminals on their somata and dendrites.

2.1.4 Projections to MSO

Principal bipolar and multipolar neurons receive input from the two ears on oppositely directed dendrites (Stotler 1953; Perkins 1973). The lateral dendrites receive input from spherical bushy cells in the ipsilateral AVCN while the medial dendrites receive input from the spherical bushy cells of the contralateral AVCN. AVCN spherical bushy cell axons form long fingerlike synaptic terminals on the somata and proximal dendrites of MSO principal cells with the characteristic features of high security synapses, i.e., synaptic terminals whose activity has a high probability of causing the postsynaptic cell to fire. These features include: a relatively large area occupied on the postsynaptic neuronal soma, the presence of multiple synaptic specializations, a general isolation of the synaptic terminals by glial ensheathment, round synaptic vesicles, asymmetric synaptic specializations and an association with coated vesicles (Schwartz 1972).

Marginal cells receive input from the ear on the side of the nucleus where they are located; lateral marginal cells from the ipsilateral ear and medial marginal cells from the contralateral ear. Marginal cells also receive collaterals of axons from principal neurons of the medial nucleus of the trapezoid body (MNTB) which project to the LSO. Marginal cells receive a higher percentage of synaptic terminals with symmetrical synaptic specializations and flat vesicles than do principal cells (Schwartz 1978, 1980). Marginal cells have also been shown to receive a higher percentage of terminals which accumulate GABA or GLY (Schwartz 1980, 1983), or have GABA or GLY immunoreactivity (Schwartz and Yu 1986). Some terminals may colocalize GABA and GLY immunoreactivity. Adams and Mugnaini (1990) showed the presence of large numbers of terminals on cat MSO marginal cells which immunostain for glutamate decarboxylase (GAD), the key enzyme in the synthesis of GABA. They also report a population of terminals on marginal cells which immunostain for leucine-enkephalin which is apparently different from the GAD(+), GABA(+) terminals.

The transmitter of the spherical bushy cell terminals has not yet been identified. Several studies provide conflicting data about whether the transmitter is either aspartate (ASP) or glutamate (GLU). Studies of uptake of tritiated amino acids suggest that it is neither ASP nor GLU (Schwartz 1985). Biochemical analysis of the bushy cell areas of AVCN reveal high concentrations of ASP and GLU suggesting that ASP or GLU could be the bushy cell transmitter (Godfrey et al. 1977). The input to rostrocaudally elongated neurons is not known.

In the cat, a minor input to principal MSO neurons is from collaterals of axons of other MSO neurons (Ramón y Cajal 1899; Elverland 1978; Schwartz 1984b). These connections may provide the anatomical substrate for complex processing. For example, the distribution of MSO axon collaterals to nearby MSO neurons could provide for inhibitory side bands around an excitatory center for select patterns of stimuli. Collaterals from a neuron's own axon could enhance or limit its response.

MSO neurons may also receive descending input from the IC (Elverland 1977; Kiss and Majorossy 1983), although most descending axons from the IC and LL seem to target periolivary cell groups (Spangler and Warr 1991). Cant (1991) reports a projection from the LNTB to the MSO.

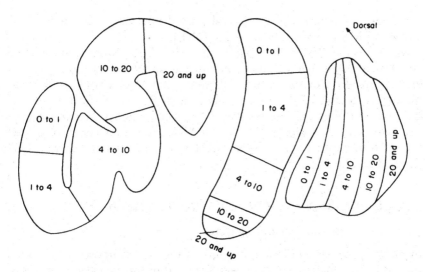

FIGURE 4.6. Tonotopic organization of the cat SOC from Figure 21, Guinan, Norris, and Guinan (1972), reprinted with permission of Gordon and Breach Science Publishers Ltd.

Simple differences in the connections to different MSO and LSO neuronal types can explain differences in their function. Some MSO and LSO principal cells are most sensitive to particular interaural phase differences, others to interaural intensity differences (Masterton et al. 1975; Heffner and Heffner 1986; Yin and Chan 1988). Single MSO neurons receiving high security synaptic input from spherical cells in both AVCNs are particularly suited to perform interaural phase analysis. MSO multiplanar principal cells with dendrites oriented in all planes may also be good candidates for detectors of intensity. AVCN spherical cell axons distribute in horizonal sheets (Scheibel and Scheibel 1974) in a tonotopic manner across the MSO (see Fig. 4.6) (Tsuchitani and Boudreau 1967; Guinan, Norris, and Guinan 1972). The flattened disc-shaped dendritic arbors of MSO principal cells are oriented parallel to the sheets, but the multiplanar cells are in contact with several sheets covering a range of best frequencies which will tend to be excited together by increasingly intense sounds. MSO principal cells are involved in both sound detection and its localization (Masterton et al. 1975; Tsuchitani 1978). As we learn more about the detailed chemistry and anatomy of MSO neurons and their synaptic inputs it becomes clear that there is still more that needs to be learned to understand the complex functions they perform.

2.1.5 Projections and Course of Axons From MSO

Almost all MSO neurons project to the ipsilateral IC; a few project to the contralateral IC (Adams 1979; Glendenning et al. 1981). Aitkin and Shuck (1985) report that MSO axons provide the predominant input to low-frequency neurons in the lateral central nucleus of the cat IC. Collaterals of MSO axons innervate the DNLL.

Injections in cat MSO of a plant lectin (*Phaseolus vulgaris* leucoagglutinin) (PHA-L), which is transported anterogradely in neurons and allows the visualization of the entire cell, show that thick axons to IC send collaterals to DNLL which terminate as a single tier. The arbors end in isolated territories that are separated from one another. Results suggest that horizontal tiers of afferent fibers in DNLL are a mosaic of restricted axonal arbors rather than uniform sheets from divergent axonal arbors (Henkel 1989).

2.2 Lateral Superior Olivary Nucleus (LSO)

2.2.1 Location and General Features

In many species the LSO is a prominent nucleus composed of a folded sheet of cells which usually appears S-shaped in transverse section (Figs. 4.3, 4.4, and 4.5). In prosimian primates (e.g., loris and galago) it is a large, relatively well-defined nucleus lacking the S shape of carnivores and rodents, but still convoluted. It remains well defined and unchanged in cytology but does not increase in size in monkeys and anthropoid primates (apes and humans), despite the large increase in brainstem size. Thus the LSO comes to constitute only a small percentage of the SOC in anthropoid primates (Moore and Moore 1971). There is a tonotopic distribution of best frequencies across the LSO with the highest frequencies in the medial limb and the lowest in the lateral limb (Fig. 4.6) (Guinan, Norris, and Guinan 1972; Tsuchitani 1977; Sanes, Merickel, and Rubel 1989).

2.2.2 Histology and Cytoarchitectonic Features

Neurons of the LSO are less densely packed in their folded sheet than the neurons of the MSO. Like the MSO, long axes of LSO principal neurons are generally oriented across the width of the cell sheet. Marginal cells are oriented parallel to the curving surfaces of the cell sheet at the ends of the dendrites of the principal cells oriented across the sheets. Myelinated fibers encircle the LSO and many enter the nucleus through the dorsal or ventral hilus. There is also some variation in cell size across the LSO in some species studied (unpublished data).

2.2.3 Morphology of Individual LSO Neuronal Types

At least five neuronal types have been identified in the LSO of the cat (Helfert and Schwartz 1986) and gerbil (Helfert and Schwartz 1987). Two classes of neurons are arranged so that their dendrites span the sheet of cells. One type are discoid principal cells (Fig. 4.7A) which appear fusiform shaped in transverse sections (Ramón y Cajal 1899; Scheibel and Scheibel 1974). Recently, a second type of cell with dendrites spanning the cell sheet has been recognized (Helfert and Schwartz 1986, 1987). These neurons are multiplanar (Fig. 4.7B). They have no preferential planar orientation of their dendrites. A third class of neurons, the marginal cells (Fig. 4.7C), is found at the edges of the cell sheet. Most of their dendrites are spread over the surface of the nucleus, with only an occasional dendrite extending into the sheet of cells. Two additional classes of smaller neurons are found distributed within the folded cell sheet: type 5 neurons (Fig. 4.7D) and small neurons (Fig. 4.7E and F). Small neurons have been shown by a variety of retrograde transport studies to be the group of lateral olivocochlear neurons which project beneath the inner cochlear hair cells. OC neurons in the gerbil preferentially accumulate D-aspartic acid (DASP), (Schwartz, Helfert, and Ryan 1986; Helfert, Schwartz, and Ryan 1988). In the guinea pig LSO single OC neurons contain both cacitonin gene-related peptide (CGRP) and enkephalin (Tohyama et al. 1990).

Principal neurons of the LSO appear fusiform in transverse section, but have a disc-shaped dendritic tree with the disc oriented in the rostrocaudal plane. In gerbil, the width of the dendritic arbor of these cells shows a systematic variation. Low-frequency neurons in the lateral LSO have broader dendritic arbors along the tonotopic axis than high-frequency neurons located medially (Sanes et al. 1990). In cat and gerbil a high percentage (>65%) of the perikaryal surface of principal and multiplanar neurons is apposed to synaptic terminals, while the small olivocochlear and type 5 cells have <35% of their surface so apposed (Helfert and Schwartz 1986, 1987). Primate material has not yet been studied electron microscopically. In paraffin or frozen sections primate LSO neurons appear small, oval and lightly stained (Moore and Moore 1971).

Helfert et al. (1989) report that in an immunocytochemical study two-thirds to three-quarters of guinea pig LSO neurons are lightly GLY(+). A similar proportion are strongly GABA(+). All the (+) neurons are fusiform or oval. The GABA(−) and GLY(−) neurons vary in shape from fusiform to polygonal. This suggests that at least some (−) neurons belong to different classes than the (+) neurons. Polygonal appearance in these preparations could be interpreted as indicative of a multiplanar character. One-half of LSO marginal neurons are GABA(+) or GLY(+). Most, if not all, neurons colocalize GABA and GLY. Small neurons throughout the middle and medial limbs are unlabeled by either GLY or GABA antibodies. The cell bodies of the vast majority of LSO neurons are contacted by only one or two GABA(+) terminals. Most LSO perikarya are surrounded by a large number of GLY(+) terminals, although some GLY(+) somata receive noticeably fewer GLY(+) terminals. A study of uptake of tritiated [H^3]GABA and GLY by slices of cat LSO shows a similar distribution of GLY(+) terminals on LSO neurons and

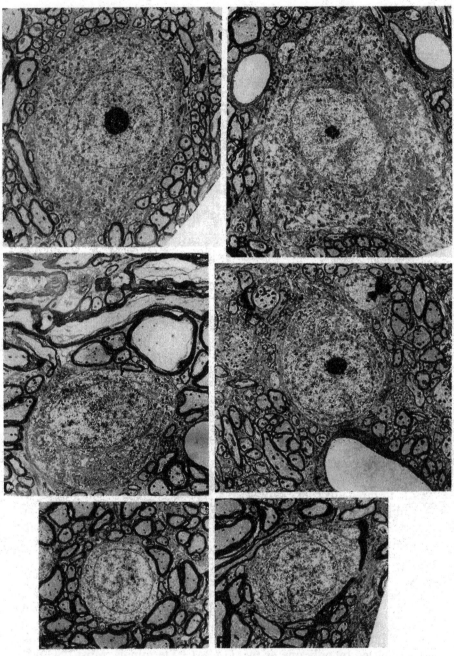

FIGURE 4.7. Electron micrographs illustrate the five cell types in the cat LSO in sections of mixed aldehyde fixed, osmicated tissues. (Figures 3C, 6B, 8, 9D and E, and 10 in Helfert and Schwartz (1986) with permission of Wiley-Liss, a division of John Wiley and Sons, Inc. (A) A fusiform principal cell surrounded

a similar smaller number of GABA(+) terminals (Schwartz 1984a, 1985). No cell bodies are labeled by uptake.

2.2.4 Projections to LSO

LSO principal neurons receive input from spherical neurons of the ipsilateral AVCN primarily via axodendritic synapses on their dendrites and input from principal neurons of the ipsilateral MNTB mostly on their somata and proximal dendrites (Spangler, Warr, and Henkel 1985; Cant and Casseday 1986). Cant (1991) suggests that the AVCN spherical neurons projecting to LSO include the same spherical bushy neurons in the rostral AVCN which project to MSO neurons, but also include spherical bushy cells in the more posterior portion of AVCN (AP) which do not project to MSO. The principal neurons of the MNTB receive their major input from the globular cells of the contralateral AVCN whose axons traverse the trapezoid body in the ventral acoustic stria. MNTB principal cell axons form multiple finger-like terminals on LSO principal neurons. These projection patterns differ significantly from the binaural input to MSO neurons and are consistent with the idea of segregated pathways associating interaural intensity analysis with the LSO and interaural phase analysis with the MSO (Warr 1982; Glendenning et al. 1985).

By studying physiologically characterized neurons filled with HRP, Friauf and Ostwald (1988) identified ipsilateral globular/bushy cells of VCN located in the area of the entrance of the cochlear nerve (neurons with "primary-like" responses and a spontaneous firing rate below 10 spikes/s) as the cells which project via the ventral acoustic stria to LSO fusiform cells. Axon terminal varicosities from the neurons in ipsilateral AVCN are distributed primarily to more distal dendrites of LSO fusiform cells (bat, *Eptesicus fuscus,* Zook and DiCaprio 1988). Presumed multipolar/stellate neurons of the contralateral PVCN ("on" responders with little spontaneous activity) project to the ipsilateral LSO (rat, Friauf and Ostwald 1988). In cat, Cant (1991) reports a projection from the poster-

---

(FIGURE 4.7. *Continued*) by an almost continuous zone of unmyelinated processes (formerly Figure 3C); (B) A multiplanar neuron surrounded by a similar distribution of synaptic terminals to those on principal cells (formerly Figure 6B); (C) a class five neuron similar in size, shape, cytoplasmic features, and orientation to a small principal neuron, but less than 50% of its somatic surface is apposed to presynaptic terminals (formerly Figure 10); (D) a marginal neuron is characterized by its position adjacent to the longitudinally sectioned axons encircling the LSO (asterisks), and the relatively small percentage of its surface apposed to synaptic terminals (formerly Figure 8); E and F) small neurons illustrating the high degree of nuclear infolding seen in small cells (Formerly Figures 9D and 9F).

odorsal portion of AVCN (PD), an area which contains globular bushy and large multipolar cells. According to Warr (personal communication 1989) neurons in the rat interstitial nucleus project to the LSO, mainly contralaterally, but also ipsilaterally. Also in the rat there is a strong topographic projection of VNTB neurons to the LSO (Warr and Spangler 1989). Inhibitory input from the MNTB seems to be directed primarily to cells in the medial limb of LSO, while cells receiving excitatory input from the two ears are more frequently located laterally in the LSO (Caspary and Faingold 1989). The calbindin immunoreactive axons of MNTB neurons have been shown to form a dense plexus around unlabeled somata in the LSO, especially in the medial to middle limb (Matsubara 1990).

2.2.5 Projections from LSO

Neurons of the LSO project to the IC (Elverland 1978; Glendenning and Masterton 1980; Glendenning et al. 1981; Glendenning and Masterton 1983; Glendenning et al. 1985). The projections are bilateral and asymmetrical. Most cells in the medial limb of LSO project to the contralateral IC, while most cells in the lateral limb of LSO project to the ipsilateral IC (Glendenning 1987). Saint-Marie et al. (1989) have shown that the projection from the cat LSO can be separated, immunocytochemically, into three components: (1) a crossed, GLY(−) projection; (2) an uncrossed, GLY(+) projection; and (3) an uncrossed, GLY(−) projection. Additional evidence suggests that the terminal fields of the two uncrossed projections may distribute differently within the IC. GLY(+) or GLY(−) projection neurons, crossed or uncrossed, do not differ in the size, shape, or location of their somata. However, most GLY(−) neurons are heavily encrusted with GLY(+) endings; GLY(+) neurons have 40 to 60% fewer of these endings. GLY(−) neurons located in the lateral limb have fewer GLY(+) perisomatic endings than those in the medial limb. Few projection neurons are GABA(+), and GABA(+) perisomatic endings are rare in the LSO. Thus, there is a heavy uncrossed projection from the cat LSO to the IC that may be glycinergic and inhibitory. Furthermore, there is a bilateral projection that is not glycinergic or GABAergic, which may be excitatory. Glendenning and Baker (1991) suggest that the contralateral projection may be glutamatergic or aspartergic.

2.3 Medial Nucleus of the Trapezoid Body (MNTB)

2.3.1 Location and General Features

The MNTB is a group of cells ventromedial to the MSO located among the fibers of the trapezoid body just lateral to the VIth nerve tract. Because of its prominence in a number of species, the MNTB is often considered one of the main SOC nuclei. Its size seems to vary with the frequency range of the animal's hearing, being largest in species with hearing in the high-frequency range, i.e., 20–80 kHz, such as the cat and some rodents. Its size seems to co-vary with that of the LSO, being large in species with large LSOs and small in species with small LSOs such as prosimians, monkeys, apes, and humans (Moore and Moore 1971). The name, MNTB, was assigned to this nucleus because of its anatomical location. Recent findings suggest that the MNTB is really a periolivary group, as are the other trapezoid body nuclei. As other periolivary cell groups supply input to the main SOC nuclei, so the MNTB supplies a major input to the LSO, relaying information from the contralateral CN. The LSO and MNTB have been linked as a single functional unit (Glendenning et al. 1985).

2.3.2 Histology and Cytoarchitectonic Features

In most primates studied (insectivores, prosimians, and monkeys) MNTB neurons are round to oval cells with darkly staining cytoplasm and a few coarse Nissl granules located near the cell and nuclear membranes (Moore and Moore 1971). In gibbon and chimp, the cells are round to oval but lightly stained and dispersed among trapezoid body fibers rather than found together in a compact group (Moore and Moore 1971). In the human (Moore and Moore 1971), MNTB cells are scattered and only occasionally coalesce to form a definable group.

2.3.3 Morphology of Individual MNTB Neuronal Types

Three cell types have been described in the MNTB on the basis of size and shape: principal neurons, multipolar (or stellate) neurons and elongate neurons (Taber 1961; Rasmussen 1967; Morest 1968b) (Fig. 4.9). The principal neurons of the MNTB are the most numerous cell type. They have a spherical (about 20 μm in cat) or oval soma with an eccentric nucleus, fine Nissl substance, 2–4 short, slender, and highly branched dendrites forming tufts. The principal neurons are, overall, somewhat flattened in the horizontal plane, parallel to the trapezoid fibers. Secondary dendrites of principal cells are spiny.

In cat, elongate neurons have oval perikarya that are slightly smaller than the principal neurons. The elongate neurons have centrally located nuclei, 4–5 dendrites covered with short, filiform appendages extended in a dorsoventral direction that sometimes span the entire width of the nucleus, and an axon which may form a short collateral before leaving the nucleus (Morest 1968b). Secondary dendrites of these cells extend in several planes, but the overall dendritic field is elongated.

Stellate neurons described by Morest (1968b) probably correspond to the multipolar cells of Taber (1961) and Rasmussen (1967). In Nissl stained sections of cat the multipolar cell soma is about 25 by 35 μm. The soma contains large amounts of cytoplasm with a darkly stained tigroid Nissl pattern and a nucleus which is usually centrally located. Long thick dendrites with small thin spines (many fewer than on other MNTB cell types) radiate in various directions, sometimes extending beyond the abducens nerve root or into the dorsomedial periolivary nucleus.

In guinea pig, Helfert et al. (1989) report that MNTB principal and elongate neurons are intensely GLY(+). Rare unlabeled neurons are typically polygonal. Small numbers of GLY(+) terminals are observed on MNTB perikaryal surfaces. Most MNTB neurons are GABA(−), but an occasional polygonal neuron is lightly GABA(+). Dupont et al. (1990) also report GABA(+) MNTB neurons. There are numerous GABA(+) terminals on most MNTB neurons. In cat and gerbil MNTB principal neurons are also reported to be intensely GLY(+) (Adams and Wenthold 1987; Wenthold et al. 1987). MNTB neurons in cat are also intensely immunoreactive for calbindin (CaBP+) (Matsubara 1990). Matsubara (1990) has suggested that CaBP+ is specific for "relay" neurons in systems requiring rapid, reliable transmission of impulses with minimal local processing or integration. MNTB principal neurons certainly receive secure synaptic input (see below).

2.3.4   Projections to MNTB

Principal cells of the cat MNTB receive their major input from the globular cells of the contralateral AVCN whose axons traverse the trapezoid body in the ventral acoustic stria and terminate in a calyx surrounding

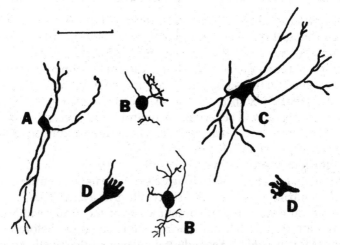

FIGURE 4.8. Diagramatic representation of MNTB cell types seen in transverse section based on rapid Golgi-impregnated cells described by Morest (1968a) from a 14 day old cat: (A) elongate cell, note the dorsoventral orientation of the dendrites; (B) principal cells, note the round cell bodies and the tufted dendrites; (C) stellate or multipolar cell; (D) calyceal endings of the kind found around principal cells. Bar = 100 μm.

the principal cell somata (Morest 1968a,b). In rat, the projection is from contralateral globular/bushy cells of VCN (Friauf and Ostwald 1988). The calyx terminals in MNTB arise from thick axons of globular cells of the AVCN (Tolbert and Morest 1978). Neural input to MNTB principal neurons is most likely the primary-like output of units in the caudal AVCN (Tsuchitani 1978). Since principal cells are the preponderant cell type they have been correlated with the 71% of MNTB units producing primary-like output (Tsuchitani 1978). MNTB units producing primary-like discharges also produce spike potentials of complex waveform similar to those produced by units in the anterior area of the AVCN defined by Brawer, Morest, and Kane (1974) as AA-AVCN; a positive presynaptic potential precedes the unit's postsynaptic potential by approximately 0.5 msec (Li and Guinan 1971). Seventeen percent of MNTB units produce off-type and 3% produce chopper-type responses (Tsuchitani 1978). Guinan, Norris, and Guinan (1972) suggested that similarities in the distribution of units producing off-type discharge and the distribution of MNTB stellates mean that stellates produce the off-type response. MNTB stellate cells receive somatic and dendritic contacts from fine and medium-sized axons of caudal AVCN neurons. Multipolar cells of caudal AVCN may produce chopper-type discharges that serve as neural inputs to MNTB stellates (Tsuchitani 1978).

Somatostatin-positive terminals in rat MNTB originate from somatostatin-immunoreactive (SS-IR) cells in or adjacent to the medial portion of the DNLL (Morley 1985).

Recently, Kinney, Ottoson, and White (1990) have reported a relatively high level of opiate receptors as determined by [³H]naloxone binding in the MNTB of human fetuses and infants, although any association with specific cell types was beyond the scope of the study.

2.3.5 Projections and Course of MNTB Axons

[³H]Leucine injections into cat MNTB and HRP injections into LSO and VNLL showed topographic projections of MNTB principal cells to LSO fusiform neurons (by passing ventral, dorsal, and through the MSO), and also to cells in the VNLL, DMPO, and VMPO (Spangler, Warr, and Henkel 1985). Immunocytochemical data from the guinea pig suggest that MNTB axons distribute terminals to the SPN (Helfert et al. 1989). Stellate neurons project to the IC (Adams 1979). Pharmacologic as well as immunocytochemical and lesion studies support the idea that the major MNTB projection is probably glycinergic (Moore and Caspary 1983; Bledsoe et al. 1990).

Neurons in the MNTB projecting primarily to the cochlea also send collaterals mainly to the ipsilateral CN (Winter, Robertson, and Cole 1989).

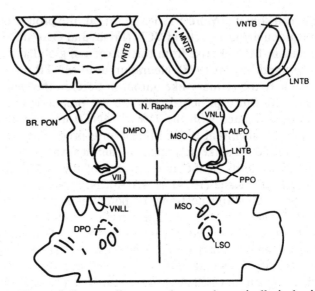

FIGURE 4.9. The periolivary cell groups shown schematically in horizontal sections of the rat brainstem. Redrawn from Figure 2 in Spangler et al. (1987), reprinted by permission of Wiley-Liss, a division of John Wiley and Sons, Inc. The most ventral section is shown at the upper left. Rostral is at the top.

2.4 Periolivary Nuclei and Cell Groups

Depending on the species, several groups of cells around the MSO and LSO are sufficiently well-defined to be given names. In cats and a number of rodents these include a lateral nucleus of the trapezoid body (LNTB) which is located near the ventral border of the brainstem lateral and slightly ventral to the LSO, a ventral nucleus of the trapezoid body (VNTB) which lies close to the edge of the brainstem ventral and medial to the MSO, a dorsal medial periolivary nucleus (DMPO) lying dorsal and medial to the MSO, and several other distinguishable groups of cells.

2.4.1 Superior Paraolivary Nucleus (SPN) or Dorsomedial Periolivary Nucleus (DMPO)

In rat and mouse a cell group called the superior paraolivary nucleus (SPN) is recognizable dorsomedial to the MSO (Harrison and Feldman 1970; Ollo and Schwartz 1979). It contains dark-staining, very large multipolar neurons (Ollo and Schwartz 1979). In some older studies cells in this group have been identified as comprising the MSO or medial accessory olive (MAO). It appears to correspond to the DMPO identified in cat and bat, a region which is composed of very large multipolar neurons with large nuclei and darkly staining, clumped Nissl granules (Zook and

Casseday 1982a). In the gerbil, the DMPO is clearly defined by immunoreactive staining with the calcium binding proteins parvalbumin and
calbindin, which stain different neuropil elements (Schwartz and Eager
1992).

In guinea pig, some large (25–35 μm) polygonal neurons are GLY(+)
and are contacted by a moderate number of GLY(+) terminals (Helfert
et al. 1989). Some unlabeled large polygonal neurons receive a larger
number of GLY(+) terminals. Light GLY(+) labeling was noted in some
smaller oval or fusiform neurons (20–25 μm long) which are contacted
by a variable number of GLY(+) terminals. Somatic staining of large
neurons is light with GABA antibodies, although a small number of 12–
20 μm fusiform and oval neurons are moderately GABA(+). Perisomatic
and peridendritic terminal GABA(+) labeling was sparse.

In the human, periolivary cells form an almost complete sphere around
the less well-defined MSO and LSO so that separate subnuclei are not
distinguishable (Moore 1987a). However large AChE(+) multipolar cells
are present medially and ventrally in the periolivary region and would
appear to be olivocochlear (OC) neurons (Moore 1987a). Cells that are
". . . medium sized, palely staining cells with a fine dispersed 'gossamer'
Nissl substance" are present rostrally in the human periolivary sphere
and seem to correspond to cells with similar morphology in the cat which
project to the IC (Moore 1987a).

2.4.2   Lateral Nucleus of the Trapezoid Body (LNTB)

2.4.2.1   Location and General Features

In cats and rodents cells of several types are found dispersed among the
fibers running along the edge of the brainstem lateral and ventral to the
LSO. The rostral and caudal boundaries of this nucleus vary somewhat
between species and may be hard to define between the LNTB and anterior and posterior periolivary nuclei (see Fig. 4.8).

2.4.2.2   Histology and Cytoarchitectonic Features

In cat, Adams (1983a) distinguishes a group of medium-sized cells with
coarse, darkly staining Nissl substance located ventral to the LSO which
send axons to the ipsilateral CN. They appear to constitute a major cell
type of the LNTB. In Protargol material these cells can be seen to receive
ring-type and/or diminutive endbulb type endings (Adams 1983a). They
can be distinguished from another group of OC cells in this location which
receive few axosomatic contacts. Adams (1983a) reports that similar cells
dorsal and anterior to the LSO also send axons to the ipsilateral CN. In
the mustache bat, *Pteronotus parnellii,* the LNTB is found only in the
anterior part of the SOC complex and lies within the lateral part of the
trapezoid body as it emerges from the CN. The LNTB contains large or
giant multipolar neurons with dendrites often oriented perpendicular to

trapezoid body fibers as well as a few medium-sized multipolar neurons. In *Pteronotus* the LNTB is distinguished from the ventral periolivary nucleus (VPO), a cell group scattered within the trapezoid body beneath the LSO more caudally (Zook and Casseday 1982a). These cells merge with the VMPO medially. Included in this group are large multipolar neurons similar to VMPO neurons and small-to-medium neurons with elongate bodies whose long axes are often oriented parallel to trapezoid body fibers.

2.4.2.3 Morphology of LNTB Neurons

Ciriello, Caverson, and Park (1986) have identified cells in the cat LNTB and VNTB and dorsolateral to the LSO which are immunoreactive for adrenalin- and noradrenalin-synthesizing enzymes. Cells in the LNTB of the deer mouse *Peromyscus bairdi* are cholinesterase positive (Ross 1969), but rat LNTB cells are reported to be negative for cholinesterase staining (Brown and Howlett 1972).

A majority of the somata in the guinea pig LNTB are moderately GABA(+), with a variable number of GABA(+) terminals on both labeled and unlabeled somata (Helfert et al. 1989). There are few GLY(+) neurons.

2.4.2.4 Projections to LNTB

The LNTB is innervated by cells from ipsilateral AVCN, but the connections are not well understood (Harrison and Irving 1966a,b; Warr 1966; Harrison 1978). PVCN octopus cells project onto globular neurons ventral to the LSO (Adams 1983b), the region Adams (1983a) has also referred to as LNTB.

2.4.2.5 Projections from LNTB

At least one major cell group in the LNTB projects to the cochlea and cochlear nucleus. This lateral olivocochlear projection is discussed in Chapter 7. Cant (1991) reports a projection to the MSO from LNTB which she suggests may be a source of some GLY(+) terminals in the MSO, although the precise LNTB cell type is unclear. Adams and Mugnaini (1990) demonstrated GAD(+) staining in cells ventral to the LSO and suggested that they might be the source of GAD(+) terminals in principal olivary nuclei such as those around cells at the margins of MSO.

2.4.3 Ventral Nucleus of the Trapezoid Body (VNTB)

2.4.3.1 Location and General Features/Histology

In cat, Adams (1983a) identifies a medial group of small cells with darkly staining Nissl substance projecting to the CN. These cells lie largely ventral to the MNTB and comprise much of what is generally considered the VNTB or the medial preolivary cell group. In the bat, *Pteronotus parnellii,* the cells are described as densely packed, small elongate or irregular shaped, with little cytoplasm and clumped Nissl granules (Zook and Casseday 1982a).

2.4.3.2 Morphology of VNTB Neurons

In the guinea pig VNTB GLY(+) fusiform and polygonal neurons (20–30 μm in length) are found among unstained neurons of similar shapes and sizes (Helfert et al. 1989). Smaller oval neurons (12–15 μm) are also GLY(+). In guinea pig (Thompson, Cortez, and Lam 1985) and gerbil (Schwartz and Yu 1986) some moderately GABA(+) and a few intensely GABA(+) neurons are observed in both VNTB and LNTB. The extent of GABA and GLY colocalization has not yet been reported. There are a large number of GABA(+) terminals in these regions.

VNTB neurons are cholinesterase positive in both mouse and rat (Ross 1969; Brown and Howlett 1972).

2.4.3.3 Projections to VNTB

Cells in the VNTB are known to receive inputs from contralateral CN (Warr 1972, 1982) and from IC (Rasmussen 1964; Faye-Lund 1986).

2.4.3.4 Projections from VNTB

Various VNTB neurons have reciprocal connections with both VCN and IC. Some cells in the VNTB have been identified as belonging to the medial olivocochlear efferent system and shown to project to both CNs (Robertson, Anderson, and Cole 1987; Helfert, Schwartz, and Ryan 1988). In the cat, collaterals of VNTB axons reach periolivary areas containing lateral olivocochlear (LOC) neurons (Spangler, Warr, and Henkel 1985). VNTB neurons in the rat also show a strong topographic projection to the LSO (which in rat contains LOC neurons) bilaterally, as well as to the LNTB (Spangler and Warr 1991).

2.4.4 Undefined or Dispersed Periolivary Cell Groups

In addition to the cells that form distinct nuclear groups several periolivary cell groups have been identified in terms of their cytologic features and projection patterns. In cat, Adams (1983a) identifies three groups: (1) large OC cells located medial to the MSO identified by retrograde staining methods (HRP from cochlea, Helfert, Schwartz, and Ryan 1988; selective uptake and transport of tritiated nipecotic acid ([³H]NIP), Ryan and Schwartz 1986; fluorescent traces, Winter, Robertson, and Cole 1989) and by acetyl cholinesterase staining (Osen and Roth 1969; Helfert and Schwartz 1986, 1987); (2) small peri-LSO neurons projecting to the cochlea similarly identified by retrograde labeling and AchE staining; and (3) medium-sized multipolar neurons with lightly staining, flocculent Nissl substance located in a region near the ventral border of the LSO (some are also dorsal to MSO) labeled retrogradely by IC HRP injections both ipsilaterally and contralaterally and not labeled by CN injections (Adams 1983b). The medium-sized multipolar neurons are distinguished from OC cells by the presence on the somas of endings containing rings of neurofilaments. The rings of neurofilaments within synaptic terminals are visualized directly in Protargol stained material and can be detected electron microscopically.

In the guinea pig a small number of GLY(+) neurons were observed near the dorsal hilus of the LSO (Helfert et al. 1989). A small population of GLY(+) cells was also observed immediately outside the medial limb of the LSO, near the dorsal pole of the MSO and the lateral tip of the SPN. There were very few GLY(+) terminals on these neurons. In the cat and gerbil a group of large round neurons ventral and caudal to the LSO are GLY(+) (Adams and Wenthold 1987).

2.4.4.1 Projections to Other Periolivary Areas

Category II neurons (with "on" responses and little spontaneous activity) located in the PVCN (presumably multipolar/stellate cells) project via the intermediate acoustic stria and innervate mainly contralateral periolivary regions. Ipsilaterally, the LSO and the SPN were innervated by some of the category II neurons (Friauf and Ostwald 1988).

The DLPO receives input from the ipsilateral PVCN via the intermediate acoustic stria (van Noort 1969; Warr 1969; Adams and Warr 1976) and the trapezoid body (Warr 1969, 1972; van Noort 1969). Most fibers in the intermediate acoustic stria (IAS) are octopus cell axons (Osen 1972). A smaller number of fibers in the IAS are from other neurons in caudal PVCN and neurons scattered through more rostral PVCN (Adams and Warr 1976). Warr (1982) also reports that IAS axons innervate the anterior lateral periolivary nucleus (ALPO), a region which may correspond to the NCAT in bat.

2.4.4.2 Projections from Periolivary Areas

Correlated studies of retrograde HRP and anterograde radiolabeled protein from the CN in the cat have shown that the lateral and dorsal periolivary nuclei project mainly ipsilaterally, while the medial periolivary nuclei project bilaterally with a contralateral bias. The projections are significant in size and widespread in their distribution in the CN (Spangler et al. 1987). Different periolivary cell groups have similar projection patterns to given CN regions in autoradiograms. MNTB cells which project to the CN are not principal cells and are located along the abducens rootlet or in a cell group at the caudal end of the MNTB (Spangler et al. 1987).

The largest projections to the CN from the periolivary regions come bilaterally from the LNTB and VNTB (Winter, Robertson, and Cole

1989). VNTB cells project bilaterally throughout the rostrocaudal extent of CN with fewer cells sending projections to the ipsilateral DCN (Adams 1983a). In the tree shrew the LNTB projects diffusely to the entire CN but only ipsilaterally, while the VNTB projects bilaterally mainly to the DCN and PVCN (Covey, Jones, and Casseday 1984). There is also a projection to the CN from the DMPO bilaterally (Winter, Robertson, and Cole 1989).

The VMPO and VLPO appear to project only to the midbrain (Adams 1983a; Spangler and Warr 1991). Some cells in VNTB and other periolivary nuclei have a small projection to the VNLL (Glendenning et al. 1981). The INLL is a major target of cells in the LNTB in the bat; however only about 1% of all the cells identified as projecting to INLL were located in the cat LNTB (Glendenning et al. 1981). In the cat Adams (1979) identified periolivary cells bilaterally projecting to the region of the IC as being mainly ventral and medial to the MSO, including slender (<5 μm diameter) elongate neurons, large (>25 μm) multipolar neurons with long dendrites and intermediate-sized cells.

The most rostral periolivary group, the NCAT, contains neurons which are known to project to the medial geniculate body (MGB) (Calford and Aitkin 1983; Henkel 1983).

2.4.5 Lateral Olivocochlear (OC) Neurons

Neurons giving rise to the olivocochelar (OC) efferents are located within the gerbil LSO (Aschoff, Muller and Ott 1988; Helfert, Schwartz, and Ryan 1988) but around the periphery of the cat LSO (Warr 1975) and bat LSO (Vater and Feng 1990). There appears to be considerable variation among species as to the exact location of the lateral OC neurons. The OC neurons in and around the LSO are generally "small." In the cat their minor axis may be only 8 μm, and the cells are either elongated or multipolar, many showing a preferred orientation of their long axis parallel to trapezoid body fibers (Adams 1983a). In cat they are characterized by long, tapering dendrites and by one or two darkly staining clumps of Nissl substance (Adams 1983b; Helfert and Schwartz 1986). The general absence of retrograde labeling of these neurons from CN injections argues that they have a limited terminal distribution in the CN (Adams 1983a). These neurons have been identified by a variety of retrograde tracers from the cochlea including HRP (Helfert, Schwartz, and Ryan 1988), PHA-L (Thompson and Thompson 1987), and fluorescent dyes (Cole, Robertson, and Johnstone 1987; Ashcoff, Muller, and Ott 1988). D-aspartic acid (DASP) injected into the perilymph is accumulated and retrogradely transported by a subset of OC efferent. Since DASP is taken up by high affinity uptake mechanisms which also work for glutamic acid (GLU) and aspartic acid (ASP), DASP uptake suggests a possible modulator role or metabolic function for glutamic acid (GLU) or aspartic acid (ASP) in these neurons. The populations of OC neurons labeled selectively by [³H]NIP, a slowly metabolized GABA analog, and tritiated D-aspartic acid [³H]DASP, an ummetabolized analog of ASP and GLU, do not overlap (Helfert, Schwartz, and Ryan 1988). OC neurons also demonstrate acetylcholinesterase activity (Osen and Roth 1969; McDonald and Rasmussen 1971; Helfert and Schwartz 1987) and are immunoreactive for antibodies to calcitonin gene related peptide (CGRP) (Schweitzer et al. 1985). Their terminals beneath the inner hair cells in the cochlea have been shown to colocalize a number of substances: acetylcholine, GABA, enkephalins, and dinorphins (Altschuler, Parakkal, and Fex 1983; Fex and Altschuler 1985). In the human the LSO contains many small AChE positive neurons suggesting that many of its neurons may be related to descending pathways rather than ascending connections (Moore 1987a).

Refer to Chapter 7 for a more extensive discussion of the lateral OC neurons, including the projections of OC neurons located in periolivary regions.

2.4.6   Nucleus of the Central Acoustic Tract (NCAT)

2.4.6.1   Location and General Features

In addition to the major stream of information ascending from the ears via the CN, SOC, LL, and IC there is evidence from a variety of species for the presence of a small pathway which bypasses the IC on the way to the superior colliculus. Recently Casseday et al. (1989) presented evidence from retrograde HRP transport studies in the echolocating mustache bat, *Pteronotus parnellii parnellii,* that a group of large multipolar neurons lying between the SOC and VNLL project to the suprageniculate nucleus (located adjacent to the medial geniculate body (MGB)) and the deep layers of the superior colliculus (SC) bypassing the IC. They have called this group the nucleus of the central acoustic tract (NCAT). The pathway seen in bat apparently corresponds to the one described by Ramón y Cajal (1899) in mouse as the central acoustic tract, and further described by Papez (1929a) to be present in all animals studied to that time, i.e., human, mouse, rabbit, and cat. The location of the NCAT in bat also corresponds to the distribution of neurons posteromedial to the VNLL in cat seen by Aitkin and Philips (1984) following retrograde transport of HRP introduced into the severed brachium of the IC (i.e., labeling fibers projecting beyond the IC). Axons from this area are known to project to the MGB (Calford and Aitkin 1983; Henkel 1983).

2.4.6.2   Morphology of Neurons in NCAT

Although information about cells in this area is sparse, cells projecting to the MGB in cat have been shown to be multipolar (about 15 × 25 µm) (Aitkin and Phillips 1984). In the echolocating mustache bat they are also reported to be large multipolar neurons (Casseday et al. 1989).

2.4.6.3 Projections to the NCAT

The NCAT receives input from the AVCN, bilaterally in bats (Casseday et al. 19889) but contralaterally in cats (Warr 1966).

2.4.6.4 Projections from the NCAT

Axons of the NCAT neurons bypass the IC and project ipsilaterally to the deep and intermediate layers of the SC and the suprageniculate nucleus (Casseday et al. 1989). They apparently also innervate the medial division of the MGB bilaterally (Calford and Aitkin 1983; Henkel 1983).

3. The Lateral Lemniscus (LL) and Its Nuclei

Neurons within the LL receive, transmit and process information ascending from the ears through the CN and SOC, and some information from neurons which bypass the SOC. Other neurons in the LL receive input from the contralateral DNLL and descending input from the IC bilaterally.

In addition to the major relay stations for ascending auditory information provided by the SOC, some information passes directly from the cochlear nuclei to the inferior colliculus, traveling in the trapezoid body and lateral lemniscii. Collaterals of some of these axons may also synapse with neurons within the SOC and LL nuclei.

The LL has been studied in a number of mammalian and nonmammalian species (see Table 4.2). Most studies have identified only a dorsal (DNLL) and a ventral nucleus (VNLL) comprised of cells distributed between the fibers of the LL. A few studies, mainly in the cat, rabbit, or bat have identified a third intermediate cell group on the basis of cytoarchitectural differences or connectional relationships (see Glendenning et al. 1981 for a review of earlier literature). There is very little data available about the human (Papez 1929a,b; Ferraro and Minckler 1977; Moore 1987a,b).

3.1 Ventral Nucleus of the Lateral Lemniscus (VNLL)

3.1.1 Location and General Features

The VNLL is the most ventral group of cells lying among the fibers of the lateral lemniscus. The VNLL is a distinct group in all nonprimate mammalian species which have been studied, but it is indistinct in New World monkeys, difficult to define in apes and very poorly developed in the human (Moore 1987a). The VNLL is part of a circuit from the CN to the reticular formation which mediates the short latency acoustic startle response (Davis et al. 1982). The reflex is clearly present in primates, but its relationship to the primate VNLL is unclear.

TABLE 4.2. Animals in which the LL has been studied.

Amphibians		
Anurans		
Marine toad	*Bufo marinus*	Pettigrew 1981
Leopard frog	*Rana pipiens*	Hall and Feng 1987
Reptiles		
Lizard		
Tegu	*Tupinambis nigropuntatus*	Browner and Rubinson 1977
Turtle	*Pseudomys scripta elegans*	Kunzle 1986
Birds		
Pigeon	*Columba livia*	Arends and Zeigler 1986
		Wild 1987
Chicken	*Galus galus*	Heil and Scheich 1986
Fish		
Catfish		
Brown bullhead	*Ictalurus nebulosus*	Knudsen 1977
Channel catfish	*Ictalurus punctatus*	Knudsen 1977
A South American catfish	*Pimelodus clarias*	Knudsen 1977
Mammals		
Ferret	*Mustela putorius*	Moore 1988
Guinea pig	*Cavia porcellus*	Tokunaga 1988
Opossum	*Didelphis virginianus*	Willard and Martin 1984
Cat	*Felis catus*	Woollard and Harpman 1940
		Goldberg and Moore 1967
		Adams 1979
		Kane and Barone 1980
		Glendenning et al. 1981
		Oliver and Shneiderman 1989
Bat		
Big brown	*Eptesicus fuscus*	Covey and Casseday 1986
		Zhang, Sun, and Jen 1987
Mustached	*Pteronotus parnellii*	Zook and Casseday 1979
Rat		
Laboratory rat	*Rattus norvegicus*	Druga and Syka 1984
Monkey		
Squirrel monkey	*Saimiri sciureus*	Goldberg and Moore 1967

3.1.2 Histology and Cytoarchitectonic Features

In the cat and rodents the VNLL is normally recognized as a single group of cells of mixed character. In an echolocating bat (*Eptesicus fuscus*), a species placing special demands on its auditory system, separate regions of the VNLL can be recognized: a lateral region (VL) with columns of spherical cells parallel to the ascending LL fibers, and a medial region

(VM) of multipolar neurons (Covey and Casseday 1986, 1991). Vater and Feng (1990) report a high packing density of round cells in the medial VNLL of the horseshoe bat *Rhinolophus rouxi*, with the lateral VNLL characterized by a mixture of multipolar, elongated and round cells. Because of its orientation in this bat the lateral VNLL corresponds to the more dorsal VNLL of cats where mixed cell types are seen.

3.1.3 Morphology of Individual VNLL Neuronal Types

In the lateral region of VNLL, (VL) special to echolocating bats, spherical bushy cells with one large highly branched dendrite are present. The cells are ASP(+). In the medial region, VM, the cells are multipolar and show little dendritic branching and a primary-like pattern of response to sound.

3.1.4 Projections to VNLL

Primary input to the VNLL comes from the CN and SOC. The major projection is from the contralateral VCN via the trapezoid body, with a much smaller projection from the ipsilateral SOC and a very slight projection from the contralateral DCN via the dorsal acoustic stria (Glendenning et al. 1981; Zook and Casseday 1985; Covey and Casseday 1986). The VNLL receives no input from the contralateral nuclei of LL (Glendenning et al. 1981). Ipsilateral CN input is restricted to a small part of the ventrocaudal VNLL (Glendenning et al. 1981). In cat three times as many cells in AVCN project to VNLL as do cells in PVCN (Glendenning et al. 1981).

Octopus cells of the PVCN send thick axons in the intermediate acoustic stria to the VNLL (Van Noort 1969; Warr 1969, 1982; Adams and Warr 1976). The heaviest terminal degeneration is observed in the lateral VNLL. Adams (1983b) reports that octopus cell axons terminate in calyx endings. Collaterals of these axons provide a diffuse innervation to the interstitial nucleus of the stria of Held (immediately medial to the DCN) and to periolivary areas (Warr 1982). These axons do not appear to reach the IC. Many of the axons projecting to the VNLL from the contralateral VCN in rat send collaterals to the superior olivary complex on both sides and terminate in the VNLL (Friauf and Ostwald 1988). Multiple horizontally oriented axon arbors are described in cats and bats (Glendenning et al. 1981; Zook and Casseday 1985; Covey and Casseday 1986). In rat they arise from axons of primary-like VCN neurons, while vertically oriented axonal arbors arise from cells with "on" responses and little spontaneous activity (Friauf and Ostwald 1988). Vater and Feng (1990) report that, in contrast to the calyx endings from thick axons from the PVCN which terminate in the medial VNLL of the horseshoe bat (*Rhinolophus rouxi*), and in the columnar cell area of the bats *Eptesicus* (Covey and Casseday 1986) and *Pteronotus* (Zook and Casseday 1985), the lateral VNLL receives medium-sized (2–3.3 μm) bead-like boutons originating from collaterals of other axons arising in AVCN and PVCN and passing through the VNLL toward the IC. Terminals immunoreactive for glutamic acid decarboxylase [GAD(+)] are present in the VNLL (Roberts and Ribak 1987).

3.1.5 Projections from the VNLL

In the cat and gerbil a group of GLY(+) neurons in the VNLL which receive endbulbs of Held from MNTB neurons send large diameter axons to the IC (Adams and Wenthold 1987).

In the frequency-modulating (FM), echolocating bat, *Eptesicus fuscus,* the columnar area of cells in VNLL, 20–30 cells high, is organized in sheets that are precisely related to the tonotopic organization of the AVCN and IC. The cells receive from the AVCN (low-frequency AVCN to dorsal VNLL, high-frequency to ventral VNLL) and project to the IC (Covey and Casseday 1986, 1991). There is an over-representation of frequencies corresponding to the range of the bats FM-echolocating call. The tonotopic connections of other LL nuclei are less precise than those of the columnar area.

3.2 Intermediate Nucleus of the Lateral Lemniscus (INLL)

3.2.1 Location and General Features

The INLL is composed of sparsely distributed cells and clusters of horizontal cells among the fibers of the LL (Glendenning et al. 1981). This region corresponds to the region identified as VLLd (dorsal zone of the VNLL) by Adams (1979). No separate INLL is recognized in the human (Moore 1987a).

3.2.2 Histology and Cytoarchitectonic Features

In the cat the INLL contains multipolar and horizontal cells (Glendenning et al. 1981). In the mustache bat, *Pteronotus,* the elongate cell with dendrites oriented mediolaterally is the principal cell type (Zook and Casseday 1982a). Large globular cells and small and medium oval cells are also present in the bat.

3.2.3 Morphology of Individual INLL Neurons

The multipolar cells resemble those of DNLL in having finely dispersed Nissl substance and frequently lacking a cap of Nissl substance around their nuclei (Adams 1979). In the horseshoe bat, *Rhinolophus rouxi,* the INLL contains ovoid cells arranged in vertical columns distributed between the fibers of the LL tract (Vater and Feng 1990). The cells of INLL

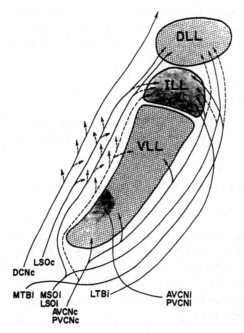

FIGURE 4.10. Schematic summary of the ascending auditory afferents to three cell groups of the right lateral lemniscal complex. Solid lines indicate major projections, dashed lines minor projections. Subscripts on labels indicate fiber origins in structures either ipsilateral (i) or contralateral (c) to lemniscal complex. For clarity, minor projections from VCN to medial border of ipsilateral VLL have been omitted. From Figure 17, page 696, Glendenning, Brunso-Bechtold, Thompson and Masterton (1981), reprinted by permission of Wiley-Liss, a division of John Wiley and Sons, Inc.

are reported to have little similarity to those of DNLL or VNLL in bat (Zook and Casseday 1979). Covey and Casseday (1986, 1991) report that in the echolocating bat, *Eptesicus fuscus,* the INLL is characterized by elongated neurons with dendrites organized in a single plane. These neurons display chopper-like response properties, are ASP(+) and are surrounded by GAD(+) terminals.

3.2.4 Projections to INLL

According to Glendenning et al. (1981) cat INLL receives afferents from some cells in the CN and SOC. There is a substantial projection from the contralateral VCN (about three quarters from AVCN and one quarter from PVCN). Input is received from the ipsilateral MSO and the LSO bilaterally. According to Warr (personal communication) INLL receives input from the lateral limb of the LSO. The INLL is the only LL nucleus to receive a substantial input from the MNTB (Glendenning et al. 1981). Zook and Casseday (1979) reported a heavy projection from LNTB to INLL in bat, but very little input (a maximum of 1%) is reported from LNTB in the cat (Glendenning et al. 1981).

3.2.5 Projections of INLL

In the bat *Eptesicus fuscus* the INLL neurons project ipsilaterally to the same regions of IC as the VNLL (Covey and Casseday 1986).

3.3 Dorsal Nucleus of the Lateral Lemniscus (DNLL)

3.3.1 Location and General Features

The DNLL is composed of neurons distributed among the fibers of the LL just ventral and caudal to the IC. Its neurons display a tonotopic organization with the most dorsal neurons most sensitive to low frequencies and successively more ventral cells most sensitive to increasingly higher frequencies (Aitkin, Anderson, and Brugge 1970; Aitkin, Irvine, and Webster 1984).

3.3.2 Histology and Cytoarchitectonic Features

Cells of the DNLL are heterogeneous.

3.2.3 Morphology of Individual DNLL Neuronal Types

A large number of cell types have been distinguished in the DNLL of the cat, varying in size, shape, orientation, and dendritic patterns. Adams (1979) noted that the characteristic medium- to large-sized cell has darkly staining Nissl substance. Smaller elongate and multipolar cells are interspersed among larger cells. On the basis of shape and diameter in Nissl material Kane and Barone (1980) recognized large round (LR), large ovoid (LO), large elongate (LE), medium round (MR), medium ovoid (MO), medium elongate (ME), small round (SR), small ovoid (SO), and small elongate (SE) cells. Most medium-sized and most LE cells are in the caudal third of the DNLL while most LO and LR cells are dorsally located in the rostral third. Moving from caudal to rostral the number of large and round types increases. Golgi material showed that LO cells have one of two orientations, horizontal or vertical. (The Golgi material showed only large and medium cell types and they are well summarized in Table 6, p. 824, Kane and Barone 1980.) There is a strong horizontal orientation of most DNLL cell bodies and dendrites. In the DNLL of the horseshoe bat Vater and Feng (1990) report loosely packed dorsoventral columns of large and small round or ovoidal cells.

Oliver and Shneiderman (1989) identified two possible cell types in the DNLL based on the ultrastructure of their nuclei and endoplasmic reticulum.

The DNLL in the human is similar in size and cytoarchitecture to that of the cat, differing mainly in an absence of the largest cell types (Moore 1987a).

Morley (1985) has demonstrated the presence of perikarya and terminals immunoreactive for somatostatin (SS) 14 and 28 in the dorsal lateral lemniscus. A group of SS-positive perikarya located either in or adjacent to the medial portion of the DNLL was identified as the source of SS-positive terminals in the MNTB. Spangler and Morley (1987) saw no labeled neurons in the nuclei of the LL, although they showed SS positive cell bodies in the central IC, lateral to the VNLL and in the lateral tegmental field of the developing cat.

3.3.4 Projections to DNLL

The DNLL receives inputs from several sources: (1) contralateral DCN (Fernandez and Karpas 1967; Adams and Warr 1976; Adams 1979), (2) ipsilateral IC (Goldberg and Moore 1967; van Noort 1969; Foster and Hall 1978; Zook and Casseday 1979), (3) contralateral DNLL (Papez 1929b; Goldberg and Moore 1967; Conlee 1979; Oliver and Shneiderman 1989), and possibly from (4) the superior colliculus (Grafova, Ottersen, and Rinvik 1978; Edwards et al. 1979). It also receives input from the ipsilateral MSO, both LSOs and the ipsilateral VNLL (Glendenning et al. 1981). Oliver, Schneiderman, and Henkel (1987) reported a heavy input from the contralateral AVCN, but found that only the low-frequency part of AVCN projected to the dorsolateral part of the DNLL.

Oliver and Shneiderman (1989) described two populations of synaptic terminals found with equal frequency mainly on the somata and dendrites of DNLL neurons: a population with round synaptic vesicles and asymmetric junctions and a population with pleiomorphic synaptic vesicles and symmetrical junctions. The pleiomorphic vesicle endings were found on axons from cells in the contralateral DNLL.

3.3.5 Projections of DNLL

The primary target of the DNLL is the IC bilaterally (Woollard and Harpman 1940; Schneiderman, Oliver, and Henkel 1988; Moore 1987a) and the contralateral DNLL (Kudo 1981). Both ultrastructurally distinct DNLL cell types identified by Oliver and Shneiderman (1989) project to the contralateral DNLL. Cells projecting to the rat IC are found in the caudal DNLL (Tanaka et al. 1985). Those to the ipsilateral IC are found centrally, while those to the contralateral IC are found medially and laterally. Some neurons projected to both ipsilateral and contralateral ICs. A different group of neurons in the rostral DNLL project to the superior colliculus (Tanaka et al. 1985). The DNLL also projects to the SOC (Van Noort 1969; Elverland 1977, 1978), other lemniscal nuclei (Adams 1979; Conlee 1979), and the CN (Kane and Finn 1977; Conlee 1979; Kane and Conlee 1979). Adams and Mugnaini (1984) report that many of the projection neurons of DNLL to IC are GABAergic. There is immunocytochemical evidence that most DNLL neurons contain GABA (Adams and Mugnaini 1984; Roberts and Ribak 1985; Thompson, Cortez, and Lam 1985; Moore and Moore 1987).

4. Functional Considerations

4.1 Sound Detection

Responses to sound include reflex behavior such as the startle response or Preyer reflex which result in alerting behavior and movements of the ear and head to facilitate detection of the location of the sound, the acoustic reflex which tenses the stapedius muscles to protect the ossicular chain from loud sounds, and conscious responses to recognize the sound so that it can be responded to appropriately. Even in environments of absolute silence the ear detects internal sounds such as respiration, circulation, mastication, etc. Fibers of the auditory nerve had spontaneous activity rates which respond with bursts of activity to sound. Cells in all three divisions of the CN receive these inputs and respond in varying ways, sending signals to neurons in the SOC and LL. While neurons in the SOC and LL are stimulated by any detectable sound, the involvement of SOC and LL neurons in reflex pathways is not clear. Reflex pathways generally appear to bypass the major nuclei of the SOC and LL. The alerting responses seem to involve the reticular formation which receives input directly from CN neurons. The stapedial reflex involves neurons located between the facial nerve nucleus and the main olivary nuclei which receive input from the CN, a posterior periolivary cell group in some species. The SOC and LL are more involved in further processing of auditory signals for purposes of sound localization, identification and selective attention.

4.2 Sound Localization

A major function of the auditory system is to allow an animal to identify the location of a sound source so that it can respond appropriately, advancing to capture a prey or with flight in the opposite direction if necessary. To localize a sound source, most mammals compare differences in the time of arrival of the signal at the two ears and its relative intensity. The neurons carrying this information from the cochlea must preserve precisely the information about time of arrival, or phase at the ear as information passes through successive levels of the auditory system, until the signals from the two ears can be compared on a single neuron. To preserve the timing information it is necessary for the neurons carrying the signal to have a very high probability of firing as soon as they are stimulated and having a high probability of causing the postsynaptic neuron to fire. Intensity differences are coded by firing rate or the number and identity of the active fibers. The synapses with a high probability of causing the postsynaptic cells to fire, e.g., calyceal endings, are referred to as high security synapses. The single neurons on which timing information from the two ears first converge are principal neurons in the MSO and LSO.

4.3 Complex Signal Processing

Although many neuronal classes in the SOC and LL are clearly projection neurons in either the ascending or descending auditory pathways, some neurons also send axon collaterals to adjacent neurons or adjacent nuclei and thus participate in local circuits. The organization of the auditory brainstem into a number of different pathways passing through the SOC and LL allows for complex processing of auditory information. The location of OC neurons in the SOC, the inputs to them from ascending and descending pathways, and their terminal distribution in the cochlea have led to the suggestion that they may be involved in facilitating detection of signals against background noise, e.g., detection of speech in noise or selective attention (Warr 1982).

4.4 Clinical Considerations

Synchronous firing of various neurons at the level of the SOC and LL are thought to be the sources of waves IV and V, respectively, of the auditory brainstem response recordings (Moller 1985). Changes in these waves, or their latencies, can be helpful in diagnosing the location of problems within the auditory pathways.

5. Conclusions/Summary

Major cell groups in the MSO, LSO, MNTB, and some periolivary areas are well characterized in some species. However, many questions remain about their chemical properties. There are also questions about the exact equivalence of cell groups between the various species. The various cell types in the lateral lemniscal nuclei are even less well defined. While neurotransmitters have been identified or implicated in some SOC and LL neurons and some synaptic sites, the transmitters are still unknown for most olivary and lemniscal neurons. Further a variety of neuromodulators and receptor types have now been recognized in the central nervous system, but their distribution in the SOC and LL are only just beginning to be studied. The details of the origins and the chemical and anatomical features of the axons contacting SOC and LL neurons are known in part for some well-characterized cell types, but not at all for some less well-known cell types. The anatomy, chemistry, and distribution of axons of SOC and LL neurons to specific cell types is also only partly known. Understanding the precise mechanisms involved at the many synaptic sites in these regions requires more information about all of these features. Defining the generality of mechanisms which have been identified to different species and classes of animals is an additional challenge. In short, we have learned a lot about the SOC and LL, but much remains to be determined.

What we now know about the anatomy and chemistry of the SOC and LL provides the basis for understanding in broad outline how sounds are detected, how they are localized in space, and how selective attention may be paid to some signals. Understanding the contribution of the lower auditory system to the performance of more subtle or complex tasks such as pattern recognition, detection of speech in noise, or the basis of certain genetic or pathologic malfunctions must await the result of continuing studies of the details of connections, neurophysiology, and neuronal chemistry among the varied cell types present in these important areas.

Acknowledgements   Much of the author's work cited in this chapter was supported by a grant from the National Institute on Deafness and Other Communication Disorders DC00132 (formerly NS09996 or NS14503). I am indebted to Dr. W. Bruce Warr, Dr. Jean Moore, and Dr. Robert Helfert for their comments and criticisms.

Abbreviations

AChE	acetylcholinesterase
ALPO	anterolateral periolivary nucleus
ASP	aspartate
AVCN	anteroventral cochlear nucleus
c	contralateral
CaBP(+)	Calbindin immunoreactive
CF	constant frequency
CGRP	calcitonin gene related peptide
CN	cochlear nucleus
CP	cerebral peduncles
CST	corticospinal tract
DASP	d-aspartic acid

DCN	dorsal cochlear nucleus
DLL	DNLL
DLPO	dorsolateral periolivary nucleus
DMPO	dorsomedial periolivary nucleus
DMSO	dorsal medial superior olive
DPO	dorsal periolivary nucleus
DNLL	dorsal nucleus of the lateral lemniscus
FM	frequency modulating
GABA	gamma amino butyric acid
GABA(+)	GABA immunoreactive
GABA(−)	not GABA immunoreactive
GLU	glutamate
GLY	glycine
GLY(+)	GLY immunoreactive
GLY(−)	not GLY immunoreactive
HRP	horseradish peroxidase
i	ipsilateral
IAS	intermediate acoustic stria
IC	inferior colliculus
ILL	INLL
INLL	intermediate nucleus of the lateral lemniscus
LL	lateral lemniscus
LNTB	lateral nucleus of the trapezoid body
LOC	lateral olivocochlear
LSO	lateral superior olivary nucleus
LTB	LNTB
MCP	middle cerebellar peduncle
MGB	medial geniculate body
MNTB	medial nucleus of the trapezoid body
MSO	medial superior olivary nucleus
MTB	MNTB
NCAT	nucleus of the central acoustic tract
NIP	nipecotic acid
NLL	nuclei of the lateral lemniscus
NTB	nucleus of the trapezoid body
OC	olivocochlear
OCB	olivocochlear bundle
PN	pontine nuclei
PO	periolivary
PPO	posterior periolivary
PVCN	posteroventral cochlear nucleus
SOC	superior olivary complex
SC	superior colliculus
SPN	superior paraolivary nucleus
SS	somatostatin

156   Ilsa R. Schwartz

TGM   upper tegmental part of the pons
VII   facial nucleus
VLL   VNLL
VMSO   ventral medial superior olive
VNLL   ventral nucleus of the lateral lemniscus
VNTB   ventral nucleus of the trapezoid body
VPO   ventral periolivary nucleus

References

Adams JC (1979) Ascending projections to the inferior colliculus. J Comp Neurol 183:519–538.

Adams JC (1983a) Cytology of periolivary cells and the organization of their projections in the cat. J Comp Neurol 215:275–289.

Adams JC (1983b) Multipolar cells in the ventral cochlear nucleus project to the dorsal cochlear nucleus and the inferior colliculus. Neurosci Lett 37:205–208.

Adams JC (1986a) Neuronal morphology in the human cochlear nucleus. Arch Otolaryngol Head and Neck Surg 112:1253–1261.

Adams JC (1986b) Cells of origin of cochlear efferents in human. ARO Absts 9:5.

Adams JC, Mugnaini E (1984) Dorsal nucleus of the lateral lemniscus; A nucleus of GABAergic projection neurons. Brain Res Bull 13:585–590.

Adams JC, Mugnaini E (1990) Immunocytochemical evidence for inhibitory and disinhibitory circuits in the superior olive. Hear Res 49:281–298.

Adams JC, Warr WB (1976) Origins of axons in the cat's acoustic strias determined by injection of horseradish peroxidase into severed tracts. J Comp Neurol 170:107–122.

Adams JC, Wenthold RJ (1987) Immunostaining of ascending auditory pathways with glycine antiserum. ARO Absts 10:63.

Aitkin LM, Kenyon CE (1981) The auditory brainstem of a marsupial. Brain Behav Evol 19:126–143.

Aitkin LM, Phillips SC (1984) Is the inferior colliculus an obligatory relay in the cat auditory system? Neurosci Lett 44:259–264.

Aitkin LM, Schuck D (1985) Low frequency neurons in the lateral central nucleus of the cat inferior colliculus receive their input predominantly from the medial superior olive. Hearing Res 17:87–93.

Aitkin LM, Anderson DJ, Brugge JF (1970) Tonotopic organization and discharge characteristics of single neurons in nuclei of the lateral lemniscus of the cat. J Neurophysiol 33:421–440.

Aitkin LM, Bush BMH, Gates RG (1978) The auditory midbrain of a marsupial: the brush-tailed possum (*Tichosurus vulpecula*). Brain Res 150:29–44.

Aitkin LM, Byers M, Nelson JE (1986) Brain stem auditory nuclei and their connections in a carnivorous marsupial, the northern native cat (*Dasyurus hallucatus*). Brain Behav Evol 29(1–2):1–16.

Aitkin LM, Horseman BG, Bush BMH (1982) Some aspects of the auditory pathway and audition in the European mole, *Talpa europaea*. Brain Behav Evol 21:49–59.

Aitkin LM, Irvine DRF, Webster WR (1984) Central neural mechanisms of hearing. In: *Handbook of Physiology*, The Nervous System, Vol III. Sensory Processes, Part 2. pp. 675–737, section eds Brookhart JM, Mountcastle VM. Vol. ed. I. Darian-Smith, exec. ed. S.R. Geiger. American Physiological Society, Bethesda.

Altschuler RA, Parakkal MH, Fex J (1983) Localization of enkephalin-like immunoreactivity in acetylcholinesterase positive cells in the guinea pig lateral superior olivary complex that project to the cochlea. Neurosci 9:621–630.

Arends JJA, Zeigler HP (1986) Anatomical identification of an auditory pathway from a nucleus of the lateral lemniscal system to the frontal telencephalon (nucleus basalis) of the pigeon. Brain Res 398:375–381.

Aschoff A, Muller M, Ott H (1988) Origin of cochlea efferents in some gerbil species. A comparative anatomical study with fluorescent tracers. Exp Brain Res 71:252–261.

Babmindra VP, Zharskaya VD (1980) [Use of antero- and retrograde horseradish peroxidase transport to analyze interneuronal connections of the bat auditory system]. Arkh Anat Gistol Embriol 78:42–48.

Babmindra VP, Zharskaya VD (1982) Use of anterio- and retrograde axonal transport of horseradish peroxidase in analyzing interneuronal connections in the bat auditory system. Neurosci Behav Physiol 12:423–428.

Belekhova MG, Zharskaya VD, Khachunts AS, Gaidaenko GV, Tumanova NL (1985) Connections of the mesencephalic, thalamic and telencephalic auditory centers in turtles. Some structural bases for audiosomatic interrelations. J Hirnforsch 26:127–152. In English.

Bishop AL, Henson Jr OW (1987) The efferent cochlear projections of the superior olivary complex in the mustached bat. Hear Res 31:175–182.

Bledsoe SC, Snead CR, Helfert RH, Prasad V, Wenthold RJ, Altschuler RA (1990) Immunocytochemical and lesion studies support the hypothesis that the projection from the medial nucleus of the trapezoid body to the lateral superior olive is glycinergic. Brain Res 517:189–194.

Borg E (1973a) Neuroanatomical study of the brain stem auditory system of the rabbit. Part I. Ascending connections. Acta Morphol Neerl Scand 11:31–48.

Borg E (1973b) Neuroanatomical study of the brain stem auditory system of the rabbit. Part II. Descending connections. Acta Morphol Neerl Scand 11:49–62.

Brawer JR, Morest, DK, Kane E (1974) The neuronal architecture of the cochlear nucleus of the cat. J Comp Neurol 155:251–300.

Brouwer B, von Walree D (1914) Uber den Hirnstamm eines Taubstummen. Folia Neurobiol 8:589–600.

Brown JC, Howlett B (1972) The olivo-cochlear tract in the rat and its bearing on the homologies of some constituent cell groups of the mammalian superior olivary complex: A thiocholine study. Acta Anat 83:505–526.

Brown MC (1985) Peripheral projections of labelled efferent nerve fibers in the guinea pig cochlea: an anatomical study. ARO Absts 8:9.

Browner RH, Rubinson K (1977) The cytoarchitecture of the torus semicircularis in the Tegu lizard, Tupinambis nigropunctatus. J Comp Neurol 176:539–557.

Browner RH, Webster DB (1975) Projections from the trapezoid body and the superior olivary complex of the kangaroo rat (Dipodomys merriami). Brain Behav Evol 11:322–354.

Brunso-Bechtold JK, Henkel CK, Linville C (1990) Synaptic organization in the adult ferret medial superior olive. J Comp Neurol 294:389–398.

Calford MB, Aitkin LM (1983) Ascending projections to the medial geniculate body of the cat: evidence for multiple, parallel auditory pathways through thalamus. J Neurosci 3:2365–2380.

Cant NB (1984) The fine structure of the lateral superior olivary nucleus of the cat. J Comp Neurol 227:63–77.

Cant NB (1991) Projections to the lateral and medial superior olivary nuclei from the spherical and globular bushy cells of the anterior ventral cochlear nucleus. In Altschuler RA, Hoffman DW, Bobbin RB, Clopton B (eds) *Neurobiology of Hearing,* Volume 2, The Central Auditory System. New York: Raven Press, (in press).

Cant NB, Casseday JH (1986) Projections from the anteroventral cochlear nucleus to the lateral and medial superior olivary nuclei. J Comp Neurol 247:457–476.

Carpenter MB, Chang L, Pereira AB, Hersh LB, Bruce G, Wu JY (1987) Vestibular and cochlear efferent neurons in the monkey identified by immunocytochemical methods. Brain Res 408(1–2):275–280.

Casey MA, Feldman ML (1985) Aging in the rat medial nucleus of the trapezoid body. II. Electron Microscopy. J Comp Neurol 232:401–413.

Caspary DM, Faingold CL (1989) Non-N-Methyl-D-aspartate receptors may mediate ipsilateral excitation at lateral superior olivary synapses. Brain Res 503:83–90.

Casseday JH, Covey E, Vater M (1988) Connections of the superior olivary complex in the rufous horseshoe bat, *Rhinolophus rouxi.* J Comp Neurol 278(3):313–329.

Casseday JH, Kobler JB, Isbey SF, Covey E (1989) Central acoustic tract in an echolocating bat: An extralemniscal auditory pathway to the thalamus. J Comp Neurol 287:247–259.

Castex A, Marchand L (1906) Etude anatomique et histologique sur la surdi-mutie. Bull Laryngol Otol Rhinol 9:81–99.

Ciriello J, Caverson MM, Park DH (1986) Immunohistochemical identification of noradrenaline- and adrenaline-synthesizing neurons in the cat ventrolateral medulla. J Comp Neurol 253:216–230.

Cole KS, Robertson D, Johnstone B (1987) Brainstem location of bilaterally projecting olivocochlear neurons in the guinea pig. ARO Absts 10:216.

Conlee JW (1979) Descending auditory projections from the inferior colliculus and nuclei of the lateral lemniscus in the cat. Ph.D. dissertation, Chicago: University of Chicago.

Conlee JW, Parks TN (1986) Origin of ascending auditory projections to the nucleus mesencephalicus lateralis pars dorsalis in the chicken. Brain Res 367:96–113.

Covey E, Casseday JH (1986) Connectional basis for frequency representation in the nuclei of the lateral lemniscus of the bat, *Eptesicus fuscus.* J Neurosci 6:2926–40.

Covey E, Casseday JH (1991) The monaural nuclei of the lateral lemniscus in an echolocating bat: Parallel pathways for analyzing temporal features of sound. J Neurosci 11:3456–3470.

Covey E, Jones DR, Casseday JH (1984) Projections from the superior olivary complex to the cochlear nucleus in the tree shrew. J Comp Neurol 226:289–305.

Davis M, Gendelman DS, Tischler MN, Gendelman PM (1982) A primary acoustic startle circuit—lesion and stimulation studies. J Neurosci 2:791–805.

Donaldson HH (1890) Anatomical observations on the brain and several sense-organs of the blind deaf-mute, Laura Dewey Bridgman. Am J Psychol 3:293–342.

Donaldson HH (1891) Anatomical observations on the brain and several sense-organs of the blind, deaf-mute, Laura Dewey Bridgman. Am J Psychol 4:248–294.

Druga R, Syka J (1984) Projections from auditory structures to the superior colliculus in the rat. Neurosci Lett 45:247–252.

Dublin WB (1976) Fundamentals of Sensorineural Auditory Pathology. Springfield, IL: Charles C Thomas.

Dupont J, Geffard M, Calas A, Aran J-M (1990) Immunohistochemical evidence for GABAergic cell bodies in the medial nucleus of the trapezoid body and in the lateral vestibular nucleus in the guinea pig brainstem. Neurosci Lett 111:263–268.

Edwards SB, Ginsburgh CL, Henkel CK, Stein BE (1979) Sources of subcortical projections to the superior colliculus in the cat. J Comp Neurol 184:309–330.

Elverland HH (1977) Descending connections between the superior olivary and cochlear nuclear complexes in the cat, studied by autoradiographic and horseradish peroxidase methods. Exp Brain Res 27:397–412.

Elverland HH (1978) Ascending and intrinsic projections of the superior olivary complex in the cat. Exp Brain Res 32:117–134.

Faye-Lund H (1986) Projection from the inferior colliculus to the superior olivary complex in the albino rat. Anat and Embryol 175:35–52.

Feng AS (1986a) Afferent and efferent innervation patterns of the superior olivary nucleus of the leopard frog. Brain Res 364:167–171.

Feng AS (1986b) Afferent and efferent innervation patterns of the cochlear nucleus (dorsal medullary nucleus) of the leopard frog. Brain Res 367:183–191.

Fernandez C, Karapas F (1967) The course and termination of the striae of Monakow and Held in the cat. J Comp Neurol 131:371–386.

Ferraro JA, Minckler J (1977) The human lateral lemniscus and its nuclei. The human auditory pathways: A quantitative study. Brain Lang 4:277–294.

Fex J, Altschuler RA (1985) Immunocytochemistry of the mammalian cochlea: results and expectations. In Drescher D (ed) Auditory Biochemistry, Springfield, Ill: Charles Thomas, pp. 5–28.

Finlayson PG, Caspary DM (1989) Synaptic potentials of chinchilla lateral superior olivary neurons. Hear Res 38:221–228.

Foster RE, Hall WC (1978) The organization of central auditory pathways in a reptile, *Iguana iguana*. J Comp Neurol 178:783–832.

Friauf E, Ostwald J (1988) Divergent projections of physiologically characterized rat ventral cochlear nucleus neurons as shown by intra-axonal injection of horseradish peroxidase. Exp Brain Res 73:263–84.

Fuller PM, Ebbesson SOE (1973) Projections of the primary and secondary auditory fibers in the bullfrog (*Rana catesbiana*). Proc Soc Neurosci 333.

Glendenning KK (1987) Asymmetries in lateral superior olive projection to inferior colliculus. ARO Absts 10:217.

Glendenning KK, Baker BN (1991) Neurochemical basis of the acoustic chiasm. ARO Absts 14:32.

Glendenning KK, Masterton RB (1980) Afferent and efferent connections of the lateral superior olivary nucleus in the cat. Anat Rec 196:63–64.

Glendenning KK, Masterton RB (1983) Acoustic chiasm: efferent projections of the lateral superior olive. J Neurosci 3:1521–1537.

Glendenning KK, Brunso-Bechtold JK, Thompson GC, Masterton RB (1981) Ascending auditory afferents to the nuclei of the lateral lemniscus. J Comp Neurol 197:673–703.

Glendenning KK, Hutson KA, Nudo RJ, Masterton RB (1985) Acoustic chiasm II. Anatomical basis of binaurality in the lateral superior olive of cat. J Comp Neurol 232:261–285.

Godfrey DA, Carter JA, Berger SJ, Lowry OH, Matschinsky FM (1977) Quantitative histochemical mapping of candidate transmitter amino acids in cat cochlear nucleus. J Histochem Cytochem 25:417–431.

Goldberg JM, Brown PB (1968) Functional organization of the dog superior olivary complex: An anatomical and electrophysiological study. J Neurophysiol 31:635–636.

Goldberg JM, Brown PB (1969) Response of binaural neurons of dog superior olivary complex to dichotic tonal stimuli: some physiological mechanisms of sound localization. J Neurophysiol 32:613–636.

Goldberg JM, Moore RY (1967) Ascending projections of the lateral lemniscus in the cat and monkey. J Comp Neurol 129:143–156.

Gorodetskaia ON, Bibikov NG (1985) Responses of auditory neurons of the medulla oblongata of the frog to presentation of tones with sinusoidal amplitude modulation. Neirofiziol 17:390–396.

Grafova I, Ottersen OP, Rinvik E (1978) Mesencephalic and diencephalic afferents to the superior colliculus and periaquaductal gray substance demonstrated by retrograde axonal transport of horseradish peroxidase in the cat. Brain Res 146:205–220.

Guinan JJ, Norris BE, Guinan SS (1972) Single auditory units in the superior olivary complex. II: Locations of unit categories and tonotopic organization. Int J Neurosci 4:147–166.

Hall JC, Feng AS (1987) Evidence for parallel processing in the frog's auditory thalamus. J Comp Neurol 258:407–419.

Harrison, JM (1978) The auditory system of the brainstem. In: Naunton, RF, Fernandez, C (eds) Evoked Electrical Activity in the Auditory Nervous System, New York: Academic Press, pp. 353–368.

Harrison JM, Feldman ML (1970) Anatomical aspects of the cochlear nucleus and superior olivary complex. In Neff WD (ed) Contributions to Sensory Physiology, New York: Academic Press, pp. 95–143.

Harrison JM, Irving R (1966a) Ascending connections of anterior ventral cochlear nucleus in the rat. J Comp Neurol 126:51–64.

Harrison JM, Irving R (1966b) Organization of the posterior ventral cochlear nucleus in the rat. J Comp Neurol 126:391–403.

Harrison JM, Warr WB (1962) A study of the cochlear nuclei and ascending auditory pathways of the medulla. J Comp Neurol 119:341–380.

Heffner HE, Heffner RS (1984) Sound localization in large mammals: localization of complex sounds by horses. Behav Neurosci 99:541–555.

Heffner RS, Heffner HE (1986) Localization of tones by horses: Use of binaural cues and the role of the superior olivary complex. Behav Neurosci 100:93–103.

Heffner RS, Heffner HS (1987) Localization of noise, use of binaural cues, and a description of the superior olivary complex in the smallest carnivore, the least weasel (*Mustela nivalis*). Behav Neurosci 101:701–708.

Heil P, Scheich H (1986) Effects of unilateral and bilateral cochlea removal on 2-deoxyglucose patterns in the chick auditory system. J Comp Neurol 252:279–301.

Helfert RH, Schwartz IR (1986) Morphological evidence for the existence of multiple neuronal classes in the cat lateral superior olivary nucleus. J Comp Neurol 244:533–549.

Helfert RH, Schwartz IR (1987) Morphologic evidence for the presence of five cell types in the gerbil lateral superior olivary nucleus. Am J Anat 179:55–69.

Helfert RH, Bonneau JM, Wenthold RJ, Altschuler RA (1989) GABA and glycine immunoreactivity in the guinea pig superior olivary complex. Brain Res 6:269–286.

Helfert RH, Schwartz IR, Ryan AF (1988) Ultrastructural characterization of gerbil olivocochlear neurons based on differential uptake of ^{3}H-D-aspartic acid and a wheatgerm agglutinin-horseradish peroxidase conjugate from the cochlea. J Neurosci 8:3111–3123.

Henkel CK (1983) Evidence of sub-collicular auditory projections to the medial geniculate nucleus in the cat: an autoradiographic and horseradish peroxidase study. Brain Res 259:21–30.

Henkel CK (1989) Axonal domains within the dorsal nucleus of the lateral lemniscus. Soc Neurosci Absts 15:746.

Henkel CK, Brunso-Bechtold JK (1990) Dendritic morphology and development in the ferret medial superior olivary nucleus. J Comp Neurol 294:377–388.

Irving R, Harrison JM (1967) The superior olivary complex and audition: A comparative study. J Comp Neurol 130:77–86.

Kane ES, Barone LM (1980) The dorsal nucleus of the lateral lemniscus in the cat: neuronal types and their distributions. J Comp Neurol 192:797–826.

Kane EC, Conlee JW (1979) Descending inputs to the caudal cochlear nucleus of the cat: degeneration and autoradiographic studies. J Comp Neurol 187:759–784.

Kane EC, Finn RC (1977) Descending and intrinsic inputs to the cat caudal cochlear nucleus: a horseradish peroxidase study. Neurosci 2:897–912.

Kinney HC, Ottoson CK, White WF (1990) 3-dimensional distribution of H-3-Naloxone binding to opiate receptors in the human fetal and infant brainstem. J Comp Neurol 291:55–78.

Kiss A, Majorossy K (1983) Neuron morphology and synpatic architecture in the medial superior olivary nucleus. Light and electron microscope studies in the cat. Exp Brain Res 52:315–327.

Knudsen EI (1977) Distinct auditory and lateral line nuclei in the midbrain of catfishes. J Comp Neurol 173:417–431.

Knudsen EI, Konishi M (1978) Space and frequency are represented separately in auditory midbrain of the owl. J Neurophysiol 41:870–884.

Kudo M (1981) Projections of the nuclei of the lateral lemniscus in the cat: an autoradiographic study. Brain Res 221:57–69.

Kudo M, Nakamura Y, Moriizumi T, Tokuno H, Kitao Y (1988) Bilateral projections from the medial superior olivary nucleus to the inferior colliculus in the mole (*Mogera robusta*). Brain Res 463:352–356.

Kunzle H (1986) Projections from the cochlear nuclear complex to rhombencephalic auditory centers and torus semicircularis in the turtle. Brain Res 379:307–319.

LaVilla I (1898) Algunos detalles concernientes a la oliva superior y focos acusticos. Revista Trimestral Micrografical 3:75–83.

Li RY-S, Guinan JJ (1971) Antidromic and orthodromic stimulation of neurons receiving calyces of Held. MIT Quart. Prog Rpt No 100 pp. 227–234.

Lindsey BG (1975) Fine structure and distribution of axon terminals from the cochlear nucleus on neurons in the medial superior olivary nucleus of the cat. J Comp Neurol 160:81–104.

Masterton B, Thompson GC, Bechtold JK, RoBards MJ (1975) Neuroanatomical basis of binaural phase-difference analysis for sound localization: A comparative study. J Comp Physiol Psychol 89:379–386.

Matsubara JA (1990) Calbindin D-28K immunoreactivity in the cat's superior colivary complex. Brain Res 508:353–357.

McDonald DM, Rasmussen GL (1971) Ultrastructural characteristics of synaptic endings in the cochlear nucleus having acetylcholinesterase activity. Brain Res 28:1–18.

Moller AR (1985) Physiology of the ascending auditory pathway with special reference to the auditory brainstem response (ABR). In: Pinheiro ML, Musiek FE (eds) Assessment of Central Auditory Dysfunction: Foundations and Clinical Correlations, Baltimore: Williams and Wilkins, pp. 23–41.

Moore DR (1988) Auditory brainstem of the ferret: sources of projections to the inferior colliculus. J Comp Neurol 269:342–354.

Moore DR, Semple MN, Addison PD, Aitkin LM (1984) Properties of spatial receptive fields in the central nucleus of the cat inferior colliculus. I. Responses to tones of low intensity. Hear Res 13:159–174.

Moore JK (1987a) The human auditory brain stem: A comparative view. Hear Res 29:1–32.

Moore JK (1987b) The human auditory brain stem as a generator of auditory evoked potentials. Hear Res 29:33–43.

Moore JK, Moore RY (1971) A comparative study of the superior olivary complex in the primate brain. Folia primat 16:35–51.

Moore JK, Osen KK (1979) The human cochlear nuclei. Exp Brain Res Suppl II:36–44.

Moore MJ, Caspary DM (1983) Strychnine blocks binaural inhibition in lateral superior olivary neurons. J Neurosci 3:237–242.

Moore JK, Karapas F, Moore RY (1977) Projections of the inferior colliculus in insectivores and primates. Brain Behav Evol 14:301–327.

Morest DK (1968a) The collateral system of the medial nucleus of the trapezoid body of the cat, its neuronal architecture and relation to the olivo-cochlear bundle. Brain Res 9:288–311.

Morest DK (1968b) The growth of synaptic endings in the mammalian brain: a study of the calyces of the trapezoid body. Z Anat Entw Gesch 127:201–220.

Morley BJ (1985) The localization and origin of somatostatin-containing fibers in an auditory brainstem nucleus. Peptides 6 (Suppl 1):165–172.

Moskowitz N (1965) Comparative aspects of the central auditory nuclei. Anat Rec 151:467.

Moskowitz N (1966) The cochlear nucleus and superior olivary complex in a lorisoid, Galago senegalensis. Anat Rec 154:478.

Moskowitz N (1969) Comparative aspects of some features of the central auditory system of primates. Annals NY Acad Sci 167:357–369.

Nakajima Y (1971) Fine structure of the medial nucleus of the trapezoid body of the bat with special reference to two types of synaptic endings. J Cell Biol 50:121–134.

Neuweiler G, Bruns V, and Schuller G (1980) Ears adapted for the detection of motion, or how bats have exploited the capacities of the mammalian auditory system. J Acoust Soc Am 68:741–753.

Nordeen KW, Killackey HP, Kitzes LM (1983) Ascending auditory projections to the inferior colliculus in the adult gerbil, *Meriones unguiculatus*. J Comp Neurol 214:131–143.

Nudo RJ, Masterton RB (1986) Stimulation-induced [14C]2-deoxyglucose labeling of synaptic activity in the central auditory system. J Comp Neurol 245:553–565.

Oliver DL, Shneiderman A (1989) An EM study of the dorsal nucleus of the lateral lemniscus: inhibitory, commissural, synaptic connections between ascending auditory pathways. J Neurosci 9:967–982.

Oliver D, Schneiderman A, Henkel CK (1987) Morphological substrates for binaural interactions in the midbrain: the dorsal nucleus of the lateral lemniscus. ARO Absts 10:218–219.

Ollo C, Schwartz IR (1979) The superior olivary complex in C57BL/6 mice. Am J Anat 155:349–374.

Olszewski J, Baxter D (1954) *Cytoarchitecture of the Human Brain Stem.* Philadelphia: Lippincott.

Osen KK (1972) Projection of the cochlear nuclei on the inferior colliculus in the cat. J Comp Neurol 144:355–372.

Osen K, Roth K (1969) Histochemical localization of cholinesterases in the cochlear nuclei of the cat, with notes on the origin of acetylcholinesterase-positive afferents and the superior olive. Brain Res 16:165–185.

Papez JW (1929a) Central acoustic tract in cat and man. Anat Rec 42:60.

Papez JW (1929b) *Comparative Neurology.* New York: Hafner Publishing Company. pp. 270–293.

Papez JW (1930) Superior olivary nucleus. Arch Neurol Chicago 24:1–20.

Perkins RE (1973) An electron microscopic study of synaptic organization in the medial superior olive of normal and experimental chinchilla. J Comp Neurol 148:387–416.

Pettigrew AG (1981) Brainstem afferents to the torus semicircularis of the Queensland cane toad (*Bufo marinus*). J Comp Neurol 202(1):59–68.

Poon PWF, Sun X, Kamada T, Jen PH-S (1990) Frequency and space representation in the inferior colliculus of the FM bat, *Eptesicus fuscus*. Exp Brain Res 79:83–91.

Ramón y Cajal S (1899) The acoustic nerve: Its cochlear branch or cochlear nerve. In: *Histologie du systeme Nerveux de l'Homme et des Vertebres.* Vol. I, pp. 774–838. (English translation by Information Center for Hearing, Speech and Disorders of Human Communications, The Johns Hopkins Medical Institutions, PB 205-473, 1967.)

Rasmussen GL (1964) Anatomical relationships of the ascending and descending auditory systems. In: Fields WS, Alford BR (eds) *Neurological Aspects of Auditory and Vestibular Disorders,* pp. 5–19. Springfield, IL: Thomas.

Rasmussen GL (1967) Efferent connections of the cochlear nucleus. In *Sensorineural Hearing Processes and Disorders,* A.B. Graham ed, Little Brown, Boston, pp. 61–75.

Richter EA, Norris BE, Fullerton BC, Levine RA, and Kiang NYS (1983) Is there a medial nucleus of the trapezoid body in humans. Am J Anat 68:157–166.

Roberts RC, Ribak CE (1987) GABAergic neurons and axon terminals in the brainstem auditory nuclei of the gerbil. J Comp Neurol 258:267–280.

Robertson D, Anderson CJ, Cole KS (1987) Segregation of efferent projections to different turns of the guinea pig cochlea. Hear Res 25:69–76.

Ross LS, Pollak GD, Zook JM (1988) Origin of ascending projections to an isofrequency region of the mustache bat's inferior colliculus. J Comp Neurol 270:488–505.

Ross MD (1962) Auditory pathway of the epileptic waltzing mouse. I. A comparison of the acoustic pathway of the normal mouse with those of the totally deaf epileptic waltzer. J Comp Neurol 119:317–339.

Ross MD (1969) The general visceral efferent component of the eighth cranial nerve. J Comp Neurol 135:453–477.

Rubel E, Durham D (1985) Afferent influences on brain stem auditory nuclei of the chicken: changes in succinate dehydrogenase activity following cochlea removal. J Comp Neurol 231:446–456.

Rubinson K, Skiles MP (1975) Efferent projections of the superior olivary nucleus in the frog, *Rana catesbiana*. Brain Behav Evol 12:151–160.

Ryan AF, Schwartz IR (1986) Nipecotic acid: preferential accumulation in the cochlea by GABA uptake systems and selective retrograde transport to brainstem. Brain Res 399:399–403.

Ryan AF, Schwartz IR, Keithley EM (1989) Collateral innervation of cochlear nucleus by lateral and medial olivocochlear neurons. ARO Absts 12:345–346.

Saint Marie RL, Ostapoff EM, Morest DK, Wenthold RJ (1989) Glycine-immunoreactive projection of the cat lateral superior olive: possible role in midbrain ear dominance. J Comp Neurol 279:382–396.

Sanes DH, Goldstein NA, Ostad M, Hillman D (1990) Dendritic morphology of central auditory neurons correlates with their tonotopic position. J Comp Neurol 294:443–454.

Sanes DH, Merickel M, Rubel E (1989) Evidence for an alteration of the tonotopic map in the gerbil cochlea during development. J Comp Neurol 279:436–444.

Scheibel ME, Scheibel AB (1974) Neuropil organization in the superior olive of the cat. Exp Neurol 43:339–348.

Schnitzler H-U, Henson Jr OW (1980) Performance of airborne animal sonar systems: I. Microchiroptera. In: Busnel RG, Fish JR (eds) *Animal Sonar Systems*. New York: Plenum Press, pp. 109–181.

Schwartz IR (1972) Axonal endings in the cat medial superior olive: coated vesicles and intracellular substance. Brain Res 46:187–202.

Schwartz IR (1977) Dendritic arrangements in the cat medial superior olivary nucleus. Neurosci 2:81–101.

Schwartz IR (1978) Differential distribution of synaptic terminal classes in the cat medial superior olive. Anat Rec 190:154.

Schwartz IR (1980) The differential distribution of synaptic terminal classes on marginal and central cells in the cat medial superior olive. Am J Anat 159:25–31.

Schwartz IR (1982) Differential tritiated amino acid labeling of synaptic terminals in the cat medial superior olivary nucleus. ARO Absts 5:21.

Schwartz IR (1983) Autoradiographic evidence that glycine labeling of synaptic terminals in the superior olivary complex has transmitter-like properties. In:

Webster WW, Aitkin LM (eds) *Mechanisms of Hearing*, Clayton, Australia: Monash Univ. Press, p. 147.

Schwartz IR (1984a) Axonal organization in the cat medial superior olivary nucleus. Contributions to Sensory Physiology, WD Neff (editor), 8:99–129.

Schwartz IR (1984b) Autoradiographic studies of amino acid labeling of neural elements in the auditory system. Intl Conf On *Auditory Biochemistry*, Absts ARO Midwinter Mtg 7:137.

Schwartz IR (1985) Autoradiographic studies of amino acid labeling of neural elements in the auditory system. In: Drescher D (ed) Auditory Biochemistry, Springfield, IL: Charles Thomas, pp. 258–277.

Schwartz IR, Eager PR (1992) Differential distribution of calcium binding proteins and neuronal surface markers and their relationship to GABA immunoreactive cells in the superior olivary complex and lateral lemniscal nuclei of the gerbil. *Abstr. ARO* Midwinter 15:59.

Schwartz IR, Helfert RH, Ryan AF (1986) Ultrastructural characterization of lateral olivocochlear efferent neurons and processes in the superior olivary complex and cochlear nucleus labeled by selective uptake of ³H-D-aspartic acid in the gerbil cochlea. Absts ARO Midwinter Mtg 9:6–7.

Schwartz IR, Yu S-M (1986) An anti-GABA antibody labels subpopulations of axonal terminals and neurons in the gerbil cochlear nucleus and superior olivary complex. Soc Neurosci Abs 12:780.

Schweizer H (1981) The connections of the inferior colliculus and the organization of the brainstem auditory system in the greater horseshoe bat (*Rhinolophus ferrumequinum*). J Comp Neurol 201:25–49.

Schweitzer LF, Lu SM, Dawburn D, Cant NB (1985) Calcitonin gene-related peptide in the superior olivary complex of cat and rat: a specific label for the lateral olivocochlear system. Neurosci Absts 11:1051.

Semple MN, Aitkin LM, Calford MB, Pettigrew JD, Phillips DP (1983) Spatial receptive fields in the cat inferior colliculus. Hear Res 10:203–215.

Shneiderman A, Oliver DL, Henkel CK (1988) Connections of the dorsal nucleus of the lateral lemniscus: an inhibitory parallel pathway in the ascending auditory system? J Comp Neurol 276:188–208.

Sidman RL, Angevine JB, Taber-Pierce E (1971) *Atlas of the Mouse Brain and Spinal Cord*. Cambridge: Harvard University Press.

Smith DJ, Rubel EW (1979) Organization and development of brainstem auditory nuclei of the chicken: Dendritic gradients in N. Laminaris. J Comp Neurol 186:213–240.

Spangler KM, Cant NB, Henkel CK, Farley GR, Warr WB (1987) Descending projections from the superior olivary complex to the cochlear nucleus of the cat. J Comp Neurol 259:452–465.

Spangler KM, Morley BJ (1987) Somatostatin-like immunoreactivity in the midbrain of the cat. J Comp Neurol 260:87–97.

Spangler KM, Warr WB (1991) The descending auditory system. In: Altschuler RA, Hoffman DW, Bobbin RP, Clopton BM (eds) *The Neurobiology of Hearing* Vol. II, New York: Raven Press, (in press).

Spangler KM, Warr WB, Henkel CK (1985) The projections of principal cells of the medial nucleus of the trapezoid body in the cat. J Comp Neurol 238:249–261.

Stotler WA (1953) An experimental study of the cells and connections of the superior olivary complex of the cat. J Comp Neurol 98:401–432.

Strominger NL (1973) The origin, course and distribution of the dorsal and intermediate acoustic stria in the monkey. J Comp Neurol 147:209–234.

Strominger NL (1978) The anatomical organization of the primate auditory pathway. In: Noback CR (ed) *Sensory Systems of Primates*, New York: Plenum, p. 53–91.

Strominger NL, Hurwitz JL (1976) Anatomical aspects of the superior olivary complex. J Comp Neurol 170:485–498.

Strominger NL, Strominger AI (1971) Ascending brainstem projection of the anterior ventral cochlear nucleus in the monkey. J Comp Neurol 143:217–241.

Strominger NL, Nelson LR, Dougherty WJ (1977) Second order auditory pathways in the chimpanzee. J Comp Neurol 172:349–366.

Taber E (1961) The cytoarchitecture of the brain stem of the cat. I. Brain stem nuclei of cat. J Comp Neurol 116:27–70.

Tanaka K, Otani K, Tokunaga A, Sughita S (1985) The organization of neurons in the nucleus of the lateral lemniscus projecting to the superior and inferior colliculi in the rat. Brain Res 341:252–260.

Thompson AM, Thompson GC (1987) Projections from PVCN to SOC in guinea pig as demonstrated by the anterograde transport of Pha-L, ARO Abs 10:215.

Thompson GC, Cortez AM, Lam DMK (1985) Localization of GABA immunoreactivity in the auditory brainstem of guinea pig. Brain Res 339:119–122.

Tohyama Y, Senba E, Yamashita T, Kitajiri M, Kumazawa T, Ohata K, Tohyama M (1990) Coexistence of calcitonin gene-related peptide and enkephalin in single neurons of the lateral superior olivary nucleus of the guinea pig that project to the cochlea as lateral olivocochlear system. Brain Res 515:312–314.

Tokunaga A (1988) Superior olivary and lateral lemniscal neurons projecting to the cochlea in the guinea pig. Neurosci Res 6:20–30.

Tolbert LP, Morest DK (1978) Patterns of synaptic organization in the cochlear nuclei of the cat. Neurosci Abs 4:11.

Tolbert LP, Morest DK, Yurgelun-Todd DK (1982) The neuronal architecture of the anteroventral cochlear nucleus of the cat in the region of the cochlear nerve root: horseradish peroxidase labelling of identified cell types. Neurosci 7:3031–3052.

Tsuchitani C (1977) Functional organization of lateral cell groups of cat superior olivary complex. J Neurophys 40:296–318.

Tsuchitani C (1978) Lower auditory brain stem structures of the cat. In: Naunton RF, Fernandez C (eds) *Evoked Electrical Activity in the Auditory Nervous System*, New York: Academic Press, pp. 373–401.

Tsuchitani C, Boudreau J (1967) Encoding of stimulus frequency and intensity by cat superior olive S-segment cells. J Acoust Soc Am 42:794–805.

van Noort J (1969) *The Structure and Connections of the Inferior Colliculus.* N.V., Netherlands: Van Gorcum and Co.

Vater M, Feng AS (1990) Functional organization of ascending and descending connections of the cochlear nucleus of horseshoe bats. J Comp Neurol 292:373–395.

Warr WB (1966) Fiber degeneration following lesions in the anterior ventral cochlear nucleus of the cat. Exp Neurol 14:453–474.

Warr WB (1969) Fiber degeneration following lesions in the posteroventral cochlear nucleus of the cat. Exp Neurol 23:140–155.

Warr WB (1972) Fiber degeneration following lesions in the multipolar and globular cell areas in the ventral cochlear nucleus of the cat. Brain Res 40:247–270.

Warr WB (1975) Olivocochlear and vestibular efferent neurons of the feline brain stem: Their location, morphology, and number determined by retrograde axonal transport and acetylcholinesterase histochemistry. J Comp Neurol 161:159–182.

Warr WB (1982) Parallel ascending pathways from the cochlear nucleus: Neuroanatomical evidence of functional specialization. Contrib Sens Physiol 7:1–38.

Warr WB, Spangler KM (1989) A novel projection of the ventral nucleus of the trapezoid body in the rat. Neurosci Abs 15:745.

Webster WR, Batini C, Buisseret-Delmas C, Compoint C, Guegan M, Thomasset M (1990) Colocalization of calbindin and GABA in medial nucleus of the trapezoid body of the rat. Neurosci Letts 111:252–257.

Webster DB, Ackermann RF, Longa GC (1968) Central auditory system of the kangaroo rat, *Dipodomys merriami*. J Comp Neurol 133:477–494.

Wenthold RJ, Huie D, Altschuler RA, Reeks KA (1987) Glycine immunoreactivity localized in the cochlear nucleus and the superior olivary complex. Neurosci 22:897–912.

Wilczynski W (1981) Afferents to the midbrain auditory center in the bullfrog, *Rana catesbeiana*. J Comp Neurol 198:421–433.

Wild JM (1987) Nuclei of the lateral lemniscus project directly to the thalamic auditory nuclei in the pigeon. Brain Res 408:303–307.

Willard FH, Martin GF (1984) The auditory brainstem nuclei and some of their projections to the inferior colliculus in the North American opossum. Neurosci 10:1203–1232.

Winter IM, Robertson D, Cole KS (1989) Descending projections from auditory brainstem nuclei to the cochlea and cochlear nucleus of the guinea pig. J Comp Neurol 280:143–57.

Woollard HH, Harpman JA (1940) The connections of the inferior colliculus and of the dorsal nucleus of the lateral lemniscus. J Anat Lond 74:441–457.

Yin TCT, Chan JCK (1988) Neural mechanisms underlying interaural time sensitivity to tones and noise. In: Edelman GM, Gall WE, Cowan WM (eds) *Auditory Function: The Neurobiological Bases of Hearing*, New York: John Wiley and Sons, pp. 385–430.

Zhang SQ, Sun XD, Jen PH (1987) Anatomical study of neural projections to the superior colliculus of the big brown bat, *Eptesicus fuscus*. Brain Res 416:375–380.

Zook JM, Casseday JH (1979) Connections of the nuclei of the lateral lemniscus in the mustache bat, *Pteronotus parnellii*. Neurosci Abstr 5:34.

Zook JM, Casseday JH (1982a) Cytoarchitecture of auditory system in lower brainstem of the mustache bat, *Pteronotus parnellii*. J Comp Neurol 207:1–13.

Zook JM, Casseday JH (1982b) Origin of ascending projections to inferior colliculus in the mustache bat, *Pteronotus parnellii*. J Comp Neurol 207:14–28.

Zook JM, Casseday JH (1985) Projections from the cochlear nuclei in the mustache bat, *Pteronotus parnellii*. J Comp Neurol 237:307–324.

Zook JM, Casseday JH (1987) Convergence of ascending pathways at the inferior colliculus of the mustache bat, *Pteronotus parnellii*. J Comp Neurol 251:347–361.

Zook JM, DiCaprio RA (1988) Intracellular labeling of afferents to the lateral superior olive in the bat, *Eptesicus fuscus*. Hear Res 34:141–147.

Zvorykin VP (1964) Morphological substrate of ultrasonic and locational capacities in the dolphin. Fed Proc Fed Am Soc Exp Biol 23:T647–T654.

5

Inferior and Superior Colliculi

Douglas L. Oliver and Michael F. Huerta

1. Introduction

In the classical neuroanatomical literature, the inferior and superior colliculi, located on the roof of the midbrain, are grouped together as the *corpora quadrigemina*. However, modern research indicates that these structures have quite different functions. The inferior colliculus is part of the auditory pathways and is involved in processing both ascending and descending information about sound. This information may concern the frequency and the monaural or binaural properties of sound. In contrast, the superficial portions of the superior colliculus are devoted to visual processing and are organized retinotopically. The deeper portions of the superior colliculus are related to initiation of eye and head movements, and it is there that information from multiple sensory systems converge, presumably to assist in the control of these movements.

Some important aspects of neural organization may be shared by the inferior and superior colliculus. Both colliculi receive inputs from a large number of sources, so that the integration of information from these sources is a major task for these midbrain centers. We would like to propose that both the inferior and superior colliculus have adopted similar anatomical mechanisms to organize their information processing. In both colliculi, inputs tend to form topographic maps. The tonotopic, retinotopic, and somatotopic maps preserve the arrangement of the information from the sensory epithelia. Many inputs are distributed across or superimposed on these maps to form locally segregated units. We propose to call these units *synaptic domains*. We define synaptic domains as *small groups of neurons and their processes that share the same population of synaptic inputs*. Such units in the colliculi are smaller than traditional cytoarchitectonic subdivisions. Depending on their location, these units are called bands, modules, or sublaminae. Because of their common inputs, all neurons within a synaptic domain would process similar types of information. Although hypotheses regarding synaptic domains are still being tested, there is already sufficient evidence to make the modular organization of the colliculi the main focus of this chapter.

2. Inferior Colliculus (IC)—A Structure for the Integration of Ascending Monaural and Binaural Pathways

Almost all ascending and descending auditory pathways synapse in the IC. Most of the monaural and binaural pathways from the lower auditory system project to the IC. Neurons in the IC send their axons rostrally to the auditory forebrain. Thus, the cellular anatomy of the IC must provide the substrate for integration of the ascending pathways on single projection neurons. This anatomy is critical for the processing of information in the ascending auditory pathway and will be the primary topic in this section of the chapter.

Studies of Golgi impregnated-material have defined the subdivisions of the IC by their cellular anatomy. A number of species, including human, have been studied with this methodology and, together, they provide a consistent picture of the mammalian IC. Golgi studies also provide a useful context for cellular analyses with modern techniques such as intracellular recording and dye injections. So, the first part of this chapter (Section 2.1) will present the cellular anatomy of the IC in relation to its subdivisions and their connections. Each subdivision does not receive the same set of inputs.

The second part of the chapter (Section 2.2) will examine the evidence for synaptic domains within the IC. The major ascending inputs to the IC form bands of axon terminals that are oriented in parallel to the dendrites of the principal cell types. One important aspect of this organization is that some banded inputs may use inhibitory transmitters. It is the overlap of banded inputs that may create synaptic domains in the IC.

The third part of the chapter (Section 2.3) will discuss how the function of efferent neurons may be related to both the synaptic domains and other intrinsic connections. There are a number of different cell types in the IC, and many may be projection neurons. In addition, the longer connections within the midbrain are important for information processing in the ascending pathways. The efferent cells from the IC may give rise to different types of parallel ascending pathways.

2.1 Subdivisions of the IC Have Unique Cellular Organizations and Connections

The IC and surrounding structures contain either tectal or tegmental neurons (Morest and Oliver 1984). Tectal neurons have frequently branched dendritic trees and usually have dendritic spines. In contrast, tegmental neurons have sparse dendritic arbors, few spines, and are typical of the reticular formation. These two cell types define the major subdivisions of the auditory midbrain. The main (tectal) subdivisions of the IC (Figs. 5.1 and 5.2) are the central nucleus, the cortex, and the paracentral nuclei (Morest and Oliver 1984). The main tegmental subdivisions are ventral to the IC and between the IC and the SC (superior colliculus). The region between the IC and the SC is called the intercollicular tegmentum.

Many of these subdivisions in the IC are similar to those recognized by Ramón y Cajal (1911). In modern studies of the IC (Morest and Oliver 1984; Oliver and Morest 1984), each subdivision is defined by a distinct neuropil and a unique set of cell types. Thus, each subdivision has axonal and dendritic arrangements that may provide different substrates for processing information about sound. When subdivisions are defined by this method, they also are found to have different combinations of inputs. In the following sections, we will discuss each subdivision, describe its neuropil and cell types, and relate the organization of its inputs to auditory information processing.

2.1.1   Central Nucleus

2.1.1.1   Cellular Organization in the Cat
(Figs. 5.1 and 5.2, L, C, M, V)

Fibrodendritic laminae define the neuropil of the central nucleus (Morest 1964; Morest and Oliver 1984). A similar region was called the ventrolateral part of the central nucleus in a previous Golgi study (Rockel and Jones 1973a). The fibrodendritic laminae are composed of principal cells (Fig. 5.3) that are disc shaped and arranged with their dendritic fields in parallel to each other and to the incoming lemniscal afferents. About 75–85% of the cells in the central nucleus are disc shaped. The remaining cells are stellate cells (Fig. 5.4). All of these cell types, together with layered axons, make up a highly structured three-dimensional neuropil that characterizes the central nucleus (Fig. 5.5).

Disc-shaped cells (Fig. 5.3) in Golgi material (Oliver and Morest 1984) are highly oriented with dendritic trees that range from 200–800 μm in length. However, dendritic fields are as narrow as 50–70 μm along an axis perpendicular to their length. Within most of the central nucleus, the disc-shaped cells are oriented rostrocaudally and also display a long axis in the dorsomedial to ventrolateral direction. Since groups of disc-shaped cells covary in the size of the dendritic trees and somata, small, medium, medium-large, and large subtypes are identified. Each of these subtypes also can be identified in Nissl-stained material and have ascending projections to the ipsilateral medial geniculate body (Oliver 1984a). Differences in some other features such as dendritic branching patterns and dendritic appendages also support the identification of more than one disc-shaped cell type (Rockel and Jones 1973a; Oliver and Morest 1984).

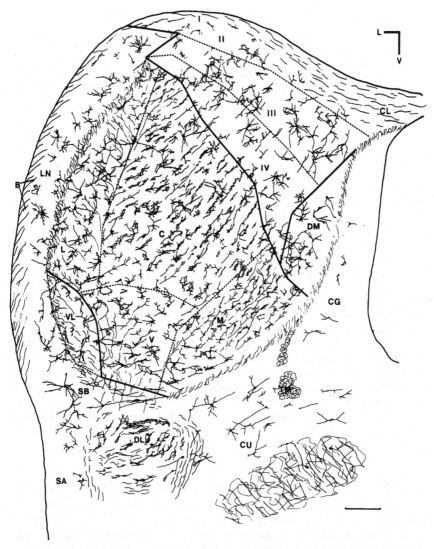

FIGURE 5.1. The inferior colliculus (IC) in Golgi-Cox material cut in the transverse anatomical plane. All major divisions of the inferior colliculus are seen. The tectal subdivisions include: Central nucleus (M, medial; C, central; L, lateral; V, ventral parts); Dorsal cortex (layers I–IV); and Paracentral nuclei (DM, dorsomedial; VL, ventrolateral; LN, lateral; B, brachium of IC). Tegmental nuclei include: SB, subcollicular tegmentum; CU, cuneiform nucleus; SA, sagulum; and CL, lateral commissural nucleus. Other nuclei at this level include: CG, central gray; DL, dorsal nucleus of lateral lemniscus. Two month-old cat. Orientation: L, lateral; V, ventral. Scale = 0.5 mm. (From Morest and Oliver (1984), reprinted by permission of John Wiley & Sons, Inc.)

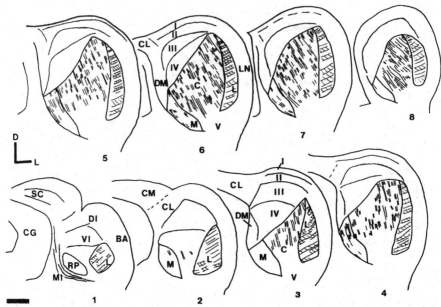

FIGURE 5.2. Subdivisions of the IC and layers of the central nucleus in Golgi-Cox preparation from a 6-week-old cat. Section 1 is at the rostral extreme of the IC where both the superior colliculus (SC) and IC are seen. At that level the rostral nucleus (RP), the nucleus of the brachium of the IC (BA), the dorsal (DI), medial (MI), and ventral (VI) intercollicular tegmentum are present. Both medial (CM) and lateral (CL) nuclei of the commissure are found. In more caudal sections, subdivisions are similar to that in Fig. 5.1 where the abbreviations are listed. Scale = 1.0 mm. (From Oliver and Morest (1984), reprinted by permission of John Wiley & Sons, Inc.)

In addition, recent immunocytochemical studies (Roberts and Ribak 1987a,b; Oliver and Beckius 1988) suggest that some disc-shaped cells may use GABA as a neurotransmitter. Undoubtedly, there are several types of disc-shaped cells, and the criteria to distinguish different functional types are still emerging.

Stellate cells (Fig. 5.4) represent the remaining neurons in the central nucleus (Oliver and Morest 1984). These are distinguished by their oval or spherical dendritic fields that often run in the mediolateral direction and cross several fibrodendritic laminae. Several types of stellate cells are found and differ in the size, shape, and branching pattern of the dendrites and in the frequency of dendritic appendages in Golgi studies (Rockel and Jones 1973a; Oliver and Morest 1984). Simple stellate cells have large, oval dendritic fields, while complex stellates have more medium-sized, spherical fields with more frequent branches and appendages. Small stellates have dendritic fields less than 200 μm in diameter. Many of these

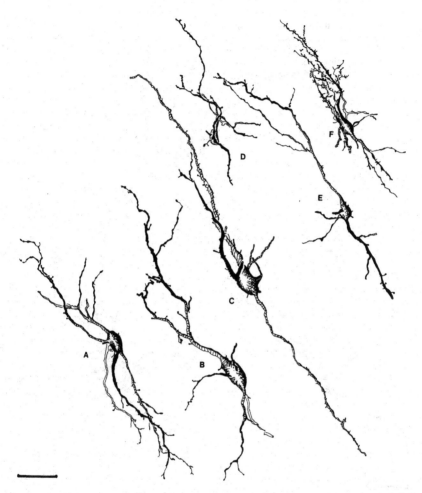

FIGURE 5.3. Disc-shaped cells in the central part of the central nucleus of the IC in Golgi-Cox material. In the transverse plane of section, these cells appear highly oriented. Large (B,C); medium-large (E); medium (A); and small (D,F). 7-week-old cat. Scale = 50 μm. (From Oliver and Morest (1984), reprinted by permission of John Wiley & Sons, Inc.)

cells have ascending projections (Oliver 1984a). Immunocytochemical data suggest that some of the stellate cells also use GABA as a neurotransmitter (Roberts and Ribak 1987a,b; Oliver, Nuding, and Beckius 1988). The largest cells in the central nucleus are GABA immunoreactive (Oliver and Beckius 1989; Oliver et al. in preparation), and these may correspond to large stellate cells.

Subdivisions within the central nucleus (Figs. 5.1, 5.2) are identified by the arrangement of the fibrodendritic laminae (Oliver and Morest 1984).

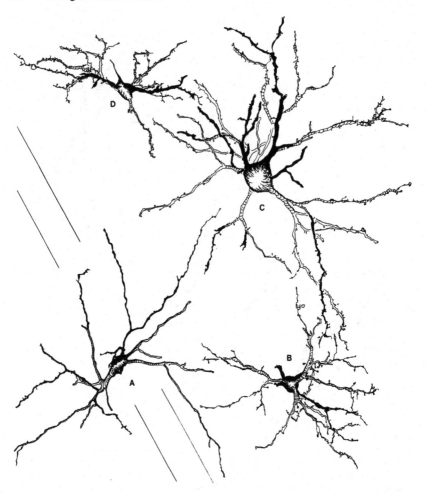

FIGURE 5.4. Stellate neurons in the central part of the central nucleus of the IC in Golgi-Cox material. These cells are from the same sections as those in Fig. 5.3. Simple stellate cells (A) have fewer dendritic branches and spines than more complex stellate cells (B,C,D). Cell C is an uncommon large stellate. Scale = 50 μm. (From Oliver and Morest (1984), reprinted by permission of John Wiley & Sons, Inc.)

These laminae may provide a structural basis for tonotopic organization (e.g., Merzenich and Reid 1974; Semple and Aitkin 1979). In *pars centralis* (*C*), the largest subdivision, the laminae are straight and run from ventrolateral to dorsomedial and within 15–20° of the sagittal plane. By comparison *pars medialis* (*M*) is smaller, but the laminae follow an orientation similar to those in *pars centralis*. The laminae near the medial edge of the IC are curved to follow the edge. In *pars lateralis* (*L*) the

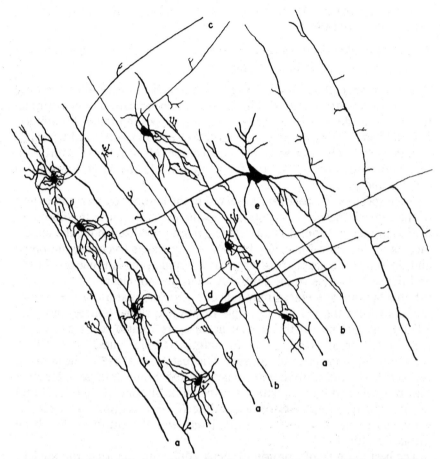

FIGURE 5.5. A simplified basic circuit of the central nucleus of the IC. Fibro-dendritic layers are composed of banded axons (a,b) and disc-shaped cells (c) that give off efferent axons. Stellate cells (d,e) have dendrites that cross the layers and have axons that give off many local collaterals but also project out of the IC. (From Oliver and Morest (1984), reprinted by permission of John Wiley & Sons, Inc.)

laminae are oriented at right angles to the laminae in the adjacent *pars centralis*. Thus, the laminae in *pars lateralis* are seen "on edge" in sections cut in the sagittal plane. *Pars ventralis* (*V*) is an interstitial zone that lacks distinct laminae because it contains few cells. It occupies the ventrolateral portion of the central nucleus where fibers of the lateral lemniscus enter.

In the cat, the packing density of neurons in Nissl stains also is used to distinguish these subdivisions (Oliver and Morest 1984). The *pars lateralis* and *medialis* are more densely packed than the *pars centralis*.

The low packing density in the central region is probably related to the larger cells found there.

2.1.1.2 Cellular Organization of the Central Nucleus Is Stable Across Species, Including Humans

When similar methods and criteria have been used, the cellular anatomy and the overall organization of the IC form a similar pattern. For example, the central nucleus contains the same neuron types in mouse (Meininger, Pol, and Derer 1986), rat (Faye-Lunde and Osen 1985), new world monkey (FitzPatrick 1975), and humans (Geniec and Morest 1971). Moreover, the subdivisions of the IC are found in roughly the same location in rodents and primates.

Some species differences are observed in the laminar pattern and extent of the central nucleus. For example, in the cat and mouse the fibrodendritic laminae are found in the same orientation throughout the subdivisions of the central nucleus. In contrast, primates (including humans) and the rat have a laminar pattern that is simpler, the laminae are flatter, and the subdivisions of the central nucleus are not clear.

Some species differences may reflect a shift or rotation of the central nucleus within the IC. For example in the tree shrew (Oliver and Hall 1978a), the laminae of the central nucleus are rotated to be parallel to the frontal stereotaxic plane. In the human (Geniec and Morest 1971), the amount of rotation is not so great. The rostrocaudal axis of the laminae is about 45° to the sagittal plane instead of 15°. Some of these differences may be explained by the relative expansion or diminution of other brain regions and resulting displacement of the IC. For example, the superior colliculus in the tree shrew is very large, and the entire IC has been displaced caudally.

The best example of a profound species difference is the central nucleus of the mustache bat (Zook et al. 1985). This may reflect a unique specialization in behavior for echolocation and doppler compensation within the 60 kHz region. Zook and co-workers (1985) suggested that fibrodendritic laminae in the 60–64 kHz region of the central nucleus are hypertrophied as compared to the normal mammalian laminar pattern.

2.1.1.3 All Types of Afferents Project to the Central Nucleus

Ascending pathways from the cochlear nucleus to the central nucleus of the IC fall into several broad categories—direct, indirect, and multisynaptic. All of the projection neurons in the cochlear nucleus receive inputs from type I spiral ganglion cells in the cochlea (see reviews by Moore and Osen 1979; Cant and Morest 1984; Chapters 2 and 3, this volume). However, the auditory nerve terminates on different cell types within the cochlear nucleus, and these cell types give rise to different efferent pathways. Thus, the pathways to the central nucleus are distinguished by both the cells of origin and the trajectory of their pathways.

Direct, monaural pathways to the central nucleus (Fig. 5.6) emerge from the stellate cells in the ventral cochlear nucleus and the fusiform and giant cells of the dorsal cochlear nucleus (Osen 1972; Adams 1979; Cant 1982; Ryugo and Willard 1985). Each of these projections terminates primarily in the contralateral central nucleus (Warr 1966; 1969; 1972; Van Noort 1969; Beyerl 1978; Roth et al. 1978; Brunso-Bechtold, Thompson, and Masterton 1981; Ryugo, Willard, and Fekete 1981; Zook and Casseday 1982; Willard and Martin 1983). The projections (Fig. 5.6) are "banded" (Oliver 1984b, 1987). Some species differences may exist (see, for example, the bat, Zook and Casseday 1987). In general, banded inputs terminate in a tightly focused manner and exhibit a topographic organization that is directly related to the tonotopic organization of the IC (Merzenich and Reid 1974; Oliver 1984b, 1987; Shneiderman, Oliver, and Henkel 1988).

Indirect, binaural pathways (Fig. 5.7) to the central nucleus are via the superior olive. These pathways originate from spherical and globular bushy cells in the anteroventral cochlear nucleus (e.g., Warr 1966, 1972; Tolbert, Morest, and Yurgelun-Todd 1982; Cant and Casseday 1986). Axons from spherical bushy cells terminate in the medial superior olive (MSO) bilaterally and in the lateral superior olive (LSO) ipsilaterally. The LSO also receives inputs from the globular bushy cells on the contralateral side via a synapse in the medial nucleus of the trapezoid body (MNTB). Periolivary nuclei receive inputs from these and possibly other cells in the cochlear nuclear complex. (For additional details on the anatomy and connectivity of cells in the cochlear nucleus and superior olivary complex, please refer to Chapters 2 and 3, this volume.) The projection neurons of the superior olive send axons to the IC (e.g., Browner and Webster 1975; Roth et al. 1978; Brunso-Bechtold, Thompson, and Masterton 1981; Willard and Martin 1983). Projections from LSO (Fig. 5.7) and periolivary nuclei to the central nucleus are bilateral, while the projection from the MSO to the central nucleus is most unilateral (Elverland 1978; Roth et al. 1978; Brunso-Bechtold et al. 1981; Glendenning and Masterton 1983; Zook and Casseday 1983; Shneiderman et al. 1988). Both MSO and LSO projections to IC are banded, tonotopic projections (Henkel and Spangler 1983; Casseday and Covey 1987; Shneiderman and Henkel 1987; Zook and Casseday 1987). In contrast, the periolivary projections to IC are "diffuse" (Adams 1983; Henkel and Spangler 1983). Diffuse projections terminate widely within the IC.

Multisynaptic pathways (Figs. 5.8 left and 5.9) to the IC include a synapse in the dorsal (DNLL), the intermediate (INLL), or the ventral nuclei of the lateral lemniscus (VNLL). These nuclei receive inputs from both the cochlear nuclei and superior olivary complex (e.g., Warr 1966, 1972; Van Noort 1969; Glendenning et al 1981; Zook and Casseday 1982; Shneiderman, Oliver, and Henkel 1988). The DNLL receives bilateral inputs from both cochlear nuclei and superior olivary nuclei. The pro-

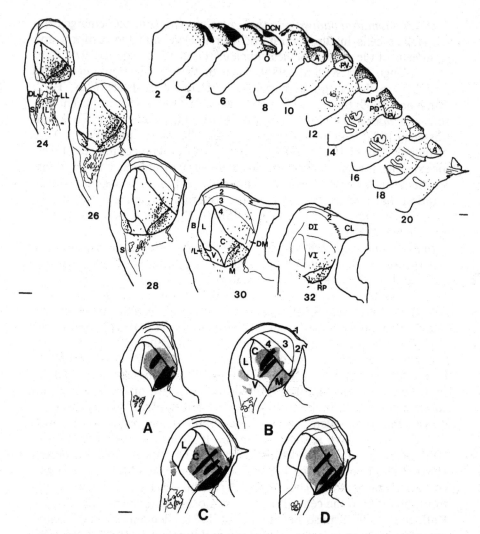

FIGURE 5.6. Axonal projections from the dorsal cochlear nucleus to the IC. Top panel shows an injection of 3H-leucine in the dorsal cochlear nucleus (DCN) and the projections to the contralateral IC. Axons terminate as bands. Labeling in posteroventral cochlear nucleus (A,O) and anteroventral cochlear nucleus (A, AP, PD, PV) also is found. Bottom panel shows a schematic of the bands of fibers from several cases. Bands terminate primarily in the central nucleus (Abbreviations as in Fig. 5.1 and 5.2). Scale bars = 1 mm. (From Oliver (1984b), reprinted by permission of John Wiley & Sons, Inc.)

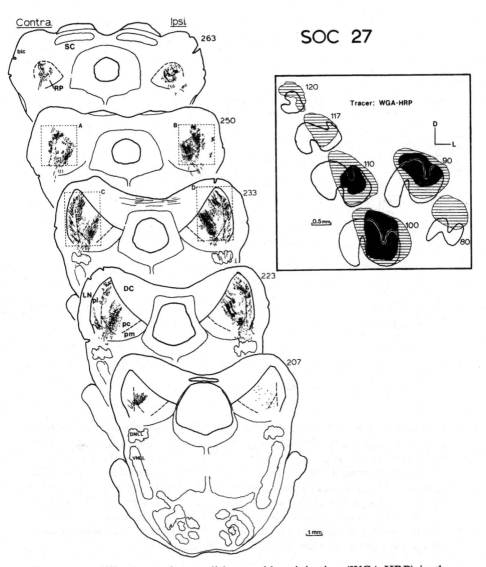

FIGURE 5.7. Wheat germ-horseradish peroxidase injection (WGA-HRP) in the lateral superior olive (see inset) and projections to the IC. Labeling in the IC is banded and confined to the central nucleus. (BIC; brachium of superior colliculus; DC, dorsal cortex; DNLL, dorsal nucleus of lateral lemniscus; LN, lateral nucleus; pc, *pars centralis* of central nucleus; pl, *pars lateralis*; pm, *pars medialis*; RP, rostral pole of IC; SC, superior colliculus; VNLL, ventral nucleus of lateral lemniscus). (From Shneiderman and Henkel (1987), reprinted by permission of John Wiley & Sons, Inc.)

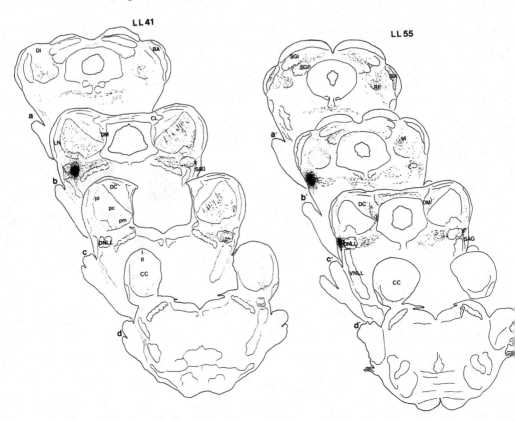

FIGURE 5.8. Projections from the dorsal nucleus of the lateral lemniscus (left) and the sagulum (right). (BA; brachium of superior colliculus; CC, caudal cortex; CL, lateral nucleus of the commissure; DC, I, II, dorsal cortex; DI, dorsal inter-collicular tegmentum; DM, dorsomedial nucleus; DNLL, dorsal nucleus of lateral lemniscus; LN, lateral nucleus; pc, *pars centralis* of central nucleus; pl, *pars lateralis*; pm, *pars medialis*; RP, rostral pole of IC; SAG, sagulum; SC, superior colliculus; SGI, *stratum griseum imtermediali*; SGP, *stratum griseum profundum*; VNLL, ventral nucleus of lateral lemniscus). (From Henkel CK, Shneiderman A (1988), reprinted by permission of John Wiley & Sons, Inc.)

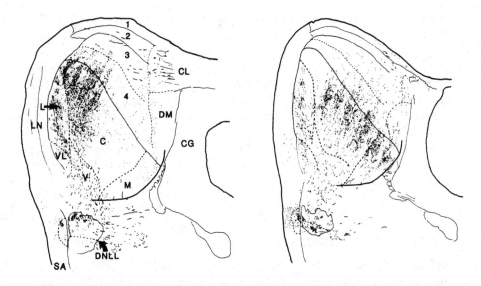

FIGURE 5.9. Enlarged view of banded afferent from dorsal nucleus of the lateral lemniscus (DNLL) as they terminate in the inferior colliculus. Scale bar = 1 mm. Central nucleus (M, medial; C, central; L, lateral; V, ventral parts); Dorsal cortex (layers 1–4); and Paracentral nuclei (DM, dorsomedial; VL, ventrolateral; LN, lateral; and CL, lateral commissural nucleus); SA, sagulum; CG, central gray.

jection of DNLL (Figs. 5.8 and 5.9) to the IC is bilateral and banded (Kudo 1981; Tanaka et al. 1985; Zook and Casseday 1987; Shneiderman, Oliver, and Henkel 1988). In the case of the VNLL, the inputs arise primarily from the contralateral cochlear nuclear complex. A diffuse projection arises, in turn, from the VNLL to the ipsilateral IC.

Other projections to the central nucleus include commissural connections (e.g., Aitkin and Phillips 1984). In most species, except possibly in primates, the central nucleus does not receive a projection from the telencephalon. (See discussion of dorsal cortex connections below.)

Subdivisions of the central nucleus may receive different sets of inputs. There may be functional differences between subdivisions of the central nucleus that reflect differences in their connections. A projection that is dense and tonopically organized in one subdivision may be absent or possibly reduced to a small number of poorly organized fibers in another subdivision. For example, the low-frequency AVCN is found to project bilaterally to the lateral, low-frequency part of the central nucleus, mostly the *pars lateralis* (Oliver 1987); whereas, the middle and high-frequency parts of the AVCN terminate only in the contralateral IC. The MSO provides a similar example. Most of the MSO encodes low or middle frequencies and projects to the ipsilateral *pars centralis* and *pars lateralis*.

However, the ventral, high-frequency, part of MSO provides only a sparse projection to the IC. Anterograde transport experiments (Henkel and Spangler 1983) have failed to reveal a projection to the *pars medialis*. But, retrograde transport experiments from *pars medialis* (Adams 1979; Brunso-Bechtold, Thompson, and Masterton 1981; Shneiderman, Oliver, and Henkel 1988) did show that a small number of cells are labeled bilaterally. The projection from the DCN provides a third example. A DCN projection to the *pars lateralis* of the central nucleus is absent in anterograde experiments (Oliver 1984b). However, retrograde experiments (Oliver 1987) suggest that a small number of cells from DCN provide sparse afferents.

2.1.2   Cortex of the IC

2.1.2.1   Cellular Organization

The cortex forms the dorsal and caudal surface of the IC (Figs. 5.1, 5.2) and is evident in all species studied with the Golgi methods (Geniec and Morest 1971; FitzPatrick 1975; Oliver and Hall 1978a; Morest and Oliver 1984; Faye-Lunde and Osen 1985; Meininger, Pol, and Derer 1986). The broad layers of the cortex are produced by the intrinsic axonal plexuses rather than the arrangement of the dendrites or the afferents (Ramón y Cajal 1911; Morest 1966; Geniec and Morest 1971; Morest and Oliver 1984). The predominate neurons are stellate cells without obvious dendritic orientations. In contrast to the central nucleus, the cortex lacks narrow fibrodendritic laminae. Although most of the neurons in the cortex are stellate cells, some cells with disc-shaped properties are also found (Oliver et al. 1991). Afferent axons in the dorsal cortex run perpendicular to the layers, not parallel as in the central nucleus.

The dorsal cortex can be divided into four layers in Golgi preparations (Morest 1966; Geniec and Morest 1971; Morest and Oliver 1984). The fibrous capsule, *layer I*, contains a few cells with flattened dendritic fields that parallel the surface of the IC. *Layer II*, in contrast, contains many small to medium-sized stellate cells. These two layers correspond to most of the "pericentral" region of the inferior colliculus (Berman 1968; Rockel and Jones 1973c). *Layer III* is delineated by a prominent fascicle of fibers dorsolaterally where fibers enter and exit the brachium of the IC. At rostral levels, the medial part of layer III is lateral to the commissure of the IC and is eventually replaced by the crossing fibers. A variety of cell types are found in layer III. *Layer IV*, that borders the edge of the central nucleus, is filled with stellate cells of all sizes. Some cells are oriented rostrocaudally/dorsoventrally or mediolaterally/horizontally. Mediolaterally oriented cells follow the border of the central nucleus. Layers III and IV largely coincide with the dorsomedial part of the central nucleus as defined by Rockel and Jones (1973a).

The *caudal cortex* forms the caudal surface of the IC. It is thinner but more elaborate than the superficial two layers of the dorsal cortex. Hence, it is not equivalent to the "pericentral" region. Instead, it may be a compressed version of the dorsal cortex. Large stellate cells are found ventrally in the caudal cortex and could be similar to large cells in the dorsal cortex.

2.1.2.2 Species Differences in the Cortex of the IC May Be Related to the Telencephalon

The size of the dorsal cortex in the IC may be correlated with the size of the auditory neocortex (Geniec and Morest 1971; FitzPatrick 1975; Oliver and Hall 1978; Morest and Oliver 1984; Faye-Lunde and Osen 1985; Meininger, Pol, and Derer 1986). The telencephalon provides the single largest source of inputs to the dorsal cortex (see below). So, an increase in the size of neocortex may produce a proportional increase in the telencephalic projections to the IC. However, there may be other factors that are related to reported differences in the size of the IC's cortex.

One specious source of differences may be the anatomical planes used to section the IC. Because of species differences in the orientation of the skull and/or body axis, the IC sectioned in a stereotaxic frontal or horizontal plane may be quite different from species to species. Sectioning effects the apparent thickness of the cortex as well as its designation as dorsal, caudal, etc. It would be preferable to have all species studied in equivalent planes, such as the anatomical transverse plane, perpendicular to the long axis of the brainstem and the floor of the fourth ventricle.

2.1.2.3 Dorsal Cortex Receives Neocortical Projections Plus Other Inputs

The dorsal cortex is distinguished from the central nucleus by its major descending input from the telencephalon. Both primary and nonprimary auditory cortex project to the cortex of the IC (Diamond, Jones, and Powell 1969; FitzPatrick and Imig 1978; Oliver and Hall 1978b; Andersen, Snyder, and Merzenich 1980b; Willard and Martin 1983; Faye-Lund 1985; Coleman and Clerici 1987; Gonzalez-Hernandez et al. 1987). The projections from the neocortex are banded and tonotopically organized, similar to those of the lateral lemniscus to the central nucleus. Projections from the primary auditory cortex extend into layer IV of the dorsal cortex (Oliver and Hall 1978b; Andersen, Snyder, and Merzenich 1980b; Luethke, Krubitzer, and Kaas 1989). In the squirrel monkey, they also continue into the central nucleus (FitzPatrick and Imig 1978). Projections originating outside of primary cortex terminate more superficially in the dorsal cortex.

A second distinguishing feature of the dorsal cortex is the absence of some lemniscal inputs. Projections from the MSO and LSO (Fig. 5.7) to the dorsal cortex are absent (Brunso-Bechtold, Thompson, and Masterton 1981; Henkel and Spangler 1983; Shneiderman and Henkel 1987; Shneiderman, Oliver, and Henkel 1988). This may mean that the dorsal cortex does not receive direct binaural inputs. Other lemniscal afferents project to the layers of the dorsal cortex incompletely. For example, the DCN (Fig. 5.6, top) and DNLL (Figs. 5.8 and 5.9) project only to layer IV and the AVCN projects only to layers II to IV (Oliver 1984b, 1987; Shneiderman, Oliver, and Henkel 1988). Even when the projections are present, they are sparse when compared to the projections to the central nucleus. In Golgi material, the lemniscal afferents are observed to give off fine collaterals that extend into the dorsal cortex while the main axonal arbors terminate in the central nucleus (Oliver and Morest 1984).

A third category of afferents to the cortex of the IC include projections from within the midbrain (Henkel and Shneiderman 1988). Nucleus sagulum, a tegmental region lateral to the DNLL, sends a heavy projection to the caudal cortex and to the superficial layers of the dorsal cortex (Fig. 5.8, right). Additional afferents to these same cortical regions arise from the retina (Itaya and Van Hoesen 1982). The superficial cortex also contains axons immuno-stained for cholecystokinin, substance P, neurotension, and metenkephalin (Adams 1985; Adams and Mugnaini 1985), but the sources of these fibers remain undetermined.

2.1.3   Paracentral Nuclei

The paracentral subdivisions in the IC surround the central nucleus and dorsal cortex. Although many of the paracentral nuclei can be distinguished by their inputs and outputs, a complete description of the connections for each nucleus is not available.

2.1.3.1   Lateral Nucleus

The lateral nucleus (Figs. 5.1, 5.2, *LN*) follows the lateral surface of the IC (Morest and Oliver 1984). This nucleus contains a mixture of tectal and tegmental cells. Since the efferent fibers of the IC enter the lateral nucleus to form the *brachium of the IC*, the fibrous capsule laterally is easily distinguished from the more cellular dorsal cortex. The lateral nucleus corresponds, in part, to the "external nucleus" (Berman 1968; Rockel and Jones 1973a).

The somatosensory system contributes a significant number of afferents to the lateral nucleus. The spinal cord, the dorsal column nuclei, and the somatosensory association cortex each project to the lateral nucleus (Schroeder and Jane 1971; Oliver and Hall 1978b; RoBards 1979; Aitkin, Kenyon, and Philpott 1981; Morest and Oliver 1984; Coleman and Clerici 1987). Projections from auditory cortex also pass through the lateral nucleus and may terminate there. Unlike the central nucleus and cortex, the lateral nucleus does not receive direct projections from the lateral lemniscus (e.g., Rockel and Jones 1973a; Oliver and Morest 1984). Nonetheless, the lateral nucleus receives inputs from the central nucleus as the efferent axons pass through into the brachium (Kudo and Niimi 1980). These intracollicular connections may be important for auditory processing within the lateral nucleus.

2.1.3.2   Dorsomedial Nucleus

The dorsomedial nucleus (Figs. 5.1, 5.2 *DM*) is between the central gray, central nucleus, and deep dorsal cortex (Morest and Oliver 1984). It is most prominent rostrally and is distinguished by large multipolar cells. This nucleus is not equivalent to the "dorsomedial part of the central nucleus" described by Rockel and Jones (1973a). Although the dorsomedial nucleus is identified in the human (Geniec and Morest 1971), it is regarded as more of a tegmental nucleus in that species. Afferents to the dorsomedial nucleus arise from the lateral lemniscus and auditory cortex (Morest and Oliver 1984). Projections from the sagulum and the nuclei of the lateral lemniscus (Fig. 5.8) also enter the dorsomedial nucleus (Whitley and Henkel 1984; Henkel and Shneiderman 1988). In addition, axons enter via the medial intercollicular tegmentum, but the source of these axons is not known (Morest and Oliver 1984).

2.1.3.3   Nucleus of the Rostral Pole

The nucleus of the rostral pole (Fig. 5.2, *RP*) is rostral to the body of the IC and the medial part of the central nucleus in particular (Morest and Oliver 1984). Many high-frequency afferents terminate in the rostral nucleus (Figs. 5.6–5.8) including those from the contralateral cochlear nuclear complex (Oliver 1984b, 1987). However, these projections do not appear to be banded. The rostral pole contains many neurons with projections to the superior colliculus (e.g., Edwards et al. 1979). Some of the neurons in the rostral pole are immunopositive for somatostatin (Spangler and Morley 1987).

2.1.3.4   Ventrolateral Nucleus

The ventrolateral nucleus (Figs. 5.1, 5.2, *VL*) contains a mixture of large and small stellate cells and is ventral to the *pars lateralis* of the central nucleus (Morest and Oliver 1984). This nucleus receives nontonotopic inputs from the lemniscus including projections from the cochlear nucleus (Oliver 1984b).

2.1.3.5   Commissural Nucleus

The commissural nucleus (Figs. 5.1, 5.2, *CL*) contains cells that lie amongst the fibers of the commissure (Herrara, Smith-Agreda, and Morera 1984; Morest and Oliver 1984; Herrara, Del Campo, and Smith-Agreda 1987; Herrara, Del Campo, and Orts 1988). The nucleus does not receive lemniscal afferents. Presumably the neurons receive synaptic inputs from the fibers in the commissure.

2.1.4 Tegmental Nuclei Separate the IC and the Superior Colliculus (SC)

Tegmental nuclei separate the tectal regions of the midbrain. Between the IC and SC, the *medial, lateral, ventral, and dorsal intercollicular tegmentum* (Fig. 5.2, *VI, MI, DI*) are found (Morest and Oliver 1984). To an extent, these regions surround the rostral pole nucleus. The *cuneiform nucleus* (Fig. 5.1, *CU*) and the *subcollicular nucleus* (Fig. 5.1, *SB*) are tegmental nuclei directly beneath tectal components of the IC. Finally, the *nuclei of the brachium of the IC* (Fig. 5.2, *BA*) also are considered tegmental nuclei.

2.2 Synaptic Domains in the IC

Although the subdivisions of the IC may have different combinations of afferents, *each nucleus may contain more than one functional type of neuron*. Thus, the subdivisional organization may be inadequate to explain the responses of single cells. In the central nucleus, for example, a number of binaural response types are reported. (For recent reviews, see Irvine 1986 and Vol. 2 of this series.) In addition, similar response types are segregated (Roth et al. 1978; Semple and Aitkin 1979). These findings suggest that small groups of neurons may receive similar inputs while other neurons in the same subdivision receive different inputs. If the small groups of disc-shaped cells in separate layers each receive a unique set of inputs, this could provide a structural basis for heterogeneous responses to sound.

2.2.1 Bands of Afferents as Demonstrated in Experimental Material

Lesions of hindbrain nuclei or their ascending tracts first demonstrated the banded inputs to the central nucleus of the cat (Morest 1964; Goldberg and Moore 1967; Rockel and Jones 1973c; Oliver and Morest 1984). A bilateral banded pattern of degeneration is produced by lesions in the superior olivary complex. Bands containing thick, degenerating fibers, oriented parallel to the fibrodendritic laminae, are separated by areas containing degenerating fine fibers. Goldberg and Moore (1967) reported that a lesion of the DNLL produced a fine, diffuse pattern of degeneration on the ipsilateral side, and a more banded, coarse degeneration on the contralateral side.

Studies using anterograde transport methods reveal the specific banded projections to the central nucleus from the cochlear nuclei, superior olivary complex, and DNLL (Kudo 1981; Henkel and Spangler 1983; Oliver 1984b, 1987; Shneiderman and Henkel 1987; Shneiderman, Oliver, and

Henkel 1988). Bands labeled with these methods are usually 200 μm wide. (See, for example Figs. 5.6–5.8.) However, bands from AVCN are twice that size. In most cases, distinct, heavily labeled bands are separated by interband areas of light labeling. Although the banded afferents are quite distinct from diffuse projections, some projections are more sharply banded than others. This is particularly evident for some bilateral projections (Shneiderman and Henkel 1987; Shneiderman, Oliver, and Henkel 1988). The number of labeled bands is determined by the size of the injection site. Small injections label one or two bands, while larger injections label more bands.

Bands in the IC also are demonstrated in experiments that label metabolically active regions with 2-deoxyglucose (Ryan, Woolf, and Sharp 1982; Servière, Webster, and Calford 1984; Webster, Servière, and Brown 1984; Webster et al. 1984). Tonal stimulation of the experimental animal produces bands in the IC that resemble the bands in anatomical experiments. Since the bands are likely to be produced by activation of axonal endings (Nudo and Masterton 1986), their appearance resembles the afferents to the IC rather than the location of the disc-shaped cells in the central nucleus.

2.2.2 Excitatory and Inhibitory Bands

Banded axons whose synapses depolarize neurons in the IC may form *excitatory bands*. This may be the most common mode of termination in the IC for axons from the lower brainstem. Direct evidence of excitation was obtained when the dorsal acoustic stria was electrically stimulated and monaural units in the central nucleus were activated (Semple and Aitkin 1980). Indirect evidence from anatomical studies shows that the synaptic connections made by the axons from DCN and AVCN in the IC are consistent with excitatory activity (Rockel and Jones 1973b; Oliver 1984b; 1985; 1987). Axonal endings from the cochlear nucleus contain round synaptic vesicles (Fig. 5.10) and make asymmetric synaptic junctions primarily on dendrites and cell bodies. These endings could account for ¼ to ⅓ of the endings with round synaptic vesicles in the central nucleus. Endings from the superior olivary complex have not been studied at the EM level, but many are expected to possess a similar morphology.

Inhibitory bands may be formed by banded afferents that use inhibitory neurotransmitters. Axons from the DNLL and the ipsilateral LSO are likely sources. Most cells in the DNLL contain gamma-aminobutyric acid (GABA) (Adams and Mugnaini 1984; Thompson, Cortez, and Lam 1985; Moore and Moore 1987; Roberts and Ribak 1987a). Since the DNLL relays monaural information from the cochlear nucleus and binaural information from the superior olivary complex (Shneiderman, Oliver, and Henkel 1988), its banded, tonotopic inputs to the IC could be related to

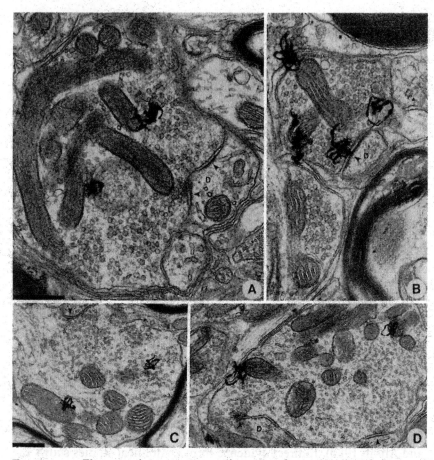

FIGURE 5.10. Electron microscope autoradiographs of axonal endings from dorsal cochlear nucleus in the inferior colliculus labeled with 3H-leucine. Silver grains (black) show radioactive endings whose cell bodies are in the cochlear nucleus. These endings contain round synaptic vesicles and make asymmetric synaptic contacts (*arrowheads*) on dendrites (*D*). Vesicles in *A* are large and round; vesicles in *B-D* are small and round. Scale bars = 0.5 μm. Size of A = B; C = D. (From Oliver (1985) reprinted by permission of John Wiley & Sons, Inc.)

both types of information. The morphology of the endings from DNLL with pleomorphic vesicles and symmetrical synapses (Fig. 5.11) is consistent with an inhibitory function (Shneiderman and Oliver 1989). This projection from DNLL contributes about ⅓ of the endings with pleomorphic vesicles in the central nucleus.

The projection from the ipsilateral LSO also may form inhibitory bands in the IC. About half of the cells in the LSO that project to the ipsilateral IC are immunoreactive with antibodies to glycine (Hutson 1988; Saint

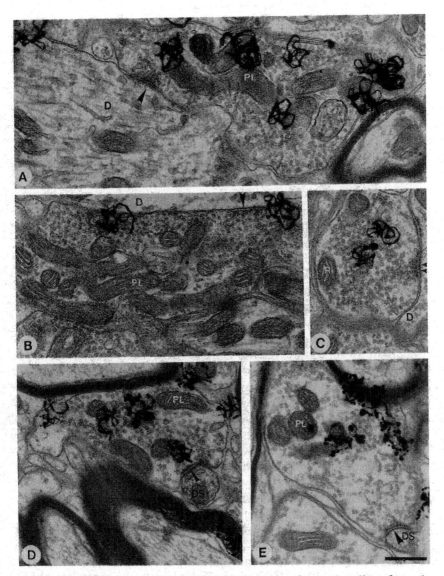

FIGURE 5.11. Electron microscope autoradiographs of axonal endings from the dorsal nucleus of the lateral lemniscus that synapse in the contralateral inferior colliculus. As in Fig. 5.10, labeled endings are indicated by silver grains. However, in this case most of the endings contain pleomorphic synaptic vesicles (*PL* in *A*, *B*, *D*, *E*) and make symmetrical synaptic contacts (*arrowhead*) on dendrites (*D*) or dendritic spines (*DS*). A small number of endings from the dorsal nucleus contained round vesicles (*R* in *C*). (From Shneiderman and Oliver (1989), reprinted by permission of John Wiley & Sons, Inc.)

Marie et al. 1988). The remaining cells are unreactive with antibodies against the putative inhibitory transmitters. Since both the ipsilateral and contralateral projections of the LSO are banded, it is possible that the axons from the ipsilateral projection contribute inhibitory bands while the axons from the contralateral projection contribute excitatory bands.

2.2.3   Synaptic Domains May Be Created by Overlapping Bands

The evidence for banded afferents to the IC suggests that different afferents will overlap to form *synaptic domains*. These unique functional zones require that bands from some sources overlap, while others do not. There are too many inputs to the central nucleus for each to have a completely separate target. When bands in the IC are observed in degeneration material after a unilateral lesion, the bands on one side of the IC appear to interdigitate with those on the opposite side (Oliver and Morest 1984). In a more recent experiment, the LSO on each side was injected with two different anterograde tracers (Shneiderman and Henkel 1987). The projection from the ipsilateral LSO is largely complementary to the projection from the contralateral LSO. These data suggest that the potentially excitatory and inhibitory bands from the LSO do not overlap.

The three-dimensional (3-D) shape of the banded afferents may be an important aspect of the synaptic domains. In single experimental cases, the bands appear to be parts of longitudinal sheets that extend in the rostrocaudal dimension. However, there has been little study of their morphology in 3-D. On the basis of current data, one may speculate that each synaptic domain does not occupy an entire fibrodendritic lamina. For example, bands may not extend throughout the entire rostrocaudal length of the central nucleus. Some bands may be located more dorsally along a fibrodendritic lamina than another. Since bands from different sources may vary in their 3-D morphology, it is not possible to predict *the shape* of the synaptic domains. It is also not possible to predict precisely *how many* banded afferents will overlap in each domain.

Synaptic domains could provide a structural basis for some of the most common neural responses in the IC (see review by Irvine 1986, and Volume 2 of this series). One may predict at least three types of synaptic domains that could be related to particular inputs: the contralateral LSO type, the MSO type, and the monaural type. Within the domain dominated by the projection from the contralateral LSO, one might expect to encounter cells that are excited by sounds in the contralateral ear and suppressed by sounds in the ipsilateral ear (contralateral-E, ipsilateral-I; the EI response). Within the domain dominated by the projection from MSO, one may find cells that are excited by sounds in either ear (the EE response). Within the monaural domain, there may be cells dominated by monaural inputs from the contralateral cochlear nuclei (the EO response).

Neural responses in the IC may not be simple replicates of those in the lower brainstem (e.g., Irvine 1986; Kuwada, Stanford, and Batra 1987), and this has implications for the neural substrate. The synaptic domains in the IC are probably based on combinations of two or more inputs. For example, an input that is compatible with the contralateral LSO input is the potentially inhibitory projection from the contralateral DNLL. These inputs could form a single synaptic domain. Another example are the inputs from the ipsilateral AVCN to the low-frequency IC and the ipsilateral MSO. Ipsilateral excitation from AVCN is compatible with the EE response predicted for a domain created by MSO axons.

2.2.4   Disc-Shaped Cells Are Inside a Single Synaptic Domain

The synaptic domain hypothesis implies that different cell types within the IC will respond predictably to the synaptic inputs in each type of domain. Most cells in the central nucleus are disc-shaped and have relatively narrow dendritic fields compared to the banded afferents (50 μm vs 200 μm). These dendritic fields may easily fit within a single synaptic domain. Several disc-shaped cells, side by side, may sit in the same domain, and nearby similar cells may reside in another domain. Within a single domain, the disc-shaped cells may receive large numbers of synaptic inputs on both dendrites and, in some cases, cell bodies (Oliver 1984b, 1987; Shneiderman and Oliver 1989). The heavy concentration of similar synaptic inputs on the cells within the synaptic domain suggests that these cells within the same domain would have similar binaural response properties as well as similar tuning (bandwidth) characteristics. The cells in an adjacent domain could have different response properties.

2.2.5   Stellate Cells Receive Inputs from More Than One Synaptic Domain

The stellate cells in the central nucleus and dorsal cortex of the IC may have a different synaptic organization than the disc-shaped cells. Stellate cells may synthesize information from several synaptic domains. Since the dendritic fields are oriented perpendicular to the fibrodendritic laminae, each stellate cell may receive inputs from more than one synaptic domain. This structure suggests that different parts of the dendritic field may receive inputs from different populations of synaptic endings. Thus, the response properties of the stellate cells, in general, may not be dominated by a single brainstem nucleus. Neurons in the IC with nonlinear phase-frequency functions may reflect such convergence (Kuwada, Stanford, and Batra 1987; Batra, Kuwada, and Stanford 1989; Stanford et al. in preparation).

Although synaptic domains may provide a global structure related to response properties in the IC, other differences between cells could exist. For example, some cells in the central nucleus receive many axosomatic synapses while others do not (Oliver 1984a). In some cases, the axosomatic inputs originate from the cochlear nuclei (Oliver 1984b, 1987). These and other differences in synaptic organization could be related to the complex interactions of excitatory and inhibitory inputs observed in intracellular recordings in the IC (Nelson and Erulkar 1963; Kuwada et al. 1980; Oliver et al. 1986; Kuwada et al. in preparation).

2.3 Functional Organization of the Efferent Neurons May Reflect Both Synaptic Domains and Intrinsic Connections

Intrinsic connections within subdivisions of the IC also may contribute to the inputs present in the synaptic domains. Axons intrinsic to the IC also may provide connections between subdivisions of the IC. Both aspects of intrinsic connections are important for the organization of the efferent pathways.

2.3.1 Cells in the IC Have Complex Local Axonal Collaterals

Many, if not all, of the neurons in the IC give off axonal collaterals within the IC. Recent studies (Kuwada et al. 1980; Oliver et al. 1991) using intracellular injections of HRP are consistent with earlier data from Golgi impregnations (Morest 1964, 1966; Geniec and Morest 1971; Rockel and Jones 1973a; Oliver and Morest 1984). Most axons give off collaterals with a widespread distribution within the IC that lack a specific orientation. For example, one injected cell (Fig. 5.12) is a stellate cell from the central nucleus whose dendrites could cross several fibrodendritic laminae. This neuron possesses an axonal arbor with over 2000 terminal boutons. Besides the local collaterals, the main stem of its axon enters the brachium of the IC. Thus, some efferent cells in the IC may contribute heavily to the synaptic plexus within the IC. Most IC neurons observed do not have an axon this complex although the widespread distribution is typical of stellate cells in the central nucleus and most cells in the dorsal cortex. In contrast to these cells, injected disc-shaped cells from the central nucleus have highly oriented axons (Oliver et al. 1991). These axons are rostrocaudally oriented and parallel the dendritic orientation.

A complex network of local axons may contribute to neural processing within the IC. Within the central nucleus, axons may establish connections within a synaptic domain or contribute to several domains. Within the dorsal cortex, the combination of banded afferents and widespread local collaterals may form a complex reticular network. Since recent studies (Roberts and Ribak 1987b; Oliver, Nuding, and Beckius 1988; Oliver and Beckius 1989) showed that cells in the IC may use GABA as a neurotransmitter, some of these local axonal collaterals may be inhibitory.

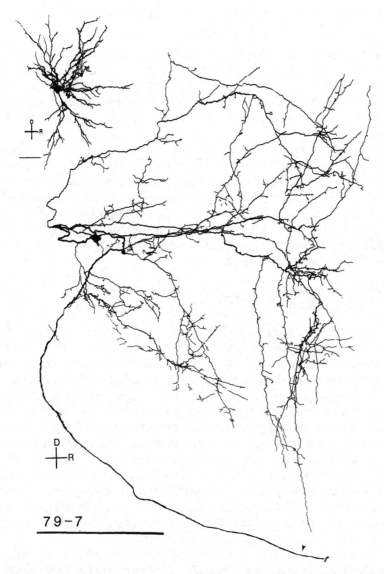

FIGURE 5.12. A stellate cell from the central nucleus of the inferior colliculus with an axon that terminates locally and also projects to another part of the nervous system. The main part of the axon (*arrowhead*) courses laterally to enter the brachium of the inferior colliculus. Scale at upper left = 100 μm. Scale at lower left = 500 μm. (After Oliver et al. (1991), reprinted by permission of John Wiley & Sons, Inc.)

2.3.2   Local Axons from the IC Form Intracollicular and Intercollicular Connections

Axons from neurons in the IC may form important connections between subdivisions. For example, fibers from the central nucleus project to the lateral nucleus as they travel towards the brachium (Andersen et al. 1980; Kudo and Niimi 1980). Some neurons in the central nucleus also send axons to the opposite colliculus via the commissure (Aitkin and Phillips 1984).

Axons from the IC also provide important connections to adjacent midbrain regions outside of the IC proper. Portions of the midbrain tegmentum, such as the cuneiform nucleus and the sagulum, may receive auditory inputs from neurons in the IC (Henkel and Shneiderman 1988; Oliver et al. 1991) and from neurons in the auditory cortex (Oliver and Hall 1978b). The deep layers of the superior colliculus also receive inputs from the lateral nucleus, superficial dorsal cortex, and rostral pole nucleus of the IC (Edwards et al. 1979).

2.3.3   Parallel Pathways Ascend from the IC to the Medial Geniculate Body

Multiple cell types within a single subdivision provide parallel ascending pathways to the thalamus. Both disc-shaped and stellate cells project to the medial geniculate body (Oliver 1984a). In the central nucleus, large, medium, and small disc-shaped neurons are labeled after injections of HRP in the medial geniculate body. It is not yet evident whether the axons of these cells terminate within the same subdivision of the thalamus. It is also not known to what extent the information from synaptic domains in the IC is transferred to the thalamus. Although binaural and monaural bands in the ventral division of the medial geniculate (Middlebrooks and Zook 1983) and primary auditory cortex (Imig and Adrián 1977) have been mapped using interaural intensity differences, their connections to the IC have not been determined. Multiple cell types with ascending connections could allow several types of information to be transmitted from the central nucleus of the IC.

Several types of efferent pathways may transmit auditory information from the IC to the thalamus in parallel (Oliver and Hall 1978a; Andersen et al. 1980; Oliver 1982; Calford 1983). The *central pathway* originates in the central nucleus and terminates in the ventral division of the medial geniculate body. The *pericentral pathways* originate in the cortical and paracentral subdivisions of the IC. The deep dorsal, dorsal, and ventrolateral nuclei of the medial geniculate body are the targets of these pathways. The *lateral tegmental system* (Morest 1965) begins in the nucleus sagulum, the posterior deep layers of the superior colliculus, and the cuneiform nucleus. This tegmental system terminates in the dorsal division and suprageniculate nucleus of the medial geniculate body. In addition, the *widespread pathway* originates from neurons throughout the IC and lateral tegmentum. These widespread neurons each send an axon to the medial division of the medial geniculate body.

Each of these pathways continues to the neocortex after a synapse in the medial geniculate body (see Chapter 6, this volume). The central pathway continues to the primary auditory cortex, while the pericentral and lateral tegmental pathways continue to nonprimary areas that are successively further away from the primary cortex. More importantly, each of the cortical targets of these ascending pathways provide descending cortical inputs to the IC (Oliver and Hall 1978b).

3. Superior Colliculus (SC)—Morphological Substrates for Integration of Sensory and Motor Systems

The superior colliculus is an important participant in orientation behavior. While the superior colliculus is often considered a component of the visual system, this structure also integrates information of multiple sensory modalities as well as certain types of motor information (recently reviewed in: Kaas and Huerta 1988; Dean, Redgrave, and Westby 1989; Robinson and McClurkin 1989; Sparks and Hartwich-Young 1989). On the basis of cell morphology, the superior colliculus comprises two major regions: a superficial region and a deeper region (Section 3.1). Connectional, physiological and behavioral data show that these regions are also functionally distinct (Section 3.2). Each region can be further subdivided into layers (Section 3.3). Finer subdivisions of these layers may be related to the functional organization of the SC (Section 3.4).

3.1 Neuron Morphology of the SC Defines Two Major Regions

The superior colliculus wraps around the dorsolateral aspect of the central gray to form the anterior tectum of the midbrain. Based on the distribution of different cell types, the superior colliculus can be parcelled into a dorsal, or superficial, region and a ventral, or deeper region (Fig. 5.13; Edwards 1980). Neurons in the superficial region of the superior colliculus are more uniform in size than are neurons of the deeper collicular region, with no superficial neuron being as large as the largest of neurons found in the deeper region (Sterling 1971; Tokunaga and Otani 1976; Edwards 1980; Norita 1980; Laemle 1981, 1983; Huerta and Harting 1984a; Moschovakis and Karabelas 1985). Besides these general features of collicular neurons which are shared by all mammals thus far studied, morphological details, and morphological classification schemes, of these neurons vary across species, especially with regard to the deeper collicular region.

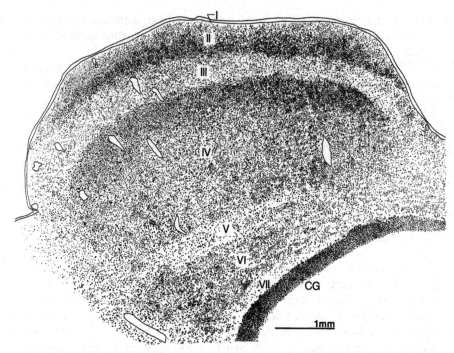

FIGURE 5.13. Drawing of a Nissl-stained transverse section through the superior colliculus of a macaque. Numerals indicate layers, CG, central gray; dorsal is to the top, medial to the right. (From Kaas and Huerta (1988), reprinted by permission of John Wiley & Sons, Inc.)

Superficial neuron types include small and medium stellate granule cells, horizontal cells, narrow field vertical cells, wide-field vertical cells, and piriform cells in the cat (Sterling 1971; Schiller 1984). In new world monkeys (*Saimiri*), old world monkeys (*Macaca*), and humans, these same cell types populate the superficial superior colliculus; in addition, primates, like some other mammals, also have marginal cells (Laemle 1981). These various cell types are differentially distributed in the superior colliculus, with marginal cells situated at the most superficial margin, and with concentrations of horizontal, small stellate, narrow field vertical, piriform, wide field vertical and medium stellate cells found in progressively deeper regions of the superficial collicular region (Sterling 1971; Laemle 1981; Schiller 1984).

Neuronal types of the deeper collicular region in the cat include large multipolar, medium multipolar, small multipolar, medium vertical, small vertical, tufted, and horizontal cells (Norita 1980; Moschovakis and Karabelas 1985). In squirrel monkeys (*Saimiri*), rhesus monkeys (*Macaca*) and humans, the deeper collicular region is populated by vertical, pyramidal, inverted pyramidal, stellate, and horizontal cell types of various sizes (Laemle 1981; Moschovakis, Karabelas, and Highstein 1988; Ma et al. 1990). As with the superficial region, there is some evidence that these cell types are differentially distributed in the deeper collicular region; in the cat, for example, the large neurons are concentrated in lateral parts of the deeper superior colliculus (Norita 1980). The accumulating evidence further suggests that the different cell types in both the superficial and deeper collicular regions can be correlated with different connections and with different chemical constituents (e.g., Mize, Spencer, and Sterling 1982; Huerta and Harting 1984a; Moschovakis and Karabelas 1985; Moschovakis, Karabelas, and Highstein 1988).

3.2 Each Region of the SC Has a Unique Combination of Connections

The superficial region of the superior colliculus participates exclusively in vision functions, whereas the deeper region is involved with processing vision-related as well as somatosensory-, auditory- and motor-related information (recently reviewed in: Kaas and Huerta 1988; Dean, Redgrave, and Westby 1989; Robinson and McClurkin 1989; Sparks and Hartwich-Young 1989). Connectional differences underlie these functional differences, with the superficial region having connections only with components of the visual system, and with the deeper region having connections with many diverse neural structures (Fig. 5.14).

3.2.1 The Superficial Region Connects with the Visual System

The retinal ganglion cells are the major source of input to the superficial region in mammals, including cats and primates. In cats, where the retina has been most extensively studied, there are at least three ganglion cell types, X, Y, and W, which are distinguished on the basis of morphological and physiological characteristics (for review see Stone 1983). While all three types innervate the dorsal lateral geniculate nucleus, the superior colliculus of the cat appears to receive most, if not all, retinal input from only the Y-cells and W-cells, with all of the Y- and half of the W-retinocollicular cells also innervating the thalamus via an axon collateral (McIlwain and Lufkin 1976; McIlwain 1978; Bowling and Michael 1980; Wassle and Illing 1980; Ogawa and Takahashi 1981). Since X-cells are considered to be important in high spatial resolution of visual stimuli, the lack of X-cell input to the superior colliculus is consistent with the interpretation that this midbrain structure participates in other aspects of visual function. The large receptive fields and the lack of specificity for visual stimulus features exhibited by superficial collicular neurons (McIlwain and Buser 1968; Berman and Cynader 1972; Cynader and Berman 1972; Goldberg and Wurtz 1972) suggest that the superior col-

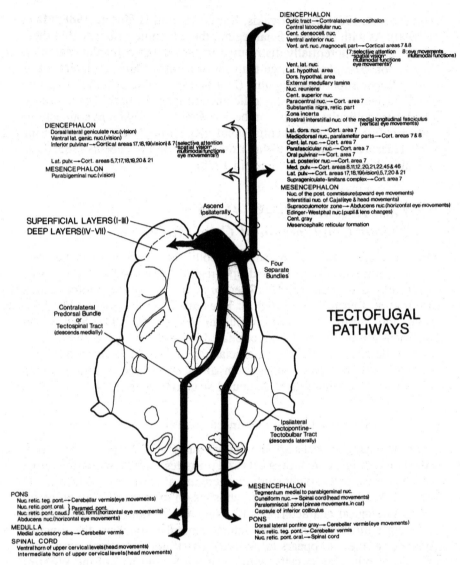

DIENCEPHALON
Optic tract→Contralateral diencephalon
Central latocellular nuc.
Cent. densocell. nuc.
Ventral anterior nuc.
Vent. ant. nuc.,magnocell. part→Cortical areas 7 & 8
(7:selective attention   8:eye movements
"spatial vision"   multimodal functions)
Vent. lat. nuc.
Lat. hypothal. area
Dors. hypothal. area
External medullary lamina
Nuc. reuniens
Cent. superior nuc.
Paracentral nuc.→Cort. area 7
Substantia nigra, retic. part
Zona incerta
Rostral interstitial nuc. of the medial longitudinal fasciculus
(vertical eye movements)
Lat. dors. nuc →Cort. area 7
Mediodorsal nuc., paralamellar parts→Cort. areas 7 & 8
Cent. lat. nuc.→Cort. area 7
Parafascicular nuc.→Cort. area 7
Oral pulvinar→Cort. area 7
Lat. posterior nuc.→Cort. area 7
Med. pulv.→Cort. areas 8,11,12, 20,21,22,45 & 46
Lat. pulv→Cort. areas 17,18,19(vision),5,7,20 & 21
Suprageniculate-limitans complex→Cort. area 7
MESENCEPHALON
Nuc. of the post. commissure(upward eye movements)
Interstitial nuc. of Cajal(eye & head movements)
Supraoculomotor zone→ Abducens nuc.(horizontal eye movements)
Edinger-Westphal nuc.(pupil & lens changes)
Cent. gray
Mesencephalic reticular formation

DIENCEPHALON
Dorsal lateral geniculate nuc.(vision)
Ventral lat. genic. nuc.(vision)
Inferior pulvinar→Cortical areas 17,18,19(vision) & 7(selective attention
"spatial vision"
multimodal functions
eye movements?)
Lat. pulv.→Cort. areas 5,7,17,18,19,20 & 21
MESENCEPHALON
Parabigeminal nuc.(vision)

Ascend
Ipsilaterally

SUPERFICIAL LAYERS(I-III)
DEEP LAYERS(IV-VII)

Four
Separate
Bundles

Contralateral
Predorsal Bundle
or
Tectospinal Tract
(descends medially)

TECTOFUGAL
PATHWAYS

Ipsilateral
Tectopontine-
Tectobulbar Tract
(descends laterally)

PONS
Nuc. retic. teg. pont.→Cerebellar vermis(eye movements)
Nuc. retic. pont.oral. | Paramed. pont.
Nuc. retic. pont. caud. | retic. form.(horizontal eye movements)
Abducens nuc.(horizontal eye movements)
MEDULLA
Medial accessory olive→Cerebellar vermis
SPINAL CORD
Ventral horn of upper cervical levels(head movements)
Intermediate horn of upper cervical levels(head movements)

MESENCEPHALON
Tegmentum medial to parabigeminal nuc.
Cuneiform nuc.→ Spinal cord(head movements)
Paralemniscal zone(pinnae movements, in cat)
Capsule of inferior colliculus
PONS
Dorsal lateral pontine gray→Cerebellar vermis(eye movements)
Nuc. retic. teg. pont.→Cerebellar vermis
Nuc. retic. pont.oral.→Spinal cord

FIGURE 5.14. Diagram of the axonal targets of neurons in superficial layers (open arrows) and of neurons in deeper layers (black arrows) of the superior colliculus in macaques. The functions and the efferent connections of some of these targets are also indicated. (From Huerta and Harting (1984b), reprinted by permission of Elsevier Trends Journals.)

liculus plays a greater role in defining the location of a stimulus than in defining the identity of a stimulus.

In carnivores and primates, the retina projects throughout the horizontal expanse of the contralateral superficial collicular region to produce a topographic map of visual space (Figs. 5.15, 5.16; reviewed in Huerta and Harting 1984a; Kaas and Huerta 1988). A more restricted ipsilateral retinocollicular projection is present in most mammals (Harting and Guillery 1976; Pollack and Hickey 1979; Florence, Connely, and Casa-

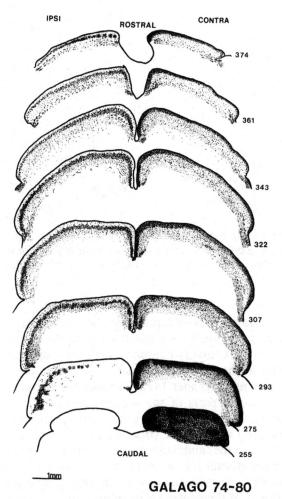

GALAGO 74-80

FIGURE 5.15. Drawing of the distribution of anterogradely transported label (stipple) in a rostral to caudal series of frontal sections through the superior colliculus of the prosimian primate *Galago* following unilateral intraocular injection of tritiated amino acid. CONTRA, contralateral to injection; IPSI, ipsilateral to injection. Dorsal to the top of each section, lateral to the left and right.

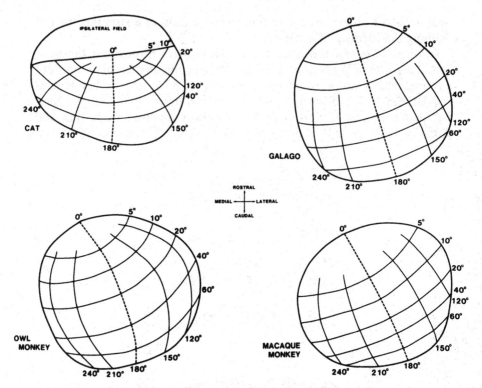

FIGURE 5.16. Schematic rendition of the topographic representation of contralateral visual space in polar coordinates from the fixation point for three primates and a carnivore. (From Kaas and Huerta (1988), reprinted by permission of John Wiley & Sons, Inc.)

grande 1986). In carnivores and primates the most rostral part of the superior colliculus that receives input from both the contralateral and the ipsilateral retinae contains neurons which are responsive to visual stimuli located near the center of the visual field, but in the contralateral hemifield (Berman and Cynader 1972; Cynader and Berman 1972). Progressively more caudal parts of the superior colliculus house neurons which are responsive to stimuli located in progressively more peripheral parts of contralateral visual space (Berman and Cynader 1972; Cynader and Berman 1972). In the caudal extreme of the superior colliculus, which receives only contralateral retinal input in carnivores and primates, the visual space which is represented is so peripheral as to impinge only on the most nasal part of the contralateral retina.

In addition to receiving visual information directly from the retina, the superficial collicular region receives projections from the ventral lateral geniculate nucleus (or pregeniculate nucleus in primates), the parabigeminal nucleus, pretectal nuclei, and layer V of numerous visual cortical areas (Edwards et al. 1979; Graham, Lin, and Kaas 1979; Kawamura and Konno 1979; Fries 1984). Moreover, in studies addressing the issue, such nonretinal visual projections to the superior colliculus appear to be topographically organized, and in register with the topography of the retinocollicular projection (McIlwain 1973).

Just as the superficial collicular region is the recipient of exclusively visual information, neurons of this collicular region project only to nuclei which are considered components of the visual system (Fig. 5.14). The vision-related targets of the superficial region include the parabigeminal nucleus, the dorsal lateral geniculate nucleus, the ventral lateral geniculate (or pregeniculate) nucleus, pretectal nuclei, and the lateral posterior and pulvinar nuclei of the thalamus (Graham 1977; Harting et al. 1980).

3.2.2 The Deeper Region Connects to Visual, Auditory, Somatosensory and Motor Systems

Like the superficial region, the deeper collicular region receives inputs from vision-related structures. The major subcortical source is the ventral lateral geniculate nucleus (Swanson, Cowan, and Jones 1974), whereas cortical sources include several extrastriate visual areas (Kawamura and Konno 1979; Fries 1984). These inputs appear to be distributed topographically, thus forming an orderly representation of visual space in the deeper region of the superior colliculus. Indeed, physiological studies reveal that the topography of the representation of visual space that is present in the superficial colliculus is also present in the deeper region (e.g., Stein, Magalhaes-Castro, and Kruger 1976). Specifically, central visual space is represented rostrally in the deeper superior colliculus, peripheral visual space is represented caudally, superior visual space is represented medially, and inferior visual space is represented laterally in the deeper superior colliculus. The representation of visual space in the superficial region of the superior colliculus is said to be in spatial register with that of the deeper collicular region.

Unlike the superficial region, which only receives inputs from vision-related structures, the deeper colliculus receives inputs from structures that participate in a variety of sensory and motor functions. Auditory-related subcortical structures which project to the deeper region include the rostral pole nucleus of the inferior colliculus, the nucleus of the brachium of the inferior colliculus, the dorsomedial periolivary nucleus, nuclei of the trapezoid body, and the ventral nucleus of the lateral lemniscus (Edwards et al. 1979). Several auditory cortical areas also project to the deeper region in cats and monkeys (Kawamura and Konno 1979; Fries 1984; Meredith and Clemo 1989). Anatomical and physiological studies indicate that in cats the most important cortical source of auditory information to the superior colliculus arises in the auditory subregion of the anterior ectosylvian cortex (Meredith and Clemo 1989); comparable analyses have not been carried out in primates. Results from experiments using dichotic or free-field stimuli suggest that auditory space is represented topographically in the deeper superior colliculus; this representation is, at least grossly, in register with the map of visual space (Wise and Irvine 1983, 1985; Hirsch, Chan, and Yin 1985; Middlebrooks and Knudsen 1987).

Somatosensory inputs to the deeper superior colliculus arise from the spinal cord, the dorsal column nuclei, principal and spinal trigeminal nuclei, and several areas of cortex (Edwards et al. 1979; Stein, Spencer, and Edwards 1983; Fries 1984). In cats, the fourth somatosensory cortical area projects most heavily to the superior colliculus, and appears to play a prominent role in modulating somatosensory responses of deep collicular neurons (Stein, Spencer, and Edwards 1983; Clemo and Stein 1986). The tactile representation of the body is represented topographically in the deeper superior colliculus, and is said to be in register with the map of visual space. Thus, the rostral part of the deeper colliculus contains neurons responsive to tactile stimulation of the face, while progressively caudal parts contain neurons responsive to stimulation of the forelimb and hindlimb (Stein, Magalhaes-Castro, and Kruger 1976).

Electrophysiological studies indicate that about half of the cells in the deeper layers of the cat's superior colliculus respond to stimuli of more than one sensory modality; the majority of such multimodal sensory neurons respond to auditory stimuli as well as visual and/or somatosensory stimuli (Meredith and Stein 1986). Although beyond the scope of this chapter, behavioral studies indicate that the multisensory integration present at the single unit level is also exhibited at the organismal level (Stein et al. 1989).

In addition to connections with sensory-related structures, the deeper superior colliculus also receives inputs from structures which are thought to participate in motor mechanisms, particularly oculomotor mechanisms. Among such sources of input are the nucleus of the posterior commissure, the deep cerebellar nuclei, nucleus propositus hypoglossi, and several cortical areas which, in monkeys, include the frontal eye field and the supplementary eye field (Edwards et al. 1979; Hartwich-Young and Weber 1986; Huerta, Krubitzer, and Kaas 1986; Stanton, Goldberg, and Bruce 1988; Huerta and Kaas 1990). Deep collicular neurons distribute axons to targets widely dispersed in the neuraxis (Fig. 5.14). Most of these targets are known, or thought, to participate in some aspect of motor function; in primates such targets are implicated in oculomotor function, while in cats the axonal targets of the deeper collicular region are implicated in pinnae and head movements, as well as in eye movements (Harting 1977; Edwards and Henkel 1978; Henkel and Edwards 1978; Harting et al. 1980; Huerta and Harting 1982a,b; also see Huerta and Harting 1984a and Huerta, Krubitzer, and Kaas 1986). Descending tectofugal pathways reach spinal cord and reticular formation nuclei, the inferior olivary complex, vestibular nuclei and nucleus prepositus hypoglossi. Ascending tectofugal axons distribute to mesencephalic structures and to many thalamic nuclei, including intralaminar and juxtalaminar nuclei.

3.3 Nissl-Cytoarchitectonics and Myeloarchitectonics Define Layers Within the Superficial and Deeper Regions of the SC

On the basis of Nissl-stained material and patterns of myelination, the superficial and deeper collicular regions can be further divided into seven layers (Fig. 5.13). The layers have descriptive names and are also numbered from superficial (layer I) to deep (layer VII) positions. Layers which are mostly fibrous alternate with cellular layers, so that layers I, III, V, and VII are made up mostly of fibers, while layers II, IV, and VI are cellular. This laminar organization is similar in a wide variety of mammals, including cats and monkeys (Kanaseki and Sprague 1974; Huerta and Harting 1984a; Wiener 1986; Kaas and Huerta 1988).

Together, layers I, II, and III correspond to the superficial collicular region described in the preceding section; these are commonly referred to as the superficial layers. Collectively, layers IV, V, VI, and VII correspond to the deeper region described in the preceding section and are called the deeper layers; alternatively, layers IV and V are called the intermediate layers while layers VI and VII are called the deep layers (Huerta and Harting 1984a). Presently, layers IV–VII will be considered together as the deeper layers.

3.4 Subdivisions of Layers May Help Explain Functional Organization in the SC

The superficial collicular layers are concerned only with visual functions and have relatively few afferent and efferent connections; the deeper layers participate in seemingly more complex sensorimotor functions and have many afferent and efferent connections. Reflecting these differences between the superficial and deeper layers is evidence that suggests that the manner in which connections are organized within the superficial layers is also simpler than that in the deeper layers. Perhaps related to this connectional organization is the distribution of certain chemical markers which also differs in the superficial layers compared to the deeper layers.

3.4.1 Superficial Layers Are Divided into Sublaminae

3.4.1.1 Sublaminae Defined by Chemoarchitecture and Connections

Layer II is divisible into sublaminae on the basis of Nissl-staining patterns, with each sublayer being continuous in the rostral-caudal and medial-lateral dimensions and with sublaminae stacked upon each other in the same way that the collicular layers are stacked upon each other. Sublaminae in layer II, and in other superficial layers, are also apparent on the basis of the distribution of certain chemical markers, including choline acetyltransferase, [Met^5]-enkephalin, and glutamate decarboxylase (Graybiel, Brecha, and Karten 1984; Lu et al. 1985; Hall et al. 1989; Hashikawa 1989).

While chemoarchitectural studies have not revealed sublaminae in superficial collicular layers in primates (Graybiel 1979; Kaas and Huerta 1988; Wallace 1988), such sublaminae have been observed in the rat based on immunochemical staining of fibers with antibodies to substance P, calcitonin gene-related peptide, and Leu-enkephalin (Miguel-Hidalgo et al. 1989). In addition to these peptidergic fibers forming different sublaminae within the superficial layers of the rat superior colliculus, analysis of similarly stained cell bodies reveals that these, too, are differentially distributed in sublaminae of the superficial layers (Miguel-Hidalgo et al. 1989).

In a variety of mammals, including carnivores and primates, sublaminae of the superficial layers are formed by the distribution of particular populations of afferent axons (reviewed in Huerta and Harting 1984a). Thus, the contralateral retina and the contralateral parabigeminal nucleus project to a superficial sublamina in layer II in cats (Graybiel 1975, 1978a; Harting and Guillery 1976), while visual cortical area 17 projects slightly deeper, and areas 18 and 19 project to an even deeper sublamina within layer II (Updyke 1977; Mize 1983; but see Behan 1984). Similar sublaminae are formed in primates (Fig. 5.17).

Finally, sublaminae of the superficial layers are also formed by the distribution of collicular neurons which send their axons to particular targets (Fig. 5.17). As with sublaminae formed by chemoarchitecture and afferent axons, sublaminae formed by specific colliculofugal populations are present in a variety of mammals (Kawamura and Kobayashi 1975; Robson and Hall 1977; Raczkowski and Diamond 1978; Graham and Casagrande 1980; Kawamura et al. 1980; Benevento and Standage 1983; Huerta and Harting 1983).

3.4.1.2 Functional Implications of a Sublaminar Organization

While morphological and connectional data clearly indicate that the superficial collicular layers comprise sublayers, the functional correlates of this type of organization are less clear. For example, anatomical data in cats suggest that the retinal input which terminates in the superficial sublamina of layer II would influence mostly neurons of that sublamina, and would exert little influence on the activity of neurons situated in deeper sublaminae. This hypothesized differential influence would have functional significance because the neurons which project to the dorsal lateral geniculate are concentrated in the superficial part of layer II and the neurons which project to the lateral posterior nucleus mostly occupy

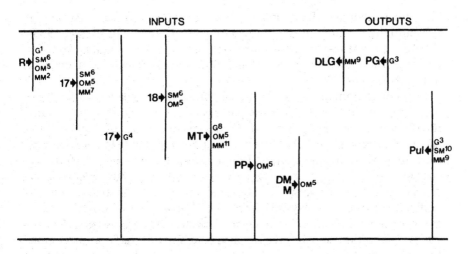

FIGURE 5.17. The approximate dorsal-ventral locations of sublaminae (vertical lines) formed by various afferent (inputs) and efferent (outputs) connections in layer II of the superior colliculus (the dorsal and ventral borders of which are represented by horizontal lines) for several primates. Arrows represent the direction of the connection, with rightward arrows indicating sublaminae formed by axons which arise in the indicated source (R, 17, 18, MT, PP, DM, M), and with leftward arrows indicating sublaminae formed by the cells of origin of axons reaching the indicated targets (DLG, PG, Pul). R, contralateral retina; 17, cortical area 17; 18, cortical area 18; MT, middle temporal cortical area; PP, posterior parietal cortical area; DM, dorsal medial cortical area; M, medial cortical area; DLG, dorsal lateral geniculate nucleus; PG, pregeniculate nucleus; Pul, pulvinar nuclei. Dorsal to the top, ventral to the bottom. (From Kaas and Huerta (1988), reprinted by permission of John Wiley & Sons, Inc.)

the deeper part of layer II (Huerta and Harting 1984a). Thus, visual information conveyed through layer II would form two parallel and separate pathways, with retinal input influencing neurons which send information to one thalamic nucleus and visual cortex input influencing neurons which send information to a separate thalamic nucleus.

A recent electrophysiological analysis, however, indicates that such retinal input strongly influences cells in both superficial and deeper sublaminae of layer II (Berson 1988). Moreover, that study revealed that most of the superficial collicular neurons which receive that retinal input are also directly activated by visual cortical axons, thus suggesting that

"... sublaminar specificity in afferent input to the SGS (layer II) may be substantially smeared at the postsynaptic level" (p. 1869, Berson 1988). Whether such smearing also occurs with other inputs, such as from the parabigeminal nucleus or from the ventral lateral geniculate nucleus, is not known.

In other mammals, however, the sublaminar organization may have more prominent functional consequences. Thus, in tree shrews (*Tupaia*) the sublaminar segregation of retinal and visual cortical afferents and the sublaminar segregation of colliculogeniculate and colliculopulvinar neurons are similar to that in cats (Graham and Casagrande 1980). In accordance with these connectional differences, electrophysiological criteria can be used to distinguish the neuronal population situated in the superficial sublamina of layer II from the neuron population situated in the deeper sublamina of that layer (Albano, Humphrey, and Norton 1978). In addition, sublaminae can also be distinguished in squirrel monkeys (*Saimiri*) on the basis of connectional and physiological criteria (Kadoya, Wolin, and Massopust 1971; Huerta and Harting 1983; 1984a, 1984b). Nevertheless, while these observations suggest that the sublaminar organization of afferent axons underlies the segregation of physiological cell types, this suggestion is thus far based on correlative evidence. Clearly, future studies must examine whether causal relationships are the bases for such correlations.

3.4.2   Deeper Layers Are Divided into Modules

3.4.2.1   Modules Defined by Chemoarchitecture and Connections

In the superficial collicular layers, the differential distribution of chemical markers defines sublaminae; most of these same markers are also differentially distributed in the deeper layers. And, in the deeper layers, as in the superficial layers, staining variations are often confined to particular dorsal-ventral positions within a layer (e.g., in the superficial or deep part of a layer). Still, the chemoarchitecture in the deeper layers is more complex than that of the superficial layers. In contrast to the more or less continuous medial-lateral and rostral-caudal distribution of chemical markers in particular sublaminae of the superficial layers, many chemical markers are distributed discontinuously and repeatedly in these dimensions in the deeper collicular layers. Such patterns have been referred to as patchy, lattice-like, or modular.

Among the chemical markers which are distributed in a modular manner are acetylcholinesterase, choline acetyltransferase, hexokinase, cytochrome oxidase, succinate dehydrogenase, nicotinamide adenine dinucleotide phosphate-diaphorase, substance P, calcitonin gene-related peptide, and Leu-enkephalin (Graybiel 1978b, 1979; Graybiel, Brecha, and Karten 1984; Sandell 1984; Harvey and MacDonald 1985; Illing and Graybiel 1985, 1986; Beninato and Spencer 1986; Wallace 1986; 1988;

Wiener 1986; Hall et al. 1989; Hashikawa 1989; Miguel-Hidalgo et al. 1989).

The distribution of afferent axons from particular sources is similarly modular in the deeper collicular layers (reviewed in Huerta and Harting 1984a). In cats, for example, inputs from the spinal trigeminal nucleus, the hypothalamus, the nucleus of the posterior commissure, the substantia nigra, the deep cerebellar nuclei, the dorsal column nuclei, the spinal cord, and several cortical areas each distribute to patches within layer IV (Graybiel 1978c; Huerta, Frankfurter, and Harting 1981; Huerta and Harting 1982c; Kawamura et al. 1982; Flink, Wiberg, and Blomqvist 1983; Stein, Spencer, and Edwards 1983; Segal and Beckstead 1984; Wiberg and Blomqvist 1984; Illing and Graybiel 1985; 1986; Rieck et al. 1986; Harting et al. 1988). A patchy distribution of afferent axons has also been demonstrated in primates to arise in the substantia nigra (Fig. 5.18) and several regions of cortex, including the frontal eye field and the supplementary eye field (Goldman and Nauta 1976; Kunzle, Akert,

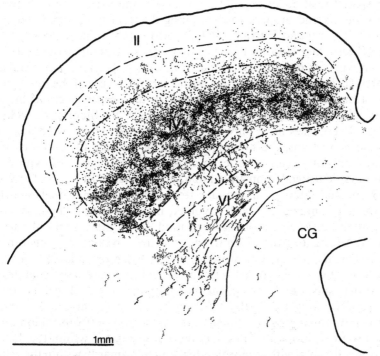

FIGURE 5.18. Patchy distribution of the nigrocollicular projection in the prosimian primate *Galago*. Distribution of anterogradely transported label (stipple) in a transverse section through the superior colliculus following injection of horseradish peroxidase conjugated to wheat germ agglutinin within the ipsilateral substantia nigra. Dorsal to the top, medial to the right.

and Wurtz 1976; Jayaraman, Batton, and Carpenter 1977; Leichnetz et al. 1981; Huerta, Krubitzer, and Kaas 1986; Huerta and Kaas 1990; Huerta, Van Lieshout, and Harting 1990). Each of these afferents is said to form an input module (Huerta and Harting 1984a,b). Neurons in the deeper layers of the superior colliculus which project their axons to particular targets are also distributed in a modular fashion. In the cat, such neurons include those which project to the spinal cord, the inferior olivary complex, and the spinal trigeminal complex (Huerta, Frankfurter, and Harting 1981; Huerta and Harting 1982a, 1984a). These are referred to as output modules (Huerta and Harting 1984a,b). Whether, and to what extent, other colliculofugal neurons form output modules remains to be elucidated.

3.4.2.2 Functional Implications of a Modular Organization

It has been hypothesized that particular input modules spatially overlap with particular output modules in the deeper collicular layers, and, obversely, that those input modules are spatially segregated from other output modules, which would overlap with a different set of input modules. Such spatial relationships could facilitate synaptic interactions between certain inputs and outputs while keeping different throughput channels separate (Huerta and Harting 1984a,b). Moreover, it has been hypothesized (Huerta and Harting 1984a,b) that particular input modules either overlap with, or remain segregated from, other specific input modules, and that such spatial relationships would form the basis of the integration of information which is known to occur in the deeper layers of the superior colliculus (Sparks and Hartwich-Young 1989; Stein et al. 1989).

These hypotheses have only begun to be tested. Thus, by comparing, in adjacent sections, the modular pattern of staining for acetyl cholinesterase with the locations of each of several input modules (with each input module being demonstrated in a separate case), Illing and Graybiel (1985) concluded that input modules from the presylvian frontal eye fields and the substantia nigra, both involved in oculomotor functions, converge upon acetyl cholinesterase-rich modules in the superficial part of layer IV. A later study using similar approaches, revealed that other input modules, which presumably arose from somatosensory-related or vision-related structures, overlapped with acetyl cholinesterase-poor modules in the deeper collicular layers (Illing and Graybiel 1986). Together, these observations suggested to those authors that, particular sensory (from vision- and somatosensory-related sources) and motor (from oculomotor-related sources) input modules may remain segregated within the deeper superior colliculus, with the former distributed primarily within the acetyl cholinesterase-poor zones and the latter distributing to the acetyl cholinesterase-rich zones (Illing and Graybiel 1986).

Illing and Graybiel's (1986) interesting hypothesis has recently received direct support from experiments carried out by Harting and Van Lieshout

(1991). Specifically, by visualizing two separate input modules in the same section, input modules from the cholinergic pedunculopontine nucleus and from the motor-related substantia nigra were found to overlap extensively within the middle of the dorsal-ventral extent of layer IV. Furthermore, somatosensory-related inputs from the spinal trigeminal nucleus were found to interdigitate with nigral inputs (Harting and Van Lieshout 1991).

To the extent that modules of the deeper collicular layers and sublaminae of the superficial layers are defined on the basis of chemoarchitecture and connectivity, the functional implications of each type of laminar subdivision are similar. In general, such subdivisions potentially afford opportunities for the integration of certain types of information as well as allowing the segregation of other types of information. The specific question of which particular types of information are being integrated or segregated in which collicular module, however, is only beginning to be examined. Although the answer to this question will probably be complex, it will undoubtedly significantly shape our understanding of the function of the superior colliculus.

4. Summary

The auditory midbrain is divided into tectal and tegmental structures. The primary tectal structure is the inferior colliculus which is subdivided into a central nucleus, cortex, and paracentral nuclei. These each are characterized by a unique neuropil and collection of cell types, and each part can be further subdivided into specific nuclei. In general, the central nucleus is closely related to the ascending auditory pathways while the other parts are less so. For comparison, the cortex receives major descending inputs from the telencephalon in addition to some ascending inputs. Although the subdivisions of the inferior colliculus are important, they are too large to account for the physiological responses of cells in each subdivision. Functional zones called synaptic domains are proposed to account for cells with different responses to acoustic stimuli. Synaptic domains represent zones with segregated ascending inputs from different brainstem sources. Synaptic domains include both excitatory and inhibitory inputs. Local axonal collaterals from neurons within the inferior colliculus also contribute to the synaptic domains. Neurons in the inferior colliculus will be heavily influenced by their location in one or more synaptic domain and transmit this integrated input to the medial geniculate body.

The superior colliculus is subdivisible into a superficial region and a deeper region on the basis of neuron morphology. Cytoarchitecture analysis reveals that the superficial region comprises three layers and that the deeper region comprises four layers. The superficial layers are connected only with vision-related neural structures, while the deeper layers are connected with neural structures known to participate in a variety of visual, auditory, somatosensory and motor functions. Such connections correspond to the visual function served by the superficial layer and the sensorimotor function carried out by the deeper layers. Different types of synaptic domains distinguish the superficial and deeper layers of the superior colliculus. In the superficial layers, synaptic domains may be represented as sublayers. In the deeper layers, the synaptic domains may be represented as modules.

Acknowledgments. This work includes the efforts of many individuals. We are indebted to Drs. W.C. Hall, J.K. Harting, C.K. Henkel, J.H. Kaas, and D.K. Morest for their contributions. We especially wish to thank L.T. Andrus, G. Beckius, S.D. Flynn, N. Julian, C. Krevolin, R.R. Morest, and D.P. Van Lieshout for technical assistance. Supported by NIH Grants DC00189 and NS25874.

References

Adams JC (1979) Ascending projections to the inferior colliculus. J Comp Neurol 183:519–538.

Adams JC (1983) Cytology of periolivary cells and the organization of their projections in the cat. J Comp Neurol 215:275–289.

Adams JC (1985) Patterns of immunostaining with antisera to peptides in the auditory brainstem of cat. Soc Neurosci Abstr 11:32.

Adams JC, Mugnaini E (1984) Dorsal nucleus of the lateral lemniscus: A nucleus of GABAergic projection neurons. Brain Res Bull 13:585–590.

Adams JC, Mugnaini E (1985) Distribution of cholelcystokinin-like immuno-reactivity in the brainstem auditory system. In: Neuronal Cholecystokinin. Ann NY Acad Sci 448:563–565.

Aitkin LM, Phillips SC (1984) The interconnections of the inferior colliculi through their commissure. J Comp Neurol 228:210–216.

Aitkin LM, Kenyon CE, Philpott P (1981) The representation of the auditory and somatosensory systems in the external nucleus of the cat inferior colliculus. J Comp Neurol 196:25–40.

Albano JE, Humphrey AL, Norton TT (1978) Laminar organization of receptive-field properties in tree shrew superior colliculus. J. Neurophysiol 41:1140–1164.

Andersen RA, Snyder RL, Merzenich MM (1980) The topographic organization of corticocollicular projections from physiologically identified loci in the AI, AII, and anterior cortical fields of the cat. J Comp Neurol 191:479–494.

Andersen RA, Roth GL, Aitkin LM, Merzenich MM (1980) The efferent projections of the central nucleus and the pericentral nucleus of the inferior colliculus in the cat. J Comp Neurol 194:649–662.

Batra R, Kuwada S, Stanford TR (1989) Temporal coding of envelopes and their interaural delays in the inferior colliculus of the unanesthetized rabbit. J Neurophysiol 61:257–268.

Behan M (1984) An EM-autoradiographic analysis of the projection from cortical areas 17, 18, and 19 to the superior colliculus in the cat. J Comp Neurol 225:591–604.

Benevento LA, Standage GP (1983) The organization of projections of the retino-recipient and nonretino-recipient nuclei of the pretectal complex and layers of the superior colliculus to the lateral pulvinar and medial pulvinar in the macaque monkey. J Comp Neurol 218:307–336.

Beninato M, Spencer RF (1986) A cholinergic projection to the rat superior colliculus demonstrated by retrograde transport of horseradish peroxidase and choline acetyltransferase immunohistochemistry. J Comp Neurol 253:525–538.

Berman AL (1968) A Cytoarchitectonic Atlas with Stereotaxic Coordinates. Madison: The University of Wisconsin Press.

Berman N, Cynader M (1972) Comparison of the receptive-field organization of the superior colliculus in Siamese and normal cats. J Physiol (Lond) 223:363–389.

Berson DM (1988) Convergence of retinal W-cell and corticotectal input to cells of the cat superior colliculus. J Neurophysiol 60:1861–1873.

Beyerl BD (1978) Afferent projections to the central nucleus of the inferior colliculus in the rat. Brain Res 145:209–223.

Bowling DB, Michael CR (1980) Projection patterns of single physiologically characterized optic tract fibers in cat. Nature 286:899–902.

Browner RH, Webster DB (1975) Projections of the trapezoid body and the superior olivary complex of the kangaroo rat (Dipodomys merriami). Brain Behav Evol 11:322–354.

Brunso-Bechtold JK, Thompson GC, Masterton RB (1981) HRP study of the organization of auditory afferents ascending to the central nucleus of inferior colliculus in cat. J Comp Neurol 197:705–722.

Calford MB (1983) The parcellation of the medial geniculate body of the cat defined by the auditory properties of single units. J Neurosci 3:2350–2364.

Cant NB (1982) Identification of cell types in the anteroventral cochlear nucleus that project to the inferior colliculus. Neurosci Lett 32:241–246.

Cant NB, Casseday JH (1986) Projections from the anteroventral cochlear nucleus to the lateral and medial superior olivary nuclei. J Comp Neurol 247:447–476.

Cant NB, Morest DK (1984) The structural basis for stimulus coding in the cochlear nucleus of the cat. In: Berlin CI, ed. Recent developments in Hearing Science. San Diego: College Hill Press, pp. 373–421.

Casseday JH, Covey E (1987) Central auditory pathways in directional hearing. In: Yost WA, Gourvitch G, (eds): Directional Hearing. New York: Springer-Verlag, pp. 109–145.

Clemo HR, Stein BE (1986) Effects of cooling somatosensory cortex on response properties of tactile cells in the superior colliculus. J Neurophysiol 55:1352–1368.

Coleman JR, Clerici WJ (1987) Sources of projections to subdivisions of the inferior colliculus in the rat. J Comp Neurol 262:215–226.

Cynader M, Berman N (1972) Receptive-field organization of monkey superior colliculus. J Neurophysiol 35:187–201.

Dean P, Redgrave P, Westby GWM (1989) Event or emergency? Two response systems in the mammalian superior colliculus. Trends in Neurosci 12:137–147.

Diamond IT, Jones EG, Powell TPS (1969) The projection of the auditory cortex upon the diencephalon and brain stem in the cat. Brain Res 15:305–340.

Edwards SB (1980) The deep cell layers of the superior colliculus: Their reticular characteristics and structural organization. In: Hobson JA, Brazier MAB, (eds) The Reticular Formation Revisited, New York: Raven Press, pp. 193–209.

Edwards SB, Henkel CK (1978) Superior colliculus connections with the extra-ocular motor nuclei in the cat. J Comp Neurol 179:451–468.

Edwards SB, Ginsburg CL, Henkel CK, Stein BE (1979) Sources of subcortical projections to the superior colliculus in the cat. J Comp Neurol 184:309–330.

Elverland HH (1978) Ascending and intrinsic projections of the superior olivary complex in the cat. Exp Brain Res 32:117–134.

Faye-Lund H (1985) The neocortical projection to the inferior colliculus in the albino rat. Anat Embryol 173:53–70.

Faye-Lund H, Osen KK (1985) Anatomy of the inferior colliculus in rat. Anat Embryol 171:1–20.

FitzPatrick KA (1975) Cellular architecture and topographic organization of the inferior colliculus of the squirrel monkey. J Comp Neurol 164:185–208.

FitzPatrick KA, Imig TJ (1978) Projections of auditory cortex upon the thalamus and midbrain in the owl monkey. J Comp Neurol 177:537–556.

Flink R, Wiberg M, Blomqvist A (1983) The terminations in the mesencephalon of fibres from the lateral cervical nucleus. An anatomical study in the cat. Brain Res 259:11–20.

Florence SL, Connely M, Casagrande VA (1986) Ocular dominance columns and retinal projections on New World spider monkeys. J Comp Neurol 243:234–248.

Fries W (1984) Cortical projections to the superior colliculus in the macaque monkey: A retrograde study using horseradish peroxidase. J Comp Neurol 230:55–76.

Geniec P, Morest DK (1971) The neuronal architecture of the human posterior colliculus. Acta Otolaryngol (Suppl) 295:1–33.

Glendenning KK, Brunso-Bechtold JK, Thompson GC, Masterton RB. (1981) Ascending auditory afferents to the nuclei of the lateral lemniscus. J Comp Neurol 197:673–703.

Glendenning KK, Masterton RB (1983) Acoustic chiasm: Efferent projections of the lateral superior olive. J Neurosci 3:1521–1537.

Goldberg JM, Moore RY (1967) Ascending projections of the lateral lemniscus in the cat and monkey. J Comp Neurol 129:143–156.

Goldberg ME, Wurtz RH (1972) Activity of superior colliculus in behaving monkey. I. Visual receptive fields of single neurons. J Neurophysiol 35:542–596.

Goldman PS, Nauta WJH (1976) Autoradiographic demonstration of a projection from prefrontal association cortex to the superior colliculus in the rhesus monkey. Brain Res 116:145–149.

Gonzalez-Hernandez TH, Meyer G, Ferres-Torres R, Castañeyra-Perdomo A, Delgado MdMP (1987) Afferent connections of the inferior colliculus in the albino mouse. J Hirnforsch 3:315–323.

Graham J (1977) An autoradiographic study of the efferent connections of the superior colliculus in the cat. J Comp Neurol 173:629–654.

Graham J (1982) Some topographical connections of the striate cortex with subcortical structures in *Macaca fascicularis*. Exp Brain Res 47:1–14.

Graham J, Casagrande VA (1980) A light microscopic and electron microscopic study of the superficial layers of the superior colliculus of the tree shrew (*Tupaia glis*). J Comp Neurol 191:133–151.

Graham J, Lin C-S, Kaas JH (1979) Subcortical projections of six visual cortical areas in the owl monkey, *Aotus trivirgatus.* J Comp Neurol 187:557–580.

Graybiel AM (1975) Anatomical organization of retinotectal afferents in the cat: An autoradiographic study. Brain Res 96:1–23.

Graybiel AM (1978a) Satellite system of the superior colliculus: The parabigeminal nucleus and its projections to the superficial collicular layers. Brain Res 145:365–374.

Graybiel AM (1978b) A stereometric pattern of distribution of acetylcholinesterase in the deep layers of the superior colliculus. Nature 272:539–541.

Graybiel AM (1978c) Organization of the nigrotectal connection: An experimental tracer study in the cat. Brain Res 143:339–348.

Graybiel AM (1979) Periodic-compartmental distribution of acetylcholinesterase in the superior colliculus of the human brain. Neurosci 4:643–650.

Graybiel AM, Brecha N, Karten HJ (1984) Cluster-and-sheet pattern of enkephalin-like immunoreactivity in the superior colliculus of the cat. Neurosci 12:191–214.

Hall WC, Fitzpatrick D, Klatt LL, Raczkowski D (1989) Cholinergic innervation of the superior colliculus in the cat. J Comp Neurol 287:495–514.

Harting JK (1977) Descending pathways from the superior colliculus: An autoradiographic analysis in the rhesus monkey (*Macaca mulatta*). J Comp Neurol 173:583–612.

Harting JK, Guillery RW (1976) Organization of retinocollicular pathways in the cat. J Comp Neurol 166:133–144.

Harting JK, Huerta MF, Frankfurter AJ, Strominger NL, Royce GJ (1980) Ascending pathways from the monkey superior colliculus: An autoradiographic analysis. J Comp Neurol 192:853–882.

Harting JK, Huerta MF, Hashikawa T, Weber JT, Van Lieshout DP (1988) Neuroanatomical studies of the nigrotectal projection in the cat. J Comp Neurol 278:615–631.

Harting JK, Van Lieshout DP (1991) Studies of the nigrotectal projection in the cat: Spatial relationship of nigral and trigeminal axons within the intermediate gray. J Comp Neurol 305:543–558.

Hartwich-Young R, Weber JT (1986) The projection of the frontal cortical oculomotor areas to the superior colliculus in the domestic cat. J Comp Neurol 253:342–357.

Harvey AR, MacDonald AM (1985) The development of acetylcholinesterase activity in normal and transplanted superior colliculus in rats. J Comp Neurol 240:117–127.

Hashikawa T (1989) Regional and laminar distribution of choline acetyltransferase immunoreactivity in the cat superior colliculus. Neurosci Res 6:426–437.

Henkel CK, Edwards SB (1978) The superior colliculus control of pinna movements in the cat: Possible anatomical connections. J Comp Neurol 182:763–776.

Henkel CK, Shneiderman A (1988) Nucleus sagulum: Projections of a lateral tegmental area to the inferior colliculus in the cat. J Comp Neurol 271:577–588.

Henkel CK, Spangler KM (1983) Organization of the efferent projections of the medial superior olivary nucleus in the cat as revealed by HRP and autoradiographic tracing methods. J Comp Neurol 221:416–428.

Herrara M, Del Campo FS, Smith-Agreda V (1987) The commissural nucleus of the inferior colliculus in the rabbit. J Hirnforsch 28:671–683.

Herrara M, Del Campo FS, Orts AP (1988) Cytoarchitecture and axonal systems in the commissural nucleus of the inferior colliculus in the albino rat. A Golgi study. J Hirnforsch 29:165–174.

Herrara M, Smith-Agreda V, Morera C (1984) The commissural neurons of the inferior colliculus of the rat. Acta Anat 119:174–178.

Hirsch JA, Chan JCK, Yin TCT (1985) Responses of neurons in the cat's superior colliculus to acoustic stimuli. I. Monaural and binaural response properties. J Neurophysiol 53:726–745.

Hubel DH, LeVay S, Wiesel TN (1975) Mode of termination of retinotectal fibers in macaque monkey: An autoradiographic study. Brain Res 96:25–40.

Huerta MF, Harting JK (1982a) Tectal control of spinal cord activity: Neuroanatomical demonstration of pathways connecting the superior colliculus with the cervical spinal cord grey. In: Kuypers HGJM, Martin GF, (eds) Descending Pathways to the Spinal Cord, New York: Elsevier, pp. 687–773.

Huerta MF, Harting JK (1982b) Projection of the superior colliculus to the supraspinal nucleus and the cervical spinal cord grey of the cat. Brain Res 242:326–331.

Huerta MF, Harting JK (1982c) The projection from the nucleus of the posterior commissure to the superior colliculus of the cat: Patch-like endings within the intermediate and deep grey layers. Brain Res 238:426–432.

Huerta MF, Harting JK (1983) Sublamination within the superficial grey layer of the squirrel monkey: An analysis of the tectopulvinar projection using anterograde and retrograde transport methods. Brain Res 261:119–126.

Huerta MF, Harting JK (1984a) The mammalian superior colliculus: Studies of its morphology and connections. In: Vanegas H (ed) Comparative Neurology of the Optic Tectum, New York: Plenum, pp. 687–773.

Huerta MF, Harting JK (1984b) Connectional organization of the superior colliculus. Trends Neurosci 7:286–289.

Huerta MF, Kaas JH (1990) Supplementary eye field as defined by intracortical microstimulation: Connections in macaques. J Comp Neurol 293:299–330.

Huerta MF, Frankfurter A, Harting JK (1981) The trigeminocollicular projection in the cat: Patch-like endings within the intermediate grey. Brain Res 211:1–13.

Huerta MF, Krubitzer LA, Kaas JH (1986) Frontal eye field as defined by intracortical microstimulation in squirrel monkeys, owl monkeys, and macaque monkeys: I. Subcortical connections. J Comp Neurol 254:415–439.

Huerta MF, Van Lieshout DP, Harting JK (1991) Nigrotectal projections in the primate Galago crassicaudatus. Exp Brain Res 87:389–401.

Hutson KA (1988) Connections of the auditory midbrain: Efferent projections of the dorsal nucleus of the lateral lemniscus, the nucleus sagulum, and the origins of the GABAergic commissure of Probst. Doctoral dissertation. Tallahassee, FL: Florida State University.

Illing R-B, Graybiel AM (1985) Convergence of afferents from frontal cortex and substantia nigra onto acetylcholinesterase-rich patches of the cat's superior colliculus. Neurosci 14:455–482.

Illing R-B, Graybiel AM (1986) Complementary and non-matching afferent compartments in the cat's superior colliculus: Innervation of the acetylcholinesterase-poor domain of the intermediate gray layer. Neurosci 18:373–394.

Imig TJ, Adrián HO (1977) Binaural columns in the primary field (A1) of cat auditory cortex. Brain Res 138:241–257.

Irvine DRF (1986) The auditory brainstem: In: Ottoson D, (ed) Progress in Sensory Physiology, Vol. 7. Berlin: Springer-Verlag.

Itaya SK, Van Hoesen GW (1982) Retinal innervation of the inferior colliculus in rat and monkey. Brain Res 233:45–52.

Jayaraman A, Batton RR, Carpenter MB (1977) Nigrotectal projections in the monkey: An autoradiographic study. Brain Res 135:147–152.

Kaas JH, Huerta MF (1988) The subcortical visual system of primates. In: Steklis HD, Erwin J, (eds) Comparative Primate Biology: Neurosciences, New York: Alan R. Liss, pp. 327–391.

Kadoya, S, Wolin LR, Massopust Jr LC (1971) Photically evoked unit activity in the tectum opticum of the squirrel monkey. J Comp Neurol 142:495–508.

Kanaseki T, Sprague JM (1974) Anatomical organization of pretectal nuclei and tectal laminae in the cat. J Comp Neurol 158:319–338.

Kawamura K, Konno T (1979) Various types of corticotectal neurons of cats as demonstrated by means of retrograde axonal transport of horseradish peroxidase. Exp Brain Res 35:161–175.

Kawamura S, Kobayashi E (1975) Identification of laminar origin of some tectothalamic fibers in the cat. Brain Res 91:281–285.

Kawamura S, Fukushima N, Hattori S, Kudo M (1980) Laminar segregation of cells of origin of ascending projections from the superficial layers of the superior colliculus. Brain Res 184:486–490.

Kawamura S, Hattori S, Higo S, Matsuyama T (1982) The cerebellar projections to the superior colliculus and pretectum in the cat: An autoradiographic and horseradish peroxidase study. Neurosci 7:1673–1689.

Kudo M (1981) Projections of the nuclei of the lateral lemniscus in the cat: An autoradiographic study. Brain Res 221:57–69.

Kudo M, Niimi K (1980) Ascending projections of the inferior colliculus in the cat: An autoradiography study. J Comp Neurol 191:545–566.

Kunzle H, Akert K, Wurtz RH (1976) Projection of area 8 (frontal eye field) to superior colliculus in the monkey. An autoradiographic study. Brain Res 117:487–492.

Kuwada S, Stanford TR, Batra R (1987) Interaural phase-sensitive units in the inferior colliculus of the unanesthetized rabbit: Effects of changing phase. J Neurophysiol 57:1338–1360.

Kuwada S, Yin TCT, Haberly LB, Wickesberg RE (1980) Binaural interaction in the cat inferior colliculus: Physiology and anatomy. In: Van Den Brink G, Bilsen FA (eds) Psychophysical, Physiological, and Behavioral Studies in Hearing. Delft: Delft Univ. Press, pp. 401–411.

Laemle LK (1981) A Golgi study of cellular morphology in the superficial layers of the superior colliculus in man, *Saimiri*, and *Macaca*. J Hirnforschung 22:253–263.

Laemle LK (1983) A Golgi study of cell morphology in the deep layers of the human superior colliculus. J Hirnforschung 24:297–306.

Leichnetz GR, Spencer RF, Hardy SGP, Astruc J (1981) The prefrontal corticotectal projection in the monkey: An anterograde and retrograde horseradish peroxidase study. Neurosci 6:1023–1041.

Lu SM, Lin C-S, Behan M, Cant NB, Hall WC (1985) Glutamate decarboxylase immunoreactivity in the intermediate grey layer of the superior colliculus in the cat. Neurosci 16:123–131.

Luethke LE, Krubitzer LA, Kaas JH (1989) Connection of primary auditory cortex in the new world monkey, *Saguinus*. J Comp Neurol 285:487–513.

Ma TP, Cheng H-W, Czech JA, Rafols JA (1990) Intermediate and deep layers of the macaque superior colliculus: A Golgi study. J Comp Neurol 295:92–110.

McIlwain JT (1973) Topographic relations in projection from the striate cortex to superior colliculus of the cat. J Neurophysiol 36:690–701.

McIlwain JT (1978) Cat superior colliculus: Extracellular potentials related to W-cell synaptic interactions. J Neurophysiol 41:1343–1358.

McIlwain JT, Buser P (1968) Receptive fields of single cells in cat's superior colliculus. Exp Brain Res 5:314–325.

McIlwain JT, Lufkin RB (1976) Distribution of direct Y-cell inputs to the cat's superior colliculus: are there spatial gradients? Brain Res 103:133–138.

Meininger V, Pol D, Derer P (1986) The inferior colliculus of the mouse. A Nissl and Golgi study. Neurosci 17:1159–1179.

Meredith MA, Clemo HR (1989) Auditory cortical projection from the anterior ectosylvian sulcus (field AES) to the superior colliculus in the cat: An anatomical and electrophysiological study. J Comp Neurol 289:687–707.

Meredith MA, Stein BE (1986) Visual, auditory, and somatosensory convergence on cells in the superior colliculus results in multisensory integration. J Neurophysiol 56:640–662.

Merzenich MM, Reid MD (1974) Representation of the cochlea within the inferior colliculus. Brain Res 77:397–415.

Middlebrooks JC, Knudsen EI (1987) Changes in external ear position modify the spatial tuning of auditory units in the cat's superior colliculus. J Neurophysiol 57:672–687.

Middlebrooks JC, Zook JM (1983) Intrinsic organization of the cat's medial geniculate body identified by projections to binaural response-specific bands in the primary auditory cortex. J Neurosci 3:203–224.

Miguel-Hidalgo J-J, Senba E, Matsutani S, Takatsuji K, Fukui H, Tohyama M (1989) Laminar and segregated distribution of immunoreactivities for some neuropeptides and adenosine deaminase in the superior colliculus of the rat. J Comp Neurol 280:410–423.

Mize RR (1983) Patterns of convergence and divergence of retinal and cortical synaptic terminals in the cat superior colliculus. Exp Brain Res 51:88–96.

Mize RR, Spencer RF, Sterling P (1982) Two types of GABA-accumulating neurons in the superficial gray layer of the cat superior colliculus. J Comp Neurol 206:180–192.

Moore JK, Moore RY (1987) Glutamic acid decarboxylase-like immunoreactivity in brainstem auditory nuclei of the rat. J Comp Neurol 260:157–174.

Moore JK, Osen KK (1979) The human cochlear nuclei. In: Creutzfeld O, Scheich H, Schreiner C (eds.) Exp Brain Res (Suppl II): Hearing Mechanisms and Speech. New York: Springer, pp. 36–44.

Morest DK (1964) The laminar structure of the inferior colliculus of the cat. Anat Rec 148:314.

Morest DK (1965) The lateral tegmental system of the midbrain and the medial geniculate body: Study with Golgi and Nauta methods in cat. J Anat (Lond) 99:611–634.

Morest DK (1966) The cortical structure of the inferior quandrigeminal laminae of the cat. Anat Rec 154:389.

Morest DK, Oliver DL (1984) The neuronal architecture of the inferior colliculus in the cat: Defining the functional anatomy of the auditory midbrain. J Comp Neurol 222:209–236.

Moschovakis AK, Karabelas AB (1985) Observations on the somatodendritic morphology and axonal trajectory of intracellularly HRP-labeled efferent neurons located in the deeper layers of the superior colliculus of the cat. J Comp Neurol 239:276–308.

Moschovakis AK, Karabelas AB, Highstein (1988) Structure-Function relationships in the primate superior colliculus. I. Morphological identity of efferent neurons. J Neurophysiol 60:232–262.

Nelson PG, Erulkar SD (1963) Synaptic mechanisms of excitation and inhibition in the central auditory pathway. J Neurophysiol 26:908–923.

Norita M (1980) Neurons and synaptic patterns in the deep layers of the superior colliculus of the cat. A Golgi and electronic microscopic study. J Comp Neurol 190:29–48.

Nudo RJ, Masterton RB (1986) Stimulation-induced [-14C]2-deoxyglucose labeling of synaptic activity in the central auditory system. J Comp Neurol 245:553–565.

Ogawa T, Takahashi Y (1981) Retinotectal connectivities within the superficial layers of the cat's superior colliculus. Brain Res 217:1–11.

Oliver DL (1982) A Golgi study of the medial geniculate body in the tree shrew (*Tupaia glis*). J Comp Neurol 209:1–16.

Oliver DL (1984a) Neuron types in the central nucleus of the inferior colliculus that project to the medial geniculate body. Neurosci 11:409–424.

Oliver DL (1984b) Dorsal cochlear nucleus projections to the inferior colliculus in the cat: A light and electron microscopic study. J Comp Neurol 224:155–172.

Oliver DL (1985) Quantitative analyses of axonal endings in the central nucleus of the inferior colliculus and distribution of 3H-labeling after injections in the dorsal cochlear nucleus. J Comp Neurol 237:343–359.

Oliver DL (1987) Projections to the inferior colliculus from the anteroventral cochlear nucleus in the cat: Possible substrates for binaural interaction. J Comp Neurol 264:24–46.

Oliver DL, Beckius G (1989) GABA immunocytochemistry and speculation on the inhibitory circuits in the inferior colliculus of the cat. Assoc Res Otolaryngol, Midwinter Meeting *12*:218.

Oliver DL, Hall WC (1978a) The medial geniculate body in the tree shrew, *Tupaia glis*. I. Cytoarchitecture and midbrain connections. J Comp Neurol 182:423–458.

Oliver DL, Hall WC (1978b) The medial geniculate body of the tree shrew, *Tupaia glis*. II. Connections with the neocortex. J Comp Neurol 182:459–494.

Oliver DL, Morest DK (1984) The central nucleus of the inferior colliculus in the cat. J Comp Neurol 222:237–264.

Oliver DL, Nuding SC, Beckius G (1988) Multiple cell types have GABA immunoreactivity in the inferior colliculus of the cat. Soc Neurosci Abstr 14:490.

Oliver DL, Winer J, Beckius G, Saint Marie R. Morphology of GABAergic neurons in the inferior colliculus of the cat, (in preparation).

Oliver DL, Kuwada S, Batra R, Stanford TR, Henkel C (1986) Structural components of binaural information processing in the auditory midbrain. Physiology and anatomy of HRP-injected cells in the cat and gerbil. Soc Neurosci Abstr 12:1271.

Oliver DL, Kuwada S, Yin TCT, Haberly L, Henkel CK (1991) Dendritic and axonal morphology of HRP-injected neurons in the inferior colliculus of the cat. J Comp Neurol 303:75–100.

Osen KK (1972) Projections of the cochlear nuclei on the inferior colliculus in the cat. J Comp Neurol 144:355–372.

Pollack JG, Hickey TL (1979) The distribution of retino-collicular axon terminals in rhesus monkey. J Comp Neurol 179:587–602.

Raczkowski D, Diamond IT (1978) Cells of origin of several efferent pathways from the superior colliculus in *Galago senegalensis*. Brain Res 146:351–357.

Ramón y Cajal S (1911) Histologie de Système Nerveux de l'Homme et des Vertébrés. Vol. 2. Madrid: Instituto Ramón y Cajal, 1972 reprint.

Rieck RW, Huerta MF, Harting JK, Weber JT (1986) Hypothalamic and ventral thalamic projections to the superior colliculus in the cat. J Comp Neurol 243:249–265.

RoBards MJ (1979) Somatic neurons in the brainstem and neocortex projecting to the external nucleus of the inferior colliculus. Anatomical study in the opossum. J Comp Neurol 184:547–566.

Roberts RC, Ribak CE (1987a) GABAergic neurons and axon terminals in the brainstem auditory nuclei of the gerbil. J Comp Neurol 258:267–280.

Roberts RC, Ribak CE (1987b) An electron microscopic study of GABAergic neurons and terminals in the central nucleus of the inferior colliculus of the rat. J Neurocytol 16:333–345.

Robinson DL, McClurkin JW (1989) The visual superior colliculus and pulvinar. In: Wurtz RH, Goldberg ME (eds) The Neurobiology of Saccadic Eye Movements, New York: Elsevier, pp. 337–360.

Robson JA, Hall WC (1977) The organization of the pulvinar in the grey squirrel (*Sciurus carolinensis*). 1. Cytoarchitecture and connections. J Comp Neurol 173:355–388.

Rockel AJ, Jones EG (1973a) The neuronal organization of the inferior colliculus of the adult cat. I. The central nucleus. J Comp Neurol 147:22–60.

Rockel AJ, Jones EG (1973b) Observations on the fine structure of the central nucleus of the inferior colliculus of the cat. J Comp Neurol 147:61–92.

Rockel AJ, Jones EG (1973c) The neuronal organization of the inferior colliculus of the adult cat. II. The pericentral nucleus. J Comp Neurol 149:301–334.

Roth GL, Aitkin LM, Andersen RA, Merzenich MM (1978) Some features of the spatial organization of the central nucleus of the inferior colliculus of the cat. J Comp Neurol 182:661–680.

Ryan AF, Woolf NK, Sharp FB (1982) Tonotopic organization in the central auditory pathway of the mongolian gerbil: A 2-deoxyglucose study. J Comp Neurol 207:369–381.

Ryugo DK, Willard FH (1985) The dorsal cochlear nucleus of the mouse: A light microscopic analysis of neurons that project to the inferior colliculus. J Comp Neurol 242:381–396.

Ryugo DK, Willard FH, Fekete DM (1981) Differential afferent projections to the inferior colliculus from the cochlear nucleus in the albino mouse. Brain Res 210:342–349.

Saint Marie RL, Ostapoff E-M, Morest DK, Wenthold RJ (1989) A glycine-immunoreactive projection of the cat lateral superior olive: Possible role in midbrain ear dominance. J Comp Neurol 279:382–396.

Sandell JH (1984) The distribution of hexokinase compared to cytochrome oxidase and acetylcholinesterase in the somatosensory cortex and superior colliculus of the rat. Brain Res 290:384–389.

Schiller PH (1984) The superior colliculus and visual function. In: Brookhart JM, Mountcastle VB, Darian-Smith I, Geiger SR (eds) Handbook of Physiology, Section 1: The Nervous System; Vol. III, Sensory Processes, Part 1, Bethesda: American Physiological Society, pp. 457–506.

Schroeder DM, Jane JA (1971) Projection of dorsal column nuclei and spinal cord to brainstem and thalamus in the tree shrew, *Tupaia glis.* J Comp Neurol 142:309–350.

Segal RL, Beckstaed RM (1984) The lateral suprasylvian corticotectal projection in cats. J Comp Neurol 225:259–275.

Semple MN, Aitkin LM (1979) Representation of sound frequency and laterality by units in central nucleus of cat inferior colliculus. J Neurophysiol 42:1626–1639.

Semple MN, Aitkin LM (1980) Physiology of pathway from dorsal cochlear nucleus to inferior colliculus revealed by electrical and auditory stimulation. Exp Brain Res 41:19–28.

Servière J, Webster WR, Calford MB (1984) Isofrequency labeling revealed by a combined [^{14}C]2-deoxyglucose, electrophysiological, and horseradish peroxidase study of the inferior colliculus of the cat. J Comp Neurol 228:463–477.

Shneiderman A, Henkel CK (1987) Banding of lateral superior olivary nucleus afferents in the inferior colliculus: A possible substrate for sensory integration. J Comp Neurol 266:519–534.

Shneiderman A, Oliver DL (1989) EM autoradiography study of the projections from the dorsal nucleus of the lateral lemniscus: A possible source of inhibitory inputs to the inferior colliculus. J Comp Neurol 286:28–47.

Shneiderman A, Oliver DL, Henkel CK (1988) The connections of the dorsal nucleus of the lateral lemniscus. An inhibitory parallel pathway in the ascending auditory system? J Comp Neurol 276:188–208.

Spangler KM, Morley BJ (1987) Somatostatin-like immunoreactivity in the midbrain of the cat. J Comp Neurol 260:87–97.

Sparks DL, Hartwich-Young R (1989) The deep layers of the superior colliculus. In: Wurtz RH, Goldberg ME (eds) The Neurobiology of Saccadic Eye Movements, New York: Elsevier, pp. 213–255.

Stanton GB, Goldberg ME, Bruce CJ (1988) Frontal eye field efferents in the macaque monkey: II. Topography of terminal fields in midbrain and pons. J Comp Neurol 271:493–506.

Stein BE, Magalhaes-Castro B, Kruger L (1976) Relationship between visual and tactile representations in cat superior colliculus. J Neurophysiol 39:401–419.

Stein BE, Spencer RF, Edwards SB (1983) Corticotectal and corticothalamic efferent projections of SIV somatosensory cortex in cat. J Neurophysiol 50:896–909.

Stein BE, Meredith AM, Honeycutt WS, McDade L (1989) Behavioral indices of multisensory integration: Orientation to visual cues is affected by auditory stimuli. J Cog Neurosci 1:12–24.

Sterling P (1971) Receptive fields and synaptic organization of the superficial gray layer of the cat superior colliculus. Vis Res Suppl 3:309–328.

Stone J (1983) Parallel Processing in the Visual System: The Classification of Retinal Ganglion Cells and its Impact on the Neurobiology of Vision, New York: Plenum.

Swanson L, Cowan WM, Jones EG (1974) An autoradiographic study of the efferent connections of the ventral lateral geniculate nucleus in the albino rat and the cat. J Comp Neurol 156:143–163.

Symonds LL, Kaas JH (1978) Connections of striate cortex in the prosimian *Galago senegalensis*. J Comp Neurol 181:477–512.

Tanaka K, Otani K, Tokunaga A, Sugita S (1985) The organization of neurons in the nucleus of the lateral lemniscus projecting to the superior colliculi in the rat. Brain Res 341:252–260.

Thompson GC, Cortez AM, Lam DM (1985) Localization of GABA immuno-reactivity in the auditory brainstem of guinea pigs. Brain Res 339:119–122.

Tigges J, Tigges M (1981) Distribution of retinofugal axon terminals in the superior colliculus of squirrel monkey. Invest Ophthalmol Vis Sci 20:149–158.

Tigges M, Tigges J (1970) The retinofugal fibers and their terminal nuclei in *Galago crassicaudasus* (Primates). J Comp Neurol 138:87–102.

Tokunaga A, Otani K (1976) Dendritic patterns of neurons in the rat superior colliculus. Exp Neurol 52:189–205.

Tolbert LP, Morest DK, Yurgelun-Todd DA (1982) The neuronal architecture of the anteroventral cochlear nucleus of the cat in the region of the cochlear nerve root: Horseradish peroxidase labelling of identified cell types. Neurosci 7:3031–3052.

Ungerleider LG, Desimone R, Galkin TW, Mishkin M (1984) Subcortical projections of area MT in the macaque. J Comp Neurol 223:368–386.

Updyke BV (1977) Topographic organization of the projections from cortical areas 17, 18 and 19 onto the thalamus, pretectum, and superior colliculus in the cat. J Comp Neurol 173:81–122.

Van Noort J (1969) The Structure and Connections of the Inferior Colliculus. An Investigation of the Lower Auditory System. Assen: Van Gorcum.

Wall JT, Symonds LL, Kaas JH (1982) Cortical and subcortical projections of the middle temporal area (MT) and adjacent cortex in galagos. J Comp Neurol 211:193–214.

Wallace MN (1986) Spatial relationship of NADPH-diaphorase and acetylcho-linesterase lattices in the rat and mouse superior colliculus. Neurosci 19:381–391.

Wallace MN (1988) Lattices of high histochemical activity occur in the human, monkey, and cat superior colliculus. Neurosci 25:569–583.

Warr WB (1966) Fiber degeneration following lesions in the anterior ventral cochlear nucleus of the cat. Exp Neurol 14:453–474.

Warr WB (1969) Fiber degeneration following lesions in the posteroventral cochlear nucleus of the cat. Exp Neurol 23:140–155.

Warr WB (1972) Fiber degeneration following lesions in the multipolar and globular cell areas in the ventral cochlear nucleus of the cat. Brain Res 40:247–270.

Wassle H, Illing R-B (1980) The retinal projection to the superior colliculus in the cat: A quantitative study with HRP. J Comp Neurol 190:333–356.

Willard FH, Martin GF (1983) The auditory brainstem nuclei and some of their projections to the inferior colliculus in the North American opossum. Neurosci 10:1203–1232.

Wise LZ, Irvine DRF (1983) Auditory response properties of neurons in deep layers of cat superior colliculus. J Neurophysiol 49:674–685.

Wise LZ, Irvine DRF (1985) Topographic organization of interaural intensity difference sensitivity in deep layers of cat superior colliculus: Implications for auditory spatial representation. J Neurophysiol 54:185–211.

Zook JM, Casseday JH (1982) Origin of ascending projections to inferior colliculus in the mustache bat, Pteronotus parnelli. J Comp Neurol 207:14–28.

Zook JM, Casseday JH (1987) Convergence of ascending pathways at the inferior colliculus of the mustache bat, Pteronotus parnelli. J Comp Neurol 261:347–361.

Zook JM, Winer JA, Pollak GD, Bodenhamer RD (1985) Topology of the central nucleus of the mustache bat's inferior colliculus: Correlation of single unit properties and neuronal architecture. J Comp Neurol 231:530–546.

6

The Functional Architecture of the Medial Geniculate Body and the Primary Auditory Cortex

JEFFERY A. WINER

1. Introduction

A signal achievement in contemporary neurobiology is the ability to trace the successive synaptic links along which neural signals travel, from receptor to cerebral cortex. This knowledge has been instrumental to understanding the nature and degree of signal transformation at various central targets, and it is a prerequisite to formulating any theory of hierarchical signal processing and transformation. This strategy of analyzing the serial signal transformations at successive synaptic stations has yielded new insights into the organization of central visual (Kuffler 1953; Hubel and Wiesel 1962) and somatic sensory pathways (Mountcastle and Henneman 1949; Mountcastle 1957). An amalgam of physiological, anatomical, pharmacological, and behavioral approaches have identified multiple sensory channels that ascend in parallel from the periphery towards higher integrative stations, then provide important descending influences to further modulate the subcortical output (Diamond 1983).

While synaptic stations analogous to those for vision and somatic sensation have been delineated for the central auditory pathways, the serial transformation of receptive field organization from one nucleus to another is not well understood, and the data derived from functional probes is at best selective and incomplete. Thus, there is a reasonably secure correlation in the cochlear nucleus between a particular architectonic subdivision (Brawer et al. 1974; Cant 1992, Chapter 3), the pattern of input arising from the cochlea (Ryugo and Fekete 1982), the type(s) of neuron postsynaptic to these axons (Brawer and Morest 1975), the temporal discharge pattern of these neurons (Rhode et al. 1983; Rhode 1992), and their targets in the auditory brain stem (Cant 1982). The same cannot be said for the medial geniculate body and the primary auditory cortex, where the relations among these properties is, at best, obscure. Nevertheless, it is possible to identify, using a combination of structural, connectional, and immunocytochemical methods, a systematic pattern of relations that may be useful in framing more explicit and testable physiological and functional hypotheses about the operations of the auditory forebrain.

Current models of sensory processing usually have some provision for the control of afferent transmission from descending pathways. Such circuitry might subserve different levels of attention, behavioral readiness, gain control, selective signal amplification or suppression, temporal delay, binaurality, tonotopic organization, or the integration of information across different modalities. The highly selective patterns of connectivity between the inferior colliculus, medial geniculate complex, and auditory cortical areas support such a conclusion, though they provide little evidence as to the physiological actions of these feedback circuits. Thus, the prominent lemniscal pathway ascending from the central nucleus of the inferior colliculus to the ventral nucleus of the medial geniculate body conveys information to layers III and IV in primary auditory cortex (AI). This information, in turn, becomes part of several, divergent pathways that influence different facets of auditory forebrain function (Fig. 6.1). For example, some layer IV cells in primary auditory cortex project ipsilaterally to nonprimary auditory cortical fields (Winguth and Winer 1986), while many layer III neurons send their axons into the corpus callosum to influence the opposite hemisphere (Imig and Brugge 1978; Code and Winer 1985). The discharge patterns of neurons in layer V that project to the inferior colliculus may be affected by ascending influences since many such cells with large apical dendrites probably receive thalamic afferents on their trunks (Games and Winer 1988), much as do neurons in layer VI whose axons end in different parts of the medial geniculate complex (Kelly and Wong 1981; Winer and Peterson 1988).

One goal of this review is to emphasize that the discharge patterns of comparatively modest numbers of auditory thalamic neurons have important consequences for vast territories of the auditory forebrain, including those devoted to limbic and autonomic responses which can be affected by acoustic input and can participate in behavioral responses to sound. A second objective is to clarify the connectional and functional affiliations between auditory centers of the pons, midbrain, and diencephalon, loci that are strongly influenced by descending projections of cortical origin. With respect to the medial geniculate body, these projections are often reciprocal and they may well equal or exceed, in size and scope, those of midbrain origin. Hence, any more refined view of auditory thalamocortical organization must identify the logical relations among these structures, and specify the functional correlations entailed by these cellular arrangements.

2. The Medial Geniculate Body

In all mammals and most vertebrates the medial geniculate body is an obligatory synaptic station for auditory information directed towards the telencephalon. In many species the medial geniculate body is an oval

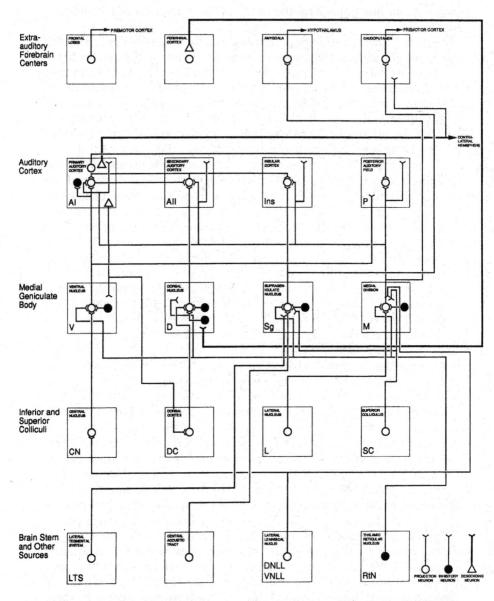

FIGURE 6.1. Synthesis of the major components of the thalamocortical auditory system, with special emphasis on the connectional relations between the thalamus, cortex, and forebrain as well as midbrain and selected brain stem centers. This composite guide is based largely on work in the cat, and there are many interspecific parallels. The figure and the text reinforce three major conclusions about auditory forebrain organization. These are (1) that there is a primary lemniscal pathway originating in the cochlear nucleus and continuing to the primary auditory cortex, and probably beyond, in which there is a systematic representation

mass protruding from the lateral edge of the brain stem at the junction of the midbrain and diencephalon. The principal boundaries in most tetrapods are the anterior surface of the superior colliculus (caudally), the lateral geniculate body (dorsally), the hippocampus (laterally), portions of the intralaminar thalamic nuclei (ventrally), and the medial lemniscal and thalamofugal axons (anterolaterally). While it is sometimes referred to as a nucleus, the medial geniculate body consists of several nuclei, each with a unique pattern of midbrain afferents (Table 6.3) and cortical targets (Table 6.4) as well as a distinct neurochemical organization (Table 6.1; Winer and Larue 1988), and physiology (for a review see Winer 1985b). The observations that follow refer to the adult cat unless otherwise specified.

2.1 Cytoarchitecture and Neuronal Organization

In a Nissl-stained coronal section midway through the medial geniculate complex, three regions are conspicuous and comprise the divisions of the medial geniculate body. The largest of these, the ventral division, constitutes about half of the structure and contains a densely packed aggregate of medium-sized and small neurons, with the former predominating. In the dorsal division, the neurons have slightly smaller somata and are more dispersed, while the medial division has a broad range of somatic sizes, including the largest cells in the auditory thalamus, and an even lower density of neurons. The dorsal and medial divisions each comprise about one-quarter of the auditory thalamus. In both the ventral and dorsal divisions, regional variations in somatic size and density imply that each contains subsidiary nuclei, a view confirmed by regional differences in cell form, midbrain input, cortical connections, and intrinsic organization. The medial geniculate body has a dense array of blood vessels comparable to that in the lateral geniculate body and an overall level of vascularity about 40% higher than that of the globus pallidus (Hough and Wolff 1939).

Historically, several alternative cytoarchitectonic schemes have been proposed; the most popular of these (Fig. 6.2B; Rioch 1929; see also Rose

---

(FIGURE 6.1. *continued*) of the epicritic dimensions of tonotopy, binaurality, amplitude, or related variables; (2) that there is a parallel, nonprimary pathway arising in the midbrain tegmentum, dorsal division of the medial geniculate body, and nonprimary auditory cortex, and beyond, in which broad tuning curves and nontopical representations predominate; (3) that an even more broadly distributed set of auditory affiliations links the auditory forebrain with cortical and subcortical components of, on the one hand, the limbic forebrain and associated autonomic centers and, on the other hand, with elements of the motor system that must organize behavioral responses to biologically significant sound and species-specific communication.

TABLE 6.1. Summary of the types of neurons in the principal nuclei of the medial geniculate body.

Division	Nucleus	Type of neuron	Somatic size	Somatic shape	Dendritic branching pattern	Projection to cortex	Possible transmitter[d]
Ventral	Ventral[a]	1. Large bushy cell	15 × 18 μm	Oval or oblate	Tufted	Yes	Glutamate (Glu) or aspartate (Asp)
		2. Small stellate cell	10 × 12 μm	Flask-shaped	Radiate	No[c]	Glutamic acid decarboxylase (GAD) or GABA
Dorsal	Dorsal[b]	3. Principal stellate cell	15 × 25–30 μm	Round or oblate	Radiate	Yes	Unknown
		4. Principal bushy cell	15 × 20–25 μm	Triangular or flattened	Tufted	Yes	Unknown
		5. Small stellate cell	12 × 18 μm	Elongated and flattened	Radiate	No	GAD/GABA
		6. Large stellate cell	15 × 25–30 μm	Elongated or flask-shaped	Radiate	No	GAD/GABA
	Suprageniculate	7. Large principal stellate cell	20 × 30 μm	Spherical	Radiate	Yes	Glu/Asp
	Posterior limitans	8. Small stellate cell	10 × 15 μm	Flask-shaped	Radiate	No	GAD/GABA
		9. Elongate principal cell	15 × 20–25 μm	Flattened	Stellate	Yes	Glu/Asp
Medial	Medial	10. Small stellate cell	≈10 × 12 μm	Oval or elongated	Stellate	?	Unknown
		11. Medium-sized stellate cell	25 × 35 μm	Round or oblate	Radiate	Yes	Glu/Asp
		12. Tufted cell	20 × 40 μm	Elongated or rectangular	Tufted	Yes	Glu/Asp
		13. Elongated cell	25 × 30–35 μm	Slightly elongated	Radiate	Yes	Unknown
		14. Large, weakly tufted cell	35 × 40 μm	Oval	Tufted	Yes	Glu/Asp
		15. Small stellate cell	15 × 20–25 μm	Triangular	Radiate	No	GAD/GABA

[a] The *pars ovoidea* of the ventral division has a comparable set of cell types and a different pattern of neuropil architecture (Morest 1964).

[b] Local differences between dorsal division nuclei are omitted here (for details see Winer and Morest 1983b, 1984).

[c] Some neurons whose somatodendritic form resembles that of stellate cells are labeled after horseradish peroxidase injections of AI (Winer 1984e).

[d] GAD, the metabolic precursor of GABA, is here considered to mark the same neuronal populations as do antisera to GABA, the presumptive transmitter. Based on immunocytochemical work in the rat for glutamate and aspartate (Larue and Winer, unpublished observations) and on studies of GAD in the rat (Winer and Larue 1988) and of GAD and GABA in the cat (Huchton et al. 1991).

and Woolsey 1949) recognized only principal (including both ventral and dorsal divisions) and magnocellular (medial) divisions, and was based primarily on Nissl and myelin preparations and the selective retrograde reaction of medial geniculate body neurons to cortical lesions. The present, tripartite scheme is derived in part from these techniques, and from reliance on Golgi-impregnated material, tissue prepared for electron microscopy, axoplasmic transport studies of midbrain, thalamocortical, and corticothalamic input, the patterns of immunostaining to a variety of antisera, and upon comparative neuroanatomical and developmental considerations, as well as data from physiological and behavioral perspectives.

2.1.1 Ventral Division

In the ventral division, the medium-sized and small neurons in Nissl material (Fig. 6.3) correspond to particular types of cells in Golgi preparations (Fig. 6.4). The medium-sized neuron has an almond-shaped soma from whose poles primary dendrites arise and branch profusely, forming complex tufts whose orientation, in transverse sections, is along a predominantly dorsomedial to ventrolateral axis, imposing a strongly polarized appearance to the neuron (see Table 6.1). This dendritic orientation confers a laminar structure upon the ventral division because long rows of these neurons form regularly ordered, conspicuous fibrodendritic laminae whose arrangement conserves and reflects the pattern of afferent input arising from the central nucleus of the inferior colliculus; such a projection undoubtedly represents the regular arrangement of frequency-specific discharge patterns across the ventral division (cf. 2.2.1 and 2.6.1). The thick, well-myelinated axons of the bushy tufted neurons have few local branches and project, in turn, to layers III and IV (Table 6.2) in the primary auditory cortex, across which an orderly arrangement of frequency (cf. 3.6.1 and 6.) recapitulates the pattern in ventral division laminae.

The laminar dendritic arrangement in the lateral part of the ventral division (Fig. 6.5A) is less evident in its medial sector, the *pars ovoidea,* where the irregularity and whorled texture of the fibrodendritic plexus obscures the regular, laminated pattern in the lateral sector. Many axons of brachial and extrabrachial origin traverse the *pars ovoidea.* Nonetheless, the same types of neurons are present, although the dendritic domains of tufted neurons in the *pars ovoidea* tend to be less polarized and, consequently, slightly larger.

The small stellate cells represent 30–40% of the ventral division neurons and are distinguished by their tiny, flask-shaped soma, sparse, slender dendrites and a thin, unmyelinated axon with robust local collaterals. These intrinsic branches divide to form a delicate, peridendritic plexus among the tufted neurons, and the terminal axonal field from each local circuit cell is probably confined to relatively few fibrodendritic laminae (Morest 1965a, 1971).   (*Text continues on page 244*).

TABLE 6.2. Summary of the types of neurons in the primary auditory cortex (AI).

Layer	Type of neuron	Somatic size[a]	Somatic shape	Size of dendritic field[a]
			Structure	
I[b]	1. Horizontal cell	10 × 15 μm	Round or flask-shaped	175 × 350 μm
	2. Small cell	20–25 × 20 μm	Round or flask-shaped	80 × 100 μm
	3. Medium-sized cell with short axon	30 × 40 μm	Flask-shaped	300 × 300 μm
	4. Large cell with short axon	40 × 50 μm	Bottle-shaped	290 × 200 μm
II[c]	5. Small pyramidal cell	15 × 15–20 μm	Flask-shaped or piriform	450 × 350 μm
	6. Medium-sized pyramidal cell	15–20 × 20–25 μm	Flask-shaped or piriform	550 × 300 μm
	7. Bipolar cell	15 × 20 μm	Piriform	550 × 80 μm
	8. Bitufted cell	15 × 20 μm	Piriform	550 × 80 μm
	9. Small smooth or sparsely spinous multipolar cell	10 × 10–12 μm	Oval	80 × 80 μm
	10. Spinous multipolar cell	10–20 × 12–30 μm	Oval or piriform	80 × 80–100 to 400 × 200 μm
	11. Large sparsely spinous multipolar cell	15 × 18 μm	Oval	300 × 200 μm
	12. Tufted multipolar cell	12 × 18 μm	Oval	300 × 180 μm
	13. Extraverted multipolar cell	15 × 20–25 μm	Round or multiangular	450 × 300 μm
III[d]	14. Small pyramidal cell	10 × 15 μm	Flask-shaped or triangular	500 × 150 μm
	15. Medium-sized pyramidal cell	15 × 25 μm	Triangular	800 × 200–300 μm
	16. Large pyramidal cell	20 × 30 μm	Triangular	900 × 350–500 μm
	17. Spinous star pyramidal cell	18 × 20 μm	Round	800–1200 × 250 μm
	18. Smooth star pyramidal cell	20 × 20 μm	Round	600 × 400 μm
	19. Tufted or bitufted cell	15 × 25–30 μm	Oval or flask-shaped	350–400 × 100–200 μm
	20. Sparsely spinous stellate cell	10–12 × 20–25 μm	Oval	250–350 × 200–250 μm
	21. Small varicose stellate cell	12 × 16 μm	Flask-shaped	200 × 200 μm
	22. Bipolar cell	15 × 25 μm	Flask-shaped	200 × 600 μm
	23. Medium-sized multipolar cell	15–20 × 15–25 μm	Round or oval	250–350 × 300–500 μm
	24. Small smooth stellate cell	10–15 μm	Drumstick-shaped	100 × 75 μm

TABLE 6.2 *Continued*

	Connectivity				Neurochemistry
Intrinsic	Cortico-cortical	Commis-sural	Cortico-thalamic	Cortico-collicular	Possible transmitter
Layer II	Unknown	No	No	No	γ-aminobutyric acid (GABA) or glutamate (Glu) or aspartate (Asp)
Layer I	Unknown	No	No	No	GABA
Layer II	Unknown	No	No	No	GABA
Layers II, III	Unknown	No	No	No	GABA
Layer III	All	No	No	No	Glu/Asp
Layers I–III	All	No	No	No	Glu/Asp
Layer III	All	No	No	No	Glu/GABA
Layer III	All	No	No	No	GABA
Layers Ib, II; III (?)	No	No	No	No	GABA
Layers Ib, II, III	Possibly	No	No	No	GABA
Layers I, II, IIIa	Possibly	No	No	No	GABA
Layers II, III	No	No	No	No	GABA
Unknown	Possibly	No	No	No	GABA
Layers II, IIIa, IIIb, IV	All	Yes	No	No	Glu/Asp
Layers IIIa, IIIb, IVa, IVb	All	Yes	No	No	Glu/Asp
Layer III	All	Yes	No	No	Glu/Asp
Layers II, IIIa, IIIb	Unknown	Possibly	No	No	Glu/Asp
Uncertain (cf. 17)	Unknown	Possibly	Yes	No	Glu/Asp
Layers II, IIIa, IIIb	Possibly	Yes	No	No	GABA or Glu/Asp
Layers IIIa, IIIb, IVa	No	No	No	No	GABA
Layers IIIa, IIIb	No	Unknown	No	No	GABA
Unknown	No	Unknown	No	No	GABA or Glu/Asp
Layer IIIa	Possibly	Unknown	No	No	GABA or Glu/Asp
Layer IIIa	No	No	No	No	GABA

TABLE 6.2. *Continued*

Layer	Type of neuron	Somatic size[a]	Somatic shape	Size of dendritic field[a]
			Structure	
IV[e]	25. Neurogliaform cell	20 × 15 μm	Round	80 × 100 μm
	26. Small tufted neuron	10 × 10 μm	Round or oblate	80 × 240 μm
	27. Medium-sized tufted cell	12 × 12 μm	Round or flask-shaped	100 × 250–350 μm
	28. Large tufted cell	12 × 16 μm	Oblate or flask-shaped	120 × 350–400 μm
	29. Large multipolar cell	15 × 15–25 μm	Round	300 × 450 μm
	30. Spiny stellate cell	8 × 10 μm	Round	300 × 300 μm
	31. Double bouquet cell	10 × 15 μm	Elongated	50–75 × 500 μm
V[f]	32. Medium-sized pyramidal cell	25 × 40 μm	Pyramidal	300 × 1400 μm
	33. Large pyramidal cell	30 × 50 μm	Pyramidal	400 × 1600 μm
	34. Star pyramidal cell	30 × 25 μm	Oval or oblate	400 × 1200 μm
	35. Inverted pyramidal cell	30 × 40 μm	Pyramidal	400 × 600 μm
	36. Medium-sized multipolar cell	20 × 20 μm	Oval or oblate	300 × 200 μm
	37. Large multipolar cell	35 × 35 μm	Oval	400 × 400 μm
	38. Fusiform pyramidal cell	25 × 40 μm	Oval	200 × 500 μm
VI[g]	39. Small pyramidal cell	13 × 17 μm	Triangular	175 × 450 μm
	40. Medium-sized pyramidal cell	17 × 28 μm	Triangular	210 × 580 μm
	41. Fusiform pyramidal cell	10 × 25 μm	Fusiform	180 × 620 μm
	42. Inverted pyramidal cell	20 × 30 μm	Triangular	225 × 500 μm
	43. Small multipolar cell	12 × 17 μm	Round or fusiform	240 × 260 μm
	44. Medium-sized multipolar cell	16 × 18 μm	Multiangular	320 × 270 μm
	45. Large multipolar cell	20 × 35 μm	Round or oval	400 × 500 μm
	46. Bipolar cell	15 × 35 μm	Fusiform	200 × 560 μm
	47. Horizontal cell	35 × 15 μm	Fusiform	600 × 130 μm

TABLE 6.2 *Continued*

	Connectivity				Neurochemistry
Intrinsic	Cortico-cortical	Commis-sural	Cortico-thalamic	Cortico-collicular	Possible transmitter
Layer III	Possibly	No	No	No	GABA
Layers III, IV	Possibly	No	No	No	GABA or Glu/Asp
Layers III, IV, perhaps others	Possibly	No	No	No	GABA or Glu/Asp
Layers IIIb, IV, perhaps others	Possibly	No	No	No	GABA or Glu/Asp
Layers II, III, IV	Possibly	No	No	No	GABA or Glu/Asp
Layers III, IV	Possibly	No	No	No	GABA
Layers II, III, IV, V	Possibly	No	No	No	GABA
Layers V, VI	Possibly	Possibly	Yes	Yes	Glu/Asp
Layer V	Possibly	Possibly	No	Yes	Glu/Asp
Layers Va, Vb, VI	Possibly	Possibly	No	Possibly	Glu/Asp
Layers Vb, VI	Possibly	Yes	No	No	GABA or Glu/Asp
Layers IV, Va	Possibly	No	No	No	GABA
Layers IVb, Va, Vb	Possibly	No	No	No	GABA
Layer V	Possibly	Yes	No	No	Glu/Asp
Layer VI	AII	Yes	Yes	Yes	Glu/Asp
Layer VI	AII	Yes	Yes	Yes	Glu/Asp
Unknown	Possibly	Possibly	Possibly	No	Glu/Asp
Layers V, VI	Yes	Yes	No	No	Glu/Asp or GABA
Layer VI	No	No	No	No	GABA
Layers V, VI	Unknown	No	No	No	GABA
Layers V, VI	Unknown	No	No	No	GABA
Layers V, VI	Unknown	No	No	No	GABA
Layer VI	Unknown	No	No	No	GABA

[a] Width by height.
[b] From Sousa-Pinto et al. (1975), Winer (1986), and Prieto and Winer (unpublished observations).
[c] From Winer (1985a).
[d] From Winer (1984b,c).
[e] From Winer (1984a).
[f] From Winer (1992).
[g] From Prieto and Winer (1991).

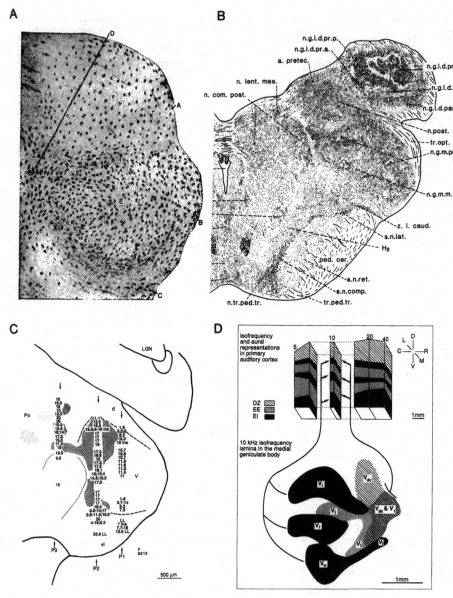

FIGURE 6.2A–D. Caption on pages 234–235.

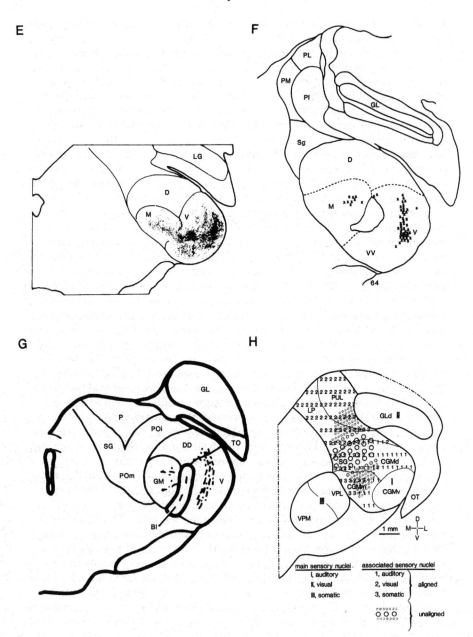

FIGURE 6.2E–H. Caption on pages 234–235.

FIGURE 6.2. Historical and methodological perspective on medial geniculate body subdivisions in carnivores. This figure reproduces material from the original sources; only the lettering has been changed for purposes of uniformity and clarity; each figure is presented on the right side for comparative purposes. Reproduced by permission. A. From Ramón y Cajal ([1911], p. 285). A Nissl preparation midway through the medial geniculate body. (*A*) superior lobe (dorsal division in the present account). (*B*) inferior lobe (ventral division, *pars lateralis*). (*C*) suprapeduncular nucleus. (*D*) medial or deep nucleus (medial division). (*E*) accessory ovoid nucleus (*pars ovoidea*). B. From Rioch ([1929], p. 105). A toluidine-blue stained section from an adult dog; × 5.6. Only two parts of the medial geniculate body are recognized: a principal nucleus (*n.g.m.pr.*) that includes both the ventral and the dorsal divisions, and a *pars magnocellularis* that refers to the large-celled portion of the medial division (*n.g.m.m.*). A territory that may correspond to the suprageniculate nucleus (above the pointer for *n.g.m.pr.*) is especially prominent. C. From Morel and Imig ([1987], p. 133). Distribution of retrogradely labeled thalamocortical neurons and density of labeled neurons (*stippled zones*) in combination with electrophysiological mapping in the medial geniculate body (*columns of numbers*) and in primary auditory cortex. Horseradish peroxidase (0.03 μl, 20% solution) was injected at one locus in the 17 kHz representation in AI. (*P1-P3*) thalamic electrode penetrations; *dashed* or *dotted lines*: limits of thalamoacoustic tract; (*d*) dorsal cap nucleus (perhaps superficial dorsal nucleus); *LGN*, lateral geniculate nucleus; *Po*, lateral part of the posterior group of thalamic nuclei; (*V*) ventral nucleus (probably corresponding to parts of the *pars lateralis* of the ventral division [cf. panel H]); (*vl*) ventrolateral nucleus; (*ins*) insecure response to tone; (*LL*) long-latency response (>50 msec); when the frequency range limits are more than one octave, the upper and lower limits are given; (*F*) denotes the anteroposterior level of the section. D. From Middlebrooks and Zook ([1983], p. 219), showing (*upper part*) the primary auditory cortex and adjoining dorsal auditory zone across which are arrayed isofrequency representations (5...40 [kHz]), and orthogonal to which are aural response bands (*stippled*). A single, 10 kHz isofrequency contour has been isolated for illustrative purposes. The auditory thalamic origins (*lower part* of the figure) of the projections to the 10 kHz region reveal that excitatory-excitatory (*EE*) or excitatory-inhibitory (*EI*) aggregates of thalamocortical cells of origin have physically segregated aural targets in the auditory cortex that conserve this representation. The entire 10 kHz representation in the ventral division has been flattened on the plane of the page, showing that V₁ (the *pars lateralis*) and Vₒ (the *pars ovoidea*) have independent aural representations, while V_dc (dorsal cap) neurons project towards the dorsal zone, whose cells have different physiological properties than do AI cells. These thalamic sources overlap spatially only in the rostral of V₁ (*lower right*). Modified with permission. E. From Kudo and Niimi ([1980], p. 548) showing the distribution of [³H]leucine terminal labeling in the ventral (*V*) and medial (*M*) divisions of the medial geniculate body after an injection in the central nucleus of the inferior colliculus. Note especially the orientation of ventral division labeling, and the differences in the disposition of silver grains in the lateral and more medial territories of the ventral division, respectively. The dispersed distribution of terminals within the medial division is also noteworthy, as is the comparatively much lighter input. (*D*) dorsal division of medial geniculate body; (*LG*) lateral geniculate body. F. From Niimi and Matsuoka ([1979], p. 10, panel 64). To show the distribution of retrogradely labeled cells in the medial geniculate body after an injection of horseradish peroxidase in AI. A prominent band of labeled cells (*crosses*) is present midway through the mediolateral axis of the ventral division (*V*), and the injection site (not shown) is between the anterior and posterior ectosylvian sulci. Fewer labeled cells are found in its magnocellular division (compare with Fig. 6.1C). (*D*) dorsal principal part of the medial geniculate body; (*GL*) lateral geniculate nucleus; (*M*) magnocellular part of the medial geniculate body; (*PI*) inferior pulvinar nucleus; (*PL*) lateral pulvinar nucleus; (*PM*) medial pulvinar nucleus; (*Sg*) suprageniculate nucleus; (*V*) main laminated portion of ventral principal part of the medial geniculate body; (*VV*) ventromedial portion of ventral principal part of the medial geniculate body. G. From Diamond et al. ([1969], p. 317). The distribution of degenerating corticothalamic axons terminating in various parts of the medial division as revealed by the Nauta-Gygax method following an ablation midway between the anterior and posterior ectosylvian sulci. The ensuing axonal debris involves mainly the ventral division along the bulk of its rostrocaudal extent, with far lighter degeneration in the medial division and virtually none in the dorsal division. Moreover, the degeneration forms a band-like array reminiscent of the pattern of thalamocortical cells of origin (panels C,F) and of tectothalamic terminal fields (panel E). (*BI*) brachium of the inferior colliculus; (*DD*) deep dorsal nucleus of small-cell division of medial geniculate nucleus; (*GL*) lateral geniculate nucleus; (*GM*) large-cell division of medial geniculate nucleus; (*P*) pulvinar; (*POi*) intermediate division of the posterior group of the thalamus; (*POm*) medial division of the posterior group of the thalamus; (*SG*) suprageniculate nucleus; (*TO*) optic tract. H. Schematic view of cytoarchitectonic subdivisions of the medial geniculate complex and adjoining thalamic territories related to physiological and functional data. The *main sensory nuclei* are the thalamic representatives of the lemniscal pathway, and they share a more or less common set of attributes such as small, nonoverlapping, exclusively auditory receptive fields, large-caliber afferent axons, brain stem or cortical affiliations primarily with one area, and a particular neuronal organization in which neurons with a bushy tufted architecture predominate. The *aligned cell groups* are specific to one modality or another, but they often have larger or overlapping (though in general still modality specific) receptive fields, receive input from several brain stem sources while projecting to more than one cortical field, and tend to have many more types of neurons than the main sensory nuclei; their dendritic arbors are usually less tufted than those of their main sensory counterparts. The *unaligned sensory nuclei* represent a region that is ill-defined in an architectonic sense, including as it does neurons from adjoining visual, somatic sensory, and auditory representations. These neurons often are responsive to input from more than one modality, they have large, broadly-tuned receptive fields, project to more than one cortical architectonic target, receive brain stem input from many sources, and contain many stellate neurons with irregularly radiating dendritic fields. (*CGMd*) dorsal division of the medial geniculate body; (*CGMm*) medial division of the medial geniculate body; (*CGMv*) ventral division of the medial geniculate body; (*GLd*) lateral geniculate complex; (*LP*) lateral posterior nucleus; (*PUL*) pulvinar; (*VPL*) ventroposterolateral thalamic nucleus; (*VPM*) ventroposteromedial thalamic nucleus.

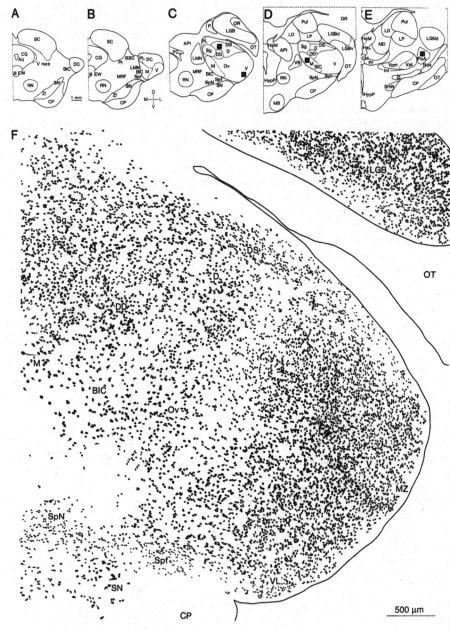

FIGURE 6.3A–F. Caption on page 238.

G   H

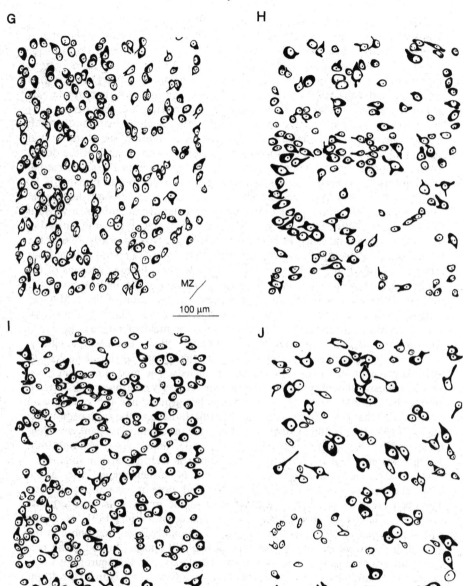

I   J

FIGURE 6.3G–J. Caption on page 238.

FIGURE 6.3. The main cytoarchitectonic subdivisions of the medial geniculate complex recognized in this account. A. From the posterior pole of the medial geniculate body, where only the caudal dorsal nucleus is present. Protocol for panels A-E: Nissl preparation; planachromat, N.A. 0.13, × 20. All abbreviations appear on the accompanying list. B. At about the caudal one-quarter of the medial geniculate body; the ventral division has begun, and assumes its characteristic position, and the medial division is at its largest relative to other divisions at this level and it is criss-crossed by brachial fibers. C. Slightly rostral to the midpoint of the medial geniculated body, showing the maximum size of the ventral division (V), the full range of the cytoarchitectonic differentiation of the dorsal division, (D, DS, Sg, and so on), and the relative regression in size of the medial division (M). The neurons in this section appear at higher power in panel F. D. At a level where less than one-quarter of the medial geniculate body lies rostral to this section. All subdivisions of the auditory thalamus are prominent; rostral to this level, thalamofugal axons begin to obscure the regular laminar organization that prevails more caudally. F. Near the rostral pole, beyond nearly 95% of the medial geniculate complex and showing the relations of the rostral pole nucleus (PoA) to the overlying lateral geniculate body and the adjoining neurons of the ventrobasal complex (Vpm, Vpl). Note the proximity of the thalamic reticular nucleus (RtN). F. Major divisions and nuclear configuration of the medial geniculate body at a level corresponding to panel C above. The high density of ventral division neurons is striking, especially in the most lateral part; the dorsal division is clearly heterogeneous, both in terms of cell size and packing density, ranging from medium-sized, scattered cells in the superficial dorsal nucleus to much larger neurons in the suprageniculate nucleus. The medial division has a very low density of neurons, many of which are relatively enormous and scattered among the brachial axons. Other, much smaller groups, like the ventrolateral nucleus (VL) that lies entirely within the ventral division or the suprapeduncular nucleus (SpN) that is affiliated with the posterior intralaminar system, are identified readily. Planapochromat, N.A. 0.32, × 125. Panels (G-J) neurons from selected nuclei at higher power. Protocol for panels G-J: planapochromat, N.A. 1.32, × 2000. G. Ventral division (V) neurons. Note their regular orientation, bimodal distribution of cell size (cf. Table 6.1) and more or less uniform density. The open areas are filled with the dendrites of the bushy cells, and terminal processes of tecto- and corticogeniculate afferent axons. (Winer 1991). H. In the rostral pole nucleus (PoA), the neurons are considerably larger than those in the ventral division proper; the distribution of size still appears bimodal, the predominant orientation is along a lateral-to-medial axis, and the neuropil is dominated by thalamofugal and corticofugal afferents. I. Cytoarchitecture of the dorsal nucleus. These neurons have a broad range of shape and size; a few of the cells are as large as those in the ventral division, but many are much smaller. Except for a few large neurons oriented mediolaterally, the cells seem scattered in small clusters without any more ordered arrangement. Some of the smaller neurons have more elaborate nuclear infoldings than any others in the medial geniculate body. J. Medial division neurons show a very broad range of size and include some of the smallest as well as the largest neurons in the medial geniculate body, not to mention other thalamic nuclei. Some cells have dendrites oriented parallel to the axons of brachial afferents, and very few cells have pronounced nuclear infoldings.

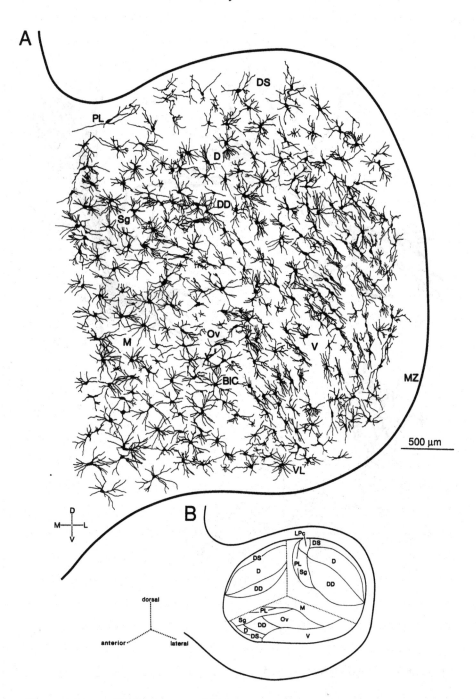

FIGURE 6.4A,B. Caption on page 243.

C

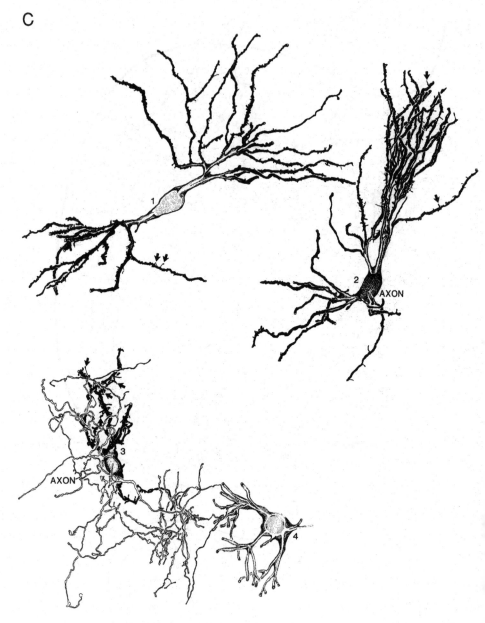

10 μm

FIGURE 6.4C. Caption on page 243.

D

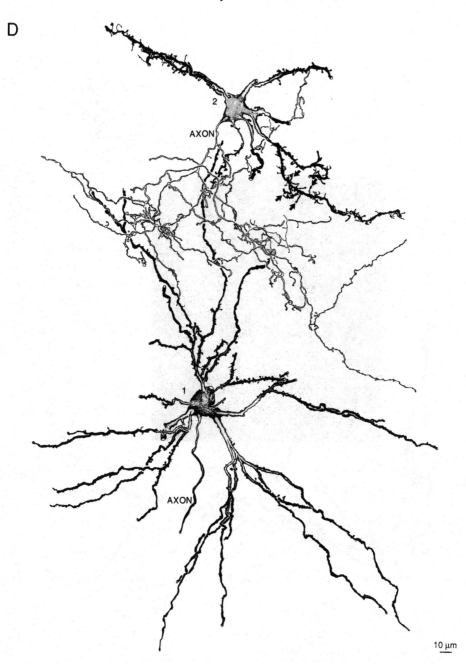

AXON

1

AXON

10 μm

FIGURE 6.4D. Caption on page 243.

E

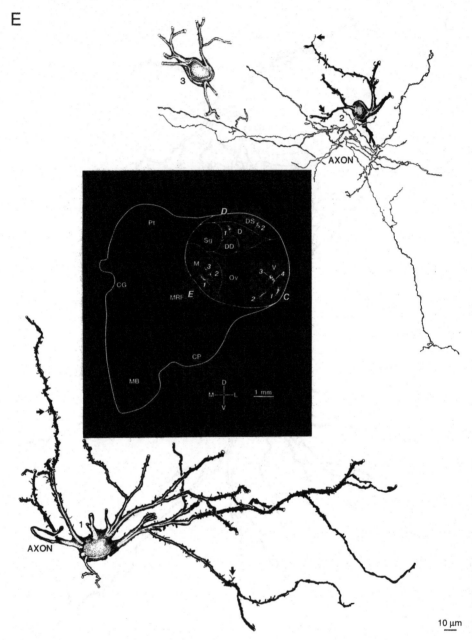

10 μm

FIGURE 6.4E. Caption on page 243.

FIGURE 6.4. Principal features of the neuronal architecture of the medial geniculate body. A. Cartoon showing the main differences between divisions; bushy tufted principal cells arranged in parallel fibrodendritic laminae are prominent in the ventral division (V), neurons with a radiating branching pattern are seen in the dorsal division nuclei (DS, D), and medial division neurons (M) express a variety of forms and a wide range in size. Golgi-Cox impregnation, planachromat, N.A. 0.35, × 200. B. Axonometric view of the major nuclear boundaries in the medial geniculate complex after dissection of the hippocampus and optic tract. Reproduced with permission. C-E. Characteristic neuronal types in various medial geniculate divisions from Golgi material. Protocol for panels C-E: rapid Golgi preparations, 41-day-old cat, planapochromat, N.A. 1.32, × 2000. C. From the ventral division, showing representative bushy neurons (1, 2). The main dendritic trunks arise at the somatic poles and, within 50 μm, branch profusely to fill a narrow volume of neuropil. Dendritic appendages are moderate in length and simple in structure (arrows) relative to those of Golgi type II cells (see below). The AXON is not impregnated beyond its initial segment; even in immature specimens, extensive axonal branching is unusual. (3) A Golgi type II cell, whose definitive features are an extensively branched local AXON, small soma, limited number of dendrites, elaborate dendritic appendages (arrows), and small dendritic field; note as well the orientation of this cell with respect to the nearby principal cell (4), demonstrating that interneurons are aligned with some elements of the fibrodendritic laminae. D. Neurons from the dorsal division. (1) A tufted neuron from the dorsal nucleus shows the large, nonoriented dendritic domains typical of such cells; several large trunks project irregularly beyond the section. Most appendages are confined to the intermediate dendrites, and the spines tend to be short and bulbous. (2) A large Golgi type II cell from the superficial dorsal nucleus; this type of cell is rarely impregnated in the medial geniculate body. It has a well developed AXON with many local trunks, which vary in diameter from quite slender up to 3 μm thick, and some of the branches form a robust local plexus and make complex endings, many of which run parallel to the cells' dendritic axis and are orthogonal to the weakly laminated arrangement typical of this nucleus. The dendrites have a complex, irregular configuration, sometimes winding about each other or projecting irregularly; their appendages (arrows) are highly developed and the most extensive of any in the auditory thalamus. E. From the medial division. (1) A magnocellular neuron (cf. Table 6.1) from the medial division, with long, sparsely branched dendrites, a myelinated AXON, and an orientation both parallel and at right angles to brachial axons. This neuron has a considerable number and a wide range of dendritic appendages, some quite long and elaborate (arrows). The irregularity of the dendritic domain can be judged by the many cut profiles (open endings); only one of the many secondary branches ends in the plane of the section. (2) Tiny, stellate Golgi type II cells with delicate dendrites are often impregnated in the medial division (see Fig. 6.3J). Their AXON is, in general, not as elaborately developed as those of comparable cells in the ventral and dorsal divisions, and their dendritic appendages are neither as widespread (arrows) nor as numerous. In fact, the dendrites of this neuron are a fraction of the diameter of those of related ventral and dorsal division cells. (3) elongate neuron soma. (Inset) locus of neurons.

While their dendritic domains are generally much smaller than those of bushy neurons, some stellate cells have dendrites far longer than might be expected on the basis of their somatic size or their appearance in Golgi impregnations; many such cells are also immunoreactive for glutamic acid decarboxylase in the rat (Winer and Larue 1988; cf. 2.5.2) or γ-aminobutyric acid (GABA) in the cat (Larue et al. 1992).

2.1.2 Dorsal Division

In contrast to the relative morphological homogeneity of the ventral division, dorsal division neurons are much more diverse. Five distinct nuclei are recognized, each with a different architectonic, connectional, and neurochemical arrangement. The dominant neuronal type throughout the dorsal division is the stellate neuron with a dichotomous, as opposed to a tufted, branching pattern. While significant numbers of bushy cells also exist, the expression of this trait is typically much weaker than in the ventral division, and there is a corresponding absence of any widespread laminar organization.

Principal stellate neurons have many more primary trunks than ventral division tufted neurons. These branches radiate more or less evenly to fill a sphere, and it is not unusual to find considerable overlap among adjoining neurons that obscures any obvious laminar pattern. The form, number, and position of dendritic appendages is reminiscent of that in the ventral division: they are of modest length (in adult material), sparse to moderate in number, and they are concentrated along the intermediate dendritic trunks.

The bushy principal neuron occurs chiefly in the superficial dorsal, dorsal, and deep dorsal nuclei in the dorsal division. It is often impregnated in the superficial dorsal nucleus, where it is as common as radiate neurons. In the dorsal and deep dorsal nuclei, cells with more spherical dendritic domains and a radiate pattern of branching prevail. The axon, like that of the radiate cell, is of medium size and usually myelinated; few local branches occur, even in immature animals. There is little preferred orientation among the dendritic arbors of bushy cells, and these tufts may be arranged at right angles to one another, especially in the dorsal nucleus. In the superficial dorsal nucleus these neurons sometimes have a mediolateral orientation that confers a laminar texture whose regularity is interrupted by the spherical dendritic arbors of the stellate neurons.

A third prominent cell type is the small stellate interneuron, which in most respects resembles its ventral division counterpart. This neuron represents about one-third of the population in the dorsal division. Their oval- or drumstick-shaped soma gives rise to various dendritic arrangements, ranging from simple radiate patterns to more polarized configurations. The very thin axon is always unmyelinated and has a rich local plexus of extremely delicate branches, many of which project beyond the limits of 120–200-μm-thick rapid Golgi preparations and have an irregular distribution within the section. These axonal branches tend to be more numerous and elaborate than those of comparable cells in the ventral and medial divisions. A larger putative interneuron exists in the dorsal nuclei, though in small numbers (Winer and Morest 1984).

The suprageniculate nucleus of the dorsal division forms a conspicuous oval mass interposed between the smaller cells and more diverse neuronal populations of the dorsal division (dorsolateral border), the posterior limitans nucleus (medial border), the medial division (ventral boundary), and the lateral posterior thalamic nucleus or pulvinar (anterodorsal border), to which its neurons are related structurally (Ogren and Hendrickson 1979). The suprageniculate nucleus is distinguished from the dorsal division territories by the comparatively larger size and the lower density of its neurons. These stellate cells have a simple, dichotomous branching pattern and large dendritic fields filling much of the nucleus; the thick axon has no local branches and is heavily myelinated.

A much smaller interneuron is also present in the suprageniculate nucleus. It has a modest dendritic field and poorly developed arbors, while the axon has fewer local branches than do comparable neurons in other dorsal division nuclei.

The posterior limitans nucleus of the dorsal division forms a slender shell on the dorsomedial face of the medial geniculate complex, between the brachium of the superior colliculus (dorsally) and the medial division and lateral mesencephalic nucleus (ventromedially); it is present only in the caudal one-third of the auditory thalamus. The major type of neuron, the elongated principal cell, has long, sparsely branched dendrites whose major axis is parallel to the plane of the nucleus. Their form resembles that of neurons in the interlaminar, as opposed to cells in the main sensory, thalamic nuclei (Winer et al. 1988). A small stellate cell with an intrinsic axon is also impregnated. The existence of a limitans nucleus has been disputed in insectivores and the rat and confirmed in rabbit, cat, monkey, and humans (Moryś et al. 1987).

2.1.3 Medial Division

The architecture of medial division neurons is the most diverse among the parts of the medial geniculate complex. In spite of this, the cells are characteristically thalamic in their form, closely resembling such neurons in related, nonauditory thalamic sensory nuclei (Winer and Morest 1983b). In Nissl preparations this division has the largest neurons and lowest packing density in the auditory thalamus, which reflects the many brachial and extrabrachial axons that traverse it (cf. 2.4). This highly reticulated neuropil texture and the intermixture of the many types of resident neurons has confounded any further regional architectonic parcellation of the medial division.

Four types of large neuron and one kind of smaller cell are routinely impregnated. Of the former, stellate and tufted neurons predominate, and elongated neurons are less common. The medium-sized stellate neurons have irregularly oriented dendritic domains of which one arbor is often much more developed at the expense of another. Tufted cells have less bushy but longer arbors than those of comparable ventral and dorsal division neurons, and the same dendritic asymmetry noted above. Elongated cells have smaller dendritic domains than tufted cells and these often follow the trajectory of brachial axons. The magnocellular neurons have modest dendritic arbors, a simple stellate branching pattern and, like the other three types of large neuron, a myelinated axon without local branches. A much smaller and rarer stellate neuron with a correspondingly limited dendritic field and a locally ramifying axon also occurs.

2.2 Connections of the Auditory Thalamus

A striking feature of the cochlear sensory epithelium is the orderly arrangement of inner and outer hair cells with respect to their characteristic frequency and the physical arrangement of the basilar membrane that supports this tonotopic organization (Békésy 1960). A primary attribute of this pathway is its divergence at early (that is, at cochlear nuclei) and subsequent stages of synaptic processing, so that multiple, independent tonotopic maps are created (Bourk et al. 1981; Rhode 1992). There is evidence for such representations among successive brain stem auditory nuclei, including the cochlear nuclei, the superior olivary and trapezoid body nuclei, the several subdivisions of the lateral lemniscus, and the inferior colliculus (Irvine 1986). In at least some of these structures, for example, the cochlear nuclei, local patterns of intrinsic connections between tonotopic nuclei provide for further regional interactions between spatially segregated representations of frequency (Wickesberg and Oertel 1988), besides the actions of local circuits confined to the nucleus of origin (for example, the granule cell system of the dorsal cochlear nucleus; see Mugnaini et al. 1980). A second property of this ascending pathway is the segregation of monaural and binaural channels to particular neuronal subsets in a nucleus (Calford 1983). Since the medial geniculate complex has both a tonotopic arrangement and a representation of different aural types, these must reflect the patterns of brain stem input as well as intrinsic processing. An appreciation of these connections is also critical for understanding the singular arrangement in different cortical fields of best frequency, sharpness of tuning, and aural representation that, at least to a first approximation, might embody the physiological attributes of different thalamic nuclei that project upon cortex (Imig and Morel 1983; Schreiner and Cynader 1984).

2.2.1 Brain Stem and Subcortical Afferents

The principal input to the ventral division (Table 6.3) arises from the central nucleus of the ipsilateral inferior colliculus (Andersen et al. 1980b; Rouiller and de Ribaupierre 1985). These axonal endings supply both the lateral, low-frequency and more medial, high-frequency parts of the ventral division, entering the auditory thalamus ventromedially from the brachium of the inferior colliculus and terminating along the proximal dendrites of bushy principal neurons and upon Golgi type II cell dendrites (Ramón y Cajal 1911; Morest 1964; Jones and Rockel 1971). Their arrangement within the ventral division is topographic, such that tonotopically related parts of the midbrain and thalamic isofrequency representations are systematically connected along a fibrodendritic domain.

Among the most prominent neurons in the central nucleus of the inferior colliculus is a principal cell with a polarized, disc-shaped dendritic domain that contributes to the laminar structure of the central nucleus (Oliver and Morest 1984) and whose axon projects to the medial geniculate body (Oliver 1984). Such cells could subserve the orderly arrangement of best frequency within the central nucleus; their Q_{10} dB values and monaural and binaural response profiles (Aitkin 1986) are comparable to those in the ventral division of the medial geniculate body (Aitkin and Webster 1972; Aitkin et al. 1981; Calford and Webster 1981). Several types of central nucleus neurons in addition to the disc-shaped cells may also terminate in the ventral division (Oliver 1984). It is unknown whether each variety has the same postsynaptic target or ultrastructure (see Winer 1991). In the mustached bat, studies of diffusion-filled horseradish peroxidase-labeled tectothalamic axons suggest that the form of axons terminating in different subdivisions may be different, even when the inferior colliculus tracer deposits are confined largely to a single physiologically defined architectonic midbrain subdivision (Wenstrup and Winer 1987).

The main nonauditory projections arise from the thalamic reticular nucleus (Rouiller et al. 1985) and ventrolateral medullary nucleus (Kamiya et al. 1988). The former projection is common to all the main sensory thalamic nuclei, and may regulate electroencephalographic spindling among these cells, or alter the excitability of principal neurons to cortical feedback and, at least in some thalamic nuclei, provide surround inhibition (Shosaku et al. 1989). There is evidence for a segregation of auditory and other modalities within the thalamic reticular nucleus. These representations embody an even finer organization, such that topographically arranged sensory thalamic subdivisions project to topographically arranged reticular nucleus subdivisions, while nontopographic thalamic and reticular representations are likewise interrelated (Crabtree and Killackey 1989; Conley et al. 1991).

The brain stem connections of the dorsal division contrast with and are more diverse than those of the ventral division, and the strength of

TABLE 6.3. Ascending input to the medial geniculate body.

Medial Geniculate Body Targets

Brain Stem Sources		VENTRAL DIVISION			DORSAL DIVISION					MEDIAL DIVISION
		V	Ov	Ds	D	DD	Sg	Vl	PL	M
Inferior Colliculus	CN	strong	strong	—	—	unknown	—	—	unknown	strong
	Dm	—	strong	—	moderate	strong	—	—	unknown	light
	DC	—	—	—	strong	—	—	moderate	—	—
	LN	—	—	—	—	—	—	moderate	—	moderate
Brain Stem	LTS	—	—	—	moderate	strong	strong	unknown	unknown	strong
	SOC	—	—	—	—	—	—	—	unknown	moderate
	other	[1]	—	—	—	[1] [2] [3]	[2] [4] [2] [5]	—	—	[2] [6] [7] [8] [9]

Strength of Projection

strong — moderate — light — absent — unknown

The sources of the projection appear on the left-hand side, their target(s) on the upper axis. The strength of the input is indicated by the size of the symbol. The numbers in boxes refer to references given below the table. For abbreviations, see the List of Abbreviations. (See 2.2.1 for details.)

[1] From the thalamic reticular nucleus (Rouiller et al. 1985).
[2] From the ventrolateral medullary nucleus (Kamiya et al. 1988).
[3] From the nucleus sagulum (Aitkin et al. 1981).
[4] From the superior colliculus (Morest and Winer 1986).
[5] From the nucleus of the brachium of the inferior colliculus and the deep layers of the superior colliculus (Calford and Aitkin 1983).
[6] From the ventral nucleus of the lateral lemniscus (Whitley and Henkel 1984).
[7] From the spinal cord (Jones and Burton 1974; Berkley 1980).
[8] From the vestibular nuclei (Roucoux-Hanus and Boisacq-Schepens 1977; Blum et al. 1979).
[9] From the superior colliculus (Graham 1977).

the different projections reflects this variability (Table 6.3). Thus, the dorsal and deep dorsal nuclei receive input from the dorsomedial part of the central nucleus (Calford and Aitkin 1983) while substantial non-lemniscal pathways arising in the lateral tegmental system of the midbrain terminate within the deep dorsal and suprageniculate nuclei (Morest 1965b). Other input originates from extraauditory brain stem sites such as the pathway between the nucleus sagulum and the dorsal nucleus (Aitkin et al. 1981; Hutson 1988) and superior colliculus input to the suprageniculate nucleus (Calford and Aitkin 1983; Morest and Winer 1986). The latter circuit could have a role in the coordination of visual, motor, and acoustic representations of three-dimensional space, or in motor activity such as the accurate spatial pursuit of moving objects which would require interactions and correlations across modalities. The large input from auditory nuclei of the inferior colliculus aligns portions of the dorsal division with the ventral division and suggests that some parts of the lemniscal system may be considered functionally as parallel, since the pathways established at the level of the midbrain (and, by extension, in medullary auditory nuclei as well) have specific thalamic targets whose representation is also conserved in the cerebral cortex (see Niimi and Matsuoka 1979; Oliver 1984; Morest and Winer 1986). It is uncertain whether there is a point-to-point topographic relationship between the inferior colliculus or other brain stem nuclei that project to the dorsal division.

The perirhinal cortex has widespread input to several subdivisions of the medial geniculate complex. Its principal targets are the suprageniculate nucleus, the caudal parts of the nuclei of the dorsal division, and the caudal extremity of the dorsal division (Witter and Groenenwegen 1986). Besides the density of this input—which is perhaps the single heaviest afferent projection to these nuclei—these findings establish a critical subcortical linkage between elements of the limbic system (see Nieuwenhuys et al. 1988/1989) and auditory thalamic centers whose output might well be modified by the cortex during ongoing behavior (Ryugo and Weinberger 1976, 1978; Ryan et al. 1984). These limbic and parahippocampal inputs, in turn, receive input from regions near the subparafascicular and suprapeduncular nuclei (Room and Groenenwegen 1986) which would then propagate influences from these posterior intralaminar thalamic nuclei toward the limbic forebrain (see Witter et al. 1989 for a review).

In some respects, the suprageniculate nucleus appears to be a promising candidate for a multimodal sensory-motor posterior thalamic relay center (a more complete account appears in Winer and Morest 1983a). A variety of converging influences arise from the superior colliculus, substantia nigra, zona incerta, thalamic reticular nucleus, and cerebral cortex to terminate within it (Norita and Katoh 1987, 1988), and this singular pattern of afferent input distinguishes it from the remainder of the dorsal division. Since in the mustached bat (Kobler et al. 1987) and the rat (Kurokawa et al. 1990) suprageniculate neurons project, in turn, towards the frontal lobes, this could subserve a global role in the sensory control of the long term planning of motor strategies.

Input ascending to the medial division obeys different rules of organization than do afferents terminating in other medial geniculate divisions. First, such projections have unusually diverse origins; second, any particular projection is rarely as large as those to the ventral or dorsal divisions; third, there is no evidence that inputs have a topographic distribution within the medial division; fourth, nonauditory afferents predominate; and fifth, the synaptic pattern of brain stem input probably differs from that to the ventral division, where a relatively small number of afferent sources terminates on a comparatively homogeneous set of postsynaptic cells, while in the medial division input from many sources may end upon a structurally heterogeneous neuronal population.

Nearly every inferior colliculus subdivision, with the possible exception of the dorsal cortex (Table 6.3), projects to the medial division (Kudo and Niimi 1980). Other, primarily auditory input arises from the vicinity of the superior olivary complex (Henkel 1983) and from the ventral nucleus of the lateral lemniscus (Whitley and Henkel 1984). The chief (though much smaller) nonauditory projections originate in the spinal cord (Jones and Burton 1974; Berkley 1980), the vestibular nuclei (Roucoux-Hanus and Boisacq-Schepens 1977; Blum et al. 1979), and the superior colliculus (Graham 1977).

2.2.2 Auditory Forebrain Targets

The logic of tectothalamic connections is conserved in the pattern of thalamocortical connectivity (Table 6.4). Thus, the output of the ventral division is predominantly to the primary auditory cortex (Rose and Woolsey, 1949; Winer et al. 1977), to the anterior auditory field (AAF) adjoining it rostrally (Andersen et al. 1980a), and to portions of the posterior (P) auditory field in the anteroventral part of the posterior ectosylvian gyrus (Imig and Morel 1984). Each of these areas is characterized by neurons with high Q_{10} dB values and a clear gradient of isofrequency representation (Merzenich et al. 1975; Reale and Imig 1980). The topography of thalamocortical input is in close register with the physiological arrangement of best frequency within the ventral division and in AI. Thus, neurons in low-frequency ventral division fibrodendritic laminae project to caudal AI sectors (Imig and Morel 1984), where the apical, low-frequency sector of the cochlea is represented (Woolsey and Walzl 1942).

As might be expected on the basis of its varied brain stem input and its architectonic diversity, the cortical targets of the dorsal division include several different fields (Niimi and Naito 1974; Winer et al. 1977), each with a cytoarchitecture distinct from that of AI (Rose and Woolsey 1949) and characteristic patterns of corticothalamic (Table 6.4), commissural (Table 6.5), ipsilateral corticocortical (Table 6.6), and corticofugal (Table 6.7) connectivity. In contrast to the pathway between ventral

TABLE 6.4. Auditory thalamocortical and corticothalamic projections and extrathalamic input.

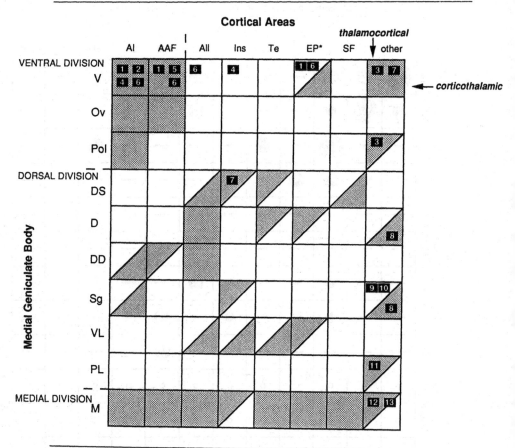

The left-hand side of the table shows the subdivisions of the medial geniculate body; the upper part, their cortical target or the source of corticothalamic projections. The *key* is in the upper right corner: the top half of each box, if filled, denotes a thalamocortical projection, the bottom half a corticothalamic one; if the box is filled completely, the projection is reciprocal. The numbers in boxes refer to extrathalamic sources of targets or input. (See 2.2.2, 2.2.3, and 3.3.3 for details.)

[1] From the lateral hypothalamic area (Sakal et al. 1983).

[2] From the claustrum (Neal et al. 1986; Rouiller et al. 1989a).

[3] From the banks of the ectosylvian visual area (Olson and Graybiel 1987).

[4] From the nucleus of the ventromedial mesencephalic tegmentum (Scheibner and Törk 1987; cf. Footnote 6).

[5] From the nucleus of the brachium of the inferior colliculus (Rouiller et al. 1989a).

[6] From the locus coeruleus, nuclei of the raphe, ventromedial mesencephalic tegmentum, posterior part of the lateral hypothalamus, and basal forebrain (Rouiller et al. 1989a).

[7] From the ventromedial tegmentum (Scheibner and Törk 1987).

[8] From the perirhinal cortex (Room and Groenenwegen 1986).

[9] To the putamen (Russchen 1982).

[10] To the caudate nucleus (Hu and Jayarman 1986).

[11] To the auditory cortex (Winer 1984d).

[12] Includes somatic sensory (Jones and Powell 1973) and auditory cortex (Winer et al. 1977; Niimi and Matsuoka 1979).

[13] To the amygdala (Russchen 1982).

* EP refers to the posterior and ventral posterior subdivisions of the posterior ectosylvian gyrus; see Table 6.5 for more specific data on other posterior ectosylvian territories.

TABLE 6.5. Commissural connections of auditory cortical fields.

Termination

area / Origin	AI	AAF	AII	SF	Ins	Te	P	VP	EPD	EPI	EPV	EPP
AI	strong	moderate	moderate	absent	absent	absent	moderate	absent	absent	absent	absent	absent
AAF	moderate	strong	light	absent	absent	absent	moderate	unknown	unknown	unknown	unknown	unknown
AII	absent	absent	strong	absent	moderate	absent	absent	absent	absent	moderate	absent	absent
SF	absent	absent	moderate	moderate	absent	absent	absent	absent	absent	absent	absent	absent
Ins	absent	absent	absent	absent	absent	absent	absent	absent	absent	absent	absent	absent
Te	absent	absent	absent	absent	absent	absent	absent	absent	absent	absent	absent	absent
P	absent	absent	light	absent	absent	light	moderate	absent	absent	absent	absent	absent
VP	absent	absent	light	absent	absent	light	absent	moderate	absent	absent	absent	absent
EPD	absent	absent	absent	absent	absent	absent	absent	moderate	absent	absent	absent	absent
EPI	unknown	unknown	unknown	unknown	unknown	unknown	unknown	unknown	unknown	unknown	unknown	unknown
EPV	unknown	unknown	unknown	unknown	unknown	unknown	unknown	unknown	unknown	unknown	unknown	unknown
EPP	unknown	unknown	unknown	unknown	unknown	unknown	unknown	unknown	unknown	unknown	unknown	unknown

Strength of Projection

strong	moderate	light	absent	unknown
● (large)	● (medium)	● (small)	—	○

The source of these projections in one hemisphere is on the left-hand side, the target at the top; the relative size and density of the input is proportional to the size of the symbols. (See 3.3.1 for details.)

TABLE 6.6. Corticocortical connections of auditory cortical fields.

The table reads from the sources (on the left) to the targets (on the top). Left-hand groupings: **Auditory** (AI–EPP), **Motor** (PS–7p), **Visual** (19–EVA). Symbol key — strong: ⬤, moderate: ●, light: •, absent: —, unknown: ○.

area	AI	AAF	AII	SF	Ins	Te	P	VP	EPD	EPI	EPN	EPP	other	
AI	╲	⬤	⬤	○	●	●	⬤	⬤	—	●	—	●	○	
AAF	⬤	╲	⬤	⬤	—		⬤	●	●	—	—	—	○	
AII	⬤	○	╲	●	●	—	⬤	⬤	—	⬤	⬤	—	●	EVA
SF	⬤	⬤	●	╲	●	—	⬤	●	⬤	⬤	—	●	●	EVA
Ins	⬤	⬤	•	•	╲	○	—	—	●	●	⬤	○	⬤	DLS EVA
Te	⬤	⬤	⬤	•	○	╲	○	○	⬤	⬤	⬤	⬤		PS
P	⬤	⬤	⬤	—	—	●	╲	⬤	—	●	●	—	○	
VP	⬤	⬤	⬤	—	—	●	○	╲	⬤	●	⬤	—	●	EVA
EPD	○	○	⬤	●	⬤	○	○	●	╲	—	—	○	⬤	EVA
EPI	○	○	○	○	⬤	○	○	●	⬤	╲	●	○	⬤	PS
EPV	○	○	●	○	●	○	○	—	●	⬤	╲	○	⬤	EVA
EPP	○	○	○	○	○	○	○	●	—	—	—	╲	⬤	DLS EVA
PS	○	○	○	○	○	—	○	—	—	○	○	○	⬤	DLS PLS
6m	○	○	○	○	○	○	○	—	—	—	—	—	⬤	DLS
7p	○	○	○	○	○	○	○	—	—	—	○	○	⬤	DLS EVA
19	○	○	○	○	○	○	○	—	—	—	○	○	⬤	DLS EVA
20a	○	○	○	○	○	○	○	—	—	—	○	○	⬤	DLS EVA
20b	○	○	○	○	○	○	○	—	●	⬤	○	○	⬤	DLS EVA
21a	○	○	○	○	○	○	○	●	—	—	○	○	⬤	DLS
21b	○	○	○	○	○	○	○	—	—	—	○	○	⬤	DLS
35/36	○	○	○	○	○	○	○	⬤	—	⬤	○	○	⬤	DLS
CG	○	○	○	○	○	○	○	—	—	⬤	○	○	⬤	DLS EVA
ALLS	○	○	○	○	○	○	○	—	—	—	○	○	⬤	DLS
PLLS	○	○	•	○	○	○	○	—	—	●	○	○	⬤	PS EVA
VLS	○	○	○	○	○	○	○	—	—	—	○	○	⬤	DLS
DLS	○	○	○	○	○	○	○	—	—	—	○	○	—	
PS	○	○	○	○	○	○	○	—	—	—	—	○	⬤	DLS
EVA	○	○	○	○	○	○	○	⬤	⬤	⬤	—	○	○	

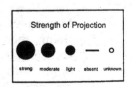

Strength of Projection

⬤ strong ● moderate • light — absent ○ unknown

The table reads from the sources (on the left) to the targets (on the top); corticocortical input to extraauditory cortical fields appears in the last column next to the abbreviations on the right-hand side. The diagonal line indicates that intrinsic corticocortical connections do not appear here, though they exist (for example, see Matsubara and Phillips 1988). The symbols denote the size and strength of the projection pattern. (See 3.3.2 for details.)

division and the cortical areas in the AI/AAF/P superfield, where sharply tuned thalamic neurons project to cortical cells with comparable tuning (see above), these nonprimary projections arise from thalamic nuclei whose Q_{10} dB values are typically much lower (Aitkin and Prain 1974; Calford and Webster 1981), and whose cortical targets likewise have only a modest degree of tonotopic organization (Reale and Imig 1980; Schreiner and Cynader 1984).

A further distinction between the ventral division—AI axis and the cortical projections of the dorsal division is that the former has few specific targets and more or less complete reciprocity of thalamocortical and corticothalamic projections (Table 6.4), while the dorsal division output is much more widely distributed and has a less regular arrangement of thalamocortical and corticothalamic reciprocity, except in a few instances (in the rat; Winer and Larue 1987). Such arrangements suggest that parallel pathways evident in the auditory thalamus (Winer and Morest 1983b; Winer 1985b, 1991) are probably present in the midbrain (Aitkin 1986; Shneiderman and Oliver 1989), and in yet more caudal brain stem loci (Brawer et al. 1974; Warr 1982).

In spite of the diversity of dorsal division projections to the cortex, many nuclei have specific targets. Thus, the superficial dorsal nucleus projects chiefly to the insular (Ins) and temporal (Te) fields adjoining the pseudosylvian sulcus (Raczkowski et al. 1976), while the deep dorsal nucleus sends axons to the second auditory cortex (AII) (Winer et al. 1977) and the AAF (Andersen et al. 1980a). The suprageniculate nucleus projects upon the insular cortex and broad territories in the vicinity of the anterior ectosylvian sulcus (Roda and Reinoso-Suárez 1983).

The widespread cortical projection of the medial division reflects the diversity of brain stem afferents to it (Table 6.3), and extends influences to each subdivision of auditory cortex, including tonotopically arranged fields as well as those with a less exact representation of frequency tuning. In addition, the medial division projects broadly to cortical territories beyond the domain classically considered as auditory, and including the somatic sensory cortex (Jones and Powell 1973) and the prefrontal cortex (Avedaño and Llamas 1984). While the areal breadth of this projection implies that it may have a nonspecific (or at least nontopographic) arrangement, other evidence suggests that there may be a more specific pattern to medial division organization. Studies using two different retrograde tracers, each injected at separate cortical loci, double-label less than 20% of medial geniculate body neurons (Bentivoglio et al. 1983; Morel and Imig 1987). Insofar as this finding applies to the medial division, then perhaps different pathways overlap spatially within it but are segregated on the basis of their brain stem and cortical connectivity or the form of their postsynaptic neurons. In short, the medial division may embody more than one pathway, the terminations of whose cells of origin overlap in an architectonically diverse territory (Winer and Morest

TABLE 6.7. Corticofugal projections of auditory cortex.

area / nucleus	AI	AAF	AII	SF	Ins	Te	P	VP	EPD	EPI	EPV	EPP
Medial Geniculate Body												
V	⬤	⬤	•	○	—	—	○	○	—	○	○	○
Ov	⬤	—	•	○	—	—	○	○	—	○	○	○
DS	—	—	●	—	—	—	○	○	—	○	○	○
D	—	—	⬤	⬤	⬤	⬤	○	○	●	○	○	○
DD	—	—	⬤	●	●	●	○	○	●	○	○	○
Sg	—	⬤	●	●	•	●	○	○	●	○	○	○
VI	—	—	⬤	—	⬤	⬤	○	○	—	○	○	○
PL	—	—	—	—	•	•	○	○	●	○	○	○
M	⬤	—	●	•	●	●	○	○	⬤	○	○	○
Inferior Colliculus												
CN	•	○	●	○	—	—	○	○	—	○	○	○
DM	⬤	○	●	○	●	●	○	○	●	○	○	○
DC	⬤	○	●	○	●	●	○	○	•	○	○	○
L	—	○	●	•	—	—	○	○	•	○	○	○
LN	●	○	⬤	—	—	—	○	○	•	○	○	○
VL	—	○	—	—	—	—	○	○	—	○	○	○
VN	—	○	—	—	—	—	○	○	—	○	○	○
M	—	○	—	⬤	—	—	○	○	•	○	○	○
Cu	—	○	—	⬤	—	—	○	○	—	○	○	○
IcT	●	○	•	—	—	—	○	○	—	○	○	○
Other												
SpN	—	○	○	—	●	●	○	—	○	○	○	○
SC	—	○	●	●	○	○	○	•	●	○	○	○
PT	—	○	•	●	○	○	○	—	⬤	○	○	○
DPN	•	○	●	○	○	○	○	•	•	•	•	○
Ca	⬤	○	●	○	○	○	⬤	○	○	○	○	○
Pu	⬤	○	●	○	○	○	⬤	⬤	○	○	○	○
LAm	○	○	○	○	○	○	⬤	⬤	○	○	○	○

Origin appears at the top. Left-hand margin groups: Medial Geniculate Body, Inferior Colliculus, Other (Termination).

Strength of Projection

⬤ strong   ● moderate   • light   — absent   ○ unknown

The origins appear at the top of the table and the left-hand side denotes the targets at auditory thalamic, midbrain, and other levels. The size and density of the input is scaled to the symbols. (See 3.3.5 for more details.)

1983b). Since AI receives such inputs, then even its sharply tuned neurons may receive nonspecific influence, or perhaps some subpopulation of medial division neurons may have Q_{10} dB values similar to those of sharply tuned cells that terminate in layers IIIb and IV (cf. 3.2).

2.2.3   Nonauditory Forebrain Targets

Many thalamic influences propagated to the cortex ultimately reach subcortical sites via corticofugal pathways (Table 6.6). However, other, subcortical thalamofugal targets also exist. These arise principally from the suprageniculate nucleus and the medial division and end in, respectively, the caudate nucleus (Hu and Jayarman 1986) and in the amygdala and putamen (Russchen 1982). Here, they may interact with motor responses to auditory stimuli (Clarey and Irvine 1986b) or participate in acoustically modulated autonomic or behavioral changes related to visceral sensibility (LeDoux et al. 1984).

2.3   Intrinsic Organization

Electron microscopic analysis of neural circuitry in the ventral division of the medial geniculate body has revealed an orderly synaptic sequence and stereotyped spatial arrangement of extrinsic afferent and interneuronal connections within the fibrodendritic laminae. Experimental degeneration studies show that medium-sized axons 1–5 μm in diameter and of inferior colliculus origin are presynaptic to dendrites and axons. Often, the presynaptic and postsynaptic elements are enveloped by fine astroglial processes, the resulting aggregate of axons and other processes surrounding a principal cell's dendrite being defined as a synaptic nest, glomerulus, or island (Jones and Rockel 1971). This arrangement is a common constituent of neuropil organization in the visual (Ohara et al. 1983) and somatic sensory (Ohara et al. 1989) thalamus of cats and primates. The preterminal axons form asymmetric endings containing spherical synaptic vesicles, and each terminal segment may have more than one ending in a particular nest. Many such terminals are evident along the main dendritic trunks. The smaller caliber (0.5–1 μm in diameter) and the largest (5 μm) axons are unaffected by midbrain lesions. The small stellate Golgi type II cell (Table 6.1) also receives afferent axons from the inferior colliculus; and their axon, in turn, is presynaptic in the glomerulus to bushy cell dendrites (Morest 1975). In addition, the interneuronal dendrites are presynaptic to these dendrites. Thus, axodendritic and dendrodendritic terminals, both of local origin, contribute small, flat vesicle-containing profiles to the neuropil; the endings originating from Golgi type II axons may be smaller and have flatter, smaller vesicles than those from dendrites (Morest 1971). Not all such axonal profiles with flattened vesicles are of local origin. The thalamic reticular nucleus, all or most of whose neurons are GABAergic (Houser et al. 1980), sends a significant number of such terminals to the ventral and dorsal divisions. In the rat ventral division, these form symmetric endings with flattened vesicles along primary and much thinner dendritic profiles (Montero 1983).

A third presynaptic profile is represented by the abundant corticogeniculate axons. These terminate on intermediate or distal dendritic segments of principal neurons and have round (Jones and Powell 1969) or sometimes flattened (Morest 1975) synaptic vesicles with associated asymmetric membrane morphology. Such axons typically enter the medial geniculate from its rostrodorsal pole to descend among the fibrodendritic laminae (Rouiller and de Ribaupierre 1990), while brachial axons enter ventromedially (Morest 1965a; Winer 1985b; for a different arrangement in the mustached bat, see Wenstrup and Winer 1987 and Fig. 6.5B).

Little is known about the ultrastructure of either neurons or axons in the dorsal or the medial divisions. In the former, the neuropil contains many extremely fine (0.2–0.5 μm in diameter) unmyelinated axons and a moderate number of thicker, myelinated axons. The thinnest axons probably arise from the fine branches of Golgi type II cells (Winer and Morest 1983b) or may represent the nonauditory brain stem afferents of limbic origin, some of which are serotonergic and of very small caliber (Dahlström and Fuxe 1964; Ungerstedt 1971; cf. 2.4.4.). In the medial division, the large preterminal fascicles of brachial axons passing toward the ventral division dominate the neuropil, and any local plexus of fine axons attributable to Golgi type II cells is correspondingly much less evident (Winer and Morest 1983a).

2.4 Neurochemistry

The development of antisera that selectively identify epitopes and the transmitter-specific labeling of neurons with tritiated amino acids have been essential tools in deciphering neural circuitry and in deriving a more complete picture of thalamic and cerebral cortical phylogeny. For example, the number of Golgi type II neurons in these structures, as identified in Golgi preparations, has long been interpreted as sparse in rodents, marsupials, and insectivores, and comparatively abundant in carnivores and monkeys. While this appears to be the case in the thalamus, it is somewhat surprising, then, to find that in the auditory cortex of the rat (Winer and Larue 1989), cat (Prieto et al. 1990), mustached bat, and the monkey (Winer, unpublished observations), a comparable laminar pattern (and perhaps a similar proportion) of GABAergic neurons occurs in each, while in the medial geniculate body there are dramatic species differences (Winer and Larue 1988; Huchton et al. 1991; Winer 1991; Winer et al. 1992). This suggests that phylogenetic interpretations about

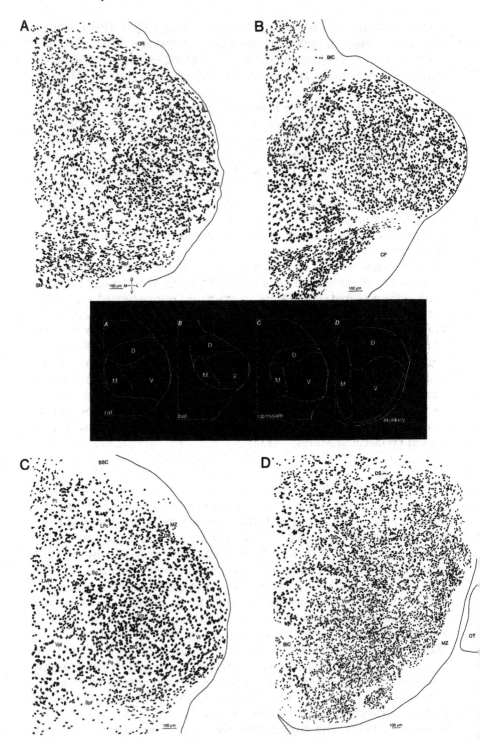

neural circuitry must be made with some caution and that a particular nuclear pattern may differ radically between species. At the same time, identification of a neuron as GABAergic does not preclude the possibility that it may project more remotely, perhaps while still having local actions (Vincent et al. 1983).

2.4.1 Tectothalamic Transmitter Candidates

Many types of inferior colliculus neurons project to the medial geniculate body (Oliver 1984), and these different cells may use a variety of trans-

---

FIGURE 6.5. Comparative cytoarchitectonic views of the medial geniculate body in the rat, bat, opossum, and monkey. A. Transverse section midway through the rat medial geniculate body. Most ventral division cells are oriented mediolaterally, except those at the perimeter of the ventral division, which are much more vertical. The dorsal and the medial divisions are reduced relative to the cat, though most of the same types of neurons can be identified in each. Protocol for panels A-C: planapochromat, N.A. 0.65, × 500; for panel D: planachromat, N.A. 0.32, × 320. B. In the mustached bat, the ventral division, especially its medial sector, is enormously enlarged compared to the rat, and the lateral-most sector (*Vl*) has a much more conspicuous laminar arrangement in Golgi preparations than the medial part (*Vm*). The dorsal division nuclei of the rat, in contrast to those in the cat, are much smaller, while retaining most of the types of neurons identified in the cat. Medial division (*M*) neurons are not conspicuously larger than cells in other parts of the medial geniculate body. Note the atypical location of the brachium of the inferior colliculus, which enters from the dorsomedial aspect of the medial geniculate body, distinguishing this bat from many other mammals. C. In the opossum, the medial division (*M*) is a mere fraction of its usual size in most placental mammals, yet all the subdivisions recognized in the cat are present, albeit often reduced to tiny nuclei less than 100 μm wide. The most unusual consequence of this is the enormous relative expansion of the ventral division, especially the *pars ovoidea* (*Ov*), which may represent as much as three-fourths the volume of the medial geniculate body. The nearby posterior intralaminar nuclei (*PIN*) are extraordinarily well developed, perhaps exceeding those in the bat. D. In the macaque monkey, the principal subdivisions noted in the cat are well delineated. Thus, the ventral division neurons situated more laterally often form laminae-like vertical arrangements, while more medial cells in this division (*V*) are separated by substantial bands of fibers more dispersed and devoid of any obvious orientation; populations of small and medium-sized neurons occur. The packing density of dorsal division neurons is much lower, and the nuclear configuration is well defined both on cytoarchitectonic grounds and fiber architecture. Thus, the suprageniculate nucleus (which lies dorsomedial to the other dorsal division nuclei and is not shown here) contains large, deeply staining cells that are distinguished from neurons in the adjoining dorsal (*D*) and deep dorsal (*DD*) nuclei on the basis of their larger size, more diverse shapes, lower packing density and the high local concentration of fibers. Medial division (*M*) neurons, in contrast, are even larger and more dispersed than suprageniculate nucleus cells, they embody a wider range of morphological diversity and are scattered among large expanses filled with brachial fibers.

mitters. Many inferior colliculus neurons show aspartate- or glutamate-like immunoreactivity (Winer, unpublished observations) or are GA-BAergic (Oliver et al. 1992). In the rat, the ventral division receives a cholinergic input of midbrain origin (Levey et al. 1987), while in the cat the number of choline acetyltransferase-immunopositive boutons is greatest in the dorsal division (Fitzpatrick et al. 1989), suggesting that many of the lateral tegmental cells of origin (Morest 1965b) may be cholinergic. The neurochemical identity of other tectothalamic pathways is unknown.

2.4.2   Thalamic Transmitter Candidates

Many of the largest, and some of the smaller, neurons in each medial geniculate body subdivision show aspartate- or glutamate-like immunoreactivity in the rat (Popowits et al. 1988) and cat (Winer, unpublished observations). It is unknown whether these cells project to the cortex (Tables 6.2, 6.4) or the nature of regional differences in their number or concentration. Some of the immunonegative neurons may use GABA (see below) or a different transmitter.

The proportion of GABAergic neurons and puncta (axon terminals) is species-specific for the limited phylogenetic series so far available. In the rat, only about 1% of auditory thalamic neurons are immunopositive for glutamic acid decarboxylase (Winer and Larue 1988), while in the cat (Rinvik et al. 1987; Winer 1991) and monkey (Smith et al. 1987) the proportion is much larger, perhaps 25% (Fig. 6.6). In the mustached bat, the number is less than 1% (Winer et al. 1992) and in the opossum the ratio is probably comparable to that in the rat and bat (Penny et al. 1984). Many, but not all auditory thalamic GABAergic neurons have small, drumstick-shaped somata and thin or medium-sized dendrites.

The distribution and density of GABAergic puncta are different among medial geniculate divisions, and these patterns, if not the absolute values, are conserved in the species where quantitative estimates exist. Thus, in the cat ventral division, the puncta are medium-sized to small, and GA-BAergic neurons typically receive a moderate number of axosomatic contacts (Huchton et al. 1991; Larue et al. 1992). In contrast, dorsal nucleus puncta are far smaller and more granular in form and less than half as dense as those in the ventral division; endings upon immunopositive or immunonegative neurons are correspondingly rarer. The medial division has a few fine and many coarse or complex puncta, some of which are much larger than those in other divisions, with an average density nearer to the ventral than to the dorsal division value; some cells receive a few GABAergic endings, others are so heavily encrusted as to stand in relief, whether or not they are immunopositive.

Certain physiological consequences may follow from these patterns. Assuming that no GABAergic neurons from the vicinity of the dorsal nucleus of the lateral lemniscus or the inferior colliculus project upon the medial geniculate body, then perhaps the intrageniculate processing of ascending (or, for that matter, descending) input is unique for each division and reflects the activity only of Golgi II cells (Table 6.1) and GABAergic input from the thalamic reticular nucleus (Table 6.3). Since the dorsal division has far fewer puncta than the ventral division in the rat, bat, cat and monkey, it is unlikely that local circuits alone could explain the wide array of effects on temporal discharge, interneuron-mediated inhibition, or on large scale excitatory-inhibitory oscillations, given the presumably selective nature of the GABAergic influence on principal cells and interneurons. However, without iontophoretic and intracellular studies, the precise nature of the neuronal circuits serving these arrangements is unknown. Comparisons between studies using gamma-aminobutyric acid and glutamic acid decarboxylase should be made with some caution since they do not invariably identify the same neuronal populations, at least in the retina (Hurd and Eldred 1989).

In the scenario for GABAergic modulation of the excitability of auditory thalamic neurons described above, the implicit presumption is that the ionic actions of GABA are directed principally toward chloride-dependent, GABA$_A$ receptor-mediated channels whose actions on the postsynaptic membrane are usually construed as temporally rapid and limited spatially to a comparatively small (and critically important) domain on the cell surface. However, insofar as these GABAergic actions could separately or in parallel entail the activation of postsynaptic GABA$_B$ receptors, an entirely different set of consequences might prevail, including slow and long-lasting changes in membrane resting potential that could exert a preparatory or 'priming' role for bursting discharge of thalamocortical relay neurons through actions upon second messenger systems, as demonstrated in other main thalamic sensory relay nuclei (McCormick and Feeser 1990; Crunelli and Leresche 1991). While the presence and internuclear distribution of GABA$_A$ and GABA$_B$ receptors remains to be determined in the medial geniculate body, the possibility of a dual mode of activation of thalamocortical neurons raises intriguing functional scenarios, not to mention the refinement in excitability that such an arrangement might impose on the physiological responses of Golgi type II cells (Morest 1971, 1974, 1975), which are thought, as a class, to be GABAergic (Huchton et al. 1991).

2.4.3 Corticothalamic Transmitter Candidates

Little is known about the neurochemical identity of these neurons. Nevertheless, four independent lines of indirect evidence each suggest that glutamate or aspartate probably play a role in this system, much as they do in other modalities (Baughman and Gilbert 1981; Fonnum 1984; Takeuchi 1987). The first (though not the most compelling) data is the abundance of layer VI neurons in AI showing glutamate- or aspartate-like

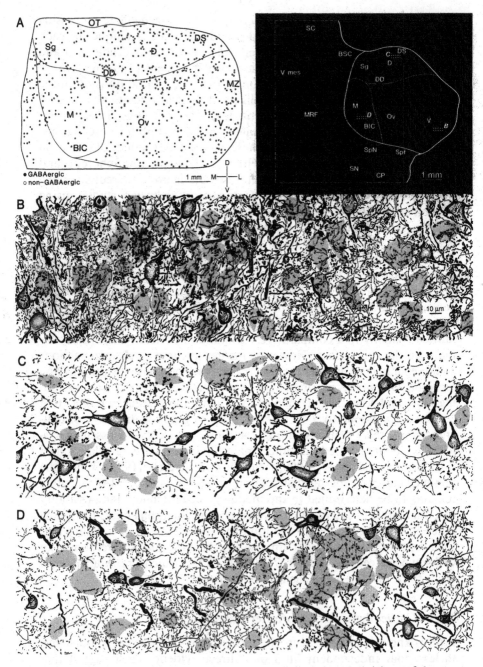

FIGURE 6.6. GABAergic neurons and axon terminals in subdivisions of the medial geniculate body. A. The distribution of GABAergic and immunonegative cells in a 1-μm-thick resin-embedded semi-thin section stained with toluidine blue. Only neurons with a nucleolus are included. There are significant differences in the number of such cells between divisions. See the text for a more detailed

immunoreactivity (Fig. 6.15F). A second and less direct measure is the decrement in biochemically assayed levels of glutamate in nonauditory thalamic sensory nuclei after cortical ablation or undercutting of the white matter to damage corticofugal axons (Fonnum et al. 1981). The third and most direct evidence pertains specifically to the nonauditory sensory and motor corticofugal projection systems, which arise in layers V and VI, and among whose connectionally identified neurons many show glutamate- or aspartate-like immunoreactivity (Giuffrida and Rustioni 1989). Finally, spinal injections of [^{3}H]D-aspartate retrogradely mark corticofugal projection neurons (Rustioni and Cuénod 1982). By inference, injections of [^{3}H]leucine in AI should therefore be expected to label a population of synaptic profiles in the ventral division of the medial geniculate body that contain round synaptic vesicles and make asymmetric synapses mainly onto the distal dendrites of principal bushy neurons. This view may be somewhat simplistic since other sensory corticofugal pathways (some of which could represent distinct neuronal populations in AI; see Games and Winer 1988 for data in the rat) might have more than one morphological type of such axon, a pattern that prevails in the lateral geniculate body (Robson 1983, 1984), and because certain axonal endings that are believed to be of cortical origin have synaptic profiles containing flattened vesicles (Morest 1975). In other sensory corticothalamic path-

---

(FIGURE 6.6. *continued*)   analysis. Planapochromat, N.A. 0.32, × 200. (B–D). GAD-immunoreactive neurons and associated puncta (axon terminals) from the ventral, dorsal, and medial divisions, respectively. *Inset*: locus of panels (B–D). Protocol for panels B–D: planapochromat, N.A. 1.32, × 2000. B. Ventral division. This part of the medial geniculate body is distinguished from the dorsal and medial divisions by its extraordinarily dense plexus of GAD-immunoreactive axon terminals, and by the prominence of preterminal fibers which often run parallel to the axis of fibrodendritic laminae and, by extension, to the main axis of isofrequency representation (see Fig. 6.2C,D). In addition, the puncta are unusually coarse compared to those in other divisions. Most of the GAD-positive neurons (*fine stippling*) have an oval or drumstick-shaped perikaryon from which two or three slender primary dendrites issue, and they are smaller than non-GABAergic cells (large, *solid gray* profiles). C. Dorsal division. In the dorsal nucleus, the concentration of puncta is far lower than in the ventral division, and few endings terminate directly on the somata of immunoreactive cells. The latter often have dendrites oriented preferentially along a lateral-to-medial axis, suggesting that a subset of bushy cells may, in fact, be immunopositive. D. Medial division. The puncta here are more numerous than those in the dorsal division and range from extremely fine to very coarse and large, and some form axosomatic endings on immunonegative principal cells. In addition, some thick (2–3 μm in diameter; *solid black*) GAD-positive axonal trunks run parallel with axons in the brachium of the inferior colliculus. Most of the immunoreactive neurons are small cells about 10–12 μm in diameter (Table 6.1), though much larger cells are seen on occasion (Winer 1991).

ways, the N-methyl-D-aspartate receptor appears to have a prominent role in the control of corticofugal excitability which, after thalamic reticular nucleus lesions, can trigger a depolarization in thalamic relay cells (Deschênes and Hu 1990).

2.4.4 Other Putative Transmitter Candidates

If the criterion for subdividing the auditory thalamus was only minute differences in form among the neuronal populations, one might understandably have little confidence in such an endeavor. However, the unique patterns of ascending and descending connections and the differential distribution of a large array of neurochemical compounds independently confirm the functional nuclear diversity of the auditory thalamus, even if they cannot yet resolve the nature of this diversity or relate it to the behavior of single neurons acting in parallel physiological circuits.

The rat medial geniculate body has a small absolute number and a regionally distinct distribution of catecholaminergic and serotonergic terminals. Thus the *pars marginalis* (perhaps corresponding to the ventral division) receives low levels of the former, and very low-to-low levels of the latter. In contrast, the *pars centralis* (perhaps corresponding to the *pars ovoidea* and/or the medial division) has very low catecholamine levels, and scattered, very thin serotonergic endings (Fuxe 1965). Iontophoretically applied L-glutamic acid excites medial geniculate cells, while brain stem stimulation reduces glutamate-evoked discharges. Iontophoresis of 5-hydroxytryptamine bimaleinate depressed most (79%) and excited a few (5%) medial geniculate cells, while L-noradrenaline inhibits many (62%) cells, excites a few (11%), and on others (27%) has no effect (Tebēcis 1967). Much of the noradrenergic input arises from the locus coeruleus (Cooper et al. 1982). Noradrenergic or cholinergic input may alter thalamocortical excitability through actions on specific potassium currents that change membrane permeability and thus block the rhythmic oscillatory discharges characteristic of sleep or drowsiness (McCormick 1989).

Both noradrenaline and serotonin modulate the hyperpolarization-activated cation current in brain tissue slices of guinea-pig or cat lateral and medial geniculate bodies. This could, in turn, reduce the actions of hyperpolarizing inputs, weaken the burst discharges characteristic of thalamic relay neurons in states of behavioral quiescence, and slightly potentiate single spike discharge—all of which may be prerequisite to increased vigilance associated with waking (McCormick and Pape 1990).

Many medial geniculate neurons (45% of the sample) and all thalamocortical relay neurons are depolarized by the iontophoretic application of acetylcholine. Most excitatory effects take place slowly, while the time course of inhibition is faster. Compounds that bind specifically to cholinergic muscarinic or nicotinic receptors have variable effects (Tebēcis

1970a). Many excitatory cholinergic effects could be blocked by atropine, while neostigmine and eserine potentiate the excitation evoked by inferior colliculus stimulation (Tebēcis 1970b). Iontophoretic application of acetylcholine onto guinea-pig or cat medial geniculate neurons has species-specific effects. In the guinea-pig, about half the cells hyperpolarize then slowly depolarize, while in the cat almost all cells quickly depolarize, after which some hyperpolarize and/or slowly depolarize. The muscarinic hyperpolarization may facilitate burst discharges, while the slow muscarinic depolarization would facilitate single-spike discharges (McCormick and Prince 1987).

Immunocytochemical work finds comparable levels of 5-hydroxytryptamine in the ventral division of the medial geniculate body, dorsal nucleus of the lateral geniculate body, and ventral posterior thalamic nucleus; a similar result prevails for tyrosine hydroxylase. In contrast, the number of choline acetyltransferase-positive axons in the visual thalamus is 300% greater than in the other thalamic nuclei, while in the auditory thalamus the level of cholinergic immunoreactivity is significantly higher in the dorsal division and suprageniculate nucleus than in the ventral division (Fitzpatrick et al. 1989). This region is largely coextensive with the thalamic terminal zone of axons labeled by injections of tracers in the midbrain region between the brachium conjunctivum, periacqueductal gray, lateral lemniscus, and inferior colliculus (Morest 1965b). These lateral tegmental inputs may represent one limb of the ascending cholinergic limbic system that profoundly affects many basal telencephalic centers (Shute and Lewis 1967).

The rat medial geniculate body shows a moderate and homogeneous density of choline acetyltransferase immunoreactivity, while the suprageniculate nucleus has comparatively weaker immunostaining than it does in the cat (Levey et al. 1987). No immunopositive cholinergic cell bodies occur in the rat medial geniculate complex (Tago et al. 1989).

In the macaque monkey thalamus, the distribution of peptidergic immunoreactivity is confined largely to the suprageniculate and posterior limitans nuclei, where moderate numbers of cholecystokinin- and fewer neuropeptide Y-positive terminals occur, and sparse or no substance P- or somatostatin-immunolabeled axonal profiles are seen. Many of those axon terminals are thought to arise from neurons in the midbrain tegmentum (Molinari et al. 1987).

2.5 Comparative Studies of the Auditory Thalamus

Without ancestral forms common to different species, it is impossible to form unassailable conclusions about homology. Nevertheless, by establishing relations among neuronal form, function, and phylogeny, these data might be used to frame inferences about related species whose evolutionary history is known with some precision and to define species differences in operational terms. Of course, any useful scheme for the study of neural homology would have to identify nonhomologous as well as presumptively homologous arrangements.

A broad and cohesive view of auditory forebrain organization in different species is not yet available. Such a perspective would require an ambitious combination and synthesis of studies of structure, physiology, connections, neurochemistry, development, and behavior that do not yet exist for the cat, much less for other species, on which the following account is based. An account of possible parallels in nonmammalian species is available (Campbell and Boord 1974), as well as a critical review of current issues in nervous system homology (Masterton et al. 1976).

2.5.1 Comparative Anatomy in Mammals

With the arrangement of the cat medial geniculate body as a frame of reference, the findings among rodents, marsupials, insectivores, carnivores, chiropterans, and primates can be used to examine parallels and species differences. In each species a medial geniculate complex can be identified and, within it, a lemniscal target, the ventral division, is recognized. There are a number of interesting species differences with respect to the volume of the medial geniculate complex relative to that of the dorsal thalamus. Thus, in the dolphin, about 9% of the dorsal thalamus consists of the medial geniculate body, while other species have a much lower ratio (rabbit 5.4%; sheep 3.2%; cat 6.2%; monkey 2.3%; from Kruger 1959).

In some treatments of the medial geniculate complex, a two-part architectonic scheme has been proposed, consisting of a principal and an internal division in the cat (Rioch 1929), rabbit (Tarlov and Moore 1966), brush-tailed possum (Haight and Neylon 1978), and marsupial native cat (Haight and Neylon 1981). Such formulations do not recognize the dorsal division as a distinct entity. The ventral division differs between species in size, shape, and its internal arrangement. In the mustached bat it forms the largest subdivision of the auditory thalamus (Winer and Wenstrup 1992a,b; Fig. 6.5B), while in humans it is proportionally smaller (Winer 1984d).

Golgi preparations reveal species similarities and differences in neuronal structure and laminar arrangement. Thus, in the opossum, the principal bushy cells form fibrodendritic laminae and receive input from disc-shaped principal cells in the central nucleus of the inferior colliculus (Morest and Winer 1986), just as they do in the cat (Oliver and Morest 1984). However, the interlaminar neuropil in the opossum ventral division is modest compared to the cat, and the bushy cell dendrites are not as well developed, nor do they have as planar an arrangement as those in the cat (Winer et al. 1988). This is not merely a function of brain or body size, since both the mustached bat (Winer 1991; Winer et al.

1992) and humans (Winer 1984d) have the most rigidly planar and highly tufted principal neurons in this phylogenetic series, despite vast differences in the size and configuration of their ventral division laminae. Thus, in humans the neuropil is so highly developed that cellular packing density is comparatively low. In macaque monkeys the intralaminar neuropil is no more voluminous than in the cat, but the number and size of GABAergic axonal endings is striking (Winer and Larue 1992). While there is continuity between the rat (Winer and Larue 1988), cat (Winer 1991), mustached bat (Winer et al. 1992) and monkey in the shape and density of GABAergic puncta among thalamic subdivisions, there are dramatic species-specific differences in the number of GABAergic neurons: the mustached bat has less than 1%, the rat about 1%, and the cat and monkey perhaps 25–35%. In spite of specific parallels in the cytoarchitectonic organization, neuronal architecture, and the subcortical and cortical connectivity of this obligatory synaptic station, variations in intrinsic organizations must impart a particular synaptic sequence that remains to be defined with precision in each species excepting the cat. Since the number and composition of synaptic glomeruli is probably different in various species, rats having a few (Špaček and Lieberman 1974) and cats many more (Morest 1975), the substrates for the transfer of synaptic information are unlikely to be identical, even among otherwise corresponding neurons, and analogous circuits may not have homologous functions.

2.5.2 Comparative Anatomy in Nonmammalian Species

While comparatively little is known about the neural substrates for auditory processing in the fish forebrain, a thorough review is available (Northcutt 1981). In the channel catfish posterior thalamus, a central posterior nucleus consisting of large and heterogeneous neurons can be identified, and it contains at least one subdivision (Striedter 1990a). In another teleost fish, the carp, the midbrain auditory center, the torus semicircularis, projects to the central posterior nucleus, which then sends descending axons (along with other forebrain nuclei) towards the torus semicircularis (Echteler 1984). The latter projection appears to be entirely absent in mammals. Both diencephalic and telencephalic auditory centers have been studied with evoked potentials and multiunit activity. The telencephalon contains two auditory representations, one dorsocaudally, the other dorsomedially, that differ in the latency of their field potential responses, with the former responding more quickly. Acoustic and lateral line input are represented at different depths in the telencephalon. Diencephalic and telencephalic evoked potentials to paired clicks and click trains are strongly refractory, in contrast to the responses of hindbrain and midbrain neurons (Echteler 1985). The pathway between the central posterior nucleus and the telencephalon has been confirmed with tract-tracing methods in the channel catfish; there appear to be few telencephalic projections reciprocal to auditory diencephalic centers (Striedter 1990b).

In amphibians, a dorsal thalamic auditory center has been identified in leopard frogs with evoked potentials (Mudry et al. 1977) and single- or multiunit (Fuzessery and Feng 1983) recordings. Thus, the central thalamic nucleus receives input primarily from mesencephalic acoustic nuclei, while the posterior thalamic nucleus has extraauditory connections with the reticular formation, tegmentum, and ventral thalamus (Hall and Feng 1987). These patterns resemble, at least superficially, a characteristic mammalian design in which axons arising in the central nucleus of the inferior colliculus terminate in the ventral division of the medial geniculate body (Jones and Rockel 1971, cat; Oliver and Hall 1978a, tree shrew; Moore et al. 1977, hedgehog and various primates; Morest and Winer 1986, opossum and cat; Wenstrup and Winer 1987, mustached bat), while those from extralemniscal or nonauditory brain stem centers end in other portions of the medial geniculate complex (Morest 1965b; Calford and Aitkin 1983; Winer and Morest 1983b; Rouiller and de Ribaupierre 1985; Winer 1985b).

Comparable pathways exist in reptiles and birds. In crocodiles, the central nucleus of the midbrain torus semicircularis projects to the posterior part of the nucleus reuniens in the dorsal thalamus (Pritz 1974a), and thalamic axons terminate in the dorsal ventricular ridge, which is thought to be a prototypic representation of a forebrain cortical auditory center (Pritz 1974b; cf 2.5.6). In the lizard, a very similar pattern prevails, such that the torus semicircularis projects upon the *nucleus medialis* of the posterior thalamus which then sends fibers to the anterior dorsal ventricular ridge in the subcortical telencephalon (Foster and Hall 1978; Ulinski 1983). An analogous arrangement of tectothalamic afferents has been noted in turtles. In addition to the classical central auditory pathway, a second route permits auditory influences to reach the somatic sensory system and vice versa (Belekhova et al. 1985). This possible substrate for audiosomatic relations is reminiscent of the intercollicular tegmentum in mammals (RoBards et al. 1976).

The avian auditory midbrain center, the *nucleus mesencephalicus lateralis, pars dorsalis,* sends fibers to the nucleus ovoidalis in the thalamus (Karten 1967), whose axons terminate in the dorsal ventricular ridge (Karten 1968). Some parts of the avian thalamic auditory representation have an orderly arrangement of best frequency (Bigalke-Kunz et al. 1987). In the budgerigar, cytochrome oxidase histochemical staining demarcates both the nucleus ovoidalis and field 'L' in the caudal neostriatum (Brauth 1990). A thalamotelencephalic pathway exists between nucleus ovoidalis and field 'L' in the ventral archistriatum (Brauth et al. 1987).

2.5.3   Comparative Structural Studies of the Medial Geniculate Body

Some caution must be exercised in extending interspecific parallels indefinitely for several reasons. Chief among these is the lack of systematic comparative studies in which phylogeny, neuronal architecture, connectivity, neurochemistry, development, and physiology are coequal partners. Second, even if a correlation can be established with a high degree of correspondence among these measures, it remains to relate these observations to the ultrastructural arrangements in any given nucleus. Third, while some headway has been made in proposing and testing hypotheses about plausible interspecific concordances or differences in the ventral division, the other divisions are much less well known, and consequently the parallels are far less secure or extensive than they might be.

In the rat, three primary thalamic divisions corresponding to those in the cat are recognized. The ventral division has a prominent laminar arrangement with fibrodendritic architecture like that in the cat except for the orientation of laminae, which are inclined at about 45° (Patterson 1976; Winer and Larue 1987). The ventral division is the largest part, followed by the dorsal and medial divisions, respectively (Fullerton 1978). The major varieties of principal cells in each division resemble those in the cat except for the comparative simplicity of their dendritic branching (Clerici et al. 1990). Few neurons with a somatic area of less than 85 μm^2 are present in Nissl material, and such cells are rarely impregnated in Golgi preparations (Winer, Larue, and Cheff, unpublished results), a finding in accord with the dimensions and the paucity of GABAergic neurons in the rat (Winer and Larue 1988) and their much greater frequency in the cat (Rinvik et al. 1987).

In the tree shrew, tufted neurons with bushy arbors dominate the ventral division, moderately branched neurons with radiating dendritic fields are common in the dorsal division, and the medial division has a range of cell types and a regional subdivision into caudal and rostral parts. The ventral division laminae are viewed best in parasagittal sections since they fan out mediolaterally (Oliver 1982). Many of the same types of neurons identified in the cat (Morest 1964) also occur in the tree shrew, including small stellate Golgi type II cells. Thus, while presumably homologous populations of neurons can be identified in many species (Winer 1991), the fibrodendritic laminae in the lateral part of the ventral division in the transverse plane in each has a unique configuration, oriented as they are along a mediolateral axis in humans (Winer 1984d), a dorsomedial-to-ventrolateral arrangement in tree shrews (see above), a dorsolateral-to-ventromedial organization in rats (Winer and Larue 1987), and a dorsoventral disposition in the mustached bat (Winer and Wenstrup 1992a). These distinctions are much less apparent in the medial part of the ventral division, where ascending axons complicate these patterns.

In the rhesus monkey, a densely packed, small-celled ventral division is recognized in Nissl preparations, adjoining the larger, scattered neurons of the medial division, while the dorsal division contains posterodorsal and anterodorsal territories; the latter, with the suprageniculate nucleus and medial division, are in close proximity at the rostral pole of the medial geniculate complex. While the orientation of fibrodendritic laminae in the ventral division is unknown, their arrangement is said to follow the surface contour of the medial geniculate body (Burton and Jones 1976), which would align them with the pattern in the rat (Winer and Larue 1987). In the tamarin, a new world primate, the ventral division stains intensely for cytochrome oxidase and has an internal topography similar to that in the macaque, while the dorsal division is cell sparse compared to the medial division and its neurons are more lightly stained than those in the ventral division. Both the dorsal and medial divisions have a broad range of cell size (Luethke et al. 1989). In the prosimian loris, and in the marmoset, squirrel monkey, and gibbon, only a principal nucleus, with medium-sized cells, and an internal division, with cells scattered diffusely among many fibers, are identified (Moore et al. 1977). However, three divisions have been described in the squirrel monkey that appear similar to those in the cat (Jordan 1973).

The cytoarchitecture of the mustached bat's auditory thalamus resembles that of the macaque monkey in many respects; the primary difference is the relatively enormous expansion of the ventral division at the expense of the dorsal and medial divisions, the striking differences between the lateral and medial ventral division territories that have physiological significance with regard to tonotopic organization, and the relatively modest volume of chiropteran neuropil. A large territory with a nonlaminar neuronal arrangement, the rostral pole nucleus, dominates the anterior auditory thalamus in the mustached bat (Winer and Wenstrup 1992b). The significance of this territory is uncertain since it contains combination-sensitive neurons (Olsen 1986), a physiological category unknown in the cat.

2.5.4 Comparative Physiological Organization of the Medial Geniculate Body

Single unit analysis of the spatial distribution of best frequencies in the squirrel monkey reveals a sequence of lower frequencies laterally and higher tones more medially in the ventral division, extending to 30–35 kHz. Electrode penetrations that traverse what may be the ventral and dorsal divisions find a clear arrangement of tuning in the former and a reversal of, and much less regular, tonotopic pattern in the latter, where loci devoid of spontaneous or auditory activity are sometimes encountered. Tracks across the major divisions reveal a regular arrangement of best frequencies; it is noteworthy that, even in the large-celled parts of the medial geniculate body (which may correspond to the suprageniculate nucleus or medial division) an orderly progression of frequency, though less precise than that in the small-celled division, is evident (Gross et al. 1974; see also Allon et al. 1981). In the guinea-pig, a systematic representation of best frequency has been inferred from experiments with 2-deoxy-D-1[^{14}C]glucose, such that both the lateral and medial divisions have an array of low-to-high frequencies caudally in a dorsomedial-to-ventrolateral sequence, and rostrally in a ventral to dorsal arrangement (Ryan et al. 1982).

Monaural and binaural units occur in the primate medial geniculate body. Two response patterns to monaural stimuli are common: a brief suppression, then numerous spikes 100 msec later, or a faster response with little later activity. Several binaural response types are identified, including time-sensitive units (59%), binaurally facilitated cells (11%), a few neurons that are inhibited by binaural stimuli (2%), and some that are not driven binaurally (28%). Many time-sensitive cells respond also to intensity differences, but there is no apparent systematic spatial arrangement of binaural classes (Starr and Don, 1972). In the rabbit and the cat, 52% and 58%, respectively, of auditory thalamic cells are driven contralaterally; the proportion of bilaterally driven neurons is comparable (Adrián et al. 1966).

In awake squirrel monkeys, almost all single units respond to vocalizations, and more than 80% have a characteristic frequency. Natural and artificial stimuli have comparable absolute thresholds, response strength, rate-level values, and binaural responses. Most cells (76%) have monotonic rate-intensity functions, and ventral and dorsal division neurons have comparable Q_{10} dB values, ranging from 1–10 (below 5 kHz) to 10–28 (above 10 kHz) (Symmes et al. 1980). Only spontaneous rate and latency distinguish the primary divisions, medial division neurons having higher rates and longer latencies (Allon et al. 1981). Most squirrel monkey medial geniculate neurons have a modulation transfer function that is bandpass-like as opposed to low-pass or multipeaked; these properties are consistent with their vocal repertoire (Preuss and Müller-Preuss 1990). In the unanesthetized guinea-pig, auditory thalamic units show much weaker inhibition than cortical units, and respond to more parts of recorded vocalizations than do the auditory cortical neurons to which they project (Creutzfeldt et al. 1980).

As might be expected, the mustached bat, whose auditory system is enormously well-developed, has a precise arrangement with respect to best frequency among thalamic neurons, with cells in the lateral part of the ventral division representing the range from about 20–50 kHz, while the more medial regions contain neurons sharply tuned to 61–63 kHz, and still higher frequencies are found more dorsally and rostrally in the ventral division. However, tonotopy is only one, and perhaps not the dominant, feature of medial geniculate body organization, and other axes of physiological organization cross thalamic nuclear borders. Thus, thalamic input to the cortical frequency-modulated (FM-FM) area originates from several parts of the medial division and from the ventral division, while the 20–50 kHz representation may arise from a single thalamic division and terminate in one cortical area, each without specialized subregions. Other response properties, such as constant frequency or frequency modulated sounds, are spatially segregated as well, but in a rostromedial gradient that crosses nuclear borders. Thus, constant frequency units are concentrated in the middle one-third, and frequency-modulated units in the rostral one-third, of the auditory thalamus (Olsen 1986).

The results from comparative work suggest some interesting possibilities. In primates, there is evidence that each medial geniculate division has a tonotopic arrangement, while in the cat such a conclusion is less secure, particularly for the dorsal division. Besides this tonotopic arrangement, however, many units have different temporal responses to stimuli, and some of these discharge patterns resemble those of onset-type or build-up units in the VIIIth nerve (Kiang et al. 1965). Another species difference is that many neurons in each auditory thalamic subdivision in the awake monkey have comparatively sharp tuning, while those in the cat do not have such values under similar stimulus conditions; many primate neurons are also highly responsive to biologically meaningful stimuli, which are best studied in conscious preparations. In the awake guinea-pig, certain neurons synchronize their discharge to speech-like sounds (Hashimoto 1980) and both auditory thalamic and cortical cells are responsive to vocalizations, though the discharge patterns at each level were different (Creutzfeldt et al. 1980).

2.5.5 Connections of the Auditory Thalamus

Large lesions of the inferior colliculus that damage the central nucleus or its outflow cause terminal degeneration in each medial geniculate subdivision in the hedgehog, loris, marmoset, squirrel monkey, gibbon (Moore et al. 1977), and rhesus monkey (Moore and Goldberg 1966), much as they do in the cat (van Noort 1969). Studies with axoplasmic tracers in the rat reveal that portions of the intralaminar nuclei adjoining the medial geniculate complex also receive input after injections of the central nucleus of the inferior colliculus, besides extraauditory projections to the intercollicular zone, and that the terminations in the suprageniculate nucleus are much lighter than in other auditory thalamic nuclei (Le Doux et al. 1987). Retrograde transport experiments, on the other hand, show that midbrain input to much of the medial part of the dorsal division and the suprageniculate nucleus comes mainly from the dorsal cortex of the inferior colliculus and caudal intercollicular tegmentum, and that some neurons in the ipsilateral superior olivary complex project to the ventral division (Fullerton 1978).

There are some interspecific variations in the basic plan of colliculo-geniculate connections. In the tree shrew, axons from the roof nucleus (dorsal cortex) of the inferior colliculus project to the thalamic deep dorsal nucleus, and the sagulum projects to the dorsal nucleus and medial division (Oliver and Hall 1978a). Perhaps these are adaptations for the rapid visual- and acoustic-motor adjustments characteristic of the tree shrew, while motor-related and other nonauditory input to the medial division might play a role in the control of visceral reflexes and their integration with auditory cues, as in other species (LeDoux et al. 1984). By the same token, in the rabbit both the nucleus of the optic tract and the superficial layers of the superior colliculus project to the dorsal division of the medial geniculate body, while the deep superior colliculus is reported to send axons to the internal division, which may correspond to the medial and dorsal divisions as defined here (Holstege and Collewijn 1982). Such input could assist in the coordination of head, neck, eye, and pinna movements, and for the integration of visual and acoustic information for accurate spatiomotor localization.

In the mustached bat most of the essential features of the tectothalamic projection common to other mammals are conserved (Fig. 6.7). Thus, low-frequency (below 25 kHz) regions of the central nucleus of the inferior colliculus, which are dorsal and caudal, project most strongly to the lateral, laminated part of the ventral division. The neurons in the ~60 kHz constant frequency region that fills much of the dorsoposterior division of the central nucleus send their axons to the large, ovoid-shaped medial part of the ventral division, as expected, while the equally heavy input to the suprageniculate nucleus is an exception to the pattern in other species, and is conspicuous for every injection site for loci above 30 kHz. The topography of tectothalamic input for the 90 kHz representation, which arises from anterior and lateral parts of the ventral nucleus, is preserved in its projection to progressively more ventral, medial, and anterior parts of the ventral division (Wenstrup and Winer 1987). Projections from other, intermediate frequencies are largely consistent with these topographic rules (Frisina et al. 1989).

The principles of the organization of thalamocortical connections in many other species suggest parallels with the pattern in the cat (cf. 3.2.1., 3.2.2.) and monkey (see below). Thus, in the rat, neurons in the laminated ventral division, which have an orderly relation with axons from the central nucleus of the inferior colliculus, in turn project to particular sectors of primary auditory cortex, which is surrounded by nonprimary fields (see Herbert et al. 1991). The posterior and anterior parts of the ventral division project to like regions of the cortex, while more ventral thalamic neurons project to dorsal cortical loci; this implies that the main axis of frequency representation in the medial geniculate body is other than a simple mediolateral one. However, the areal distribution of thalamocortical input is much like that in the cat, with the ventral division

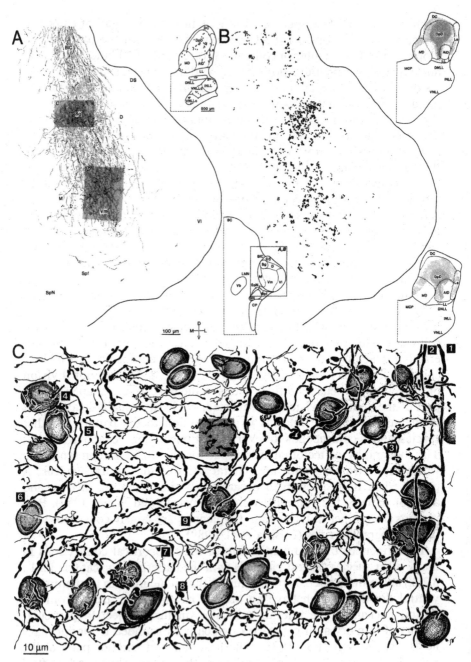

FIGURE 6.7A–C. Caption on pages 276–77.

D

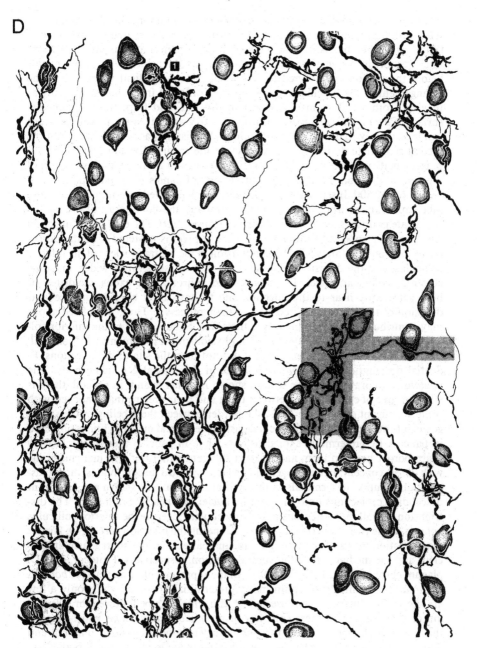

FIGURE 6.7D Caption on pages 276–77.

FIGURE 6.7. Morphology and spatial distribution of inferior colliculus axons bulk-filled with horseradish peroxidase and terminating in the medial geniculate body of the mustached bat. A. Regional distribution of preterminal axons in a 30-μm-thick section in which the reaction product was intensified with heavy metals before the section was stained for Nissl substance. A salient point is that the arrangement of the brachial axons is entirely different from the cat, entering as they do from the dorsomedial aspect in this bat, and from the ventromedial surface of the medial geniculate body in the cat. A second feature is that injections confined mainly to one inferior colliculus subdivision mark axons in each of the medial geniculate divisions—in this case, in the medial division (*M*), in the suprageniculate nucleus (*Sg*) of the dorsal division, and in the medial part (*Vm*) of the ventral division. Protocol for central parts of panels A, B: planapochromat, N.A. 0.65, × 500. Insets show, respectively, the bulk-filled inferior colliculus and other midbrain neurons (dots in *upper inset*) and the boundaries of the medial geniculate body architectonic subdivisions (*lower inset*). The core of the injection site (panel B, upper and lower insets, *dark stipple*) is centered near the dorso-posterior (approximately 60 kHz) subdivision of the central nucleus of the inferior colliculus, and the diffusion of horseradish peroxidase is much wider, invading adjoining collicular nuclei. These bulk-filled neurons have axons, dendrites, or both sufficiently near to the injection site to have been filled by intracellular diffusion of tracer or direct injection of their processes. Thus, their filled axons may contribute to the labeling in the thalamus. Subdivisions of the inferior colliculus correspond to those described in physiological experiments by Zook et al. (1985), though the present experiment did not map the site of the injections. B. Spatial distribution of *boutons terminaux* in the medial geniculate body. The *boutons* were drawn from the section shown in panel A and, in general, there is a high degree of concordance between them and their preterminal branches, with some significant exceptions. Thus, the border between the labeling in the suprageniculate nucleus (*Sg*) and the medial part of the ventral nucleus (*Vm*) is obscured in panel A by preterminal axonal trunks, while in panel B there is a relatively sharp border between them and a rather different pattern of terminal clustering, with dense, concentrated suprageniculate labeling and a discontinuous, more clustered distribution of terminals in the ventral division. Not shown are significant descending projections onto the pontine nuclei and nearby brain stem. The distribution of bulk-filled neurons and extracellular tracer in the inferior colliculus appears in the two insets (*black dots*) surrounding the injection core. C, D: Higher magnification views of the terminal labeling patterns in the suprageniculate nucleus and the medial part of the ventral division, respectively. Protocol for C,D: planapochromat, N.A. 1.32, × 2000. C. In the suprageniculate nucleus the *boutons terminaux* are embedded within many preterminal fibers, of which only a few are shown for illustrative purposes (*1–3*), and which can be followed for 500–800 μm or more; note that these emit few or no *boutons de passage*, which implies that some single fibers passing through a nucleus towards another may not end divergently in several nuclei. *Boutons terminaux* are evident on cell bodies (*4–6*) and in the neuropil, where they range in size from about 1 μm in diameter (*7*) to much larger and coarser (*8,9*). The bulk of the endings are in the neuropil, and single endings (*stippled profile*) have a relatively simple morphology. D. In the ventral nucleus (*Vm*) an entirely different architecture prevails among the endings near these much smaller neurons. Here, single terminals are far more complex projecting to area 41 (Clerici and Coleman 1990), the dorsal division to nearby nonprimary auditory cortical fields Te2 and Te3 (Arnault and Roger 1990), and the medial division to both of these and beyond. The input to nonprimary cortical areas is also organized regionally, and arises chiefly from the dorsal division; the target of the caudodorsal nucleus of the medial geniculate body includes Te2. The auditory thalamic projections onto nonprimary cortex (Patterson 1976; Arnault and Roger 1990) appear to be devoid of topography, while those from the ventral division upon Te1 have a clear topography (Scheel 1988; Roger and Arnault 1989).

The laminar distribution of ventral division input is primarily to layer IV, while axons from the medial division end in layers I and VI (Ryugo and Killackey 1974), and there is a general reciprocity of thalamocortical and corticothalamic projections, albeit with significant zones of regional discontinuity (Winer and Larue 1987). Comparable architectonic areas in the auditory thalamus and cortex and analogous patterns of thalamocortical relations are described for the guinea-pig (Redies et al. 1989a,b), squirrel and macaque monkeys (Mesulam and Pandya 1973; Burton and Jones 1976; Jones and Burton 1976), marmoset (Brysch et al. 1990), tamarin (Luethke et al. 1989), tree shrew (Casseday et al. 1976; Oliver and Hall 1978b), Virginia opossum (Kudo et al. 1986), northern native cat (Kudo et al. 1989), marmoset (Aitkin et al. 1986b), birds (B.A. Bonke et al. 1979; Brauth et al. 1987), and in the mustached bat (Olsen 1986). Among the noteworthy species differences are the comparatively more massive subcortical telencephalic projections in marsupials (Kudo et al. 1986) and the specificity of thalamocortical connections in the mustached bat, in which each electrophysiologically defined cortical area and thalamic nucleus appears to have a particular relationship (Olsen 1986) that could reflect the areal segregation of cortical and subcortical function (Suga and Horikawa 1986).

Among the synaptic targets of thalamic axons terminating in layer IV of primate area KA are small bipolar cells, large bitufted neurons, and larger multipolar cells. Each type of neuron is contacted by asymmetric, round vesicle-containing endings, at least some members of each class of neurons are GABAergic, and certain of these also contain somatostatin-

---

(FIGURE 6.7. *continued*) (*stippled profile*) but spatially segregated from one another, and they form elaborate clusters with numerous swellings and dilatations and, consequently, differ from the very fine endings that dominate the supragen-iculate nucleus. Moreover, the bulk of these endings are probably onto processes in the neuropil, since relatively few neurons appear to receive axosomatic contacts (*1–3*). The axons in this panel have been drawn selectively to show the preterminal and terminal segments on the left-hand side, and the terminal morphology on the right-hand side. From unpublished observations by Wenstrup and Winer; see also Wenstrup and Winer (1987).

or cholecystokinin-immunoreactive material (Cipolloni and Keller 1989). The morphological and neurochemical diversity of these targets suggests that, even within the lemniscal pathway terminating in layer IV, thalamic input may influence diverse postsynaptic targets with, presumably, divergent sites of action.

Knowledge of corticothalamic projections in species besides the cat is modest. In the rat, every part of the medial geniculate complex receives descending projections, but there are regional differences among thalamic subdivisions in their strength and sources. Thus, the ventral division always has the strongest critical input, and the medial division the weakest. Cortical fields that receive axons from a given thalamic nucleus tend to project most heavily onto that nucleus, less so upon others, and no part of the medial geniculate body is without cortical input. Corticothalamic projections are topographically arranged, at least in the ventral division, where labeling is concentrated in lateral and caudal zones after injections in the posterior part of the primary auditory cortical fields (Winer and Larue 1987). In the owl monkey, both AI and the rostral auditory field project to each of the three main medial geniculate body divisions, suggesting that descending connections may be more divergent than in rodents (FitzPatrick and Imig 1980). In the budgerigar, descending axons from field 'L' terminate largely in the rostromedial archistriatal nucleus and not, as one might predict, in the nucleus ovoidalis (Brauth and McHale 1988). Similarly, canary field 'L' does not project, as might be expected, to diencephalic auditory cells, but to motor-related neurons in the paleo- and hyperstriatum that function as upper motor neurons for the control of song (Kelley and Nottebohm 1979). The mustached bat has widespread corticothalamic projections that are often but not always reciprocated by thalamocortical input (Olsen 1986).

2.5.6   Comparative Thalamic Neurochemistry

While there are as yet no published studies on the thalamocortical auditory system in which candidate transmitters have been combined with tract-tracing methods in the same neurons, it is possible that the chief tectothalamic, thalamocortical, and corticothalamic transmitters are glutamate or aspartate, though others cannot be excluded. An impediment to proving this is that many neurons have significant metabolic pools of these (or perhaps other) amino acids which the antisera now available cannot distinguish from the transmitter-specific pool; hence, it is appropriate, until such specificity can be demonstrated, to consider the findings, especially those pertaining to the excitatory amino acids, with some caution (Yingcharoen et al. 1989).

In the rat (Popowits et al. 1988) and cat (Winer, unpublished observations), many neurons in each thalamic subdivision show aspartate- or glutamate-like immunoreactivity, as do neurons in the inferior colliculus and auditory cortex. This contrasts with the smaller, but still substantial number, of GABAergic cells in the rat (Mugnaini and Oertel 1985) and cat (Oliver et al. 1992) inferior colliculus, medial geniculate body (rat, Winer and Larue 1988; cat, Rinvik et al. 1987), and auditory cortex (rat, Winer and Larue 1989; cat, Figs. 6.14, 6.15G-L; Prieto et al. 1990; cf. 2.5.1). In the chicken, no GABAergic neurons are immunostained in nucleus ovoidalis, though there are abundant, immunopositive puncta (Müller 1988; Granda and Crossland 1989). A similar pattern but with a lighter density of puncta is reported in the pigeon nucleus ovoidalis (Domenici et al. 1988). In the crocodile, the presumed homologue of the medial geniculate body, the *nucleus reuniens pars centralis* (posterior) is devoid of GABAergic neurons and puncta, as are all other nuclei whose projections terminate in the telencephalon. This implies that GABA-mediated local circuits evolved after the emergence of reptiles or that some other neuroactive substance might serve a comparable role (Pritz and Stritzel 1988).

In the macaque medial geniculate body, each division also contains neurons immunopositive for parvalbumin and for calbindin. In the ventral and dorsal divisions, they were colocalized together within single cells, while in the medial (magnocellular) division they were not. Only a few cells were parvalbumin-positive in the limitans and suprageniculate nuclei, while many somata are positive for calbindin in each. Cells of different size or form were immunoreactive in the medial division, suggesting that chemically-specific subtypes of thalamocortical relay neurons might exist. The patterns of immunoreactivity for these calcium binding proteins in the primate thalamus is not indicative of their transmitter status since colocalization of GABA with either binding protein is uncertain (Jones and Hendry 1989).

Other putatively neuroactive substances in the rat medial geniculate body include acetylcholine and luteinizing hormone releasing hormone; their density appears to be highest in the dorsal division and lowest in the medial division, and it has been proposed that their presence constitutes a rostral element of the limbic brain stem (see Nieuwenhuys et al. 1988/1989). Enkephalin-like immunoreactivity is limited largely to the marginal zone along the perimeter of the medial geniculate body (Fallon and Leslie 1986), in contrast to the cat, where much of the dorsal division contains immunopositive elements (Covenas et al. 1986). There is a discrepancy between the restricted pattern of enkephalin-like immunoreactivity and the presence of immunostaining for neutral endopeptidase-24.11 (enkephalinase), which is found in many small cells throughout the rat medial geniculate complex and is believed to play a regulatory role in peptide activity (Back and Gorenstein 1990). The latter study also finds a wider distribution of enkephalin-like immunoreactivity than do others (Fallon and Leslie 1986), including substance P-immunoreactive fibers in the ventral and dorsal divisions. Still other studies describe enkephalinergic cells along the ventromedial border of the medial geniculate body, and immunopositive fibers and terminals on the dorsomedial margin (Uhl et al. 1979; see also Petrusz et al. 1985). Calretinin-immunoreactive fibers occur in the marginal zone and more sparsely along the medial half of the auditory thalamus; there are few immunostained somata (Jacobowitz and Winsky 1991). In contrast, somatostatin-like perikaryal immunoreactivity is present in many neurons in the medial one-half of the medial geniculate body, with only occasional immunopositive fibers; while its role in medial geniculate body function is obscure, it does satisfy many of the criteria for a neurotransmitter (Johansson et al. 1984).

In the little brown bat, a few cholecystokinin-immunopositive neural somata have been seen in the auditory thalamus, intermingled among immunostained axons (Cotter and Laemle 1990). While the function of cholecystokinin is unknown, it is thought to act as a neuromodulator (Bradford 1986). Since it is apparently not present in the rat (Nieuwenhuys et al. 1988/1989), its presence in the auditory thalamus of this bat may be species-specific.

2.6 Functional Organization

The topographic, reciprocal, and specific laminar projections between the thalamus and the cortex attest to their interrelatedness. A key index of this is the distinct physiological organization among different thalamic nuclei and between various cortical fields. Insofar as the relations between thalamus and cortex violate the rule of one thalamic nucleus per cortical area, then different or emergent cortical axes of representation may exist, perhaps as a consequence of divergent projections. Thus, a model that includes elements both of serial and hierarchical thalamocortical relations may prevail across parallel pathways. If the physiological representations at different levels are indistinguishable and the projections convergent, then local circuits could segregate functional subregions within larger topical arrangements. At present, there is some evidence for both types of organization.

The following brief survey concentrates on recent literature; more detailed accounts are available (Imig and Morel 1983; Imig et al. 1992).

2.6.1 Tonotopic Arrangement

There is strong evidence for a regular disposition of neurons with respect to best frequency in the ventral division, with lower frequencies represented in ventral, lateral, and caudal territories (Aitkin and Webster 1972; Calford and Webster 1981). Frequencies of 7–14 kHz fill the medial part of the ventral division, while cells tuned to still higher frequencies (above 16 kHz) are found in the most medial part of the ventral division, the *pars ovoidea* (Imig and Morel 1985b; Fig. 6.2C). In the latter, large arrays of tectothalamic axons passing toward the ventral division degrade the more orderly laminar arrangement prevailing in the lateral parts of the ventral nucleus (Morest 1964).

A second tonotopic thalamic field, in the lateral part of the posterior nucleus, is affiliated with the ventral and medial divisions and lies at the rostral pole of the auditory thalamus, beyond the thalamic territory usually construed as auditory. These neurons have short latencies, sharp tuning curves (Imig and Morel 1985a) and project to one or more of the primary cortical fields (Imig and Morel 1984). In the mustached bat, a zone presumably corresponding to this region receives afferents from physiologically identified regions of sharply tuned neurons in the inferior colliculus (Wenstrup and Winer 1987; cf. 2.5.3).

The evidence for a frequency-specific arrangement in any part of the dorsal division is weak. The tuning of single units is typically broad or multipeaked (Calford and Webster 1981), though about 12% of the units are sharply tuned; neurons in the deep dorsal nucleus respond only to frequencies above 8.5 kHz. Many suprageniculate nucleus units show long latencies, little frequency tuning, and habituate readily to repeated stimuli; a large number are not driven acoustically (Calford 1983).

There is a disagreement over the degree of medial division tonotopy. Physiological descriptions report irregular or degraded representations compared to those in the rostral pole (Imig and Morel 1985a). After injections of tracer in physiologically defined regions of the primary auditory fields, there is much more overlap of transport than would be expected in a tonotopically arranged system. However, low-frequency foci of transport were found laterally in the caudal part of the medial division, and more ventrally in the rostral half, while high-frequency label lay more medially in the caudal medial division, and more dorsally in the rostral part (Imig and Morel 1984). Physiological studies find low-frequency units concentrated in ventral and lateral parts of the medial division, and neurons responsive to higher frequencies dorsally and medially (Rouiller et al. 1989b), though the tuning of single neurons is often broad (Aitkin 1973).

2.6.2 Binaural Processing

The nuclei of the medial geniculate complex show a surprisingly uniform distribution of binaural interactions, with excitatory-excitatory interactions accounting for some 60% of the sample in each division. The chief difference between areas is the concentration of monaural units in the caudal parts of the dorsal division (Calford 1983). Click stimuli evoke short-latency or long-latency responses, with binaural input preferentially affecting the latter (Altman et al. 1970). It is not surprising that there is little in the way of any segregation of aural properties in cortical areas, such as AII (Schreiner and Cynader 1984), that are among the chief targets of the dorsal division. Perhaps the convergence of local or ipsilateral corticocortical (Winguth and Winer 1986) or of commissural projections in AII (or related fields) serves functions other than tonotopic representation. In contrast, the topographic commissural (Imig and Brugge 1978; Code and Winer 1985, 1986) and tonotopic ipsilateral corticocortical (Imig and Reale 1980) projections would tend to conserve the thalamic patterns in cortical fields with a clear organization of best frequency.

2.6.3   Other Physiological Representations

Remarkably little is known about the biophysical properties of thalamocortical relay neurons, not to mention those of Golgi type II cells. Insofar as valid generalizations can be made with respect to thalamic principal cells, the following observations may be relevant for medial geniculate body neurons. These cells exhibit two modes of discharge: single spikes or bursts of spikes. The latter reflects a low-threshold Ca^{2+} current, represents a non-linear transform of de- or hyperpolarization, and cannot respond to rates above 15 Hz. The bursting mode is hypothesized to act as a sensory filter during periods of decreased vigilance. Single spikes, in contrast, have time-, frequency-, and intensity-dependent rates, and they can follow rates beyond 100 Hz (McCormick and Feeser 1990).

A wide range of properties are characteristic of medial geniculate neurons when they are compared to one another in terms of consistency of response, rate sensitivity, novelty, and lability. As might be expected, these dimensions have a logical distribution, for example, more than 95% of ventral division neurons show consistent responses, while more than 50% so respond in the dorsal or medial divisions, where neurons responsive to each of the four dimensions noted above occur (Calford 1983). Neurons sensitive to the direction of sound movement also occur in the dorsal and ventral divisions (Altman et al. 1970). Other medial geniculate neurons (77% of a sample) respond preferentially to cubic difference tones below 10 kHz rather than either tone presented alone, and the index of synchrony was often greater for the cubic difference tones than for a pure tone at a particular characteristic frequency (Horner et al. 1983). Some 28% of a large sample of medial geniculate body neurons respond selectively to interaural phase differences, and a further 37% were sensitive to interaural intensity cues; among the latter, half represented phase differences, too. The rates change over a large range, indicating that the spatial tuning is broad (Ivarsson et al. 1988).

Besides the mediolateral arrangement of best frequency (cf. 2.6.1.), the ventral division is functionally heterogeneous along its rostrocaudal axis. Thus, units situated more caudally discharge to pure tones, are not driven by simple auditory signals, show broad tuning and comparatively weak tonotopy, and exhibit long response latencies, equal numbers of excitatory and inhibitory responses, nonmonotonic intensity functions, and little time locking to repeated clicks. In contrast, units in more rostral parts of the ventral division respond to simple auditory stimuli, have narrower tuning and a clearer tonotopy, shorter and more uniform response latencies, a predominance of excitatory responses, a tendency to have monotonic rate-intensity functions, and more time-locked discharges. These ventral division subregions project preferentially to different primary fields: the caudal part terminates largely in the posterior auditory field, while cells in the more rostral part send their axon to AI or AAF (Rodrigues-Dagaeff et al. 1989). Thus, primary fields, like the auditory thalamic subregions, may be nonequivalent in a functional sense.

Besides the mediolateral arrangement of best frequency (cf. 2.6.1.), the ventral division is functionally heterogeneous along its rostrocaudal axis. Thus, units situated more caudally discharge to pure tones, are not driven by simple auditory signals, show broad tuning and comparatively weak tonotopy, and exhibit long response latencies, equal numbers of excitatory and inhibitory responses, nonmonotonic intensity functions, and little time locking to repeated clicks. In contrast, units in more rostral parts of the ventral division respond to simple auditory stimuli, have narrower tuning and a clearer tonotopy, shorter and more uniform response latencies, a predominance of excitatory responses, a tendency to have monotonic rate-intensity functions, and more time-locked discharges. These ventral division subregions project preferentially to different primary fields: the caudal part terminates largely in the posterior auditory field, while cells in the more rostral part send their axon to AI or AAF (Rodrigues-Dagaeff et al. 1989). Thus, primary fields, like the auditory thalamic subregions, may be nonequivalent in a functional sense.

While most single units in the ventral division have narrow tuning curves in anesthetized preparations, a significant number have much broader, W-shaped response profiles in awake, freely moving animals. The breadth of these tuning curves has been interpreted as reflecting frequency-specific inhibitory or excitatory subregions within the larger receptive field (Whitfield and Purser 1972).

Electrical stimulation of the medial division activates globus pallidus units 3–9 msec later, and basal amygdaloid neurons 12–23 msec. This suggests that the conduction velocities or synaptic arrangements of these circuits are somewhat different even among limbic nuclei, since stimulation causing pallidal units to discharge inhibits amygdaloid cells. So broad a distribution of latencies could reflect more than one population of thalamofugal neurons (Wepsic and Sutin 1964).

From a neural modeling perspective, the discharge properties of single neurons in the squirrel monkey medial geniculate body have been simulated for species-specific calls using Volterra kernels. Many of the actual firing patterns are closely approximated by this method (Yeshurun et al. 1985).

2.6.4   Behavioral Studies

The classical literature devoted to the behavioral effects of medial geniculate body damage is modest given the difficulty of making selective
ablations in a fiber-rich milieu. Nevertheless, it suggests that particular
auditory thalamic lesions produce deficits specific to and selective for a
given frequency (see Neff et al. 1975 for a review). More recent studies
in dogs (Heffner and Heffner 1984) and cats (Jenkins and Masterton 1982)
indicate that these effects include both frequency discrimination and
sound localization, respectively. In the latter study, localization performance was essentially normal in an easier localization paradigm. Still larger
lesions that destroy auditory thalamic outflow and corticothalamic input
cause profound auditory deficits in the discrimination of clicks and pure
tones; even comparatively small lesions near the brachium of the inferior
colliculus produce massive auditory and somesthetic losses (Glassman
et al. 1975). In contrast, medial geniculate body lesions in rats do not
affect sound localization (Kelly and Judge 1985), and acoustically conditioned changes in heart rate in rabbits are likewise unchanged (Jarrell
et al. 1987). Acoustically mediated motor reflexes are still present after
destruction of the cerebral cortex and thalamus (Forbes and Sherrington
1914). Human psychophysical studies in which an auditory cue is used
to assist visual spatial localization and orientation suggest that visual
search strategies are strongly influenced by concurrent acoustic input
(Perrott et al. 1990).

2.6.5   Physiological Analysis of Learning and Memory

It has been known for some time that classical conditioning of the discharge rate of single auditory thalamic units was possible, and that this
manipulation can affect the temporal distribution of spike activity (Weinberger et al. 1972). In multiple unit recordings, interesting dynamic
changes in the behavior of medial division neurons are reported. These
include the long-term potentiation of amplitude increases and latency
decreases after short, high-frequency stimulation of the brachium of the
inferior colliculus (Weinberger 1982; Gerren and Weinberger 1983), and
increases in the multiple unit activity during presentation of a white noise
and shock-conditioned stimulus in cats (Ryugo and Weinberger 1978)
and rats (Edeline et al. 1988). Single neuron studies of rat medial division
receptive fields during acquisition of learning in a classical conditioning
paradigm find that 54% of the cells show plasticity of their evoked response to a conditioned stimulus, and that 75% of these changes are
frequency specific to the latter stimulus (Edeline 1990). Multiunit changes
to a conditioned stimulus appear first in the medial division, then later
in the hippocampus and auditory cortex. In studies of retention after 45
days, discriminative responses are still observed in each, which is consistent with the view that their discharge patterns remain stable (Edeline

et al. 1990). Pairing of different tones with (CS⁺) and without (CS⁻) shocks reveals two groups of medial division neurons in the rabbit, one responsive to the CS⁺ and the other to the CS⁻ condition (Supple and Kapp 1989), suggesting that medial division neurons that project to the amygdala in the rat (LeDoux et al. 1985) could play an important role in the acquisition of conditioned autonomic reflexes (Gabriel et al. 1976). On the other hand, corticothalamic input may modulate inhibition of the response to the CS⁻ (Jarrell et al. 1987). Significant differences are reported between cells in the posterior and anterior parts of the medial geniculate, with the former showing a more dramatic change from habituation through training (Buchwald et al. 1966), while neurons in the region of the rostral pole alter their discharge pattern and reduce their baseline activity to a CS⁺ during acquisition of a conditioned eyeblink response (Woody et al. 1991).

The short latency discharge of multiple unit clusters in the ventral division is unaffected by cortical cooling of AI; however, this treatment inhibits click-evoked, late-reverberatory responses but does not alter the discharge to tones. Cooling has little effect on background activity of reverberatory units, and facilitates the discharge of nonreverberatory units (Ryugo and Weinberger 1976).

In an integrated series of connectional, physiological, and behavioral studies in the rat, the fundamental neural circuitry in a pathway essential for fear conditioning has been dissected (LeDoux et al. 1984, 1990a,b). This auditory-limbic (and somato- and visceromotor) interface originates in the thalamus in the vicinity of the posterior intralaminar nuclei (see Winer et al. 1988 for a discussion of nomenclature), portions of which form the ventrolateral surface of the medial geniculate body, and in the medial subdivision of the thalamic posterior nucleus, which lies just dorsal and medial to the posterior intralaminar nuclei. The lateral amygdaloid nucleus and amygdalostriatal transition zone receive their input largely from the medial division and from the suprageniculate nucleus of the medial geniculate body, from the posterior intralaminar nuclei, and less so from the lateral posterior thalamic nucleus. The afferents to the central, medial, and basomedial amygdaloid nuclei arise solely from the medial part of the posterior group (LeDoux et al. 1990b), a nucleus considered also to have strong affiliations with the primary somatic sensory cortex (Fabri and Burton 1991). It is significant that the inferior colliculus afferents to the posterior intralaminar nuclei virtually define the extralemniscal sources of input to the auditory thalamus, including as they do the external, pericentral, and dorsal cortical regions of the inferior colliculus, and the sagulum, and cuneiform nuclei (LeDoux et al. 1990b).

Electrical stimulation of the thalamoamygdaloid pathway evokes mainly excitatory responses in amygdaloid and striatal units, many of which have low spontaneous rates. The fastest responses occur in the caudoputamen, lateral amygdaloid nucleus, and amygdalostriatal transition zone, with slower responses in the basolateral, basomedial, and central amygdaloid nuclei. Neurons in the lateral nuclei also require stronger input at higher frequency to elicit a comparable response (Clugnet et al. 1990). Lesions of the lateral amygdaloid nucleus alone abolished the capacity to form a classically conditioned fear response, while damage to the striatum or overlying cortex had no such effect (LeDoux et al. 1986b, 1990a). Electrical stimulation of the medial geniculate body in turn produces long-term potentiation among lateral amygdaloid neurons (Clugnet and LeDoux 1990). Chemical destruction of medial geniculate body neurons prevents the formation of acoustically, but not visually, mediated learned autonomic responses (LeDoux et al. 1986a). The results of the foregoing studies have a number of important implications relevant to the functions of the medial geniculate body. First, there are extensive relations with the limbic forebrain that may be specific to particular behavioral contexts and that are mediated by auditory input. Second, there are robust inputs to the corpus striatum whose role is unknown, but which in all likelihood influence the early stages of motor planning, preparation for action, and the gross postural adjustments prerequisite to the smooth and accurate execution of movement. Third, the connectional output of portions of the medial geniculate complex thought to be devoid of a clear tonotopic organization, such as the medial division and the posterior intralaminar nuclei, is equally selective and as specific as that of subdivisions whose predominant organizational features are related to tonotopic organization or a highly topographic set of projections, such as the ventral division. Fourth, the pattern of inferior colliculus input to the medial geniculate body is logically consistent and congruent with the patterns denoted in the preceding points. Fifth, these observations are consistent with the view that the medial geniculate complex embodies auditory, limbic, motor, and associational components, each of which has a distinct subset of morphological, connectional, physiological, and chemical properties.

In experiments on classical conditioning among multiunit ensembles in the rat medial geniculate complex, the CS^+ to CS^- comparison was the only one significantly different within a broad sample of cortical and subcortical sites, and the absolute change seen in the posterior nucleus is noteworthy. In general, thalamic neurons, compared to cortical cells, show greater changes in acquired responses, exhibit more generalization, and form unique learning patterns (Disterhoft and Olds 1972). Many neurons in midbrain, thalamic, and cortical auditory centers respond selectively to tonal signals with different behavioral relevance. About 20% of rat auditory thalamic and cortical units have unique discharge patterns in the delay period following the tone, suggesting that they participate in memory processes. It is noteworthy that significant changes also occur in the entorhinal cortex (Sakurai 1990; cf. 2.2.1).

3. The Primary Auditory Cortex (AI)

There is a striking laminar, areal, and neurochemical homogeneity in the arrangement of much of the neocortex (Eccles 1966; Chow and Leiman 1970; Rockel et al. 1980; Hendry et al. 1987) that can be somewhat enigmatic considering the unique peripheral connections and functions of different architectonic fields. As an example, consider the distinct roles of layer III pyramidal neurons in the primary auditory and visual cortices. In the former, many (but not all) of these cells project to the contralateral AI (Imig and Brugge 1978; Code and Winer 1985), and others to nearby homolateral auditory fields (Winguth and Winer 1986). In contrast, in area 17 only neurons adjoining the border with area 18 and representing the vertical meridian have callosal axons (Segraves and Rosenquist 1982a,b); other layer III pyramidal cells are presumably ipsilateral (or interlaminar) projection neurons and can thus affect the discharge patterns of contralateral neurons only through circuitous, polysynaptic commissural pathways outside area 17. Thus, morphologically similar neurons have distinct connections and perhaps different functional roles. Despite a basic structural arrangement that aligns it with the remainder of the neocortex, the primary auditory cortex thus has several different neuronal arrangements that distinguish it from other sensory cortical fields. Whether the nonprimary fields are likewise unique remains to be determined. Evidence supporting the proposition that functional (and, by inference, structural) differences exist among the many cortical auditory fields is the diverse regional cytoarchitectonic arrangements among them (Rose and Woolsey 1949), and the specific behavioral effects of bilateral damage to each area (Colavita et al. 1974; Neff et al. 1975).

3.1 Cytoarchitecture and Neuronal Organization

In each layer of AI it has been possible to identify sublayers, for example, layers Va and Vb, based on local differences in neuronal architecture, connectivity, and on immunocytochemical criteria. These designations should be regarded as provisional, since even more refined laminar distinctions may be necessary when more is known about the distribution of thalamic axons within layers III and IV (Table 6.8), or about intralaminar connections or receptive field arrangements. By analogy with the primary visual cortex, where axons from thalamic laminae representing X- or Y-cells terminate in different sublaminae of layer IV (Humphrey et al. 1985), perhaps in AI the thalamic afferents representing monaural or binaural subsystems have specific laminar or morphological arrangements. If so, they remain to be identified; if not, then they further distinguish AI from area 17. A parallel question is the nature of the laminar arrangement of thalamocortical connectivity in the second auditory cortical area, AII (Fig. 6.8), whose neurons receive information from homo-

TABLE 6.8. Laminar terminations and origins of main projection systems in cat primary auditory cortex (AI) and relations with GABAergic axon terminals (puncta).

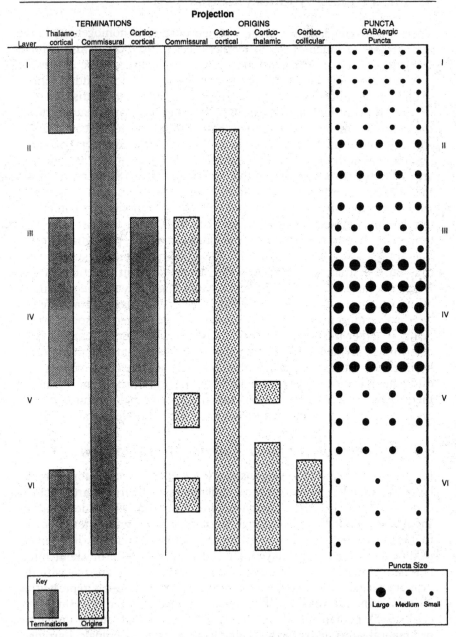

The left-hand one-third shows the laminar distribution of afferent input onto the layers in AI from the auditory thalamus (see 3.2.1 and 3.2.2), contralateral AI (see 3.3.1), and ipsilateral AII and other fields (see 3.3.2). The middle one-third shows the laminar sources of commissural (see 3.3.1), corticocortical (see 3.3.2), corticothalamic (see 3.3.3), and corticocollicular (see 3.3.4) projections. The right-hand side is a schematic representation of the size and density of GABAergic puncta. It is noteworthy that the layers with the largest such endings are the target of all three types of afferent input (left-hand side), while the layers containing the cells of origin for the main sources of corticofugal output have a different, much lighter pattern. (From Priero et al. 1992b).

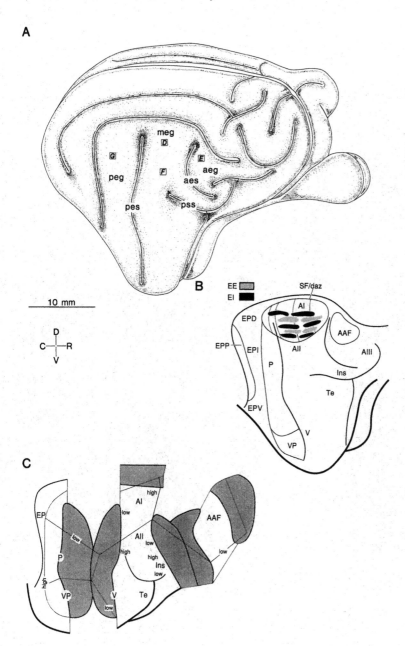

FIGURE 6.8A–C. Caption on page 291.

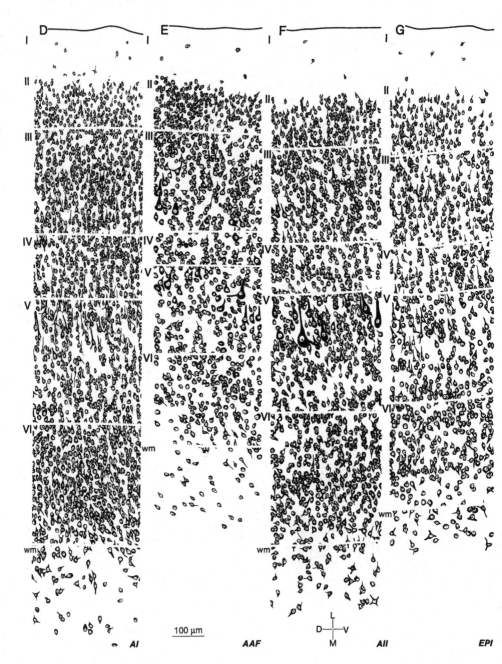

FIGURE 6.8D–G. Caption on facing page.

lateral AI cells (Tables 6.6, 6.8) and participate in much larger ipsilateral corticocortical as well as contralateral networks.

3.1.1   Layer I

The molecular layer is remarkable for its few neurons, the corresponding richness 'of the neuropil development, and for its neurochemical homogeneity, which is unique among the cortical layers (cf. 3.4). The bulk of its connections are with the apical dendrites of deeper-lying neurons or within layer I.

The absence of layer I pyramidal cells is a clue to its functional organization since more than 90% of the resident neurons are GABAergic in the rat (Winer and Larue 1989). Many dendritic profiles in this 150-μm-thick layer are from pyramidal neurons in layers II–V, suggesting that possible distal inhibition onto these elaborate arbors from intrinsic layer I neurons may have a modulating effect upon neurons that participate in the corticocortical, commissural, corticothalamic, corticostriatal, cor-

---

FIGURE 6.8. Topography of auditory cortical fields in the cat. A. Schematic lateral view of the major gyri and sulci. B. Areal organization of the main auditory architectonic regions on the lateral cortical convexity. In AI, the disposition of isofrequency contours (*thin vertical lines*) and of aural subregions (varieties of *stippling*) is shown schematically. The subdivisions of the posterior ectosylvian gyrus are also shown. C. Tonotopic arrangements in different auditory areas. The direction of the progression from low-to-high frequencies is indicated. Buried, sulcal cortex is also revealed in this flattened surface map. Redrawn from Reale and Imig (1980). D-G. Cytoarchitectonic arrangements in fields (C) AI, (D) AAF, (E) AII, and (F) EPI. Protocol for panels D-G: planachromat, N.A. 0.65, $\times$ 500, Nissl-stained 30-μm-thick celloidin-embedded sections with no correction made for shrinkage. C. AI is characterized by an overall high packing density that is most pronounced among the neurons in layers II and IV, the much lower cell density in layer Vb, and the thickness and prominence of layers V and VI. B. In AAF the neuronal density is somewhat lower than that in AI and the cells are slightly larger, the pyramidal cell populations in layers IIIa and Va have larger somata than their AI counterparts, and the cell-poor part of layer Vb is reduced. In addition, layer IV contains a significant number of pyramidal cells, unlike layer IV in AI (Winer 1984a). C. AII has a distinctive cytoarchitectonic arrangement: there are fewer of the pyramidal cells characteristic of layer III in AI (Winer 1984b), the density of neurons is more or less uniform throughout, except in layer Vb, and large or giant pyramidal neurons mark layer Va. Nevertheless, layer IV is dominated by small, round cells and the columnar arrangement evident in AI is conserved here as well. D. EPI also has a unique cytoarchitecture, in which small pyramidal cells predominate, with few medium-sized or large cells; the packing density is far less than in AI and the neurons, while smaller than AI cells, do not impart an especially granular texture to this region. Layers V and VI in particular are thinner and less elaborated than in AI. Vertical clustering, as in AI and AII, is also present here.

ticopontine, and corticocollicular projection systems. Since these complex branches often have many trunks that divide at the layer I–II boundary, these potentially Ca²⁺ channel-rich zones could amplify otherwise decremental dendritic signals, much as they are believed to do in Purkinje cells (Llinás and Sugimori 1980). Hence, for so cell-poor a layer, its influence is likely to be diverse though modest.

The neuronal architecture is more variable than might be expected, ranging from small cells with horizontal dendritic fields to much larger neurons; all of the neurons are nonpyramidal (Table 6.2). Most cell types have less than five primary dendrites, and their distal arbors are poorly developed though often quite long. Another cardinal feature of layer I neurons is the lateral stratification of their dendrites, an orientation at odds with the prevailing vertical arrangement of many neocortical cells. Finally, the axon is often unmyelinated and largely confined to layers I or II in these adult preparations.

Since a complete description of the morphology of neurons in layer I in auditory cortex is unavailable, the following classification should be regarded as preliminary and subject to revision. This account largely follows the schemes proposed in prior Golgi studies in cat AI (Sousa-Pinto et al. 1975) and is supplemented by other investigations in the rat (Winer and Larue 1989) and cat (Winer 1986; Prieto and Winer, unpublished observations) which included Golgi preparations. Among the most unusual neurons is the horizontal cell, whose sparsely branched, moderately spinous dendrites parallel the pia to run laterally for distances that no neuron, other than the horizontal cells in layer VI (Table 6.2), matches. Some of the small neurons may also have a horizontal disposition, while others span much of the depth of layer I. The medium-sized cell has a correspondingly larger dendritic domain and a simple, dichotomous branching pattern with unusual, recurving dendrites, some having surprisingly long appendages. In the rat, these neurons appear to have more appendages than do otherwise comparable cells in the cat. The largest layer I neuron has a multipolar appearance and a dendritic field that is spherical or laterally oriented; the dendrites are relatively smooth. In all likelihood, other types of neurons may exist in layer I, types that may not readily be accommodated into the provisional plan outlined above. This would include vertical cells that, on the basis of their somatic diameter, are medium-sized, but that have an entirely different orientation than other medium-sized cells. By the same token, cells of comparable somatic diameter but with very dissimilar dendritic branching patterns, the tufted and the multipolar cells, fall in the range of medium-sized neurons while probably representing distinct categories. In any event, the neuronal architecture of layer I is diverse.

The myeloarchitecture confirms certain features noted above. Thus, the upper half (layer Ia) is cell sparse, and the neuropil is correspondingly well-developed and dominated by fine myelinated and unmyelinated axons and the dendritic arbors ramifying in it. Layer Ib contains many more neurons, and the neuropil is even more complex, containing all the above components as well as axonal endings arising in deeper layers. Unmyelinated terminal axonal segments predominate, giving layer I a far paler texture than the other layers.

3.1.2 Layer II

The external granule cell layer has a wide range of both pyramidal and nonpyramidal neurons, a columnar and vertical organization that is conserved in the deeper layers, and significant neurochemical diversity. Its principal connections are with adjacent nonprimary auditory fields, and it provides local interlaminar projections within layers I–III.

Many layer II neurons are medium-sized nonpyramidal cells, and the appearance of large pyramidal cells some 400 μm below the pia denotes the beginning of layer III. Pyramidal cells are especially numerous in the deeper half (layer IIb), while the myeloarchitecture is progressively finer and lighter in the superficial half, through which the pyramidal cell apical dendrites begin to branch as they ascend towards layer I (Winer 1985a). Layer II pyramidal neurons are less well developed than those in layer III, with simpler arbors and thinner dendritic trunks, though their axons have the same pattern of recurrent collaterals as do pyramidal cells in other layers, except that they are distributed towards more superficial targets.

Nonpyramidal cells dominate layer II neuronal architecture. These neurons include small smooth or sparsely spinous multipolar cells with fine, unmyelinated, and locally branched axons, and medium-sized or large sparsely spinous multipolar cells, the latter having either extensive vertical or lateral axonal territories extending several hundred micrometers and perhaps farther since these observations come from adult specimens. Their vertical axonal domains, in particular, would be ideally arranged to influence the dendrites of pyramidal cells. A corollary of this is the columnar organization of AI so evident in Nissl preparations through layers II–VI, where vertical aggregates of cells spanning several layers are bordered by, and alternate with, 25–30-μm-wide zones of neuropil (Sousa-Pinto 1973b; Winer 1984a, 1991). The terminal axonal territories of some sparsely spinous neurons extend well past the dendritic domains. Such projections could subserve lateral interconnections among different functional cortical modules, much as they do in the primary visual (Gilbert and Wiesel 1983) and somatic sensory (DeFelipe et al. 1986) cortical fields. A type of neuron apparently unique to layer II is the extraverted multipolar cell, whose broad lateral domain and comparatively massive dendritic arbors distinguish it from the other types of neurons (Winer 1985a).

3.1.3 Layer III

The external pyramidal cell layer has a neuronal architecture as or more diverse than any other layer in AI, a correspondingly complex set of intrinsic and extrinsic connections, including relations with the auditory thalamus and ipsilateral as well as contralateral auditory cortices, and a neurochemical arrangement that sets it apart from other layers.

The pyramidal neurons, of which there are three varieties differing chiefly in size, represent the most conspicuous population in layer III. While they occur throughout layer III, they are especially common in the deeper one-half (Winer 1984b), where many of the commissural cells of origin lie (Code and Winer 1985). In myeloarchitectonic studies and in material prepared for electron microscopy, there is an abrupt change at the layer II–III junction, about 400 μm beneath the pia, where the prominent lateral axonal plexus in layer II is replaced by long vertical columns of myelinated and unmyelinated axons, between which are columns of dendrites arising from deeper-lying neurons. This arrangement and layer III extend to some 900 μm beneath the pia. Layer III proper represents about one-quarter of the thickness of AI.

The distinction between small, medium-sized, and large pyramidal neurons is not based merely on considerations of size but has implications for their synaptic arrangements and interlaminar connections as well. Thus, small deep-lying pyramidal cells have apical dendritic arbors that may barely reach layer II, segregating them from the latter influence or those arising in layer I, each of which probably includes substantial GABAergic populations. By the same token, large pyramidal cells might be expected to receive input from the full panoply of intrinsic, ipsilateral corticocortical, commissural, and (perhaps both specific and nonspecific) thalamic sources, suggesting that their receptive field organization may differ in some fundamental ways from that of other neuronal types with more limited dendritic domains and, by inference, a more limited set of afferent inputs, though the synaptic arrangements of only a few types of auditory cortex neurons is known. One feature that aligns each of the pyramidal cell populations (and many of the nonpyramidal cells, for that matter) is the local arrangement of their axons: with few exceptions, and excluding any corticofugal branch, they project principally towards the supragranular layers and have widespread lateral, and presumably excitatory, influences (Table 6.2) that could affect relatively distant cortical territories. These observations must be qualified since the full intercortical trajectory of axons cannot be revealed with Golgi preparations in adult animals, and in immature specimens it is a reasonable supposition that their connections are necessarily incomplete. Intracellular injection experiments that fill single neurons with horseradish peroxidase have so far yielded views of the distribution of their axonal branches that are comparable to the results from Golgi material (Mitani et al. 1985), but neither picture can be considered complete when, in the somatic sensory cortex, such cells with ipsilateral intercortical projections of several millimeters have been demonstrated (DeFelipe et al. 1986).

In contrast to the pyramidal neurons, the size and influence of nonpyramidal neurons argue that their actions are properly considered as more local, at least within the constraints noted above. Some neurons, like the sparsely spinous stellate cell, probably exert their primary effects within layer III, and their axonal trajectory is parallel to the apical dendrites of pyramidal cells. The largest nonpyramidal cells have dendritic fields up to 500 μm, but most are far smaller. The neurogliaform neurons have fine, wavy dendrites and an axon largely confined to the immediate vicinity of the limited, spherical dendritic arbor (Fig. 6.9[5]). A small, axonless granule cell with a tiny soma and fine, sparsely branched dendrites has also been reported (Meyer et al. 1984).

3.1.4 Layer IV

The internal granule cell layer has an almost completely nonpyramidal neuronal population, and a pattern of connectivity dominated by thalamic, corticocortical, and intrinsic input. It receives projections from, but does not contribute in any substantial way to, the commissural system. Only about 250 μm thick, it represents some one-eighth of the cortical depth and is conspicuous between the pyramidal cell-rich populations in layers III and V.

The vertical arrangement of neuronal cell bodies is evident in layer IV and is parallel to dendritic bundles ascending through it as well as the coarse, unmyelinated terminal processes of large axons presumed to be of thalamic origin by virtue of their size and shape (Winer 1984c, Fig. 6.1), as well as their close resemblance to such robust terminals in other sensory cortical fields (Ferster and LeVay 1978; Landry and Deschênes 1981). A prominent, laterally arranged axonal plexus can be traced to the recurrent intracortical branches that arise from more superficially situated pyramidal cells, then ascend to re-enter layer IV. Other intrinsic axons are those from layer IV neurons.

Of the six main types of layer IV neurons, four have the granular and supragranular layers as their main axonal target, while only two varieties project towards the infragranular layers (Winer 1984a; Table 6.2). This is consistent with the finding that layers V and VI have far fewer GABA-immunoreactive axon terminals than do layers I–IV (Peterson et al. 1990), and it supports the view that some GABAergic neurons in layers V and VI may project towards more superficial layers.

Many of the small or medium-sized layer IV neurons have strongly tufted and spatially restricted dendritic arbors, while among the large multipolar neurons the arbors are often much simpler and correspondingly larger, implying that the different cell types each have specific spatial

I

II

III

IV

V

VI

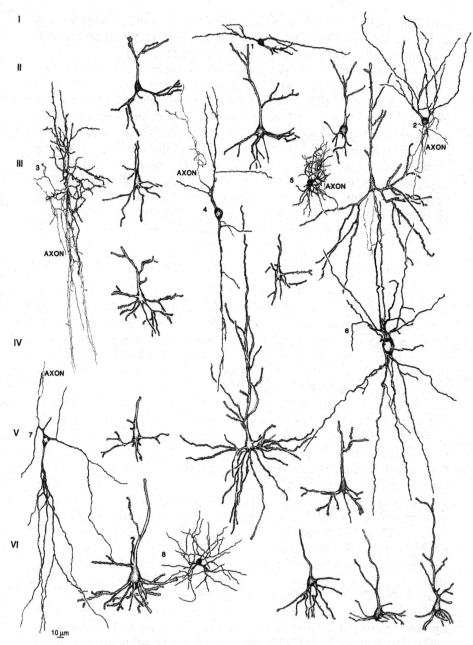

10 µm

FIGURE 6.9. Representative AI neurons from Golgi preparations, with emphasis on the varieties of nonpyramidal neurons. Planapochromat, N.A. 1.32, × 2000, Golgi-Cox method, 140-µm-thick sections. From Prieto et al. (1992a). (*1*) a horizontal cell with widely-ramified, sparsely spinous lateral dendrites confined entirely to layer I; see Table 6.2, No. 1. (*2*) an extraverted multipolar cell in layer II whose apical dendrites extend into layer I and whose *AXON* projects towards domains within layer IV. There is a further suborganization within layer IV where the smallest neurons are in the superficial half (layer IVa), the larger ones in the deeper half (IVb). By analogy with the other, pyramidal cell-dominated layers, perhaps the largest, deepest-lying nonpyramidal cells have the most distant corticocortical targets, while more superficially situated neurons have a more limited, local pattern of connectivity (see Games and Winer 1988 for data from the rat bearing on this point).

A variety of tufted cells and multipolar neurons are the dominant neuronal classes, and most of these cells have smooth, aspinous dendrites. Spiny stellate neurons are much less common, and they are among the smallest somata in layer IVa. A much larger neuron, also found in layer IVa, is the double bouquet cell, whose vertically oriented dendritic arbors reinforce the columnar architecture of the IIIb–IVa border, and whose axon has an extensive vertical domain and many lateral side chains that branch at regular intervals.

3.1.5 Layer V

The internal pyramidal cell layer contains a broad morphological array both of pyramidal and nonpyramidal neurons and is the source of a diverse set of connections that include the ipsilateral nonprimary auditory cortex, the contralateral AI, the medial geniculate body, the inferior colliculus, and perhaps other brain stem targets. It also receives a significant number of commissural and corticocortical axon terminals, but

---

(FIGURE 6.9. *continued*) layer III; see Table 6.2, No. 13. (*3*) double bouquet (small multipolar) cell in layer III with many dendritic appendages and a highly branched, vertically arranged axonal projection largely confined to a narrow cortical territory. The *AXON* is very slender and has several *boutons de passage*; see Table 6.2, No. 23. (*4*) layer III bipolar cells have one of the larger vertical dendritic fields of any AI cell, sending their smooth dendrites deep into adjoining layers, while the *AXON* projects towards layer II; see Table 6.2, No. 22. (*5*) Neurogliaform cells have thin, delicate dendrites with a few appendages and a very fine, locally projecting axon. Their dendritic domain is among the smallest of any AI cell; see Table 6.2, No. 25. (*6*) Large multipolar cells in layer IV have smooth, well developed apical and basal arbors that can reach layers III and V, respectively, and which branch sparsely; see Table 6.2, No. 29. (*7*) Among the most rarely impregnated neurons in AI, the smooth inverted pyramidal cell has several striking features, including dendrites that are almost entirely aspinous and unusually slender, poorly developed "basal" dendritic arbors, and an axon projecting towards layer IV. Although it is provisionally considered as a member of the class of inverted pyramidal cells, it differs from them in its smooth dendrites and comparative scarcity; see Table 6.2, No. 35. (*8*) a small multipolar cell in layer VI with smooth dendrites that radiate spherically. While there are fewer layer VI GABAergic neurons than in other layers in AI, such neurons may be immunopositive; see Table 6.2, No. 43. *Stippled profiles,* various types of pyramidal cells.

few of the latter are likely to arise from intrinsic GABAergic neurons (see Peterson et al. 1990). Beginning about 1100 μm beneath the pia, it is nearly 400 μm thick or almost one-fourth of the total cortical depth. Like layers I–IV, and VI, it contains more than one type of neuron showing glutamate- or aspartate-like immunoreactivity. Many neurons are immunopositive for gamma-aminobutyric acid, but the local arrangement of the GABAergic processes is entirely different from that in other layers, and aligns it most closely with layer VI.

Layer V cytoarchitecture and myeloarchitecture distinguish it from adjoining layers. Thus, layer IV receives a continuous band of large-caliber, myelinated afferents and has a high density of small, closely packed nonpyramidal neurons. In contrast, layer V has a cell-sparse, myelin-rich outer half (Va), and an inner half (Vb) with many medium-sized and large pyramidal cells. The layer VI border is marked by a decrease in cell size analogous to that in layer IV and less well-developed pyramidal neurons, as well as a different constellation of extrinsic projections. The columnar organization of cell bodies and dendritic bundles evident in layers II–IV is also present here, as are myelinated and unmyelinated terminal axonal profiles with a vertical orientation.

Of the seven main types of layer V cells, three are pyramidal, three are nonpyramidal, and the seventh may be construed as one or the other depending on the criteria used and their interpretation. The pyramidal neurons are unusual for their large sizes and relative dendritic complexity. Their apical dendrites extend to layers II or I, where they form more or less elaborate, curved tufts, many of which are parallel to the pia and are thus in a position to intercept comparatively fine thalamocortical axons originating in the medial division of the medial geniculate body (seen in the rat by Ryugo and Killackey 1974). As a rule, layer V pyramidal cells have simpler local axonal networks than their counterparts in layers VI, III, or II. The large dendritic trunks are selectively spinous, often with smooth expanses in layer V and many more appendages in the superficial layers, especially in layer I. Star pyramidal cells, which occur in both layers III and V, have a more radiate dendritic domain and a simpler branching pattern than do other pyramidal neurons; their large apical dendrite and the form of their axon define them as pyramidal (Winer 1992).

The inverted pyramidal cells resemble pyramidal cells in many respects and differ in others. Thus, their superior dendritic arbors are often less developed, while the apical, inverted primary dendrite may be truncated or have complex spray-like tufts. When the myelinated axon is impregnated, it has an origin, form, and trajectory much like that of pyramidal cell axons. While it therefore seems plausible to classify these cells as pyramidal, the fact that many such somata are GABAergic, often including immunopositive segments of the dendrites and axon, argues against such a conclusion (Peterson and Winer 1989; Prieto et al. 1990; cf. 3.4).

Among the nonpyramidal neuronal populations in layer V, the large multipolar cell is conspicuous for the size of its perikaryon, which exceeds that of any other cortical interneuron except the giant multipolar cell in layer VI (cf. 3.1.6.). The slender dendrites are smooth and rarely have appendages. Their sparsely branched arbors radiate across layer V in every direction. The thick, unmyelinated axon has many local branches, some of which are vertical and may terminate on the apical dendrites of pyramidal cells (Winer 1992), while others project laterally among the basal dendritic arbors to form clasp-like endings whose postsynaptic targets are unknown. These neurons are commonly found in layer Vb.

A medium-sized multipolar cell also occurs in layer V. It has an oval soma with several thin, smooth, and poorly branched dendrites that radiate spherically. The thick, unmyelinated axon forms a local collateral network with vertical ascending or descending branches; the breadth of this terminal field is smaller than that of the large multipolar neuron, and these cells are frequently seen in layer Va. A bitufted neuron, which has a soma almost equal in size to that of the large multipolar neuron, is also impregnated. The thin, vertically oriented dendrites arise at the perikaryal poles and form sparse tufts with a few appendages; the myelinated axon is sometimes impregnated.

3.1.6 Layer VI

Layer VI has been described variously as the multiform layer (Brodmann 1909) or spindle cell layer (Campbell 1905) or layer of medium-sized pyramidal and triangular neurons (Ramón y Cajal 1911). It contains the most diverse neuronal population within AI, consisting of at least nine readily recognized types of cells (Table 6.2; Prieto and Winer, in preparation). While the vast majority of these neurons are pyramidal in their morphology, the classes of nonpyramidal cell are as varied (though perhaps not so numerous) as those in the supragranular layers, where the intrinsic actions of inhibitory intracortical neurons might be considered to be more critical than in the infragranular layers, whose output is directed chiefly towards subcortical targets and whose synaptic consequences are assumed implicitly to be excitatory, as opposed to those of supragranular neurons. This view is consistent with the finding that layer VI has a smaller proportion of GABAergic neurons (Prieto et al. 1990) and significantly fewer such puncta (Peterson et al. 1990) than do supragranular layers in AI. A little less than 400 μm thick, layer VI begins about 1500 μm beneath the cortical surface; while there is a sharp border between layer VI and the white matter for most neurons, a surprising number are scattered in the neuropil far beneath layer VI, and many of these cells are GABAergic (Fig. 6.14[wm]).

In preparations stained for myelin, layer VI is virtually opaque when the fiber plexus of the supragranular layers is optimally differentiated. Nevertheless, it is possible to discern that the upper half, layer VIa, is dominated by pyramidal cells, and that it has a vertically arranged fiber architecture reinforced by the apical dendrites of pyramidal cells, while layer VIb contains polymorphic cells with a heterogenous orientation, including neurons whose longest axis is horizontal or tangential. On average, all of the types of layer VI neurons are smaller than layer V cells, with one important exception (see below).

Among the neuronal classes, three are undoubtedly pyramidal, one more type may or may not be, depending on how pyramidal cells are defined, and five types are nonpyramidal. The shape of the small and medium-sized pyramidal cells is distinct from that of their counterparts in layers II (Winer 1985a), III (Winer 1984b), IV (Winer 1984a), and V (Winer 1992) in having fewer and simpler dendritic arbors and a less polarized mode of branching that obscures the sharp transitions between cell types characteristic of neurons in other layers. Their apical dendrite rarely extends into layer III and they often tilt obliquely, in contrast to the vertical orientation of other pyramidal cells. Their axon commonly projects into the white matter.

Among the class of atypical pyramidal neurons, the vertically arranged pyramidal cell is striking with its enormously long apical and basal dendrites, which may span 600 μm or more, the upper branches extending to the base of layer IV. The proximal dendrites branch sparsely and the distal ones have appendages typical of those found on classical pyramidal cells.

The inverted pyramidal cells are also simpler than comparable cells in layer V. In particular, their dendritic arbors are smaller and single dendrites are thinner; two variants—one smooth, the other spinous—are recognized. Many (but not all) inverted pyramidal cells are GABAergic (Table 6.2; Prieto et al. 1990); it is unknown if these correspond to one subtype or include members of both. The smooth variety is virtually without appendages.

Small and medium-size multipolar cells have a simple stellate or weakly tufted mode of branching. The former have relatively small dendritic fields largely confined to a long axis of about 150 μm, though some are up to twice this size, and their dendrites are slightly varicose and entirely smooth save for one or two spines. Nothing is known of the form of their axon. While there is a considerable range in the size of the dendritic domains, all the neurons in this class share a similar dendritic configuration and have few appendages. Their dendritic orientation is likewise varied: some have a spherical arrangement, and others are polarized vertically or horizontally.

The largest and most unusual of the multipolar cells, and perhaps the rarest layer VI neuron, is the giant multipolar cell, which is *sui generis* in AI and unique to this layer. It has a spindle-shaped soma with a long axis of nearly 50 μm, whose size exceeds that of all but a few unusually large pyramidal cells in layer V. Unlike the large multipolar cell of layer V, however, this cell has a dendritic domain consistent with its somatic dimensions, and may dominate areas of 600 μm high by 400 μm wide; these dimensions are sufficient, if their orientation is appropriate, to span considerable domains within an isofrequency (Merzenich et al. 1975) or aural (Middlebrooks et al. 1980) representation, or perhaps to straddle adjoining but dissimilar representations. Their dendrites, like those of most other nonpyramidal cells, are unusually smooth.

Bipolar neurons have a fusiform soma from whose poles 1–2 slender primary dendrites arise and branch sparsely, usually dichotomously; in their main features they closely resemble bipolar cells in cat area 17, though the latter are not apparently impregnated in layer VI (Peters and Regidor 1981).

The horizontal cell completes the present survey. Like the giant multipolar cell, it is unique to layer VI, combining as it does different attributes of the bipolar and pyramidal cell classes. Like the bipolar cell, it has a highly polarized orientation, in this case parallel to the pia rather than vertical, and resembling the ovoid cell in this regard. However, its branching pattern and dendritic span resemble those of bipolar neurons much more than those of ovoid cells. The dendrites are not as rich in appendages as those of spinous pyramidal, inverted pyramidal, or spiny ovoid cells, but they have more spines than any other type of layer VI neuron and, once again, underscore the cytological diversity of this heterogeneous population.

3.2 Laminar Distribution of Afferent Input

Historically, it was unclear whether each part of the cerebral cortex received thalamic input, or if every thalamic nucleus projected to the cortex (Walker 1938; Diamond and Utley 1960). The main impediment in making this determination was the unreliability of the retrograde degeneration method, whose conclusions rested on the completeness of the terminal axotomy (as well as the preservation of the blood supply to adjoining cortical fields), and on the unambiguous interpretation of thalamic cell shrinkage, pallor, and ensuing gliosis. Where the thalamic neurons are more or less uniform in size and densely packed together, as in the ventral division of the medial geniculate or the A-laminae of the lateral geniculate body or the main parts of the ventrobasal complex, the analysis of these changes was relatively straightforward. In other parts of the auditory thalamus, such as the dorsal and medial divisions, however, the absence of any conspicuous laminar organization, variability in cell form and density, and the complex neuropil organization confounded this direct approach and led to the conclusion that either some parts of the thalamus were without a cortical projection, or, if they had such a projection, then the afferent thalamic axons branched to supply more than one cortical field, thereby preserving them from, or at least making them less susceptible to, the retrograde changes following axotomy since some part of their axonal terminal domain would survive all but the largest cortical lesions. It is now clear that only a relatively small proportion of auditory thalamic neurons possess such branched axons (cf. 2.2.2.) and that no part of the auditory cortex so far explored with contemporary axoplasmic transport methods is devoid of thalamic input (Winer et al. 1977; Niimi and Matsuoka 1979; Andersen et al. 1980a). However, the view that the pathway for the ventral division to AI is essential (since these thalamic cells degenerate after AI lesions) while that from the nonprimary thalamic nuclei to the nonprimary cortex is sustaining (since most of these neurons show only equivocal signs of retrograde degeneration even after comparatively large cortical lesions) may be valid in another sense than Rose and Woolsey (1958) intended originally. Thus, thalamic input to layers III and IV in AI arises only from the ventral division, while thalamic projections to presumably comparable layers in AII, and in the insular, temporal, and posterior ectosylvian fields, each arise from more than one auditory thalamic nucleus (Table 6.4). Other evidence favoring this conclusion is the differential behavioral effects of damage to one cortical area or another (cf. 9; Neff et al. 1975; Winer 1985b).

3.2.1   Thalamic Input to Layers III and IV

The laminar distribution of thalamocortical axons is highly conserved across species, sensory systems, and cortical areas (Jones 1985), and the main features of this pattern are evident in AI. Briefly, a topically arranged main sensory thalamic nucleus that represents a lemniscal channel (for example, the ventral nucleus of the medial geniculate body [Winer and Morest 1983a] or the dorsal nucleus of the lateral geniculate body [Stone 1983], or the ventrobasal complex [Scheibel and Scheibel 1966]) conveys afferent input to a cortical field with analogous topographic organization of a sensory surface (for example, primary auditory [Niimi and Matsuoka 1979], primary visual [Niimi et al. 1981], or primary somatic sensory [Dykes et al. 1986] cortex). Thus, auditory thalamic injections of various tracers (Peterson and Winer 1988, 1993; Fig. 6.10) or thalamic lesions and the ensuing silver-impregnated axonal degeneration (Sousa-Pinto 1973a) both reveal comparable terminal patterns: dense labeling or heavy degeneration in layers IV and IIIb, somewhat less input to layer IIIa, and alternating foci of lighter and heavier bands of transport/degeneration in 500–1000 μm-wide lateral expanses. Terminal labeling or degenerating axonal fragments are near background levels in layers I, II, V, and VI (Table 6.8). Afferents to layer IV arise from medial geniculate neurons other than those with terminations in layer I; the cells projecting to layer

I have many origins, with the medial division contributing most, while those ending in layer IV are predominantly from the ventral division (Mitani et al. 1984; Niimi et al. 1984). Projections from the rat suprageniculate nucleus terminate mainly in layers III and IV in temporal cortex, and more widely in layer I (Kurokawa et al. 1990).

The medial division has a different physiological arrangement (Aitkin 1973) and neuronal architecture (Winer 1985b) than the ventral division, and a unique set of connections with the midbrain (Morest and Winer 1986) and cortex (Andersen et al. 1980a). Since these attributes distinguish it from the lemniscal pathway and align it more closely with polysensory lemniscal adjunct channels, it is not surprising that its laminar targets are likewise unique, and that terminal labeling or degeneration in layers I and VI is marked, while that in other layers is sparse. Among the consequences of such an arrangement are (1) the possible conservation of independent thalamic (and by implication, discrete brain stem) channels in the cortex, and their potential divergence or convergence by way of corticocortical connections; (2) the creation of alternating functional subregions within AI where robust thalamic input is intercalated between zones where such influence may be much smaller; and (3) the spatial segregation of thalamic terminations upon particular postsynaptic loci, for example, apical or basal dendritic arbors, or neuronal perikarya, among others. Besides the direct effects of thalamic input upon neurons in layers III and IV, many of whose cells have local cortical connections (Table 6.2), thalamic axons terminating on the apical dendrites of pyramidal neurons must affect these cells, which in turn modify thalamic, midbrain, and brain stem auditory centers (Table 6.7). Thus, the information carried in thalamocortical axons might affect even spatially remote neuronal populations.

How far the analogy between the primary visual and auditory fields can be extended with respect to thalamic input to layer IV remains to be seen. In area 17, the unique sublaminar terminations in layer IV of X- and Y-cell axons (Humphrey et al. 1985) suggest that the segregation among afferent channels is preserved. On the other hand, there is strong evidence within any sublayer that the neurons postsynaptic to this input are varied in their form (Davis and Sterling 1979) and, by inference, their homolateral (Meyer and Albus 1981) and interlaminar (Lund et al. 1979) connections. If this applies to layer IV in AI, then thalamic influences might reach, through corticocortical connections, the anterior auditory field, the posterior auditory field, and even more remote loci (Table 6.6). Thalamic input, then, could thus diverge progressively, degrading any exact tonotopy while propagating different binaural or temporal attributes of the stimulus to converge within more remote areas (Tsuzuki and Suga 1988; Jen et al. 1989; Pantev et al. 1989). While this hypothesis has not been confirmed, it is consistent with the attributes of neurons in auditory area AII, whose binaural properties and breadth of afferent tuning are

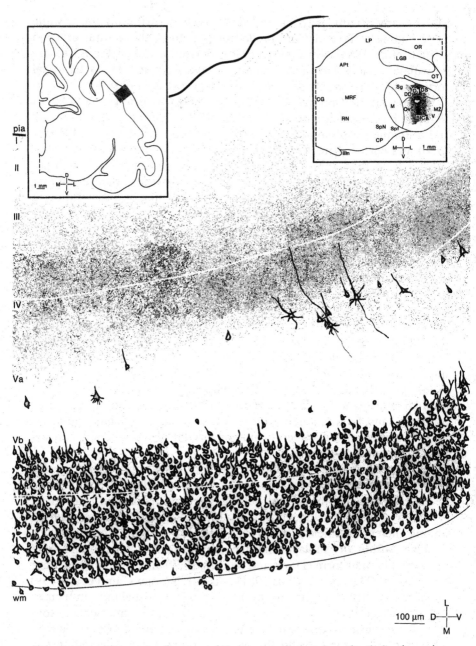

FIGURE 6.10. The major features of thalamocortical and corticothalamic projections. In this experiment, 0.25 μl of horseradish peroxide was injected into the medial geniculate complex (*inset, upper right*), centered in what is probably near the 20 kHz representation in the ventral division (see Fig. 6.2C), though portions of the dorsal division and of the medial division are also involved. The resulting anterograde transport labeled a large expanse of the cortical convexity along which

strikingly different from those of AI cells, as is the spatial organization of these dimensions (Schreiner and Cynader 1984). Such analogies cannot be pursued indefinitely, however, since the output of many layer IV cells in AI is directed primarily at cells in layers III and II (Table 6.2; Mitani et al. 1985), whose targets include both the ipsilateral corticocortical (Winguth and Winer 1986) and commissural (Imig and Brugge 1978) pathways, while the comparable neurons in layers II–IV in area 17 project to the homolateral area 18, and thus receive commissural influence, either as cells of origin or targets, only at subsequent synaptic stations (Fisken et al. 1975). It is interesting to note that area 18 receptive fields, in contrast to those in area 17, are larger, less sharply tuned, and show other traits that support the idea that their response properties reflect serial, afferent convergence (Wiesel and Gilbert 1983; see also Dreher et al. 1980). Such patterns would therefore align cells in areas 17 (Hubel and Wiesel 1959) and AI (Merzenich et al. 1975) (and, by analogy, areas 3, 1, 2; Dykes et al. 1980) with one another, and neurons in area 18 (Hubel and Wiesel 1965; see also Orban et al. 1985) and AII neurons (Schreiner and Cynader 1984) (and, again by analogy, to cells in SII; Burton et al. 1982). Further evidence supporting the linkage between architectonic areas, cell types, and connections are the different relations between lemniscal and lemniscal-adjunct systems in the thalamocortical and corticothalamic pathways (Table 6.4) and in their contrasting patterns of commissural connections (Table 6.5).

Anterograde degeneration studies of the laminar distribution of thalamic afferents to nonprimary cortex find axonal fragments in all but

---

(FIGURE 6.10 *continued*) AI is found, and from which a representative segment is illustrated (*stippled area* in *inset, upper left*). Two types of transport are evident. A fine, granular reaction product is concentrated in layers III and IV and forms a band running dorsoventrally through the central, presumably middle-frequency, parts of AI. A much smaller projection to the deepest part of layer II and to layer Ia is also present. Also labeled, though by retrograde transport, are the corticothalamic cells of origin in the superficial part of layer Va, the deep part of layer Vb, and throughout layer VI (see Table 6.8). From these and similar experiments using other tracers, several conclusions follow. These are (1) that thalamocortical input to AI involves parallel projections that terminate in different layers (and which probably arise from different nuclei) and whose density is lamina-specific; (2) that thalamocortical input is spatially reciprocated by corticothalamic neurons, the latter being more extensive in these experiments than the former; (3) that because the corticothalamic projections originate from at least two layers, and hence are likely to project convergently within the auditory thalamus, they may represent more than one descending pathway. Finally, (4) both ascending and descending systems have main and subsidiary elements segregated on the basis of their laminar origins and terminations, respectively. Planapochromat, N.A. 1.32, × 2000. From Peterson and Winer (1993).

layer II in most fields (Niimi and Naito 1974). Several types of dorsal division neurons project to nonprimary cortex (Table 6.1) and perhaps to AI (Mitani et al. 1984); their postsynaptic targets and the form and spatial distribution of their terminals are unknown.

3.2.2 Thalamic Input to Layers I and VI

The term, nonspecific, that is often applied to this pathway is a misnomer since its brain stem (Table 6.3) and cortical (Table 6.4) connections are largely parallel in some respects to those of the lemniscal pathway ascending to AI, and convergent in other ways, but hardly random or generalized. While the physiological properties of these lemniscal-adjunct neurons are conspicuously different from those of ventral division neurons (Aitkin 1973; Aitkin et al. 1981; Calford and Aitkin 1983), these attributes probably represent other facets of auditory function than orderly arrangements of frequency, aurality, or azimuth (cf. 9.).

While the functional significance of this pathway (Fig. 6.10) is obscure, some hypotheses can be offered. The terminal plexus of these axons in layer I could contact the processes of resident neurons or those of deeper-lying cells. Since many medial division neurons in the rat (Popowits et al. 1988) and cat show glutamate- or aspartate-like immunoreactivity, and can be labeled simultaneously by cortical injections of horseradish peroxidase (Larue and Winer, unpublished observations), then the thalamic projection to layer I could affect this largely GABAergic neuronal population (Winer and Larue 1989). Since the output of these neurons is directed chiefly towards layers II and III, the potential postsynaptic targets include the many GABAergic cells in these layers (Fig. 6.15H,I) and those with glutamate- (Fig 6.15B,C) or aspartate-like immunoreactivity. Thalamic input to layers I and VI is parallel to that of thalamic afferents to layers IV and III, but may arrive somewhat later because of the presumably larger caliber of lemniscal axons and the fact that they make monosynaptic connections with supragranular neurons.

The projections to layer VI might have different consequences. If the synaptic target were chiefly the corticofugal neurons (Tables 6.2, 6.4, 6.7, and 6.8), then corticothalamic cells would be affected primarily, and perhaps those with ipsilateral corticocortical projections as well (Fig. 6.3A,B). However, if GABAergic layer VI cells (Figs 6.14, 6.15; Prieto et al. 1990) are among the targets, then other scenarios might follow. More than one type of layer VI cell is GABAergic (Table 6.2; Fig. 6.15L); if any of these, like certain cells in layer VI of the primary visual cortex, have interlaminar projections to layer IV (McGuire et al. 1984), then the extralemniscal thalamic pathway could, by such input, alter the discharge of neurons in the lemniscal pathway. An indirect piece of evidence supporting this idea is the relative paucity of GABAergic axon terminals in layer VI, compared to the number in layer IV (Peterson et al. 1990).

3.2.3   Other Ascending Projections

While the laminar target of certain brain stem input to AI is unknown, a compact group of nuclei in the ventromedial midbrain tegmentum near the root of the oculomotor nerve project widely upon the cerebral cortex, including primary auditory, insular, and primary visual and motor fields. These substantial projections involve at least four tegmental nuclei in each instance, with the bulk of the auditory input arising from the rostral linear tegmental nucleus. The function(s) of this putatively dopaminergic pathway remain(s) obscure (Scheibner and Törk 1987).

3.3   Laminar Origins and Terminations of Cortical Projections

The areal and laminar relations in the auditory cortex suggest a number of functional implications that distinguish it from thalamic cellular arrangements in several ways. First, almost every cortical layer is subject to the influence of many others, and in turn influences them (Mitani et al. 1985), while particular thalamic nuclei have no known interdivisional connections with one another through local circuits (Morest 1964, 1965a) and largely parallel patterns of brain stem input (Table 6.3). Second, the relative proportion of GABAergic neurons shows enormous variability in the medial geniculate body in different species (Winer and Larue 1992), while the proportion of such cells in the cortical sensory and motor fields is highly conserved among species and between systems (Fitzpatrick et al. 1987; Hendry et al. 1987; Winer and Larue 1989). Third, the thalamic input to the auditory cortex, while large, is only one of many (Table 6.8) and if the density of such synaptic terminals in primary auditory cortex resembles the pattern in the primary visual cortex, the ratio of these synapses to the total number of layer IV synapses is surprisingly small, about 15% (LeVay and Gilbert 1976). In contrast, the corticothalamic projection is relatively enormous, and there are many instances where it is not reciprocated by a corresponding thalamocortical input (Colwell 1975; Andersen et al. 1980a; Winer and Larue 1987). Fourth, the thalamic projections are distributed largely to the cortex and some subcortical limbic and motor targets (Table 6.3), while those from the cortex reach the ipsi- and contralateral cortices, subcortical forebrain centers, thalamus, midbrain, and medullary nuclei in a highly selective yet widespread pattern (Table 6.7; reviewed by Huffman and Henson 1990).

3.3.1   Commissural Pathway

The commissural projection from AI arises in layers III, V, and VI (Table 6.8), the former contributing three-quarters of the total (Code and Winer 1985). Within layer III, about two-thirds of the retrogradely labeled neurons were pyramidal cells, and members of each class of morphologically

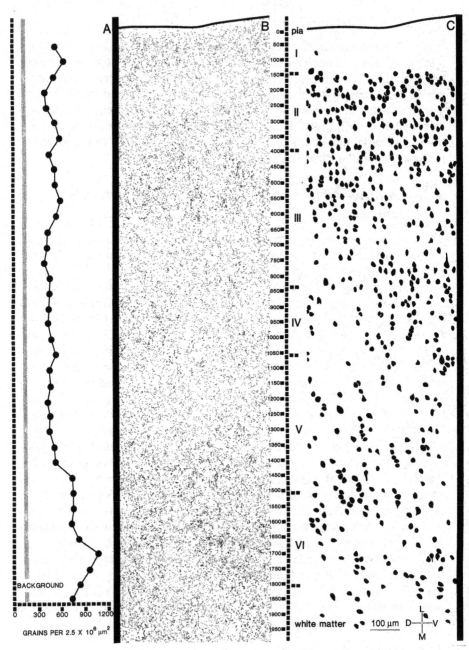

FIGURE 6.11. Commisural axon terminals in AI in a band of intense labeling after injections of [³H]amino acids in the contralateral hemisphere. The ventral division of the medial geniculate body (not shown) received significant descending input. A. Grain counts across the cortical depth; the background was derived from nearby, nonauditory cortex. B. Silver grains in AI. Protocol for panels B,C:

identified pyramidal cell were labeled (Table 6.2). In addition, some bipolar cells and certain large multipolar neurons are commissural. Since
GABAergic neurons are not thought to project commissurally, perhaps
there are subtypes of bipolar and multipolar neurons that are GABAergic
and (presumably) noncallosal, while others might show glutamate- (Fig.
6.15C) or aspartate-like (Winer, unpublished observations) immunoreactivity and could project in the corpus callosum. Analogous neurochemical and synaptic diversity have been demonstrated for bipolar cells
in the rat primary visual cortex (Peters and Harriman 1988).

In contrast to layer III, whose cells of origin are largely continuous and
terminate abruptly at the border of a focus of transport, those in layers
V and VI are more widely dispersed and apt to exceed the margins of
the heaviest concentrations of labeled layer III neurons. At least five types
of neurons in these layers project commissurally and are classified as
pyramidal; the possible exception is the inverted pyramidal cell, many
of which are commissural or GABAergic (showing immunoreactivity for
glutamic acid decarboxylase [GAD] or GABA and accumulating
[^{3}H]GABA), though it is unknown if any embody both features (Peterson
and Winer 1989).

There is a systematic topographic relation between the site of injection
in AI, the contralateral locus of commissural cells of origin, and the
position of retrogradely labeled ipsilateral thalamic neurons. Thus, injections centered at the caudal, low-frequency representation (Merzenich
et al. 1975) label commissural AI cells throughout the same contralateral
site (Fig. 6.12B, *insets*), and at a thalamic locus (Fig. 6.12D, *inset*) that
would correspond approximately to the 0.1–0.8 kHz representation in
the ventral division (Morel 1980). Analogous results have been seen with
injections at other sites along the cortical convexity and within AI (Code
and Winer 1985).

The pattern of anterograde transport to the contralateral hemisphere
is somewhat more complex. After large injections of [^{3}H]amino acids, all
layers receive substantial input (Fig. 6.11). With somewhat smaller injections, all layers still receive terminal labeling, with layer III having the
heaviest concentration, layers I, V, and VI intermediate values, while
layers II and IV receive less than half as much input as does layer III,
though still far above background (Code and Winer 1986; see also Imig
and Brugge 1978). Even when corticothalamic anterograde labeling in the
latter experiments was essentially continuous, the commissural labeling
in AI is patchy and discontinuous. However, the transport to zones intercalated between the foci of dense terminal labeling was still above

---

(FIGURE 6.11. *continued*) planapochromat, N.A. 1.32, × 2000. C. Cytoarchitecture from the 30-μm-thick frozen section autoradiograph from which the silver
grains in panel B were plotted; compare with Figure 6.8D.

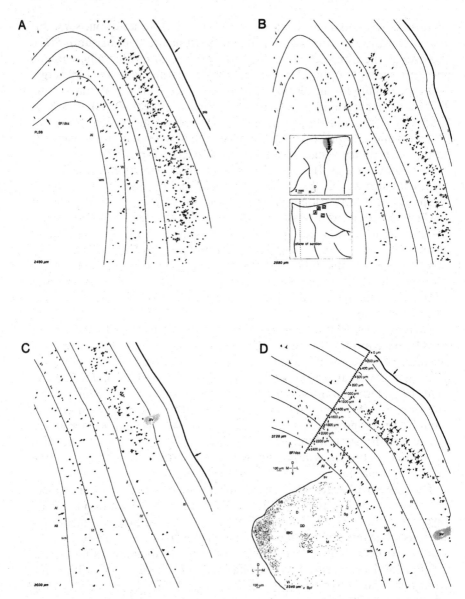

FIGURE 6.12. Laminar origins of commissural neurons in AI. Four injections of horseradish peroxidase were placed along an axis that might correspond to an isofrequency contour (panel B, *upper inset, black dots;* see also Fig. 6.8B) with the resulting spread of tracer largely confined to the low-frequency representation as assessed by the concentration of retrogradely labeled neurons in the lateral part of the ventral nucleus (panel D, *inset;* see also Fig. 6.2C). The ensuing cells that are retrogradely labeled in contralateral AI (panel B, *lower inset*) are mainly in layer III, with far fewer in layers V and VI, and very few in layer IV. Sometimes in layer III (panel C) the labeled cells form clusters, and in other sections (panel B) they are essentially continuous, while cells of origin in layer V (panel A) are

background, suggesting that the commissural system is divergent. There is some indication that the caudal, low-frequency portion of AI may receive fewer or sparser commissural terminals than other parts (Diamond et al. 1968; Imig and Brugge 1978), while other studies find a more complete representation (Code and Winer 1986). A discrepancy also exists with respect to the degree of anterograde-retrograde reciprocity between terminal fields and cells of origin, one study noting almost complete concordance between heavily labeled zones (Imig et al. 1982) while another describes many instances of nonreciprocity (Code and Winer 1986).

Even more refined patterns of interhemispheric relations between physiologically defined AI subregions may exist. Thus, zones with many ipsilateral-dominant suppression neurons have more callosal cells of origin and terminals than subareas where monaural contralateral-dominant responses are prevalent. Perhaps the rather sparse monaural contralateral and contralateral-dominant suppression area projects heavily and convergently onto limited zones where ipsilateral-dominant suppression neurons occur. In contrast, the more numerous ipsilateral-dominant suppression neurons could supply monaural contralateral neurons and contralateral-dominant suppression cells (Imig and Brugge 1978).

While the strongest commissural interconnections are homotopical, there is also significant evidence for divergence. Thus, AI and AII project to at least three other contralateral fields besides one another, and there is more than one possible exception to the rule of homotopic interhemispheric connections (Table 6.5).

3.3.2 Corticocortical Projections

Corticocortical cells of origin have more diverse laminar origins than do commissural cells. Thus, ipsilateral neurons arise in layers II–IV (Table 6.8), with occasional projections even from layer I cells, while the commissural cells reside chiefly in layer III and, to a lesser degree, in layers V and VI (Imig and Brugge 1978; Code and Winer 1985). The laminar distribution of AI neurons projecting to AII is about 35% in layer III, 30% in layer V, and 10–15% in the other layers (excepting layer I). These neurons often form topographically arranged clusters in AI (Winguth and Winer 1986) whose position could be related to the representation of best frequency in AI and AII (Reale and Imig 1980). The distribution of

---

(FIGURE 6.12. *continued*) situated chiefly in the intermediate region above and below which corticothalamic projection neurons reside (see Fig. 6.10). The commissural projection terminates abruptly at the dorsal (panel D, *upper part*) and ventral (panel C) borders of AI. While most of the cells of origin are pyramidal, a significant fraction are not (Code and Winer 1985). Distances are computed from the posterior ectosylvian sulcus and the caudal pole of the medial geniculate body, respectively. Planachromat, N.A. 0.35, × 320.

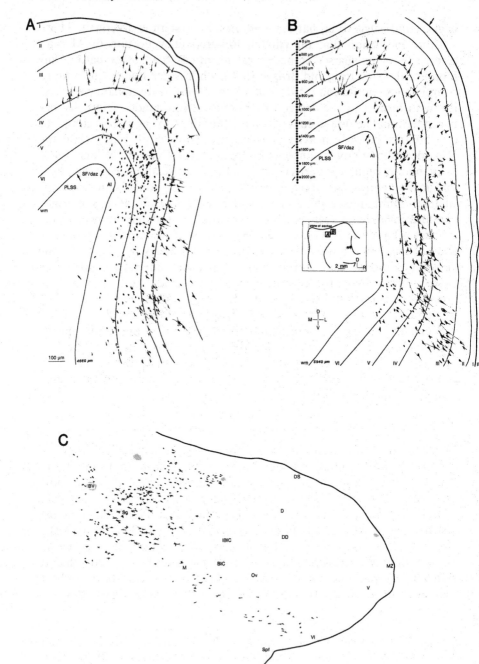

FIGURE 6.13 Caption on facing page.

thalamic cells of origin within the dorsal division was also highly specific (Fig. 6.13C; Winer et al. 1977), although the nature of any frequency-specific organization within this part of the thalamus remains uncertain (Aitkin and Prain 1974). There is evidence of strong divergence in the projections of single AI foci onto multiple targets within this field (Imig and Reale 1981).

Another feature distinguishing the corticocortical and commissural systems is the large variety of the projection neurons in the former. As might be expected from its more restricted laminar sources (Table 6.8), the commissural pathway has perhaps 10 distinct types of candidate projection neurons, while the corticocortical system involves more layers and may have 20 or more classes of neurons (Table 6.2). Thus, about half of the types of AI neurons may be part of the latter system. Both pyramidal and nonpyramidal cells are involved in the corticocortical projection (Winguth and Winer 1986) and the size range of these cells includes many of the smallest and some of the largest AI neurons (Fig. 6.13A,B). Such correlations between the form of Golgi-impregnated cortical neurons and their retrogradely labeled counterparts are reasonably secure since the extent of somatic and dendritic filling among the latter is often robust and may mark secondary and higher-order dendrites.

There is as yet no detailed study of the laminar targets of corticocortical projection neurons (Table 6.8). What data there are suggest that layers III and IV receive the bulk of this input in the cat (Diamond et al. 1968) and monkey (FitzPatrick and Imig 1980).

An orderly pattern prevails among at least some of the many corticocortical connections and can be summarized as follows: (1) powerful

---

FIGURE 6.13. Corticocortical neurons projecting from AI to AII. Two injections of horseradish perioxidase in the rostral part of AII (inset in panel B) retrogradely label neurons in layers II–VI in the caudal sector of AI (panels A,B). The distribution of thalamocortical cells of origin (panel C) reveals that only cortical dependencies of the nuclei of the dorsal division were involved, including large cells in the suprageniculate (*Sg*) and much smaller neurons in the medial limb of the dorsal nucleus (*D*), with occasional medial division (*M*) neurons. The retrograde cortical labeling is clustered in some layers (for example layers III and VI) and scattered elsewhere (layer II and in the white matter) or absent (layer I). Also noteworthy is that the cells of origin arise from adjoining nonprimary fields such as the suprasylvian fringe/dorsal auditory zone (*SF/daz*), suggesting that AII cells may receive convergent input or, alternatively, that such projections may be segregated regionally within AII, selectively within certain layers, or in their pattern of projection onto particular types of postsynaptic neurons. Even more remote projections may arise from nearby visual association cortex (*PLSS*) in the lateral suprasylvian gyrus. Both pyramidal and nonpyramidal cells are projection neurons, and more than one type of each is labeled (from experiments by Winguth and Winer [1986]). Planachromat, N.A. 0.35, × 320.

projections link tonotopically organized fields; (2) weaker, divergent projections connect tonotopic and nontonotopically arranged cortical areas; (3) corticocortical connections between different fields tend to be reciprocal, though (4) most areas have at least one nonreciprocal projection; finally, (5) some connections, especially those arising from the nonprimary fields in the caudal part of the posterior ectosylvian gyrus, are directed primarily toward limbic or extralemniscal visual pathways, where they may serve as intermodal linkages (Table 6.8; see also Bowman and Olson 1988a,b). While the current view of corticocortical auditory relations is too incomplete to infer that there is a hierarchical arrangement of homolateral projections in serially ordered sequences passing from area to area, as proposed for the visual cortex (DeYoe and Van Essen 1988; Zeki and Shipp 1988; Girard and Bullier 1989), neither can such a scenario be excluded. Thus, the corticocortical projections of AI and of the other tonotopically organized fields are both sequential (in their topographic relations with tonotopically arranged cortical fields) and divergent (by their projections to multiple fields with limited or uncertain tonotopic organization; see Imig and Reale 1980). If this view is valid, then it may be more relevant to define broad families of affiliations, for example, tonotopic-to-tonotopic, or tonotopic-to-nontonotopic, or tonotopic-to-extraauditory, with more precision before attempting to establish any hierarchy of such relations. In contrast, the projection patterns of areas adjoining the primary fields are widespread, and directed chiefly towards other polymodal and extraauditory targets (Reinoso-Suárez 1984).

As in other sensory neocortical fields (Fisken et al. 1975), there is an intricate pattern of local corticocortical connections within AI. Thus, a single tracer injection in AI labels small clusters of neurons in up to five discontinuous patches within 800–3000 μm of the deposit site. These patches are 400–500 μm wide, and include retrogradely labeled cells concentrated mainly in layers III and IV, with anterograde transport to more superficial layers. Many such patches are rostral to the injection site and at frequency representations corresponding to, or higher than, those injected. There is no obvious segregation of connections with regard to binaural aggregates of neurons (Matsubara and Phillips 1988).

Little is known of the projections between different layers in AI or the function(s) that such pathways might serve, especially with regard to the modulation of receptive field properties. In visual cortex, for example, convergent input from layer V neurons onto layer VI cells is believed to have a role in generating the elongated receptive fields characteristic of layer VI neurons (Bolz and Gilbert 1989).

3.3.3 Corticothalamic Cells of Origin

A wide variety of thalamic, midbrain, and nonauditory brain stem sites receive descending cortical projections. Such inputs often reciprocate the thalamocortical pathways, but there are many exceptions to the rule of strict cortical and thalamic reciprocity (Table 6.4). Considering these projections from area to area, it is clear that some relations that might be predicted logically indeed occur, such as from AI to the ventral nucleus of the medial geniculate body, while others that might be expected, such as from AI to the central nucleus of the inferior colliculus, are weak or absent altogether (Meredith and Clemo 1989; Tamai and Miyashita 1989; Table 6.7). Each cortical field projects to more than one thalamic, midbrain, and brain stem center, while all but a few inferior collicular and brain stem nuclei receive such input from more than one cortical area of origin. Taken together, these observations suggest that some part of the corticofugal projections may be convergent, and other parts divergent, though a more precise test of the nature of the individual connections awaits scrutiny (see Brandner and Redies 1990 for a discussion of divergent *versus* point-to-point topography of thalamocortical connections). While it embodies orderly, map-like relations, this system also serves functions other than the maintenance of tonotopy or the mirror-image connectional reciprocity that are characteristic of both lemniscal and extralemniscal pathways. It may be significant that the diversity of targets in the former is comparable to that in the latter.

The laminar sources of the corticothalamic pathway include the superficial and the deep parts of layer V and all of layer VI (Fig. 6.10). The projection arising in layer Va is relatively sparse and widely dispersed, while those from layers Vb and VI are massive and continuous, with no interruption at the layer Vb/VI border (see also Kelly and Wong 1981). Layer IV (and, to a slightly lesser degree, layer III) receive dense thalamic input through their depth (Fig. 6.10, *fine dots*), while the intercalated, neuropil-rich and cell-poor zone in layer Va is virtually free of these terminals. Many of these neurons, some of which are among the largest cells in AI, project to the inferior colliculus (Kelly and Wong 1981; in the rat, Games and Winer 1988). Other layer Va neurons are connected with the contralateral AI (Code and Winer 1985), but few of these cells have both commissural and corticothalamic axonal branches (Wong and Kelly 1981).

About 50% of layer VI cells are estimated to project to the medial geniculate body, and both small pyramidal and fusiform neurons are labeled (Kelly and Wong 1981). Medium-sized pyramidal cells in layer V are the predominant corticogeniculate neuron, though other types may occur (Table 6.2; Fig. 6.8). The substantial degree of thalamocortical-corticothalamic reciprocity suggests that areas with a comparable tonotopic or aural arrangement are most likely to be interconnected (Colwell 1975; Andersen et al. 1980a; in the rat, Winer and Larue 1987). There is evidence that the corticothalamic pathway has varied effects on the discharge of ventral division neurons, and that these descending functional influences may travel in parallel (Ryugo and Weinberger 1976; cf. 2.6.5.). Just as the thalamocortical input terminates in fields beyond those considered as auditory (Jones and Powell 1973), so do corticothalamic projections arise from remote areas such as nonprimary somatic sensory cortex (Jones and Powell 1971).

Corticothalamic axons are slender, have small *boutons terminaux* at intervals of 20–80 μm along their trajectory, follow the contours of ventral division fibrodendritic laminae, reciprocate spatially the origins of thalamocortical projections, and probably send collaterals into the thalamic reticular nucleus as they travel towards the medial geniculate body (Rouiller and de Ribaupierre 1990).

3.3.4   Corticocollicular Cells of Origin

With the possible exceptions of the anterior auditory field (AAF) and the posterior (P) and ventral posterior (VP) auditory areas, and segments of the remaining posterior ectosylvian nonprimary auditory areas whose corticofugal projections have not been investigated in detail, the other auditory cortical areas each appear to project to more than one subdivision of the inferior colliculus (Table 6.7). While these connections are large and the scope of the targets diverse, their organization is a significant departure from the pattern embodied by corticothalamic projections. In the latter, the reciprocal corticothalamic projection includes more divergent targets than the thalamic afferents. However, in the corticocollicular system, the central nucleus, the primary link of the lemniscal pathway, receives only modest or equivocal projections from AI, while the input from the other tonotopically arranged fields remains to be investigated. Indeed, the only descending input to the central nucleus of consequence seems to arise from AII (Diamond et al. 1969), a projection whose functional role is obscure. Perhaps the descending control of lemniscal input is accomplished principally through the corticothalamic pathway arising in AI and terminating in the ventral division, while the corticofugal projection to inferior colliculus subdivisions, such as the lateral nucleus, a site dominated by input from the spinal cord (Aitkin et al. 1978), allows the primary auditory cortex access to motor and somesthetic information that plays a role in the alignment of visual, auditory, and somatic sensory maps or in other polymodal functions. An analogous role has been proposed for the auditory representation in the deep layers of the superior colliculus (Wise and Irvine 1983). Some indirect evidence supporting this linkage between AI and AII and the sensory-motor system is the comparatively modest corticopontine projections (Brodal 1972) on the one hand and the strong connections of AI with the basal ganglia (Reale and Imig 1983) on the other hand. It would be interesting to know if these corticopontine targets receive convergent input from the inferior colliculus (see Hashikawa 1983; Hashikawa and Kawamura 1983), thus imposing both midbrain and forebrain influences on cerebellar representations of hearing (Huang and Liu 1985) and presumably, upon the ultimate subcortical targets of cerebellar outflow (Hikosaka et al. 1989). Such connections could propagate auditory cortical influences to non-auditory loci and, by a network of widespread connections, influence either motor planning and output (Kobler et al. 1987; Wu et al. 1989) or indirectly affect remote cortical territories whose cells may respond to, but are not driven primarily by, auditory input (Thompson and Sindberg 1960; Irvine and Huebner 1979).

A second feature distinguishing the corticocollicular projections and the corticothalamic pathway is that the former is confined to layer V (Kelly and Wong 1981), in contrast to all other cortical projection systems, which have multiple laminar origins (Table 6.8). In the rat, about 50% of layer V neurons project to the inferior colliculus, including medium-sized and large pyramidal cells (Table 6.2) whose somata form small clusters among patches of unlabeled cells; as many as 10% of the projection neurons arise in layer VI (Games and Winer 1988). It is unknown whether there is a topographic relation between the cortical origin and the collicular targets of these projections.

3.3.5 Other Corticofugal Projections

Besides the descending projections already noted, other input to a variety of brain stem or subcortical sites exists, and there is reason to believe that further study with new methods might reveal yet more such connections. While nothing is known of the laminar origins of these neurons or their form, or of the topographic pattern of their projection, they have diverse targets. Thus, neurons in the insular and temporal nonprimary fields project to the thalamic suprapeduncular nucleus, which in mammals has affiliations with the intralaminar system, and may be involved in the control of oscillatory rhythmic discharge or of polymodal thalamic neurons (Steriade and Llinás 1988). Input to the superior colliculus or pretectum arises from several nonprimary cortical fields (Meredith and Clemo 1989; Tamai and Miyashita 1989; Table 6.7) and could have a role in spatial localization of sound (Middlebrooks and Knudsen 1984, 1987; Palmer and King 1985), or in cortical and subcortical auditory-visual interactions (King and Palmer 1985). The significance of corti-copontine projections has been considered (cf. 3.3.4.) and might play a part in sound localization behavior requiring the integration of visual or vestibular cues (Wallach 1940; Lackner 1973) or perhaps modify the excitability of gamma motor neurons in cervical and cranial motor nuclei serving postural or proprioceptive reflex adjustments (Lackner and Shenker 1985; Rudell and Eberle 1985). Other motor-related connections include strong projections to the caudate nucleus and putamen arising from each of the primary auditory fields. In contrast, cortical input to the lateral amygdala may arise only from a subset of these areas (Reale and Imig 1983) and could participate in limbic-related cardiovascular and visceral adjustments during learning (LeDoux et al. 1984).

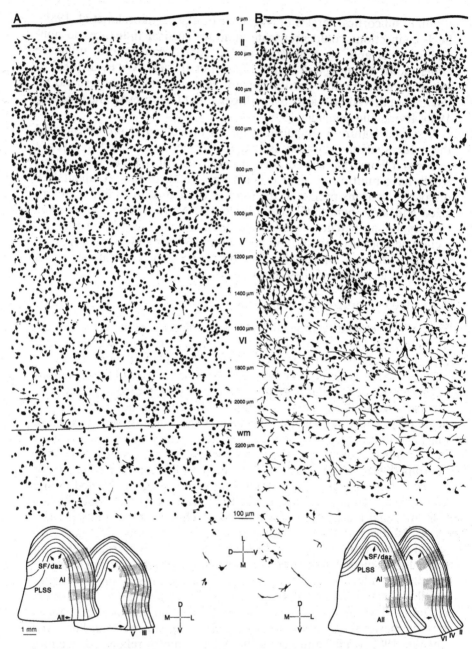

FIGURE 6.14. GABAergic and GAD-immunoreactive neurons in AI. A. GA-BAergic cells. This drawing was made by harvesting complete, immunopositive neuronal perikarya from six frames in two nearby, 50-μm-thick sections immunoreacted to demonstrate GABAergic neurons and then superimposing them (*insets*). The actual proportion of GABAergic cells, as determined in 1-μm-thick, plastic-embedded sections is about 25%, and varies somewhat between layers.

3.4 Cortical Neurochemistry

A proposition that is increasingly subject to experimental scrutiny is the control of receptive field organization in cortical neurons by local circuits as well as extrinsic connections. Since only a comparatively small fraction of cortical neurons are GABAergic, and their connections in other parts of sensory cortex appear to be highly specific, then the pattern of intracortical connections could define particular sets of relations with a specific functional role. Thus, in the visual cortex, the end-inhibition shown by layer IV neurons is thought to be a consequence of GABAergic inhibition imposed by neurons in layer VI that project upon layer IV. Other facets of receptive field organization, such as directional preference or orientation tuning, are unchanged by the reversible inactivation of layer VI neurons (Bolz et al. 1989; see also McGuire et al. 1984).

3.4.1 GABAergic Neurons and Axon Terminals

In AI there are specific and contrasting laminar patterns in the distribution of GABAergic cells (Fig. 6.14) and the smaller punctate elements that are presumed to represent axon terminals (Fig. 6.15G-L). Thus in layer I, 94.7% of the cells are GABAergic, while in layer V only 27.2% are. The average proportion of these cells, 24.6% throughout AI (Prieto et al. 1990, 1992a) is somewhat higher than the 15% estimated for the macaque striate cortex (Fitzpatrick et al. 1987) but lower than the value of 25% in the monkey precentral motor, somatic sensory, parietal association, and other cortical areas (Hendry et al. 1987). The GABAergic neurons in AI have several interesting properties: (1) each layer contains more than one type of GABAergic neuron; (2) some of the smallest and many of the largest cortical neurons are GABAergic, and all GABAergic cells are nonpyramidal; (3) of the approximately 50 types of morphologically defined AI neurons, about three-fourths have some GABAergic

(FIGURE 6.14. *continued*) The principal conclusions are that (1) every layer has a significant proportion of such cells, that (2) all the members of this class are nonpyramidal (with a possible exception, the inverted pyramidal cell, discussed in the text; cf. 3.4.1.), that (3) GABAergic cells include some of the smallest and the largest neurons in AI, as well as many medium-sized cells, and that (4) significant numbers of neurons in the white matter are also immunoreactive. Protocol for A, B: planachromat, N.A. 0.35, × 320. B. Neurons immunoreactive for glutamic acid decarboxylase. More of their proximal dendritic segments are immunostained (see, for example, layer IV or the white matter [*wm*]) but the main findings from the GABA immunostained preparations are corroborated. While many immunopositive cells have a vertical somatic orientation (for example, those in layer V), others are oriented horizontally (for example, those deep in layer VI); this implies that, besides differences in size, the GABAergic neurons represent more than one neuronal population.

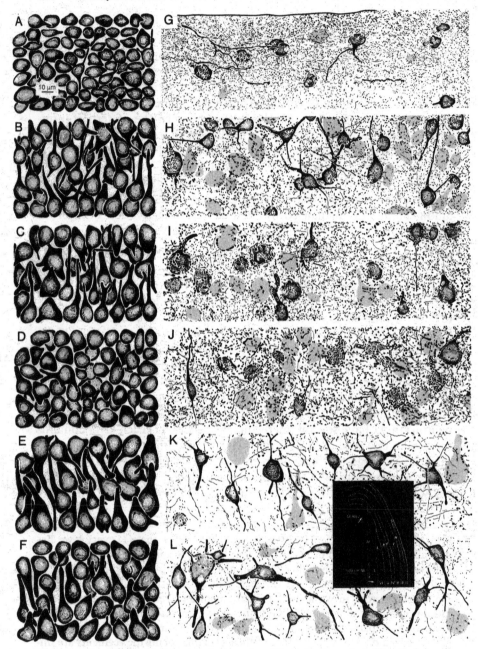

FIGURE 6.15. Views of glutamate-immunopositive cells (panels A-F) and glutamic acid decarboxylase-immunopositive (GAD) neurons and axon terminals (panels G-L) in different layers in AI (see *inset* in panels K,L). A. In layer I, the vast majority of the neurons are GABAergic, and many of the other cells are glutamatergic; no pyramidal cells are apparent. Protocol for all panels: planapochromat, N.A. 1.32, × 2000. B. In layer II, many of the glutamatergic cells are

members; (4) some six types of neurons represent the vast majority of GABAergic neurons in the primary auditory field; and (5) there is a systematic gradient of GABAergic neurons in layers III and IV along the dorsoventral axis of AI that is correlated positively with increases in the sharpness of afferent tuning (see Sutter and Schreiner 1991).

Much less is known about the form and spatial distribution of the puncta in primary sensory cortex, despite the fact that their laminar concentration is strikingly different among supragranular, granular, and infragranular layers, and even within particular layers in the monkey striate cortex (Fitzpatrick et al. 1987) and in AI as well (Peterson et al. 1990; Prieto et al. 1992b). Within AI, there are specific laminar patterns in the number and form of puncta that are noteworthy: (1) there is about a six-fold range in puncta density associated with cortical depth, with the highest values (30 puncta/100 μm^2) in layer I, the lowest (5–8/100 μm^2) in layers V and VI, and intermediate values in layers III and IV, so that there is an approximately geometric (and stepwise) numerical increase from white matter to pia; (2) except in layer I, many GABA-immuno-negative pyramidal and nonpyramidal neurons receive significant numbers of axosomatic puncta, while few neurons in layers V and VI have such accumulations; (3) size and density are distributed independently,

---

(FIGURE 6.15 *continued*) pyramidal and have a vertical arrangement. C. Layer III glutamate-positive neurons are somewhat larger than those in layer II, and probably include nonpyramidal cells. D. Glutamatergic layer IV cells are smaller than those in any layer except layer I, and are without an apical dendrite. E. Layer V glutamate-immunoreactive cells are the largest in any layer, and include several different types of pyramidal, and perhaps some nonpyramidal, cells. F. Layer VI glutamatergic cells are heterogeneous in size and shape and include some of the layer's smallest somatic profiles. G. Layer I GAD-immunopositive cells and puncta. The neurons include all of the types of layer I neurons recognized in Golgi preparations (see Table 6.2), while the puncta (*fine dots*) are numerous and among the smallest in AI. *Uniformly stippled outlines,* immunonegative cells. H. In layer II, many types of nonpyramidal cells are GAD-positive and the puncta are larger and less dense than those in layer I. I. Many GAD-immunoreactive layer III neurons receive comparatively large numbers of axosomatic puncta. J. The GAD-positive puncta in layer IV are the largest and coarsest in AI and often form clusters on the soma and proximal dendrites of immunoreactive neurons. K. In layer V, there is an abrupt decrease in the number of GAD-immunopositive puncta, especially those on the somata of neurons, with a few exceptions. The immunoreactive cells are heterogeneous in size and shape and include the smallest and largest neurons, as well as cells of intermediate size. L. Layer VI GAD-positive neurons also have varied shapes and sizes, including vertically oriented as well as horizontally arranged neurons; the density of puncta is reminiscent of the pattern in layer V, but on the whole they are finer and more delicate. From Prieto et al. (1992b).

with layer I having the highest density of fine, granular puncta, while many endings in layers III and IV are large and coarse, though small puncta are present in all but layers V and VI. Since there are no data on the intralaminar connections within AI, nor any iontophoretic studies, specific functional conclusions cannot, at present, be made except to note that the proximity of GABAergic neurons to cells that are targets of thalamic input or sources of corticocortical, corticofugal, and intrinsic connectivity suggests that they may have a role in each network.

Several of the GABA-containing neuronal populations in cat AI also colocalize one or another of the broad range of neuropeptide-immunoreactive compounds (Hendry and Jones 1991), a pattern common to nonpyramidal cells in other cortical fields (Hendry et al. 1984). This suggests that such cells might have temporally rapid as well as extended physiological actions.

3.4.2   Glutamate- and Aspartate-Like Immunoreactivity

The present impossibility of distinguishing among metabolic and transmitter-specific pools of intracellular glutamate and aspartate has confounded efforts to identify cortical and thalamic excitatory amino acid neurotransmitters with certainty. In any event, many AI cells contain glutamate-immunoreactive material (Fig. 6.15A-F). The direct demonstration of vesicular glutamate in an immonulabeling study of rat cerebral cortex using microbeads shows that, after enrichment, it is ten times more common than other amino acids and that there is little exchange or reuptake of glutamate (Burger et al. 1989).

3.4.3   Other Neurotransmitter Candidates

Many serotonin-immunoreactive pericellular axonal aggregates, known as baskets, are present in the posterior auditory field and other auditory cortical areas. These form terminal nests with elaborate varicosities about the somata and dendrites of several different types of postsynaptic neurons. The morphology of the serotonergic axons is varied, including small and large varicose endings and nonvaricose processes. Two distinct systems of serotonergic influence on cortex are proposed, one with a divergent distribution, the other directed toward particular types of postsynaptic cells (Mulligan and Törk 1988). A continuous, dense band of serotonergic immunoreactivity occurs in immature rat auditory and other cortical areas, and various sensory or motor fields have distinct patterns, suggesting that serotonergic afferents to cortex are distributed nonuniformly (D'Amato et al. 1987). The source of dopaminergic innervation of AI and in insular cortex may arise from the ventromedial tegmentum (Scheibner and Törk 1987).

3.5   Functional Neurochemistry

The iontophoretic application of muscarinic agonists has a marked effect on many facets of single unit activity of cells in the middle layers of AI.

Thus, spontaneous activity is affected in two-thirds of the neurons, and tone-evoked discharges in nearly 90%. The effects on spontaneous or evoked responses often differed in a cell. Facilitation of the on-response was a common finding, but in units with complex, tone-evoked discharges particular agonists affected different parts of the response. These findings suggest that several variables besides rate may be affected by cholinergic mechanisms (McKenna et al. 1988). In iontophoretic studies using acetylcholine release paired to specific tones, the frequency tuning of single unit receptive fields changes as a result of this pairing, many decreasing while others show frequency-specific increases or general changes in excitability to many frequencies (Metherate and Weinberger 1989).

Many cortical units (77% of the sample) in awake, paralyzed, artificially respired cats show frequency-specific modulation of discharge to iontophoretic or micropressure injection of muscarinic cholinergic agonists, and response increases or decrements were observed at different frequencies (McKenna et al. 1989). The effects of application of anticholinergic compounds were comparable to those of acetylcholine or acetyl-β-metacholine and cause response shifts in best frequency by depressing the latter while facilitating responses to nearby frequencies (Ashe et al. 1989). The source of putative cholinergic afferents is unknown. There is evidence that some aspects of associative learning are selectively blocked by the pirenzepine (M_1) but not the gallamine (M_2) subtype antagonists in the guinea-pig auditory cortex. Thus, acetylcholine facilitates rate-level functions and reduces neural thresholds (Metherate et al. 1990).

In one of the few pharmacological studies of AI, it has been proposed that two components of the middle latency auditory response evoked by brain stem stimulation are a consequence of intracortical circuitry. Thus, two potentials, Pa and Nb, are reversibly affected or eliminated by surface or intracortical GABA administration, whereas wave VI, which has a much shorter latency, is not (Knight and Brailowsky 1990).

Iontophoretic studies in monkey auditory cortex reveal that both norepinephrine and GABA inhibit vocalization-dependent discharges and sometimes produce substantial effects on the pattern of discharge. Acetylcholine produces excitation (Foote et al. 1975). Particular transmitters may thus have a specific role in vocalization, though little is known about the arrangement of these forebrain circuits.

3.6 Functional Organization

The acoustically responsive and medial geniculate-recipient cortex in the cat has an area of nearly 400 mm^2, and the bulk of the experimental work has explored the readily accessible gyral cortex and, more recently, the banks of the ectosylvian sulci using tone burst stimuli in sodium pentobarbital-anesthetized preparations. The following brief review is intended to complement the preceding, anatomical observations; more complete accounts of forebrain physiology are available (Imig et al. 1992) as are conceptual expositions of cortical function (Whitfield 1982).

3.6.1 Tonotopic Arrangement

At least seven more or less complete representations of the cochlear partition are recognized in the cerebral cortex, including areas AI, AAF, AII, AIII, P, V, VP as well as Ins, Te, and the other anterior and posterior ectosylvian fields, whose complete tonotopic arrangement is uncertain as defined by the evoked potential method using electrical stimulation of the osseous spiral lamina and single- or multiunit extracellular mapping (Woolsey and Walzl 1942; Rose and Woolsey 1949; Downman et al. 1960; Sindberg and Thompson 1962; Knight 1977). These areas are nonequivalent, differing as they do in the arrangement of frequency, latency, Q_{10} dB values, response monotonicity, degree of binaural representation, connectivity, and behavioral function. Thus, AI neurons show more monotonic responses than cells in field P, their response areas have broader thresholds at low frequencies, and shorter latencies, and there are other differences as well (Phillips and Orman 1984). In AAF the dominant input to one-third of the cells was of ipsilateral origin, in contrast to most AI cells (Phillips and Irvine 1982). The integrated Q_{10} and Q_{40} dB values for unit clusters are distributed nonuniformly within AI, with multiple peaks and valleys. The transition between AI and the overlying suprasylvian fringe cortex in terms of sharpness of tuning is gradual rather than abrupt. The sharpest tuning is in the central 2 mm of AI, with increased values dorsally and ventrally (Schreiner and Mendelson 1990). Therefore, even among the tonotopic fields, the distributions of certain functions may be different. By implication, these differences probably reflect unique thalamocortical (Table 6.4; Calford 1983) or corticocortical (Tables 6.5, 6.6; Imig and Reale 1980) input to particular subpopulations of AI cells, or significant local variations in the analysis of such information. Within AI, not all frequencies are represented equally, the upper three octaves filling a proportionally larger expanse than the lower two (Merzenich et al. 1975).

In anesthetized, quiescent preparations, the physiology of AI neurons is somewhat different. Single units show little in the way of fine-grained or systematic tonotopic spatial organization (Evans et al. 1965) despite the overall low-to-high frequency gradient from caudal-to-rostral across AI. Units have labile responses at particular frequencies, and some transient units acquire sustained discharge properties during extended recordings (Evans and Whitfield 1964). Evoked potentials to clicks in unanesthetized preparations are stereotyped, though the waveform of later events is less secure and more prone to change as a consequence of trauma (Teas and Kiang 1964).

In AII, the orderly tonotopic sequence typical in AI is degraded and highly variable, the bandwidth of cell clusters increases, there is a 10–15 dB threshold decrement, and binaural representation is much less regular and more fragmented than that in AI (Schreiner and Cynader 1984). Little is known of the detailed arrangement of best frequency in the nonprimary areas of the temporal, insular, and posterior ectosylvian fields (Sindberg and Thompson 1962).

3.6.2 Binaural Processing

Most cells in AI are driven by acoustic stimulation of both ears (Brugge et al. 1969), and some respond exclusively to binaural input while others are purely monaural (Hall and Goldstein 1968; Kitzes et al. 1980). Quantitatively, 50% of AI cells show responses to binaural stimulation, 25% exhibit suppression, and 17% have mixed responses. Studies of neuron clusters suggest that these response classes are distributed differentially in AI, with the mixed class showing a gradient of near monotonic growth with distance from dorsal AI, while ventral AI has an equal mix of the various classes (Reale and Kettner 1986). Neurons studied in radial penetrations share common aural properties (Abeles and Goldstein 1970; Imig and Adrián 1977) and frequency tuning (Phillips and Irvine 1981). Contralateral dominant summation regions have weaker corticocortical input than do the corresponding suppression regions (Imig and Reale 1981), while the pattern of callosal projection is the opposite (Imig and Brugge 1978).

The arrangement of binaural neurons across AI also has a more global organization revealed by long rows of tangential electrode penetrations. Projected on the surface of AI, these 1–2 mm long tracks reveal a systematic (though irregular) distribution of strips representing one aural pattern or another and running across AI in an approximately anteroposterior orientation, and roughly at right angles to the sequence of best frequencies. Within each of these strips, either a binaural or a monaural arrangement, which is paired with a summation or a suppression representation, predominates, though zones without such interactions (Middlebrooks et al. 1980) or clusters with nonmonotonic response classes also occur (Imig and Adrián 1977). Injections of different retrograde tracers centered at defined aural and tonotopic loci within AI label arrays of medial geniculate neurons in the ventral division described as "... class-specific projection sources ..." that form "... large slabs of projecting neurons, each sending projections to the representation in AI ..." (Middlebrooks and Zook 1983, p. 219; cf. 2.6.2. and Fig. 6.2D).

Little is known about the finer organization of binaural properties in AII except that aural subregions have irregular shapes and that each binaural response class is more strongly represented in AI (Schreiner and Cynader 1984). In even more distant cortical association areas, well outside territories classically considered as auditory (Rose and Woolsey 1949), some 70% of single units in the pericruciate, anterior lateral, and medial suprasylvian fields are driven by auditory input, while others respond to visual and somatic sensory stimulation as well. Many neurons have wide tuning curves but their thresholds are comparable to those in AI, and their latencies are similar to, or longer than, those of AI cells. Some 95% of these cells have binaural responses, and three-fourths of these show occlusion (Irvine and Huebner 1979). Some of these response properties could reflect the influences of corticocortical pathways (Paula-Barbosa et al. 1975) or overlapping thalamocortical afferents (Graybiel 1972a,b).

Besides the classically defined auditory areas and those in the adjoining, nonprimary fields, other polysensory association zones such as the anterior ectosylvian area in the periauditory sulcal cortex have been mapped and found to have auditory, visual, or a few polymodal neurons. The auditory cells are broadly tuned and may have several high-frequency tuning peaks, are often binaurally driven, and show an onset-type response (Clarey and Irvine 1986a).

3.6.3   Other Physiological Representations

The classical approach to mapping in sensory neurobiology traces a representation from point-to-point, describing the discharge properties of central neurons as a prelude to defining their role in one circuit or another, then quantifying the magnification factors and other axes indicative of connectional convergence or divergence; in the most fortuitous cases, as in the visual system, the local synaptic arrangements that subserve receptive field structure and its transformation at successive stations have been delineated (Kuffler 1953; Martin 1988). Another approach to this task would characterize the attributes of single units using nonclassical stimuli to reveal features of receptive field structure otherwise masked by the use of static or unidimensional stimuli (Allman et al. 1985). Insofar as analogous representations might play a role in processing biologically relevant sounds (Newman 1970) or in dynamic reorganizations of receptive field structure during behavior (Merzenich et al. 1984), they would be appropriate to a more refined view of the role of the auditory forebrain in the analysis of species-specific communication signals. Except in bats (Suga 1988), however, there has been little in the way of a systematic approach to this problem (see also Brugge 1975) and the bulk of the data available reflect the situation in AI.

Besides the tonotopic map in AAF, there is a further orderly arrangement of unit clusters to monaural sinusoidal and rectangular, amplitude-modulated tones; their synchronization and rate of discharge behave as a bandpass filter to changes in amplitude (Schreiner and Urbas 1986). Later work has described regional cortical differences in the modulation transfer functions in fields AI, AAF, AII, P, and VP. Thus, AAF has the largest such values, AI is about 50% of the AAF value, and AII, P, and

VP are each about 20–25%, suggesting that different fields have unique roles in the coding of complex stimuli (Schreiner and Urbas 1988). Some AI neurons responsive to the profile of the stimulus envelope also occur (Swarbrick and Whitfield 1972). Finally, many neurons outside the 2 mm-wide core of sharply-tuned AI neurons have multipeaked tuning curves with suppressive inhibitory bands interposed between their peaks (Sutter and Schreiner 1991). Responses of unit clusters to broad-band transients reveal multiple peaks and valleys within AI (Schreiner and Mendelson 1990). A systematic relationship has been found at interstimulus intervals for tone pips, white noise bursts, and short trains of clicks for most, but not all, AI cells (Hocherman and Gilat 1981). These findings each support the idea that there is a further functional segregation in AI other than tonotopy or binaurality, that the different architectonic areas have a unique and perhaps nonuniform physiological organization, and that the representation of biologically relevant stimuli may differ among cortical fields, especially with regard to segregation of spatial and temporal aspects of the acoustic signal. Analogous conclusions apply also to the broad range of physiologically defined representations within and among subregions in the visual (Tusa et al. 1981) and somatic sensory (Kaas et al. 1981) cortices.

Just as there appears to be a variable and fine-grained distribution of differential sharpness of tuning across AI (see above), so, too, are there regional differences in the spatial distribution of directionally sensitive neurons. Early studies showed that a subset of AI neurons responds selectively to sound source movement (Sovijärvi and Hyvärinen 1974). In subsequent work the arrangement of such cells has been examined with more precision. In a recording from single or multiple units at seven sites through layer IV in a 1 mm-long isofrequency strip in AI, five separate directional preferences through the frontal sound field were noted. Units with strongly nonmonotonic azimuth and level functions were exclusively highly directional; they are also more sharply tuned than are neurons with lower directional selectivity. Further, there is some evidence that neurons with comparable azimuthal and level-dependent profiles are clustered spatially, and these ensembles often have homogeneous physiological properties. A subset of highly directional neurons do not appear to encode level differences over a 40–60 dB range, and many units showed preferential spatial tuning to the central 10° of the midline (Imig et al. 1990). In certain studies about 50% of AI cells are omnidirectional (Middlebrooks and Pettigrew 1981) while in others this value is 17.6–19.9% (Rajan et al. 1990a); the proportions are much more comparable when cells with multipeaked functions are included. For most cells in the latter study, azimuthal sensitivity (to contra-, ipsi-, central-field or omnidirectional sound) remains constant across intensity. An even more refined arrangement of azimuthal selectivity has been described within AI isofrequency contours, where successive units represent a particular azimuth function, between which are zones with mixed or rapidly changing azimuthal values. The spatial distribution of omnidirectional units, in contrast, is random. Neurons in any particular isofrequency representation thus respond to azimuth in one of four patterns: (1) transitions across the frontal hemifield; (2) ensembles devoted to the contralateral or (3) ipsilateral quadrant, or (4) to stimuli centered near the midline (Rajan et al. 1990b).

Taken together, these results and those of intracortical connectional experiments (Matsubara and Phillips 1988) suggest that the local patterns of cortical microorganization could entail short-axon pathways between groups of neurons organized along an axis other than tonotopic organization. In contrast, many AII neurons with similar best frequencies have differences in receptive field size, sharpness of tuning, and in the shape of their response. Moreover, in the dorsocaudal region of AII, beneath AI, a tonotopic map is reported along a 4 mm-tall, 14 mm-long strip (Galazyuk and Volkov 1988).

Other cortical neurons that are responsive to interaural phase differences in the frequency range of 190–2400 Hz have been described. Most such cells discharge preferentially when contralateral input arrives first, and fire less when ipsilateral input is earliest. One-quarter of these cells responded to dynamically changing phase cues, and changes in intensity had little effect on discharge when results near threshold and at non-monotonic stimulus levels were eliminated (Reale and Brugge 1990). As the latter authors suggest, this would make both time and intensity (Phillips and Irvine 1981) information readily available to, and preserved with high fidelity for, auditory cortical neurons.

A steady-state response to clicks at 30–50 and 90–100 Hz in 350 ms trains has been described in auditory cortex. This response is barbiturate-sensitive and has a different form of organization in the medial geniculate body than in cortex. While the function of these potentials is unknown, their expression may reflect the state of behavioral vigilance, and it has been postulated that they are a product of cortical processing (Mäkelä et al. 1990).

A significant proportion (68%) of AI cells discharge predictably to complex stimuli, while the remainder do not respond in a regular or stereotyped way. Some of the latter are inhibited and others are excited by vocalizations or other complex sounds (Sovijärvi 1975).

3.6.4 Behavioral Studies

There is a large literature on the manifold and often inconsistent effects of damage to the various subdivisions of the auditory cortex. In many species the deficits in absolute threshold, onset detection, sensitivity to intensity or frequency modulation, or sound pattern changes, spatial localization, and binaural processing have been assessed; a comprehensive review is available for the earlier literature (Neff et al. 1975).

Massive lesions that include all of the primary and several of the non-primary areas produce no or only small losses in pure-tone thresholds despite some brief, transitory postoperative deficits. Detection of changes in sound pressure level are likewise unaffected, as is the ability to discriminate frequency. In contrast, the accurate perception of sound sequences or of signals including pairs of tones at different frequencies is impaired after large, bilateral cortical lesions. The ability to detect changes in stimulus duration is preserved after ablations limited to auditory cortex, but disturbed when damage includes the second somatic sensory area. In sound localization experiments, a variety of effects follow cortical lesions. Orientation reflexes are somewhat affected by unilateral damage, more so after bilateral ablation, while even decorticate animals retain some capacity to locate sound sources. Complete bilateral lesions severely affect the localization of a sound source. When the blood supply to various parts of AI is selectively disrupted or conserved, the effect on sound localization is frequency-specific and permanent for brief sounds, while longer tones are still localized (Jenkins and Merzenich 1984). Tests of binaural sound patterns that combine various elements of laterality and frequency in tone sequences reveal moderate to severe performance deficits. With respect to the analysis of sounds with complex or speech-like properties, the data are equivocal, suggesting some deficit after ablation of nonprimary areas.

The evidence available in other carnivores suggests that even relatively large cortical lesions do not abolish the capacity for auditory discriminations. Thus, dogs trained to make different spatial responses to low-frequency tone pairs or location cues are impaired on only one of three tasks (Stępien et al. 1990).

3.6.5 Physiological Analysis of Learning and Memory

A paradox of auditory cortex function is that its neurons must encode presumably invariant stimulus dimensions such as tonotopicity while retaining the flexibility to acquire, reify, and forget biologically relevant events on demand. Since ablation studies by definition are limited to considerations of the residual functional capacity of the nervous system, a natural alternative approach is to examine the dynamic attributes of neurons using a variety of experimental approaches, ranging from evoked potentials in anesthetized animals to single-unit studies in awake, paralyzed preparations. Among the results is that habituation, which is a hallmark of nonassociative learning, occurs in AI cells, but not in the cochlear nucleus. In an associative learning task, the pupillary conditioned response can be controlled by auditory cortex at the multiunit level, and two-thirds of a single unit sample rapidly develops plasticity to a conditioned stimulus (Weinberger and Diamond 1987). Discharge plasticity is evident both in background and stimulus evoked activity, and the conditioned response is acquired as quickly as are changes in evoked activity; background changes take somewhat longer to acquire (Weinberger et al. 1984). The genesis of conditioned responses among AI neurons in cats with chronically indwelling electrodes shows a short onset discharge elicited by the conditioned stimulus, then a decrement in this response succeeded by mild, prolonged inhibition temporally like that in the thalamus. With training, inhibition disappears and the onset burst grows in strength and consistency (Buchwald et al. 1966).

Extracellular recordings from small groups of auditory cortex units reveal that only a few cells respond as if there are strong connections between near neighbors, and that most such connections are excitatory. Pairs of neurons often receive common input, and functional connectivity can either be altered by different inputs (Frostig et al. 1983) or remain unaffected (Dickson and Gerstein 1974).

Further work in AII and the ventral ectosylvian gyrus has demonstrated that, in a model of the classical conditioning and plasticity of frequency-specific single unit receptive fields, behavior during task acquisition differs qualitatively from performance after training. These associative changes are unrelated to arousal and have led to a theory that these nonprimary fields serve as depots for the storage of biologically relevant sounds, the access to which is a function of postulated bandpass or notch filters that permit neurons to behave adaptively (Diamond and Weinberger 1989). The associative changes seen in these neurons during acquisition of a conditioned stimulus suggest that local cortical mechanisms (Diamond and Weinberger 1986), rather than changes at the level of the auditory thalamus (Ryugo and Weinberger 1978), are responsible for response plasticity.

The acquisition of a pupillary conditioned response to an auditory stimulus during learning in awake, paralyzed preparations indicates that changes in auditory cortex neurons precede those in the cochlear nucleus or in other nonauditory subcortical or cortical centers. In fact, the establishment of the conditioned response itself antedates conditioned changes in auditory neural discharges (Oleson et al. 1975). This raises the possibility that in some sense even the primary auditory cortex is an association area, or at least that it does not necessarily have temporal primacy in the acquisition process.

Rabbits that sustain auditory cortex lesions and then learn to respond to a tonal signal with a conditioned increase in heart rate are unable to eliminate the response during extinction trials. Since visual cortex ablations do not impair such extinction, it suggests a central role for corticothalamic and thalamocortical neurons in acquired visceral response control (Teich et al. 1989).

Almost all AII neurons compared to AI cells (95% *versus* 63%) show plasticity during an associative conditioning task. About 80% of AII neurons and nearly all ventral ectosylvian cells show decreased background activity during conditioning. These differences and the area-specific consequences of arousal on learning suggest that the different fields are functionally nonequivalent. Perhaps such dynamic changes reflect thalamic influences since the behavior of medial division neurons embodies some of the same properties, including response lability, seen in AII (Diamond and Weinberger 1984, 1989; cf. 2.6.5.).

Cats raised in behaviorally enriched or impoverished environments show significant changes in neuronal density, cortical thickness, and the number of cells in 1 mm² of primary and secondary visual cortex, but not in auditory or somatic sensory cortex. The lack of such changes has been attributed to developmental differences between cortical areas (Beaulieu and Colonnier 1989; cf. 5.6).

4. Comparative Studies of the Auditory Cortex

The proliferation of cortical fields with a tonotopic or other specialized representation in the rat (Horikawa et al. 1988), primates (Brugge 1982), and mustached bat (Suga 1984a,b), to mention only a few, strongly suggests that multiple areas are a common denominator of mammalian auditory cortical organization. The following sections compare the findings in various species to those in the cat.

4.1 Comparative Cortical Structure

In the rat, the primary auditory cortex (Te 1, area 41) is a typical sensory neocortical field with a thick layer I, a granular layer IV, and a well-developed layer V (Krieg 1946; Zilles 1985; Roger and Arnault 1989). The neuronal architecture resembles that in the cat, except that the number of different types of nonpyramidal neurons may be lower (Games and Winer 1988). The primary field is distinguished from caudal (Te 2) and rostroventral (Te 3) nonprimary fields that are thinner than Te 1 and have a lighter pattern of myelination. In the rhesus and squirrel monkey, AI is distinguished by its conspicuous granular cytoarchitecture, which is so great that layers II–IV appear virtually continuous, while layers V and VI are not usually divided further. The ultrastructure of layer IV in squirrel monkey primary auditory cortex shows many small nonpyramidal neuronal somata that are densely packed and may be electrotonically coupled (Smith and Moskowitz 1979). A comparable ultrastructural somato-somatic arrangement is seen in field L of the starling (Saini and Leppelsack 1977). In primates, two nonprimary areas, the parainsular and rostrolateral fields, are also identified (Jones and Burton 1976), the former including the parainsular and prokoniocortical areas described in rhesus monkeys (Pandya and Sanides 1973), while additional primary rostral and nonprimary anterolateral, caudomedial, and posterolateral fields are described in architectonic and single neuron mapping studies in the owl monkey (Imig et al. 1977). Multiple architectonic subregions have been defined within a polymodal nonprimary field in the rhesus monkey (Seltzer and Pandya 1989b).

Golgi studies reveal that many of the types of neurons in cat AI are impregnated in other species. Both spinous and smooth nonpyramidal cells occur in all layers of rabbit auditory cortex, but the wide variety of local axonal arrangements in carnivores and primates is reduced in lagomorphs (McMullen and Glaser 1982a). On the other hand, there appears to be a reasonably good interspecific concordance with respect to pyramidal cells in various species. Thus, in dog auditory cortex, inverted pyramidal cells are seen at about 1500 μm beneath the pia, and fusiform and double bouquet cells are also described (Tunturi 1971) that resemble those in the cat. In the rabbit, many of the layer III–IV nonpyramidal neurons have either vertically oriented dendritic domains or a tangential configuration relative to the cortical convexity (McMullen et al. 1984) much like that in the cat (Wong 1967; Reale et al. 1983). In the dolphin, layer I is relatively enormous, up to one-third of the cortical depth, layer II is thin but densely stained, layers III and VI are moderately thick and poorly differentiated from one another, and there is little distinct architectonic individuality among functionally distinct areas (Morgane et al. 1985). Many different local circuit neurons, including inverted pyramidal cells with long ascending axonal branches, occur, though a specific layer IV could not be identified (Ferrer and Perera 1988; see also Kesarev et al. 1977). In the mustached bat, layer I is also much enlarged compared with layer I in the cat and monkey (Winer, Bui and Larue, unpublished observations). In the starling, the microneurons that comprise the bulk of field L may be comparable to the thalamic-recipient koniocortical neurons in layer IV, while cells resembling pyramidal neurons are prevalent in adjoining, nonprimary areas (Saini and Leppelsack 1981). Field L is not homogeneous in its internal organization, and there is some evidence that its subdivisions might represent an avian homologue of specific cortical layers in mammals (Bonke et al. 1981).

4.2 Comparative Cortical Physiology

A frequency-specific organization has been found in rat temporal cortex, where narrowly-tuned units form isofrequency contours across a 3 mm–wide band, and adjoining fields have a different arrangement of characteristic frequency (Sally and Kelly 1988). In related studies, a comparable tonotopic primary field is flanked by three further fields, the posterodorsal, anterior, and anteroventral areas, each with a different sequence of characteristic frequency than in the primary field, and significantly longer latencies (Horikawa et al. 1988). In the rat, the sound-evoked potentials with the largest amplitude occur some 600 μm beneath the surface (Borbély 1970), which is consistent with the location of layers III and IV (Games and Winer 1988). Current source density analysis suggests that the waveform variance is consistent with a two-part model in which component one is generated by supragranular, and component two by infragranular, pyramidal cells (Barth and Di 1990).

Most rat AI neurons have monotonic responses to stimulus amplitude across a 5–35 dB range or more. Perhaps, as in the cat, there is a neural substrate for amplitude representation (Phillips and Kelly 1988). Progressively lower frequencies of amplitude modulation are present in the squirrel monkey inferior colliculus, medial geniculate body, and auditory cortex (Müller-Preuss 1986).

More than one systematic cortical representation of basilar membrane best frequency is found in the hedgehog (Batzri-Izraeli et al. 1990), mouse (Stiebler 1987), gray squirrel (Merzenich et al. 1976; Luethke et al. 1988), guinea-pig (Hellweg et al. 1977; Redies et al. 1989b), rabbit (Galli et al. 1971; McMullen and Glaser 1982b), owl monkey (Imig et al. 1977), and rhesus monkey (Merzenich and Brugge 1973). In most species (Fig. 6.16), a tonotopically organized primary field with sharply tuned cells abuts one or more nonprimary representations, ranging from two (rabbit and hedgehog) to four (rat and mouse) to five (guinea-pig and monkey) or more (squirrel; see Luethke et al. 1988). In the marsupial possum, only one tonotopic auditory field is reported (Gates and Aitkin 1982). In certain species, such as the marmoset, the organization of best frequencies differs from that in cat AI, such that low frequencies are situated rostroventrally while high frequencies are more caudal in the primary field. As in the cat, however, there is considerable interanimal variability in the size and position of this area (Aitkin et al. 1986a; see also Merzenich et al. 1975). A substantial area of acoustically responsive cortex in the ectosylvian region has been reported using evoked potentials in the miniature swine (Andrews et al. 1990). In sharks, an auditory-responsive telencephalic zone has been identified with click and tone-burst stimuli; while these well-defined areas are multimodal, the optimal acoustic areas were remote from fields responsive to photic or electric modalities (Bullock and Corwin 1979).

In other physiological studies, mesencephalic stimulation in encéphale isolé (spinal transection behind the first cervical root) monkeys increases the amplitude of cortical responses or facilitates auditory thalamic responses to paired click stimuli. Thus, significant nonauditory influences can alter the firing pattern of forebrain neurons (Symmes and Anderson 1967). In awake monkeys, simultaneous reticular formation stimulation and experimentally presented vocalizations affect response strength or discharge pattern in only about one-third of the units. This suggests that many units have consistent response profiles that reticular influences cannot modify in intact animals (Newman and Symmes 1974).

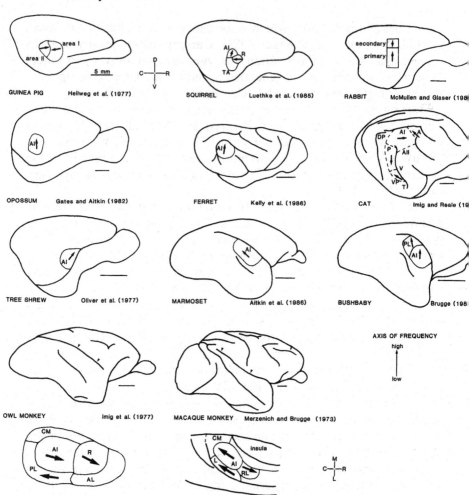

FIGURE 6.16. Auditory cortical areas and the arrangement of best frequency in eleven mammals. In owl and macaque monkeys, the auditory cortex is hidden within the lateral sulcus (*arrowheads*); it is revealed and enlarged below the lateral views. *Arrows* indicate the sequence of best-frequency representations from low-to-high. See the text for further discussion. Abbreviations: (*A*) anterior auditory area; (*AL*) anterior lateral auditory cortical area; (*AI*) primary auditory cortical area; (*AII*) secondary auditory cortical area; (*CM*) caudomedial auditory cortical area; (*DP*) dorsoposterior auditory cortical area; (*P*) posterior auditory cortical area; (*PL*) posterolateral auditory cortical area; (*R*) rostral auditory cortical area; (*RL*) rostrolateral auditory cortical area; (*T*) temporal auditory cortical area; (*TA*) temporal anterior auditory cortical area; (*V*) ventral auditory cortical area; (*VP*) ventroposterior auditory cortical area. Reproduced from Luethke et al. (1988) and modified with the permission of the author.

Exploration of auditory cortical responses in awake primates using species-specific vocalizations finds that 90% of the single units are responsive. The complex response patterns suggest that neurons are not preferentially responsive to one stimulus attribute but integrate information across a wide spectrum, where frequency, duration, or the temporal structure each may act (Newman and Wollberg 1973). The lack of any clear tonotopic arrangement and the variability in temporal discharge pattern and complex response areas of these neurons in alert animals are consistent with this idea (Funkenstein and Winter 1973). In other single unit studies of awake squirrel monkeys, single AI neurons show far sharper tuning, and the Q_{10} dB values are much narrower, than those of medial geniculate neurons; thus it is proposed that local cortical inhibitory neurons participate in this improvement in tuning (Pelleg-Toiba and Wollberg 1989). Indeed, the inhibitory responses revealed by two-tone interactions suggest that neurons in different cortical fields may have different response profiles. Thus, most such AI or rostral field neurons have strong inhibition that may span as much as four octaves, and they possess typically nonmonotonic rate-level functions, show low spontaneous rates, and have little or no response to broad band noise. Cells in other fields show very different two-tone profiles, including high spontaneous rates, excitatory responses to broad-band noise, or monotonic rate-level functions (Shamma and Symmes 1985). This would suggest that the intrinsic cortical circuits, insofar as they contribute to these unique areal patterns, are unlikely to be identical in the various architectonic fields.

In old world monkeys, the neural representation of consonant-vowel syllables has been studied in the primary auditory cortex and thalamocortical radiations in awake animals with multiple unit recordings. The main conclusions are that the pure tone elements of these stimuli can readily be related to the known features of the tonotopic map, and that interactions among formants affects the response to the syllable. The interactions are often marked by strong inhibitory effects, and analogous physiological patterns are seen in thalamocortical axons and, by inference, in thalamocortical relay cells (Steinschneider et al. 1990). These authors note that the interpretation of their data has significant bearing on the question propounded by Suga (1978) as to whether cortical neurons behave as feature detectors—units that are tuned optimally to one or another attribute of the stimulus such as frequency, phase, or amplitude—or, in contrast, if the discharge patterns of so-called nonspecialized cells that are not specifically responsive to a single axis of stimulus representation are critical. It does not necessarily follow that these theoretical formulations are incompatible, and they could coexist.

Auditory responses from neurons in cortical areas remote from the sensory fields are more robust and specific than might be expected. Thus, two-thirds of the units near the sulcus principalis in awake monkeys respond to vocalizations. Since this frontal lobe area receives no direct auditory thalamic input, corticocortical connections must link these neurons with those on the supratemporal plane (cf. 4.3.). The outstanding characteristic of their response is a preference for complex vocalizations that would predominate in social activity (Newman and Lindsley 1976). More than 50% of squirrel monkey auditory cortex neurons responding to natural calls are inhibited during phonation, suggesting that a feed forward pathway of cortical origin is operative during voluntary vocalization (Müller-Preuss and Ploog 1981).

Auditory potentials evoked from medial geniculate body stimulation after large brain stem lesions in the squirrel monkey reveal that much of the neocortex can still be activated, especially the parietal cortex and caudal prefrontal areas, and with no concomitant involvement of visual or somatic sensory cortex (Bignall and Singer 1967). Such extralemniscal pathways to the cortex have been demonstrated in the cat (Galambos et al. 1961).

Comparisons between bat species and the cat with respect to physiological organization must be made with caution for four reasons. First, the characterization of auditory cortical fields in the cat is at best incomplete, and large areas, especially in the banks of the ectosylvian cortex, remain unexplored. Second, most cat studies use pentobarbital-anesthetized preparations and nonnaturalistic pure tone or tone burst stimuli, while many bat experiments use awake, restrained preparations and more ecologically relevant stimuli. Third, bats do not represent a single, homogeneous adaptive strategy towards the central representation of hearing and the role of audition in their behavioral milieu is unlikely to be identical to that of purely terrestrial mammals. Fourth, the approach toward establishing areal subdivisions in bats has been driven largely by functional ethological constraints as opposed to considerations of tonotopic organization. Except as noted otherwise, the following conclusions apply to the mustached bat.

At least eight distinct auditory representations occur along the lateral cortical convexity. The principal representation of frequency is similar to that in the cat, with basal cochlear positions situated caudally and apical cochlear points arrayed rostrally, and the frequency range extending from less than 5 to more than 150 kHz. Among the representations, nearly 40% of the length of the sequence is devoted to the (1) Doppler-shifted constant frequency 61–63 kHz range (Suga 1984a), whose physiological fine structure is unique to each bat and sexually dimorphic (Suga et al. 1987). This area plays a key role in (2) object detection, encoding of azimuthal motion, and frequency analysis. Three subareas participate in the analysis of frequency modulated (55–60 kHz, 84–90 kHz, and 110–120 kHz) signals, frequencies which are not represented in the main sequence, but lie slightly dorsal to it. These neurons respond optimally to pulse-echo combinations with a specific echo delay, and have been characterized as range-sensitive neurons (O'Neill and Suga 1979). Further maps of (3) 61–63 kHz and 92–94 kHz include the constant frequency area intercalated between the primary sequence and the frequency-modulated area. The constant frequency neurons are responsive to two sounds whose frequencies differ by a particular interval (Suga et al. 1983). These neurons show little preferential discharge to orientation sounds, echoes, or single sounds, but respond vigorously to combination tones. They could thus encode information useful in velocity computations, azimuthal localization, and sound amplitude. The dorsal fringe area (4) consists of three sets of frequency-modulated facilitation neurons that subserve the range (time) axis; while the time axis in this area is shorter than in (2), there is no advantage in processing range information (Suga and Horikawa 1986). A neural representation of (5) frontal azimuth and velocity lies above the 61–63 kHz acoustic fovea (Suga 1984a). A (6) ventroanterior area rostral to and beneath the 90–120 kHz isofrequency representation receives projections from the frequency modulated field and responds to first and second harmonics of pulse and/or echoes. The best facilitation is between 61–62 kHz, and these cells could encode target range and velocity (Tsuzuki and Suga 1988). A (7) ventrolateral area contains cells sensitive to target filter, while the more caudal (8) ventroposterior neurons encode azimuth motion (Suga 1984a,b). Not all cortical auditory neurons are narrowly tuned; certain cells between 88–99 kHz have higher Q_{10} dB values than those of neurons in the constant frequency area and are not arranged in a clear tonotopic sequence (Asanuma et al. 1983).

It should not be surprising that different bats have evolved species-specific patterns of cortical neural representation. Thus, the little brown bat and the mustached bat both use delay lines to derive range information, but have unique pulse-echo combinations (Berkowitz and Suga 1989). The auditory cortex of the little brown bat contains, besides separate tonotopic and delay-sensitive zones, a third area where neurons sensitive both to spectrum and echo-time delay overlap. Thus, in contrast to the mustached bat, single frequency-modulated sounds may be used to process convergent information, while in the former individual cortical areas appear to be segregated (Wong and Shannon 1988).

Besides a columnar segregation of best frequency, minimum threshold, and azimuthal locus in the mustached bat (Kujirai and Suga 1983) and the big brown bat (Jen et al. 1989) there is a systematic cortical representation of auditory space that is frequency dependent and whose performance improves at higher frequencies. While single neurons in cat AI respond selectively to various domains of acoustic space ranging from omnidirectional to hemifield to axial, no orderly map of these dimensions is found across AI (Middlebrooks and Pettigrew 1981).

In the avian auditory forebrain, electrophysiological analysis of field L neurons finds that they have a large range of responses to naturalistic acoustic stimuli, though almost 50% of the cells respond to less than 10 different sounds (Leppelsack 1978), and many are driven by only one stimulus from among those of a particular class (Leppelsack and Vogt 1976). About two-thirds of low frequency field L neurons appear to encode attributes of the stimulus envelope (Hose et al. 1987).

There is a continuous tonotopic representation running through field L, the adjoining neostriatum, and in the ventral hyperstriatum (Bonke et al. 1981; Müller and Leppelsack 1985). Stimulation studies with [^{14}C]2-deoxyglucose in the parrot, duck, pigeon, gull, and other avians reveal that field L has internal subdivisions which show differential patterns of stimulus-dependent labeling, and that the exact arrangement of these subareas differs from species to species. However frequencies above 3 kHz represent less than 20% of the total spectrum (Müller and Scheich 1985). A similar, selective arrangement of binaural banding is revealed in gerbil auditory cortex by 2-deoxyglucose metabolic mapping (Caird et al. 1991).

Further support for subdividing field L is the selective distribution of spontaneous and tonal-evoked activity dependent labeling within it. Thus, L_2 has a high rate of resting activity, while frequency-modulated stimuli selectively decrease spontaneous discharge in zones adjoining responsive areas (Scheich and Bonke 1981). Within L_2 there is evidence for a representation of aural input orthogonal to the map of frequency. Thus, after unilateral cochlear ablation in chicks, [^{14}C]2-deoxyglucose labeling marks distinct bands that cross the isofrequency representation and which alternate with activity-free columns (Scheich 1983). Such an arrangement may be related to the form of aural dominance bands in the cat (Middlebrooks et al. 1980).

There is a selective distribution of certain physiological properties among single neurons within field L. Thus, L_2, whose dominant input is from nucleus ovoidalis (B.A. Bonke et al. 1979), contains cells with high spontaneous rates, simple discharge patterns, unimodal best frequency, and weak responses to biologically important calls (D. Bonke et al. 1979). Many neurons in L_1 and L_3 show low spontaneous rates, complex discharge patterns, multipeaked frequency responses, and are driven optimally by naturalistic stimuli (Scheich et al. 1979). A few neurons show frequency modulation (Leppelsack 1974).

Besides the tonotopic order and aural segregation described above for field L and its subparts as well as in the ventral hyperstriatum, a third systematic representation, relating to isointensity band width, has been delineated (Heil and Scheich 1991a). Single unit onset latencies also had a topographic distribution, decreasing at higher best frequencies; they were shortest in L_2 and longer in L_3 and L_1/ventral hyperstriatum (Heil and Scheich 1991b). Such a distribution would have important consequences for the spatial arrangement of temporal processing analogous to those found in cat auditory cortex (cf. 3.6.3.).

4.3 Comparative Patterns of Cortical Connections

The laminar distribution of thalamic input to the primary auditory field is similar to the cat in all species so far studied other than avians or reptiles, with the principal focus of transport to layers III and IV after involvement of the ventral division and lesser labeling in layers I and VI if the medial division is invaded (Sousa-Pinto 1973a; Niimi and Naito 1974). Such a pattern has been seen in the rat, though the superficial part of layer III appears to receive fewer such endings than in the cat (Patterson 1976; Vaughan 1983), and it occurs also in rhesus and squirrel monkeys (Mesulam and Pandya 1973; Jones and Burton 1976), in the tree shrew (Casseday et al. 1976; Oliver and Hall 1978b) and in the opossum (Ebner 1969). There may be areal differences in the pattern of laminar input to nonprimary areas. Thus, in the parainsular nonprimary auditory field, there is moderate input to layer IIIB, with a much smaller projection to layer IIIA and the superficial part of layer IV; in contrast, the first and second temporal areas have heavy input to layer IIIB and light labeling in layer VB, while the third temporal field is similar save for moderate layer IIIB labeling (Jones and Burton 1976); it is uncertain if there are corresponding arrangements in the cat since the laminar organization of projections to nonprimary cortex has not been studied in detail, and because correspondences between different areas in the cat and primates remain to be established. The synaptic target of thalamic input must include a substantial number of GABAergic neurons and some GABA-negative cells (Cipolloni and Keller 1989).

The thalamocortical cells of origin in the ventral division of the medial geniculate body have a systematic relation to the tonotopic or topographic subregion of the primary auditory field in each species so far studied (rat [Patterson 1976; Kelly et al. 1993]; squirrel [Luethke et al. 1988]; guinea-pig [Redies et al. 1989a]; mustached bat [Olsen 1986]; opossum [Kudo et al. 1986]; native cat [Kudo et al. 1989] and tamarin [Luethke et al. 1989]). Such experiments usually produce strips or bands of retrogradely labeled ventral division neurons, and a much less regular distribution of such cells within the medial division. This pattern has also been observed in the porpoise (Revishchin and Garey 1990) and in the brush-tailed possum (Neylon and Haight 1983). In one primate, the tamarin, neurons in the suprageniculate nucleus project to AI (Luethke et al. 1989) while in the macaque monkey its target is the granular insular cortex (Burton and Jones 1976) or the temporopolar cortex, which also receives input from the parvocellular part of the medial geniculate body and from the limitans nucleus (Morán et al. 1987). Other parts of the medial geniculate body project to the middle and caudal parts of the upper bank of the superior temporal sulcus (Yeterian and Pandya 1989). There are insufficient data to draw firm conclusions about the comparative cortical affiliations of the other parts of the primate dorsal division.

In its main outlines, the organization of thalamocortical projections in the mustached bat is readily compared to the pattern in the cat, with some significant exceptions. The chief parallel is the orderly relation between the retrogradely labeled ventral division territories and the locus of transport from physiologically defined primary cortical loci. The cortical target of the bat dorsal division is mainly to areas representing frequencies above 60 kHz or to cortical fields whose neurons respond predominantly to frequency-modulated sounds. The principal exceptions are the projection patterns of the medial division and of the suprageniculate nucleus. In contrast to the cat, in which medial division neurons have moderately heavy but widespread affiliations with each subdivision of auditory cortex and with many nonauditory fields, in the mustached bat much of the frequency representation (from below 10 kHz to more than 90 kHz) in AI receives little or no medial division input, while the frequency-modulated and constant-frequency representations adjoining the primary field receive comparatively dense input in most cases (Olsen 1986).

The projection of the cat suprageniculate nucleus is largely to the insular cortex and to the caudate nucleus and putamen (Table 6.4), while in the mustached bat, it has strong input to frequency-modulated and to constant-frequency cortical representations (Olsen 1986) as well as a large projection to portions of the frontal cortex (Kobler et al. 1987) that, in primates, might correspond to the principal sulcus, which is thought to receive auditory input only through corticocortical routes (Pandya et al. 1969; see below). The brain stem afferents to the bat suprageniculate nucleus are concentrated largely in the nucleus of the central acoustic tract, a group of large multipolar cells near the base of the ventral nucleus of the lateral lemniscus (Casseday et al. 1989). These neurons probably correspond to one or more of the subcollicular efferent pathways that arise in the midbrain tegmentum and terminate predominantly in the nuclei of the dorsal division (Papez 1929; Morest 1965b; Henkel 1983). The sources of acoustic input to these centers are not well understood. The pathway between the suprageniculate nucleus and the frontal lobes has also been demonstrated in the rat (Kurokawa et al. 1990), suggesting that it may be common to many mammals, though not to the marsupial brush-tailed possum (Neylon and Haight 1983).

Corticocortical connections in rodents and primates embody a more refined and complex picture than those in the cat, in part because of the superior sensitivity of axoplasmic transport methods over the silver degeneration studies that have dominated the classical feline literature. Nevertheless, certain basic principles are common to both newer and older studies: (1) each cortical field projects to multiple areas, some nearby, while other targets are more remote; (2) most or all such projections are reciprocal; (3) some of the homolateral projections are topographic; (4) the primary and nonprimary fields are more heavily interconnected with each other than are the nonprimary areas with themselves; and (5) the laminar origins of these projections include more layers than do the origins of commissural connections; (6) there is always access to perirhinal cortex, and hence to the limbic system of the basal forebrain. Many of these findings are derived from or have been confirmed in connectional studies in the gray squirrel (Luethke et al. 1988), tamarin

(Luethke et al. 1989), marmoset (Aitkin et al. 1986b, 1988), and in rhesus (Galaburda and Pandya 1983; Cipolloni and Pandya 1989) and owl (FitzPatrick and Imig 1980) monkeys. In the latter, the laminar origins of these projections are diverse, and their target is often layer I; few corticocortical neurons project to more than one region. The dorsolateral portions of the association cortex in the superior temporal gyrus projects more broadly to the temporal lobe and frontal lobes in the squirrel monkey than does the supratemporal plane proper (Forbes and Moskowitz 1977). In the rhesus monkey the pattern of corticocortical connections is somewhat different, with feed forward projections arising from supragranular layers and ending in and near layer IV, while reciprocal connections involve infragranular origins and terminations in layer I (see above). Other, fine-grained local intrinsic connections originate and terminate in all layers (Seltzer and Pandya 1989b).

Convergent long-axon corticocortical afferents arise from many sources and terminate within several fields along the supratemporal plane. These include projections from the vicinity of the intraparietal sulcus to nonprimary parts of the superior temporal gyrus and sulcus. Large expanses of the somatic sensory cortex, primary and nonprimary visual cortex, and inferotemporal cortex send axons to nonprimary fields of the superior temporal sulcus. These have an orderly arrangement such that visual input is segregated caudodorsally, inferotemporal afferents are concentrated rostroventrally, and somatic sensory corticocortical projections are intermediate; however, each modality projects to more than one area (Seltzer and Pandya 1978, 1989a). There is also electrophysiological evidence for corticocortical connections between the temporal lobe and areas 6 and 8, near the arcuate sulcus (Bignall 1969). These responses depend on the integrity of the medial geniculate body and of the brachium of the inferior colliculus (Bignall 1970). Lateral frontal cortical lesions impair acquisition and retention of a simple auditory discrimination (Weiskrantz and Mishkin 1958).

Extensive corticocortical connections exist between subdivisions of the rhesus monkey auditory cortex and the prefrontal cortex near the principal sulcus (Petrides and Pandya 1988). The laminar origins of these within auditory cortex are specific to a given area, with almost 75% of such neurons arising in layer III in primary fields, in contrast to the nonprimary cortex, where only 35% of the projection cells are in layer III. While there are some differences in the laminar origins of cortiofrontal projections between different modalities, the correspondence among them suggests that the convergence in the frontal lobes is a common principle (Barbas 1986). Particular architectonic fields within the superior temporal sulcus have both singular and overlapping patterns of input to broad territories of the frontal lobe ranging from area 8 (the frontal eye fields) to area 10 (in the vicinity of the principal sulcus; see Seltzer and Pandya 1989a). This, in turn, implies that uni- and polymodal influences reach areas whose primary role is response planning and control rather than sensory evaluation as such, that is, areas defined as association cortex (Rose and Woolsey 1958). Further sites of corticocortical convergence implicate the temporal pole as a target in a variety of primates (Markowitsch et al. 1985; Morán et al. 1987).

In the dog, long corticocortical projection fibers link the auditory association fields adjoining the primary auditory cortex with the frontal association cortex. Auditory- and visually-related afferents are segregated from limbic input, and there is only a weak association between the thalamic mediodorsal nucleus and its direct auditory affiliations (Stępniewska and Rajkowska 1989). Hence, auditory input to frontal lobes in carnivores is largely through corticocortical, as opposed to thalamocortical, circuits; the latter pathway appears to be more highly developed in rodents (Kurokawa et al. 1990) and the mustached bat (Kobler et al. 1987).

The interhemispheric projections in macaque monkeys reach their contralateral targets through several commissures. Thus, the entire temporal lobe contributes fibers to the anterior commissure, with the largest input originating from rostral cortical territories. Callosal axons carry information mainly from the caudal two-thirds of the temporal lobe and from the superior and inferior temporal gyri (Demeter et al. 1990; see also Cipolloni and Pandya 1985). Perhaps each component of the commissural projection system evolved at a somewhat different time and has a distinctive functional role.

Corticocortical input from the cingulate cortex vocalization areas to nonprimary auditory cortex provides limbic access to the auditory forebrain, while there is no corresponding cingulate projection onto the auditory thalamus (Müller-Preuss and Jürgens 1976). Thus, in primates, certain highly important thalamocortical projections are without a reciprocal corticothalamic input.

The organization of commissural connections is comparable in most species. The main principles subserving this arrangement are as follows: (1) the projection is topographic, at least in the primary field; (2) the cells of origin are chiefly in layer III, less so in layers V and VI; (3) the targets of this input include all contralateral layers, though the projection is strongest to layer III and weakest to layer IV; (4) terminal fields contain radially arranged, often alternating foci of high and moderate terminal density in axoplasmic transport experiments; and (5) projections are often, but not always, reciprocal, and may be divergent as well. Most of these conclusions apply to findings in the rat (Cipolloni and Peters 1979, 1983; Vaughan 1983), gray squirrel (Luethke et al. 1988), tree shrew (Cusick et al. 1985), marmoset (Jouandet et al. 1984), tamarin (Luethke et al. 1989), and rhesus monkey (Pandya et al. 1969; Cipolloni and Pandya 1985).

There are little data from rodents on the laminar origins and terminations of auditory corticofugul projections to the thalamus and brain stem. However, the distribution of terminal labeling in the rat after auditory cortex injections of horseradish peroxidase conjugated to wheat germ agglutinin is widespread, the targets including the ipsi- and contralateral medial geniculate complex, the middle and deep layers of the superior colliculus, the central gray, the nucleus reticularis cuneiformis, the red nucleus, the dorsal cortex and dorsomedial part of the central nucleus of both inferior colluculi, the vicinity of the ventral nucleus of the lateral lemniscus, the superior olivary complex, and the lateral paragigantocellular reticular nucleus (Newman et al. 1989). Other sources of corticofugal input include the perirhinal cortex (Deacon et al. 1983), which would afford limbic access to lemniscal sensory pathways. So broad a projection provides for cortical influence over almost every subcortical sensory and motor center that could play a significant role in the analysis of, or response to, auditory stimuli.

A more complete consideration of the evidence for the proposition that homologous synaptic stations exist in reptiles, birds, and mammals is available (Ulinski 1983). Briefly described, the anterior dorsal ventricular ridge of the telencephalon is a massive subcortical structure that dominates the caudolateral and periventricular portions of the forebrain. As such, it is the recipient of ascending visual, auditory, and somatic sensory (but not olfactory) input from the brain stem and thalamus which it then conveys either to the corpus striatum or to the basal part of the dorsal ventricular ridge. The former projects successively through the tectum towards the reticular formation, while the latter sends axons to the hypothalamus after having received powerful olfactory connections. Thus, fundamental and largely parallel pathways between the auditory thalamus and the striatum on the one hand, and between the auditory thalamus and the limbic forebrain, on the other, appear to be common denominators in reptiles, birds, and mammals. In the former, the anterior dorsal ventricular ridge is intercalated between thalamostriatal and thalamoautonomic linkages. In other respects, however, there are tantalizing parallels with mammals in that one branch of medial geniculate body output is directed towards subcortical sensory-motor targets (such as the caudate nucleus; see Hu and Jayarman 1986) while others have chiefly limbic affiliations (through the amygdala; see Russchen 1982).

4.4 Comparative Cortical Neurochemistry

The data available are too modest to permit firm conclusions or generalizations about a common mammalian pattern or to allow specific comparisons with the cat, save that in several species many cortical neurons are GABAergic despite interspecific differences in architectonic organization (Fig. 6.17). Further, all subregions in rat temporal cortex have many choline acetyltransferase-positive terminals in layers I, Va, and VI (Eckenstein et al. 1988). This pattern differs somewhat in the monkey,

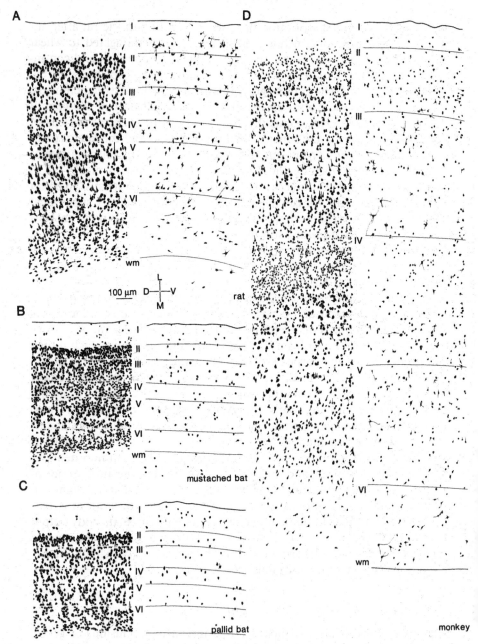

FIGURE 6.17. Some comparative views of auditory koniocortex from AI or its presumptive homologue showing the main architectonic features and the disposition of GABAergic neurons in several species. A. Cytoarchitecture of rat primary auditory cortex. The most conspicuous features are the well-developed layer II, the small size of neurons in layers IV and VI, and the relatively cell-poor layer Vb. See also Games and Winer (1988). Planapochromat, N.A. 0.35,× 320, 25-

where there are few cholinergic layer VI fibers, numerous serotonergic-immunopositive axons in all layers, sparse dopamine-β-hydroxylase-immunoreactive profiles, and more tyrosine-hydroxylase-immunostaining in rostral than caudal sections (Campbell et al. 1987). Certain thalamic-

---

(FIGURE 6.17 *continued*) μm-thick frozen sections for cytoarchitecture (*left-hand* panel) and for GAD (*right-hand* panel) immunocytochemistry. GAD-immunoreactive cells in rat primary auditory cortex show many of the features common to all mammals, including puncta-ringed but immunonegative pyramidal cells, especially prominent in layers III and V, and a heterogeneous population of GABAergic neurons in each layer. Considering the very small number of immunoreactive medial geniculate neurons in the rat, such cells are surprisingly numerous in the auditory cortex. From unpublished observations by Lin, Larue, Prieto, and Winer. B. Cytoarchitecture of the mustached bat auditory cortex, from the Doppler-shifted constant-frequency area and showing a prominent layer I, a cell-dense layer II, a thin, granule cell-dominated layer IV, and a layer VI in which many of the cells are oriented laterally. As in other species, the mustached bat has a substantial complement of cortical neurons immunoreactive for GAD, and all of these appear to be nonpyramidal. They are heterogeneous in size and shape and probably represent several types of neurons. The distribution of immunopositive puncta (not illustrated) is also consistent with the pattern in cat and rat, except that their concentration in layers III–IV appears extremely dense, while the number in layers V and VI may be smaller. From unpublished observations by Bui, Larue, Prieto, and Winer. C. The cytoarchitecture of the pallid bat's temporal cortex is similar to that of the mustached bat in some respects and different in other ways. Thus, layer II is thin but prominent, layer III has fewer neurons comparatively, and much of layer VI is dominated by horizontally arranged neurons. However, layer IV is not as conspicuous, nor is layer Vb as lightly populated. While the limits of auditory responsive cortex are unknown in this species, these observations are taken from temporal convexity cortex at a level midway through the anteroposterior extent of the thalamus. With respect to the pallid bat's GAD-immunoreactive neurons, their arrangement follows many of the same principles as in other species: they are most prominent in supra- as opposed to infragranular layers, and they represent several neuronal populations including both large and small nonpyramidal cells. The puncta appear even smaller than those in the mustached bat, many with a diameter of less than 0.5 μm, though their laminar distribution resembles that in other species. D. Cytoarchitecture of macaque monkey area TC, in which the degree of laminar differentiation is remarkably well-developed. In particular, the granule cells in layer IV merit the term, koniocortex, though larger, nonpyramidal neurons are scattered among them. The other layers are just as finely differentiated, and layer VI appears devoid of the horizontally arranged neurons that identify it in the other species included here. The GAD-immunoreactive neurons in macaque AI (area TC) are nearly as diverse as their counterparts in Nissl preparations. In particular, the granule cell and supragranular layers have many immunoreactive nonpyramidal neurons encompassing a wide range of sizes and shapes. While they may be less numerous in the infragranular layers, they are still a significant proportion of the neurons.

recipient neurons in layer IV colocalize somatostatin or cholecystokinin as well (Cipolloni and Keller 1989), suggesting that they may have a role in neural events whose time course far exceeds that of responses to GABA.

About 4–8% of the neurons in field L and the adjoining ventral hyperstriatum are GABAergic in the chicken (Müller 1988), compared to the approximately 20% in mammalian primary visual, somatic sensory (Hendry et al. 1987) and auditory (Prieto et al. 1990) cortex. More GABAergic puncta are evident in avian thalamic-recipient field L_2 than in other regions. A variety of GABAergic neurons and a dense concentration of GABAergic processes is reported in the neuropil of the pigeon ventral hyperstriatum (Domenici et al. 1988).

Iontophoretic application of a GABA antagonist to field L_2 cells does not change their response to species-specific calls, while L_1 and L_3 neurons are affected markedly. In contrast, glutamate always facilitates the evoked response (Müller and Scheich 1987a).

The aural properties of field L neurons can be modulated by anti-GABAergic compounds. Cells driven by contralateral but not ipsilateral stimulation (EO neurons) are excited by ipsilateral input during antagonist application to resemble binaurally driven cells. Half of the cells driven by contralateral and inhibited by ipsilateral input (EI) lose this inhibition with application of the antagonist, and the effect is reversible (Müller and Scheich 1987b). Microiontophoresis of acetylcholine in field L or ventral hyperstriatal neurons in awake chickens and in tissue slices commonly excites the former and inhibits the latter. In the hyperstriatum, the excitatory cholinergic effect is suppressed by indirect inhibition arising from local circuit GABAergic neurons (Müller 1987). These effects are similar to those of muscarinic receptor-dependent actions in slice preparations from guinea-pig limbic, sensory-motor, and visual cortex (McCormick and Prince 1986), and analogous changes are also seen in the visual cortex of the anesthetized cat (Sillito and Kemp 1983).

4.5 Comparative Behavioral Studies

Just as in the cat, other species show significant associative changes in the discharge pattern of auditory thalamic and cortical neurons during the acquisition and extinction of a classically conditioned task. Thus multiunit recordings from freely moving rats show different rates of response acquisition, with the posterior nucleus adjoining the medial geniculate body showing the most rapid changes, followed in turn by cortical neurons, medial geniculate body cells, and inferior colliculus neurons. Conversely, cells in the posterior nucleus are slow to extinguish, while cortical and other thalamic neurons do so more rapidly (Disterhoft and Stuart 1976).

Behavioral studies after massive cortical lesions in the rat reveal that pure tone localization is slightly but not severely impaired by lesions that cause most of the medial geniculate body and many of the adjoining thalamic nuclei to degenerate retrogradely; even in more difficult, three-choice tasks, the deficits are minimal, with the performance of controls only 5–10% superior to that of animals with comparatively massive cortical damage (Kelly and Kavanaugh 1986), in contrast to the cat, where even small lesions can have relatively specific and severe effects in similar tasks (cf. 3.6.4).

The mature auditory cortex can rearrange its tonotopic organization in response to damage limited to one part or another of the organ of Corti. Thus, in guinea-pigs with lesions to specific cochlear loci, the responses of unit clusters in auditory cortical areas that would have received this input are shifted toward the frequencies of intact, adjoining areas. In recordings made soon after such lesions, the thresholds were elevated substantially, while 5–12 weeks later, the thresholds were only 3.8 dB greater than controls. Such changes are restricted to cortical zones about 700 μm long in the rostrocaudal dimension (Robertson and Irvine 1989), a value that approximates the size of segments of single thalamocortical axon terminal fields in other pathways (Landry and Deschênes 1981), and the dimension of comparable zones of reorganization in the primate somatic sensory cortex after peripheral manipulation (Merzenich et al. 1983).

Behavioral studies of single units in awake rhesus monkeys find that the firing pattern of auditory cortical neurons is sensitive to patterns of reinforcement. In a task requiring discrimination between tonal stimuli, 25% of the single units showed stable responses in their evoked activity, generally an increase in on-excitatory response, with no change in spontaneous rate. This suggests that the receptive field organization of at least a subset of cortical neurons is malleable and contingent on experience in intact animals (Beaton and Miller 1975). An even higher proportion of units in the primary and adjoining auditory cortex of the baboon show augmented activity during an auditory task (Gottlieb et al. 1989). Monkeys performing a simple behavioral task have a larger evoked response than when the animal is not performing or when drowsy, lightly sleeping, or anesthetized (Pfingst et al. 1977). Studies of the temporal discharge pattern in the unanesthetized primate find that about half of the cells have complex, variable forms of temporal coding that led the authors to question whether a typical response pattern could be defined for central auditory neurons, at least in koniocortex and nearby supratemporal fields (Miller et al. 1974). Analogous data have prompted similar conclusions for certain cortical neurons in the cat (Goldstein et al. 1968). The intricacy of receptive field organization thus challenges the view that tonotopic representation is the appropriate or singular metric for assessing cortical organization. Thus, some 30% of cortical neurons in awake squirrel monkeys are not driven by pure tones, while 28% of the units are responsive only to one vocalization among the repertoire (Winter and Funkenstein 1973).

In another study of the behavior of single neurons in the awake primate at various levels of the central auditory pathway, cells in both the medial geniculate body and auditory cortex have lower rates of spontaneous activity than hindbrain auditory neurons and there is a systematic trend of decreasing spontaneous discharge from brain stem to cortex. In the auditory thalamus, the initial latency to discharge is higher during the performance of the behavioral task, and decreases systematically with stimulus intensity. In contrast, latency measures for cortical units are independent of stimulus intensity. Many medial geniculate units show significant changes in both early (0–75 msec) and late (75–200 msec) evoked discharges as well as in initial latency, while the response of cortical units changes less dramatically across these dimensions. At the cortex, these patterns would serve to enhance the signal-to-noise ratio of an auditory behavioral evoked response (Ryan et al. 1982).

The response lability and inherent variability of auditory cortical unit clusters in the tree shrew (Manley and Mueller-Preuss 1978a) and squirrel monkey (Manley and Mueller-Preuss 1978b) to noise or vocalizations have been proposed as obstacles to signal processing theories that would predict an invariant representation of peripheral events by central neurons. In a baboon, the response of about one-fourth of auditory cortex units is contingent on stimulus invariance (Gottlieb et al. 1989). In the rhesus monkey two-thirds of auditory cortical neurons respond preferentially to either light or sound as a behavioral cue in a motor task, and prolonged training may modify single unit discharge patterns (Hocherman et al. 1976). About 18% of rhesus monkey auditory cortex units have greater evoked discharges to stimuli that are attended to compared to nonattended stimuli (Benson and Hienz 1978). In unanesthetized rhesus monkeys performing an auditory behavioral task, some single units in AI or adjoining, tonotopically arranged fields do not show narrow tuning or there are inconsistencies or reversals in the tonotopic sequence; many nonmonotonic units are reported (Pfingst and O'Connor 1981).

Historically, the localization of primate auditory cortex on the basis of stimulation experiments and from behavior-ablation studies has engendered much controversy that reflects, at least in part, the prevailing expectations about the permanence of perceptual deficits anticipated from cerebral lesions (Heffner 1987). Compared to the results of auditory cortical ablations in the cat on sound localization, analogous lesions of large areas in the rhesus monkey have much more severe effects that are specific to audition (Wegener 1973). Thus, ablation of the rostral supratemporal cortex that causes retrograde changes only in the posterior one-quarter of the medial geniculate body, that is, chiefly in the caudal extremity of the dorsal division, produces transient deficits in auditory and visual pattern discriminations that are reversible (Wegener 1976). Monkeys cannot relearn a vowel discrimination task after lesions of primary auditory cortex, while a tone-noise distinction is reacquired readily. Damage to nonprimary cortex causes a substantial but reversible speech discrimination deficit, and neither lesion affects visual discriminations (Dewson et al. 1969). Other lesion-behavior studies of nonprimary cortex find short-term auditory memory impaired in a delayed matching-to-sample task, while sound discrimination behavior itself is relatively normal. Thus, *Cebus* monkeys are capable of sound discriminations of one octave or less, but fail in a behavioral task with a four-octave interval between signals whose discrimination requires short-term memory after association cortex is damaged (Colombo et al. 1990). Japanese macaques sustaining small, bilateral cortical ablations of primary and secondary fields show a partial loss of hearing and deficits in the capacity to discriminate species-characteristic "coo" vocalizations (Heffner and Heffner 1989). Damage to the lateral part of the monkey frontal lobes impairs auditory discrimination behavior, and the effects are marked when the lesion is remote from the principal sulcus (Gross and Weiskrantz 1962; cf. 4.3.).

Other ablation and behavioral experiments in the rhesus monkey indicate that animals with bilateral AI lesions can localize sound sources with normal acuity when only head movements are required, but cannot locate the sound source when an active motor approach response is the criterion (Heffner and Masterton 1975). Perhaps the disruption of auditory corticofugal projections to the basal ganglia (though it is unknown if these have the same arrangement in primates as they do in the cat; see Reale and Imig 1983) disconnects the motor system from access to appropriate auditory information, while the integrative capacity of intact, nonprimary fields is sufficient to localize stimuli but cannot subserve or initiate a gross motor response. Alternatively, the nonprimary areas may lack direct access to the basal ganglia that is essential for the normal performance of the primary field.

The question of whether auditory cortical lesions cause permanent hearing deficits is answered affirmatively in experiments in primates involving shock-avoidance or conditioned suppression. Performance at the most sensitive middle ranges showed 10–70 dB impairments that are unchanged 6–8 months postoperatively (Heffner and Heffner 1986a). There is evidence of a unique contribution from the left hemisphere in the discrimination of macaque vocalization, while right hemisphere lesions have no effect. Recovery of function occurs after left hemisphere lesions, but not after bilateral ablations (Heffner and Heffner 1986b).

Damage to the posterior (auditory) region of insular cortex or to the second somatic sensory cortex in rhesus monkeys disrupts postoperative auditory discrimination and causes severe gliosis in the medial geniculate body (Hörster and Ettlinger 1987). The degree to which other modalities besides audition are represented in the primate posterior insula is a matter of debate (see Friedman et al. 1986).

5. Developmental Studies of the Auditory Forebrain

The assembly of a mature and functional thalamocortical auditory system entails cellular interactions that may be typical of many species and sensory pathways. However, except in the case of the periphery (Rubel 1978) and lower brain stem (Schweitzer and Cant 1984) relatively little is known about the finer features of auditory neurogenesis, and this omission is most conspicuous in the cortex, particularly for the cat. Since a much more thorough account is available for the rat, it and other species are included in this account.

5.1 Ontogenesis of Acoustically-Evoked Potentials and Studies of Single Units

Kittens 18–24 days old show clear motor orientation responses to acoustic stimuli (Clements and Kelly 1978), suggesting that much of the essential auditory forebrain connectivity must be established by this period (cf. 5.4), though the fully adult pattern may be incomplete (cf. 5.5). In a series of evoked potential studies of auditory cortex responses to thalamic electrical stimulation (summarized in Mysliveček 1983), cats respond to tone bursts in the vicinity of AI three days postnatally and show adult-like latencies by 35 days after birth (Pujol 1972). By four days postnatally, the evoked potential is adult-like in form; before this, negative components dominate (cf. 5.3). With regard to single unit properties, latencies achieve adult values 2–4 months postnatally, as does the temporal synchronization of neural discharge; the emergence of complex, adult-like firing patterns begins in the third week and is followed by the development of rhythmic responses (König et al. 1972).

5.2 Thalamic Development

The neurons comprising the adult rat medial geniculate complex are generated between embryonic (E) days 12.5–15.5; some 75% are born sometime near day 14. This sequence partly precedes that in the dorsal cochlear nucleus and inferior colliculus; in the latter, proliferation occurs between E16–19 (Altman and Bayer 1981). By day E15 the migration of all the main sensory thalamic nuclei is largely complete (Altman and Bayer 1979a, 1988b), and by E17–18 a distinct medial geniculate complex is recognized, and some corticofugal axons may be in transit to the thalamus (Altman and Bayer 1979b). Later work describes neuroepithelial lobes which give rise to regional lobules from which, in turn, specific sensory nuclei are derived. Thus, the posterior lobule (which is the source for much of the dorsal thalamus) in the caudal lobe is the origin of the medial geniculate sublobule, which forms by eversion of the ventricular thalamic neuroepithelium (Altman and Bayer 1988a).

Further analysis reveals that different rat auditory thalamic territories are generated at specific times (Altman and Bayer 1989a,b). Thus, dorsal and ventral clusters of neurons above and below the primordial ventral division arise significantly before the latter; the locus of these clusters suggests that they may be related to the mature superficial dorsal and ventrolateral nuclei, each of which is considered as part of the dorsal division (see Winer et al. 1988). Next, on E14, most of the lateral part of the future ventral division arises, followed (on E15) by neurons in the more medial parts of both the ventral and dorsal divisions; the last-generated cells, from the medial division, also rise on E15. All parts of the medial geniculate complex have a caudal-to-rostral developmental gradient with the former neurons arising earliest (Altman and Bayer 1989b). Ventral division neurons in immature rats display a considerable degree of acetylcholinesterase staining, as does the primary auditory cortex. The significance of this transient enzymatic feature is unknown, and it is sufficiently robust to withstand cytotoxin-induced microencephalopathy (Robertson et al. 1990).

5.3 Development of Cortical Neurons

In rabbit auditory cortex the neocortical layers are visible six days after birth, when hearing begins, and after which the numbers of dendrites and branches of nonpyramidal cells in layers III and IV multiply to almost twice their adult values (McMullen et al. 1988). Many of these branches are lost subsequently and surviving dendrites mature and elaborate spines. The adult, vertical dendritic arrangement of some arbors is present from early postnatal periods; perhaps the later arrival of thalamic or commissural afferents plays a role in the development of the mature pattern in which tangential dendritic orientations are a common feature (see Wong 1967). Since many of the nonpyramidal neurons with vertical dendritic arbors are GABAergic and have an unmyelinated axon with local branches (Prieto et al. 1990), perhaps some substrates for local cortical processing appear at a comparatively early age. Because the characteristic, initially positive evoked-potential is missing or minute in 5 day-old animals, the deeper-lying pyramidal neurons presumed to be the source of this potential may be immature (König and Marty 1974). Since some of the earliest synaptic contacts in rat temporal cortex are in the marginal layer (König et al. 1975), most of whose cells will be GABAergic in adults (Winer and Larue 1989), then local cortical circuits could begin to elaborate even before specific (and perhaps nonspecific) thalamic input arrives. Because most thalamic neurons have birth dates during this period (Altman and Bayer 1989a,b), it is unlikely that they could have established cortical synapses at this time since most cortical contacts are either superficial or deep, and not at intermediate, laminar levels (König et al. 1975).

5.4 Development of Extrinsic Connections

By four days after birth, the major afferent and efferent connections of AI in the cat are established. There may be some significant departures from the adult pattern, for example, most commissural cells of origin lie in the infragranular layers, especially layer VI (Cornwell et al. 1984), while in mature animals layer VI is the source of only a fraction of this projection (Code and Winer 1985). Other work finds the bulk of commissural labeling present in layer III one day after birth (Feng and Brugge 1983). Similarly, the topographic corticofugal projection to the basal ganglia present in adults (Reale and Imig 1983) occurs in kittens, too, but is thought to be devoid of such order (Cornwell et al. 1984). As a rule, projections from the deepest cortical layers develop first, while those of the granular and supragranular layers arise somewhat later. Hence, long corticocortical connections appear to be the last to mature, while auditory corticogeniculate pathways would be expressed earliest (Payne et al. 1988).

5.5 Transient Connections of Developing Cortex

The maturation of connectivity entails certain cellular rearrangements. Thus, the commissural cells of origin, which are at first continuously distributed, later achieve a mature, clustered form, and immature callosal axon terminals have not, early in ontogeny, assumed their adult, banded distribution. Neither the time course of the development of a particular projection nor its refinement, through cell death or loss of connections, is uniform. Thus, the ontogenesis of callosal connections is delayed relative to all other corticofugal and thalamocortical projections (Feng and Brugge 1983). Perhaps this is merely a result of propinquity, or it may reflect genuine qualitative differences in the complexity of a particular pathway or the distance that its axons must travel.

In kittens less than 38 days old, there is a transitory ipsilateral corticocortical projection from AI and AII to visual areas 17 and 18; most of these cells arise from layers III, V, and VI (Innocenti and Clarke 1984) or layer II and III pyramidal cells (Dehay et al. 1988), and there is no reciprocal projection from area 17 to AI (Clarke and Innocenti 1986). While these auditory-visual pathways are eliminated in mature animals, the expression of the adult pattern implies the loss of a portion of their axon collateral system rather than the death of the parent neuron. These cells thus persist as ipsilateral corticocortical projection neurons whose connections are confined to auditory and periauditory targets (Clarke and Innocenti 1990).

5.6 Effects of Neonatal Manipulation

The consequences of peripheral damage on the properties of AI units have been studied in adult cats that sustained unilateral cochlear destruction in the first postnatal week. In AI on the same side as the intact ear, a normal tonotopic arrangement exists and thresholds are often comparable to those in the opposite hemisphere. In normal adult cats, two-thirds of AI neurons are driven by ipsilateral stimulation and these thresholds are appreciably higher than those for contralateral input. Lesioned animals also have a rearrangement of the distribution of thresholds to sound intensities above or below 20 dB, with the lowest values predominantly rostrally and caudorostrally and higher thresholds caudally and rostroventrally in AI, while in normal adults the spatial arrangement of threshold is random (Reale et al. 1987). Parallels between the effects of these manipulations and the consequences of neonatal occlusion on the arrangement of ocular dominance columns (Hubel et al. 1977) or binocular response properties (Hubel and Wiesel 1970) of visual cortex neurons suggest that similar mechanisms may be involved, and the nature of the neural adjustments in the somatic sensory cortex after peripheral digit manipulation (Merzenich et al. 1983) implies analogous modes of functional reorganization.

Neonatal atresia has comparatively modest effects on the organization of adult AI. In cats whose ligated, cut canals were cleared of debris, thresholds from the operated side are elevated but still normal, and all monaural spike count/intensity functions were unaffected. Most surprisingly, each binaural response type is present, and in numbers indistinguishable from controls. The only significant effect is a mismatch of up to 50 dB required to elicit binaural responses from cells driven by the occluded ear (Brugge et al. 1985). These changes are in striking contrast to the marked and permanent deficit in binocular tuning of area 17 cells ensuing from even relatively brief periods of monocular deprivation (Malach et al. 1984).

A morphological effect of cochlear deafferentation on nonpyramidal auditory cortex neurons in layers III and IV in rabbits is a 27% increase in dendritic length and the appearance of unusual, recurrent dendrites (McMullen et al. 1988). Pyramidal cells show a 39% decrease in dendritic spines (McMullen and Glaser 1988), suggesting that transneuronal changes after cochlear damage are not limited to the brain stem nuclei or strictly to neuronal perikarya (Powell and Erulkar 1962).

5.7 Nonauditory Responses in Auditory Cortex: Genetic Factors and Early Cochlear Destruction

While there are no direct projections from primary visual to primary auditory cortex (Clarke and Innocenti 1986; see also Jones and Powell 1970), it is nonetheless possible to record large-amplitude visual evoked potentials at moderate latencies in AI in cats after cochlear ablation and in hereditarily deaf white cats; these potentials were not present in somatic sensory cortex (Rebillard et al. 1977). There is a critical period (1–8 weeks) within which the ablation must be performed (Rebillard et al.

1980). Even in animals deaf for several years, photically-evoked potentials occur in AI (Rebillard et al. 1976).

5.8 Experimentally Induced Reorganization of Thalamic Projections

Ablations of the dorsal nucleus of the lateral geniculate body and of the superior colliculus in newborn hamsters induce retinal axons to innervate the medial geniculate complex and adjoining thalamic nuclei. These redirected axons form topographically organized projections (Frost 1981) and comprise fibers that, by reactive sprouting, occupy almost 40% of the auditory thalamus (Frost 1986). Whether retinal projections form functional synapses in the medial geniculate body (see below) as they do in the ventrobasal complex (Frost and Metin 1985), is unknown, but the ultrastructure of the experimentally induced synapses more closely resembles the normal fine structural arrangement in the medial geniculate body than that of retinal axons ending in the lateral geniculate body (Campbell and Frost 1988). This suggests that intrinsic organization exerts important local influences on the mature configuration of extrinsic afferent thalamic synapses.

In ferrets, neonatal ablation of the superior and inferior colliculi provokes retinal ganglion cell axons to innervate neurons in the ventral and dorsal divisions and cells in nearby parts of the posterior thalamus (Sur et al. 1988). These neurons, compared to typical visual thalamic X- or Y-cells, had longer latencies, larger receptive fields, and their visual responses were variable or sluggish, responding best to large, moving spots of light, all of which are consistent with the view that they represent W-like cells (Stone 1983). Recordings from AI find visual responses in neurons between 600 and 900 μm deep to the pia, and with latencies about twice as long as those in visual cortex cells. Some auditory cortex neurons have directional tuning or orientation sensitivity (Sur et al. 1988). Most auditory corticocortical and corticofugal projections are conserved in these preparations, and others that would have been eliminated, presumably as a result of competition, programmed cell death, or other effects, now persist into maturity (Pallas et al. 1990). These findings suggest that at least some permissive factors are of cortical origin and that projections induced experimentally onto AI conserve the topographic organization of the visual field (Roe et al. 1990). Indeed, in deaf humans the visually evoked N150 component is larger in temporal and frontal areas than in hearing subjects (Neville et al. 1983). By the same token, both the medial and lateral geniculate bodies of the hereditarily blind subterranean mole are strongly activated by sound, as assessed by 2-deoxyglucose autoradiograms (Bronchti et al. 1989).

6. The Human Auditory Forebrain

Two constraints impede the direct extrapolation of patterns of thalamic and cortical organization from infrahuman species readily to humans. First and foremost is the inherent unreliability of the criteria, and the second is the impossibility of using invasive contemporary methods to study presumptively homologous neuronal populations. Despite these limitations, the degree of coherence in the overall design of the mammalian auditory forebrain suggests that analogous stations exist despite uncertainties about the synaptic sequence or local functional organization in any given nucleus. For example, no conspicuous or consistent specializations for speech have been discerned in the human auditory thalamus (see Eidelberg and Galaburda 1982).

6.1 Medial Geniculate Body

The human medial geniculate body is situated between the pulvinar (dorsally) and the lateral geniculate body (laterally) and the cerebral peduncle (ventrally); it is an ovoid mass about 5–6 mm wide and 4 mm tall. The primary parts and many of the subdivisions defined in nonprimates (Ramón y Cajal 1911; Morest 1964; Winer 1985b; Morest and Winer 1986) are recognized in humans in Nissl, Golgi, and fiber stained material (Winer 1984d), though other workers neither subdivide the auditory thalamus (Andrew and Watkins 1969) nor present any rationale for proposed boundaries other than somatic size or packing density (Dewulf 1971).

In Golgi preparations, the ventral division has many medium-sized neurons with long, highly tufted dendrites whose predominant orientation is along a mediolateral axis. In their form and distribution, these cells resemble the ventral division bushy neurons, which in the cat serve as thalamocortical relay neurons, show aspartate- or glutamate-like immunoreactivities (Table 6.1), and project to the cerebral cortex (Winer 1984e). Smaller neurons with thinner, radiating dendrites are also impregnated, and in Nissl material these cells form the smaller of the bimodal peaks of somatic size in the ventral division. While the fibrodendritic laminae are oriented from dorsolateral to ventromedial, nothing is known about the frequency-specific organization of the auditory thalamus, nor are the data on retrograde thalamic degeneration after temporal lobe lesions sufficiently detailed to permit inferences about a topographic thalamic organization (Van Buren and Borke 1972). The dorsal division has a preponderance of medium-sized and small cells except in the suprageniculate nucleus, where they are much larger. While some dorsal division neurons have tufted dendritic branches, many have dendrites that radiate spherically to form rather simple arbors; the latter neuronal type is common throughout the human lateral thalamic sensory nuclei (de Courten and Garey 1982; Braak and Braak 1984). Smaller stellate cells are especially numerous in the dorsal division. In the lateral geniculate body, comparable small putative interneurons represent less than 9% of a large sample, suggesting possibly significant differences in intrinsic organization among the principal thalamic sensory nuclei (Braak and Bachmann 1985). Some medial division neurons are the largest in the auditory thalamus, and the resident neuronal population is the most diverse and resembles many of the types of cells impregnated in the cat (Winer and Morest 1983a). However, the heavy myelination of this region has frustrated any systematic survey of neuronal architecture in these adult preparations.

Two other features distinguish human medial geniculate body organization from that of carnivores or other species. First, the relative proportions devoted to particular subdivisions are different: in the cat (Morest 1964; Winer and Morest 1983b) and bat (Winer and Wenstrup 1992a,b), the ventral division is preeminent in size, followed by the dorsal and the medial divisions, while in humans the dorsal division is conspicuous, and the ventral division appears relatively reduced. Perhaps this is a consequence of the apparent phylogenetic regression in the laminar architecture of the human cochlear nuclei (Moore 1980). Second, the human neuropil is extremely well developed, and cell-free regions 100–500 μm wide, especially in the dorsal division, are common (Winer 1984d).

There is some evidence for thalamic structural asymmetry since in 7 of 10 specimens the right medial geniculate complex was larger than the left, though the difference was not statistically significant (Eidelberg and Galaburda 1982; Galaburda 1986). This is in contrast to the case of the left temporal plane, which is larger in two-thirds of the sample (Geschwind and Levitsky 1968).

Auditory thalamic neurochemical studies describe only modest histochemical staining for acetylcholinesterase compared to other thalamic nuclei, with the dorsal and medial divisions distinctly darker than the extremely pale ventral division (Hirai and Jones 1989a), a pattern consistent with the immunocytochemical results obtained with choline acetyltransferase in the cat (Fitzpatrick et al. 1989). A few tachykinin- or metenkephalin-immunoreactive axon terminals are present throughout the medial geniculate complex, concentrated primarily in the ventral division, in the suprageniculate nucleus, and in the nearby posterior thalamus, which may include parts of the dorsal division; in the latter, their concentration is somewhat heavier and more clustered (Hirai and Jones 1989b). Such a finding supports the view that this part of the dorsal thalamus is a target of convergence for nociceptive afferents (Poggio and Mountcastle 1960), a function that these axon terminals might subserve both in carnivores and primates (see also Mehler et al. 1960).

In developmental studies, the human medial geniculate is present in nine-week-old embryos, after which it begins to show a characteristic lobular organization that is partly obscured by the subsequent expansion of the pulvinar (Cooper 1948).

6.2 Auditory Cortical Fields

Many cytoarchitectonic fields have been defined along the convexity of the transverse temporal gyrus, and the notable features common to all with presumptive auditory affiliations are (1) a thick layer I; (2) an irregular texture in layer II, ". . . as if bunches of its cells were protruding into I . . ."; (3) a columnar arrangement in layer IV; (4) the prominence of layers V and VI; and (5) a radial organization spanning layers VI–II (von Economo 1929, p. 111). These attributes apply to feline AI (Rose and Woolsey 1949; Sousa-Pinto 1973b; Winer 1984a,b,c, 1985b, 1991) and to portions of the parietal lobe (Bailey and von Bonin 1951). Area TC, the primary field, is a 900–1600 mm² area of classic koniocortex along the first and second gyri of Heschl, some 3 mm thick, and it is bounded by nonprimary areas, of which four are noteworthy (von Economo 1929). Area TC corresponds to area 41 (Brodmann 1909), and the supratemporal *isocortex koniosus* (Iκst; Bailey and von Bonin 1951). Some surface estimates of areas 41 and 42 describe values of 1198–1380 mm², with a pronounced right-side areal advantage in 7 of 10 brains (Campain and Minckler 1976), while other measurements find the left supratemporal plane to be larger in 65% of brains, and the right bigger in only 11% (Geschwind and Levitsky 1968). More recent studies (Galaburda and Sanides 1980) describe two adjoining core areas (KAm and KAlt) that are bordered by five further, nonprimary fields. Thus, this arrangement resembles that in the monkey (Pandya and Sanides 1973) except that some of the auditory areas extend into the parietal operculum and inferior parietal lobule. In pigment architectonic preparations, this primary area (*area temporalis granulosa*) has unusually light aldehyde fuchsin staining and is bordered by anterior and posterior nonprimary fields (Braak 1978a,b).

There are only a few studies of the neuronal architecture of human auditory cortex, and these provide limited data on the types of neurons. Ramón y Cajal (1900) describes characteristic pyramidal neurons, an abundance of bitufted neurons with rich vertical dendritic arbors and parallel dendritic branches, and inverted pyramidal cells; with a few exceptions (for example, the giant specific cells) these and the appropriate terminal axonal patterns can be found in cat AI (Sousa-Pinto 1973b; Winer 1984a,b,c, 1985a) and, in the case of chandelier cells, the monkey (De Carlos et al. 1987). A heterogeneous population of spiny stellate neurons mingled with a large population of small pyramidal neurons is characteristic of layer IV koniocortex and parakoniocortex in humans. The stellate cells have an axon whose collaterals are mainly of the descending type but with recurrent branches, too, and some ascending intracortical axons (Meyer et al. 1989). Chandelier cells, whose axons are presynaptic to the initial axonal segment of pyramidal cells, occur in human temporal cortex, much as they do in feline AI (Kisvárday et al. 1986). This implies a considerable degree of concordance in the basic pattern of neuronal architecture between human and cat layer IV (Winer 1984a). Further Golgi studies of human temporal cortex (Ong and Garey 1990) reveal that many of the cell types described in the cat (Sousa-Pinto 1973b; Winer 1984a,b,c, 1985a, 1992), especially the populations of non-pyramidal neurons, are present in humans. This investigation of human material also finds that columnar organization in the right temporal cortex is far less prominent than in the left hemisphere. Other Golgi preparations from area *TC* on the medial part of the anterior segment of the transverse temporal gyrus show some small layer IV pyramidal cells, and no extraverted layer II neurons (Conel 1959), the latter finding in contrast to the pattern in the cat, where these cells are a hallmark of layer II organization (Winer 1985a). Perhaps the columnar organization is related to the extreme vertical polarization of many of the dendritic domains in layer IV and above (Seldon 1981).

The neurochemical organization of area 41/*TC* has received little attention. In a study of receptor binding using [³H]amino acid probes to localize the *N*-methyl-D-aspartate, quisqualate, or kainic acid receptors in the primary auditory cortex, the supra- and infragranular layers show the highest ligand binding for [³H]glycine and L-[³H]glutamate, with layer IV lower in each. [³H]kainic acid binding and quisqualate receptor binding have a similar distribution. In contrast, the primary visual cortex has a more or less homogeneous laminar distribution of ligand binding for glycine and glutamate, and a quisqualate pattern like that in the auditory cortex (Jansen et al. 1989).

The distribution of GABA immunoreactive terminals in Brodmann's area 38 essentially matches that in the cat, with those in layers I–IV most numerous, and far fewer in layers V and VI, while vertically oriented, GABAergic terminal plexuses occur in layers II and III, and a lateral concentration is prominent at the base of layer VI. Small (less than 15 μm) and large (from 15 to 25 μm major diameter) GABAergic neurons occur, the former present in each layer, the latter more common with increasing depth. Most layer II and III GABAergic neurons have a multipolar dendritic configuration, often with a vertical polarization, and some cells have bitufted dendrites; immunopositve bipolar cells are rare, and a significant number of immunoreactive neurons lie deep in the white matter (Schiffman et al 1988). These results correspond to those in the cat, with some exceptions; for example in cats the immunoreactive axon terminals have different shapes and concentrations in each layer (Table 6.8; Fig. 6.15G-L), and particular spatial arrangements with presumptive postsynaptic neurons (Peterson et al. 1990; Prieto et al. 1990; see also Winer 1986). The synaptic targets of GABAergic terminals in human temporal cortex include neuronal somata, dendritic shafts and spines and, less often, axonal initial segments (Kisvárday et al. 1990). Certain cells in the infragranular layers of human temporal cortex, including bipolar neurons, show tyrosine hydroxylase immunoreactivity; this transmitter may be implicated in psychiatric disorders (Hornung et al. 1989).

Intracellular and iontophoretic study of pyramidal neurons from human temporal lobe tissue obtained at biopsy show that the effects of GABA agonists and antagonists are consistent with the interpretation that GABA is probably a physiological transmitter. Since blocking the normal inhibitory postsynaptic potentials to these neurons causes ictal discharge, GABA could modulate some facets of the temporal organization of their output (McCormick 1989; Wuarin et al. 1990).

The neurophysiological consequences of temporal lobe stimulation or the effects of such damage on human behavior exceed the scope of this review. Suffice it to say that auditory behavioral responses are evocable from regions near the temporal pole and along large expanses of the supratemporal plane and extending deep into the parietal cortex (see Penfield and Perot 1963 and Gloor 1990). Trauma due to many causes, such as pathological changes consequent to methyl mercury poisoning, besides affecting cells in the auditory brain stem, may cause especially severe, and perhaps transneuronal, changes in the medial geniculate body and transverse temporal gyrus (Oyanagi et al. 1989).

The cortical locus of stimulation that elicits a behavioral response is largely congruent with, though smaller than, the zone from which evoked potential studies reveal neural activity. It is uncertain if such a representation extends into the insula, and potentials evoked from the parietal operculum have smaller amplitudes, but a comparable shape, to those from the first (anterior) temporal gyrus (Celesia et al. 1968). Further work localized short latency evoked potentials to the rostral and caudal parts and along the medial limb of the second (posterior) transverse temporal gyrus and adjoining gyri. Potentials evoked in the first temporal gyrus are small in amplitude, longer in latency, and more variable (Celesia and Puletti 1969). Positron emission tomography indicates that low frequencies are represented laterally, higher frequencies more medially (Lauter et al. 1985). In a more recent study of evoked potentials using stereotaxic coordinates and arterial landmarks near the medial part of Heschl's gyrus, the largest and earliest potentials are found in the posteromedial part of the gyrus (Liegeois-Chauvel et al. 1991), in close correspondence with architectonically defined koniocortical fields (Galaburda and Sanides 1980).

At least one tonotopic representation has been described along the transverse temporal gyrus using neuromagnetic techniques (Romani et al. 1982), and a second, later response with an amplitude one-third that of the primary one has been interpreted as a contribution from nearby, nonprimary auditory areas (Elberling et al. 1982; see also Pantev et al.

1989b). A compelling argument for functional subregions within a primary field or for an adjoining field is the specialized representation found for pure tone stimuli. This implies that complex (virtual pitch) stimuli have a different representation from pure-tone (spectral pitch) signals, and that the primary cortical role is analysis in the frequency domain (Pantev et al. 1989c). A systematic representation of stimulus amplitude that is independent of frequency occurs in auditory cortex (Mäkelä et al. 1987; Pantev et al. 1989a). There is evidence for a functional hemispheric asymmetry such that patients sustaining left hemisphere temporoparietal damage have deficits in temporal processing, with normal temporal perception, while right hemisphere lesions produce the opposite effect (Robin et al. 1990). Broad areas besides the auditory fields show increased regional blood flow during an auditory discrimination task. These areas include the frontal eye fields, frontal polar cortex, superior frontal areas near the midline, and Broca's area (Roland et al. 1981).

Binaural stimulation produces stronger responses than monaural stimulation, and large territories of the posterior temporal cortex show evoked potential responses to clicks (Celesia 1976). Magnetoelectroencephalographic responses have a similar contralateral preference (Tiihonen et al. 1989) and shorter latencies than for ipsilateral stimuli (Mäkelä 1988).

Using neuromagnetic recording, several interesting functional arrangements are reported in the auditory cortex. Thus, novel stimuli presented ipsilaterally evoke a response 130% stronger than contralateral, standardized stimuli (Mäkelä 1988) or more than 200% larger for tones at unexpected temporal intervals (Hari et al. 1989b; see also Kaukoranta et al. 1989). Other, event-related potentials responsive to novel acoustic stimuli require the integrity of the posterior superior temporal plane, while even substantial lateral parietal cortex damage has no effect on them. An intact association cortex is required for appropriate neurophysiological responses to such input (Knight et al. 1989a). The finding that evoked potentials in primary auditory cortex are facilitated by damage to prefrontal cortex suggests that nonauditory corticocortical input to auditory cortex may modulate its excitability (Knight et al. 1989b), besides the role of intrinsic inhibitory mechanisms (McCormick 1989).When a noise alone or a noise and a square wave with various stimulus onset configurations is presented, the amplitude or the latency of the main component of the evoked magnetic potential is differentially affected, independently of stimulus strength in the 40–70 dB range. Such mechanisms could be useful for encoding rapid phonetic changes (Mäkelä et al. 1988). Thus, the bipoles for /ka/ and /a/ are some 7 mm apart in the left hemisphere, and indistinguishable spatially in the right hemisphere (Kuriki and Murase 1989). It has been postulated that the minimal mean latency of spike timing in cat primary auditory cortex neurons is appropriate to the temporal resolution of sounds in the range of a few milliseconds to tens-of-milliseconds, which is optimal for the representation of stimuli that may be important in syndromes such as acquired word deafness (Phillips and Hall 1990; Phillips and Farmer 1990); for a clinically relevant case, see Albert and Bear (1974). There is no known correlate of such lateralization in the cat, despite the fact that hemispheric specialization for species-specific vocalizations has been reported in Japanese macaques (Heffner and Heffner 1986a).

The operations of auditory cortex are not limited strictly to hearing. Thus, damage to the transverse temporal gyrus impairs acoustic control of a spinal reflex, implying that such lesions reduce cortical facilitation or that vital corticocortical connections essential for this reflex are disrupted. Damage to other parts of the temporal, parietal, or occipital lobes had no effect on this reflex, while frontal lobe lesions had inconsistent results (Liegeois-Chauvel et al. 1989; see also Knight et al. 1989a).

There is evidence that event-related potentials show behaviorally induced plasticity, supporting the idea that their response properties selectively encode different experiences. Thus, dichotic listening to relevant and irrelevant sounds produces different responses. The discrimination of meaningful *versus* meaningless words or tones differing in duration consistently affects the second, sustained component of the evoked potential (Hari et al. 1989a).

7. Synthesis of Connections and Directions for Future Work

The rudimentary status of current knowledge of the afferent and efferent connections of the auditory forebrain is the primary obstacle to framing more explicit hypotheses about the functional anatomy of this system. Consider, for example, the question of whether the projection of inferior colliculus neurons to the thalamus is divergent, and the form that such divergence might take. It is uncertain now whether single inferior colliculus neurons have branched axons that project to more than one thalamic division, or if tectothalamic neurons in a particular nucleus have only one axonal branch but different thalamic divisional targets. Each possibility has different ramifications for thinking about thalamic operations. In the first instance, it would be essential to know if the postsynaptic targets of branched neurons were similar, both in a connectional and functional sense. The second possibility is the simplest statement of the parallel processing dogma, but leaves unresolved what is parallel (for example, aurality, tonotopy, or sharpness of tuning) or the operational meaning of divergent parallel channels. For example, if any GABAergic inferior colliculus neurons project to the medial geniculate body, how are their synaptic actions (or cellular target[s]) different from those of presumptively aspartatergic or glutamatergic tectothalamic neurons? Are the projections of particular types of midbrain neurons regionally specific?

While there may be reasons favoring one scenario over another, the lack of a more complete picture severely constrains the generality of any single finding, and makes it impossible to integrate physiological, neurochemical, and anatomical data into a coherent view.

Nevertheless, a few organizational principles of tectothalamic connections can be deduced: (1) single midbrain divisions may project to more than one thalamic division; (2) the neurons representing the lemniscal pathway in the ventral division receive the smallest number of different inputs, while the dorsal and medial divisions receive many more; (3) this implies a segregation of specific input to the ventral division and its convergence in the dorsal and medial divisions, contributing in the latter divisions to representations whose internal order is unknown.

Many of the same principles are conserved in the thalamocortical projections, and at the same time new complications emerge. For example, there are at least four cortical fields (AI, AAF, P, and VP), with narrowly tuned neurons, and which may be considered as related. If so, what different functions do they serve? Are they each a part of some enormous functional superfield which is *the* primary auditory cortex, and which, as a consequence of the massive evolutionary elaboration of the temporal lobe, now consists of four discontinuous subareas linked largely by functional thalamic affiliations and a vast network of corticocortical connections? If they are part of one superfield, then why are their physiological properties so different? Does this mean that adjacent thalamic neurons that project to different cortical fields might have different physiological profiles that would align them with their cortical targets and, by inference, unique constellations of midbrain afferents that would signify their thalamic function? If so, these differences have so far eluded structural, connectional, physiological, and neurochemical scrutiny.

The function(s) of the corticogeniculate, corticocortical, and commissural pathways are equally obscure. The laminar and areal specificity of each of these, as well as their sheer size, suggest that they must each play a particular behavioral role. Perhaps the development of pathway-specific antibodies or other functional probes will stimulate more focal inquiries.

Among the questions that might be asked are these:

- Do aurally specific inferior colliculus neurons projecting to the medial geniculate body use the same transmitter?
- What are the differences in the synaptic arrangements of monaural and binaural neurons in the auditory thalamus, both with respect to afferent pathways and intrinsic organization?
- Are there significant areal or laminar differences in the arrangement of corticocortical projections that are functionally meaningful?
- Is the laminar distribution of thalamic input to layers III and IV homogeneous, or is there a finer vertical or horizontal sublaminar arrangement analogous to that in the visual cortex?

- How do thalamic reticular nucleus inputs affect the responses of medial geniculate body neurons?

8. Synthesis of Physiology and Directions for Future Work

It is now apparent that there are auditory thalamic territories with a well-developed tonotopic organization (the ventral division and the lateral part of the posterior nucleus), regions with a weaker representation of best frequency (medial division), and nuclei where no such systematic arrangement, if it exists, has yet been discerned (the dorsal division). Corresponding representations have been described in their affiliated cortical fields. These findings raise central issues about the function of the auditory forebrain since different functional arrangements appear to have unique local patterns of representation in the thalamus and cortex. Many physiological studies have been limited largely to investigations of tonotopic organization or binaurality, usually in anesthetized preparations. Insofar as such an approach is inimical to revealing other facets of function, the present view of thalamic and cortical function is unduly narrow. It is not clear intuitively why so many serial representations of frequency or aurality are required, nor is it valid to construe thalamic or cortical centers without such maps as less important or hierarchically more advanced than those with them. The lack of such a metric for specifying function in these nontopical areas reflects present methodological and conceptual limitations, not to mention the immense physiological complexity imposed by vast local and remote patterns of interconnectedness.

In spite of these limitations, it is possible to make some general statements that may apply (in varying degrees) both to thalamus and cortex: (1) besides the tonotopic arrangement, some segregation of neurons by aurality is also evident; (2) cortical fields and thalamic nuclei with similar physiological attributes are often (but not always) interconnected; (3) the tuning curves and response profiles of neurons in species with many well-developed local circuits (for example, primates) and in those with presumably more modest interneuronal populations (such as rodents and bats) are more similar than might be expected, insofar as Golgi type II cells are predicted to have any substantial role in modulating sharpness of tuning or affecting the spatial contours of receptive fields; (4) even in areas where a topical physiological arrangement is predominant, there may be other functional subregions with nontopical or multiple representations; (5) adjoining neurons may or may not share like physiological attributes.

In the central visual and somatic sensory pathways, plausible neural circuits have been proposed to explain how receptive fields are constructed at various neural stations, and in some cases the predictions of these models have been tested experimentally (Dreher and Cottee 1975; Allard et al. 1985). In the auditory thalamus and cortex, no such models exist at present, and while the chief contributors to the excitatory region of the receptive field can be surmised logically in certain instances, the other components remain obscure, as the example of the layer III pyramidal cell in AI shows. As the recipient of strong, monosynaptic, and narrowly tuned thalamic input, this cortical cell can be expected to have a similar profile of responses. However, insofar as it is also the recipient of commissural afferents, it will reflect these inputs, too. If any ipsi- or contralateral afferents that rise from nontonotopic fields converge on this cell, then they could effectively alter the thalamic-derived tuning, or activate local circuits presynaptic to the cell. Local circuits might endow the cell with diverse responses to different stimuli, for example, through collateral excitation of GABAergic neurons that could either suppress nonspecific input or sharpen receptive field borders to amplify signal contrast or specify spike timing with enhanced precision. By virtue of its local axonal branches projecting towards the supragranular layers, this cell could facilitate or inhibit the discharge of more remote neurons, either within a specific aural or frequency domain, or it may have lateral interconnections across such borders. It can likewise affect neurons in the opposite hemisphere, perhaps in every layer, thus influencing every contralateral cortical projection system. Thus, one hemisphere might facilitate or depress transmission in the opposite thalamus during a localization task. While this is mere speculation at present, it is nonetheless amenable to testing.

Answers to the following questions might be useful:

- How are the binaural (or other functional properties) of AI cells altered by selective, reversible inactivation of afferent input?
- What effects, other than the broadening of otherwise narrow tuning curves, might the administration of GABA antagonists have on AI cells?
- In what ways do the discharge properties and ionic configuration of intrinsic neurons differ from those of projection cells?
- Is there a physiological and pharmacological basis for response plasticity or invariance in the discharge properties of auditory thalamic and cortical cells?
- Can intracellular recordings delineate patterns of synaptic relations or specify important facets of the tuning of postsynaptic potentials in cortical neurons?

9. Synthesis of Neurochemistry and Directions for Future Work

In other sensory systems, immunocytochemical methods have revealed some surprising facts, for example, the possibility that GABA is a primary afferent transmitter of avian vestibular hair and ganglion cells (Usami et al. 1989). While the study of neurotransmitter candidates in the auditory forebrain is still in its earliest stages, with respect to GABA some initial generalizations can be made: (1) particular patterns of immunoreactivity reliably differentiate the major divisions of the medial geniculate complex; (2) there are significant species differences in the proportion of thalamic GABAergic neurons, rodents and bats having relatively few and carnivores and primates many more; (3) these species differences are much less pronounced in the auditory cortex, where the number of such neurons is comparable in different species; (4) each of the cortical layers, like the subdivisions of the auditory thalamus, has a characteristic pattern of GABA immunoreactivity that distinguishes it from other layers, both qualitatively and quantitatively; (5) in both the auditory thalamus and cortex, the proportion of GABAergic cells is less than 40% (except in cortical layer I, where it is 65–90%; see Winer and Larue 1989), and many of the GABA-negative neurons contain aspartate- or glutamate-immunoreactive material; (6) there is a complex relationship between the number of GABAergic neurons and puncta in a nucleus or a layer such that they do not covary consistently in every case.

The relative paucity of GABAergic neurons in the rat and bat medial geniculate body is not especially surprising, given the comparative rarity of Golgi type II neurons. However, the conservation of specific nuclear patterns of puncta immunostaining suggests on the one hand that even the few local circuit neurons in these preparations have axonal collateral systems that may represent thousands of branches (and must therefore have been underestimated radically in Golgi material from immature specimens), or that, on the other hand, the arrangement of local patterns is, at least in some species, a consequence of extrinsic input from the thalamic reticular nucleus (or other sources). An arrangement favoring the latter conclusion is that these regional patterns of puncta immunostaining are very similar qualitatively in cats and monkeys, each of which has many more GABAergic neurons than rodents. Since the number and type of such puncta are so different in the main parts of the auditory thalamus, then within the auditory subdivision of the thalamic reticular nucleus there may be regions specific to one or another part of the medial geniculate complex. Insofar as these projections might arise from other sources, the same considerations would apply.

As noted above, equally specific conclusions can be made for the cortex. Thus, there is a sharp contrast in the number and form of GABAergic puncta in the supra- *versus* infragranular layers, while the number of GABAergic neurons also has a nonuniform laminar distribution. The sources of these synaptic endings remain largely a matter of speculation, as do the identity of the transmitter(s) for the intrinsic, corticocortical, and corticofugal systems.

Among the questions for neurochemical scrutiny are:

- Is the thalamic transmitter for axons afferent to layers III and IV the same as that for axons ending layers I and VI?
- Do GABAergic synaptic endings have the same ultrastructural target and arrangement throughout the medial geniculate complex?
- How many transmitters participate in the corticocortical and cortico-fugal systems?
- Are the targets of GABAergic terminals comparable in puncta-rich (layers I–IV) and puncta-poor (layers V–VI) layers?
- What light can iontophoretic neurochemical studies shed on the functional arrangement of the auditory thalamus and cortex?

10. Synthesis of Behavioral Studies and Directions for Future Work

Any more refined analysis of the behavioral capabilities and functions of AI must reconcile the disparate attributes of these neurons. On the one hand, they embody tonotopic and aural properties that apparently are invariant and strongly conserved. On the other hand, they (and in all likelihood many of the subcortical neurons that project to them) are capable of an enormous range of normal response lability, dramatic functional readjustment after peripheral trauma, and the capacity to remember and forget biologically significant events. Tonotopic organization could then be construed as a metric defining an axis of representation, much as the borders and local internal features of a map confer some ordinal boundaries upon otherwise randomly organized aggregates. Within the map, however, is an indefinitely large set of subsidiary and discontinuous representations that are less than complete iterations of any dimension and hence are not formally maps or, if they are, have a fundamentally different internal arrangement than do contemporary maps. The rich interconnections between different representations and within them suggests that the pattern of anatomical relations has utility only in defining global relations between or within areas. Any more precise treatment would require a focus on one area and detailed knowledge of connectivity, the basic physiological analysis of single- and multi-unit properties in anesthetized and behaving animals, pharmacological and iontophoretic study of transmitter specificity, behavioral work using conditioning techniques to probe the functional response of neurons, combined with neuropharmacological or reversible inactivation methods that reveal functional deficits with agonists or antagonists to a particular transmitter, or changes ensuing from selectively or reversibly inactivating one afferent pathway or another. This strategy would include careful ablation-behavior experiments where small lesions and specific tasks are used to define the residual capacity in a number of functional contexts.

Among the specific objectives of future studies might be these:

- How is the receptive field organization of single neurons altered in cortical areas when corticocortical afferents are selectively and reversibly inactivated?
- Are there significant differences among the primary auditory fields (AI, AAF, VP, P) with regard to their neural plasticity or behavioral capacities?
- Is there any hierarchical sequence across different auditory cortical fields with regard to signal analysis that has behavioral significance and with implications for the question of serial processing?
- What is the nature of species differences with respect to the cortical distribution of function compared with subcortical allocation of function?
- Which deficits and functions are specific to nonprimary fields, and how are these distributed within and between different architectonic areas?

Acknowledgements Many people contributed generously to the ideas contained in this review. It is a pleasure to acknowledge their thoughtful and constructive criticism, though I take responsibility for the present interpretation and expression of the ideas. I am grateful to Drs. R.A. Code, K.D. Games, B.A. Peterson, J.J. Prieto, and J.J. Wenstrup. Mr. D.T. Larue provided superb technical assistance as well as intellectual support. Ms. J.G. van de Vere, Mr. J. Shin, and Ms. D. Lambert prepared the manuscript conscientiously.

Drs. D.E. Schmechel, E. Mugnaini, and W.H. Oertel kindly donated the antiserum to glutamic acid decarboxylase, and Drs. A.J. Beitz and J.R. Clements the antisera to glutamate and aspartate; many of the immunocytochemical observations reported here depended on such cooperation. Dr. H.J. Karten made helpful suggestions on a preliminary version of the tables.

This research was supported by United States Public Health Service Grant 1 R01 NS16832-13, by a Deafness Research Foundation Grant, and by Biomedical Research Support Grants and Faculty Research Grants from the University of California at Berkeley.

This is dedicated to my mother, my father, and my two sisters.

List of Abbreviations

AAF	anterior auditory field
aeg	anterior ectosylvian gyrus
aes	anterior ectosylvian sulcus
a. pretec.	pretectal area
AID	anterolateral subdivision of the central nucleus of the inferior colliculus

ALLS	anterolateral lateral suprasylvian visual area
Aq	cerebral acqueduct
APt	anterior pretectum
Asp	aspartic acid
AI	primary auditory cortical area
AII	second auditory cortical area
AIII	third auditory cortical area
BI, BIC	brachium of the inferior colliculus
BSC	brachium of the superior colliculus
BV	blood vessel
Ca	caudate nucleus
Cb	cerebellum
CG	central gray substance *or* cingulate gyrus
CGMd	dorsal division of the medial geniculate body
CGMm	medial division of the medial geniculate body
CGMv	ventral division of the medial geniculate body
CGL	lateral geniculate body
CM	centre-médian thalamic nucleus
CP	cerebral peduncle
Cu	cuneiform nucleus of the inferior colliculus
D	dorsal nucleus of the dorsal division of the medial geniculate body *or* dorsal principal part of the medial geniculate body
d	dorsal cap nucleus
DC	caudal pole of the dorsal division *or* dorsal cortex of the inferior colliculus
DD, D_d	deep dorsal nucleus of the dorsal division of the medial geniculate body
Dm	dorsomedial subdivision of the central nucleus of the inferior colliculus
DNLL	dorsal nucleus of the lateral lemniscus
DpD	dorsoposterior subdivision of the central nucleus of the inferior colliculus
DPN	dorsolateral pontine nucleus
DS	superficial dorsal nucleus of the dorsal division of the medial geniculate body
DZ	dorsal auditory cortical zone
EE	excitatory-excitatory binaural cortical band
EI	excitatory-inhibitory binaural cortical band
EP	posterior ectosylvian cortex
EPD	dorsal part of posterior ectosylvian area
EPI	intermediate part of posterior ectosylvian area
EPP	posterior part of posterior ectosylvian area
EPV	ventral part of posterior ectosylvian area
EVA	ectosylvian visual area

GABA	gamma-aminobutyric acid
GAD	glutamic acid decarboxylase
GL, GLd	lateral geniculate nucleus
Glu	glutamate
GM	large-cell division of the medial geniculate body
H_2	field H_2 of Forel
Ha	habenula
HaI	habenulo-interpeduncular tract
HaL	lateral nucleus of the habenula
HaM	medial nucleus of the habenula
HypP	posterior hypothalamus
IBIC	interstitial nucleus of the brachium of the inferior colliculus
IC	inferior colliculus
IcT	intercollicular tegmentum
INLL	intermediate nucleus of the lateral lemniscus
Ins	insular cortical field
Int	intralaminar thalamic nucleus
L	lateral part of the central nucleus of the inferior colliculus
La	lateral nucleus of the inferior colliculus
LAm	lateral amygdaloid nucleus
lat	lateral gyrus
LD	lateral dorsal thalamic nucleus
LGBd, LGB	dorsal nucleus of the lateral geniculate body
LGBv	ventral nucleus of the lateral geniculate body
LL	lateral lemniscus
LMN	lateral mesencephalic nucleus
LN	lateral nucleus of the inferior colliculus
LP	lateral posterior nucleus
LPc	caudal division of the lateral posterior nucleus
LTS	lateral tegmental system of the midbrain
M	medial *or* magnocellular division of the medial geniculate body *or* medial part of the central nucleus of the inferior colliculus
MB	mammillary body
MD	medial subdivision of the central nucleus of the inferior colliculus
meg	middle ectosylvian gyrus
MRF	mesencephalic reticular formation
mss	middle suprasylvian sulcus
MZ	marginal zone of the medial geniculate body
n. com. post.	nuclei of the posterior commissure
n.g.l.d.m.	magnocellular layer of the lateral geniculate body
n.g.l.d.parv.	parvocellular layer of the lateral geniculate body
n.g.l.d.pr.a.	anterior principal layer of the lateral geniculate body
n.g.l.d.pr.p.	posterior principal layer of the lateral geniculate body

n.g.m.m.	magnocellular nucleus of the medial geniculate body
n.g.m.pr.	principal nucleus of the medial geniculate body
n.post.	posterior nucleus
n.tr.ped.tr.	nucleus of the transverse peduncular tract
n. lent. mes.	lentiform mesencephalic nucleus
OR	optic radiation
OT	optic tract
Ov	ovoid nucleus (*pars ovoidea*) of the ventral division of the medial geniculate body
P	posterior auditory field of the ectosylvian gyrus *or* pulvinar nucleus *or* posterior thalamic nucleus
PC	posterior commissure
ped. cer.	cerebral peduncle
peg	posterior ectosylvian gyrus
pes	posterior ectosylvian sulcus
Pf	parafascicular thalamic nucleus
PI	inferior pulvinar nucleus
PL	lateral pulvinar nucleus *or* posterior limitans nucleus
PLS	posterolateral suprasylvian visual area
PM	medial pulvinar nucleus
Po	posterior thalamic nucleus
PoA	anterior pole of the medial geniculate body
POi	intermediate division of the posterior group of thalamic nuclei
Pol	lateral part of the posterior group of thalamic nuclei *or* rostral pole of the medial geniculate body
POm	medial division of the posterior group of thalamic nuclei
PS	posterior suprasylvian visual area
pss	pseudosylvian sulcus
Pt, PT	pretectal nuclei
Pu	putamen
Pul, PUL	pulvinar nucleus
RN	red nucleus
rs	rhinal sulcus
RtN	thalamic reticular nucleus
SC	superior colliculus
SF	suprasylvian fringe cortex
SF/daz	suprasylvian fringe/dorsal auditory zone
Sg, SG	suprageniculate nucleus of the dorsal division of the medial geniculate body
SN	substantia nigra
s.n.lat.	lateral part of the substantia nigra
s.n.ret.	reticulated part of the substantia nigra
s.n.comp.	compact part of the substantia nigra
SOC	superior olivary complex

SPF, Spf	subparafascicular nucleus
SpN	suprapeduncular nucleus
SthN	subthalamic nucleus
Te	temporal cortical field
TO	optic tract
tr.opt.	optic tract
tr.ped.tr.	transverse peduncular tract
VB	ventrobasal complex
V, V_1	ventral nucleus of the ventral division of the medial geniculate body *or* ventral cortical field
V_o	ovoid part of the ventral division
VL	ventral lateral nucleus of the dorsal division of the medial geniculate body *or* ventral lateral part of the central nucleus of the inferior colliculus
Vl	lateral part of the ventral nucleus of the ventral division of the medial geniculate body
VLS	ventral lateral suprasylvian visual area
Vm	medial part of the ventral nucleus of the ventral division of the medial geniculate body
VN	ventral part of the central nucleus of the inferior colliculus
VNLLc	ventral (columnar) part of the ventral nucleus of the lateral lemniscus
VNLLd	dorsal (noncolumnar) part of the ventral nucleus of the lateral lemniscus
VP	ventral posterior part of posterior ectosylvian area
Vpl, VPL	ventroposterolateral nucleus of the ventrobasal complex
Vpm, VPM	ventroposteromedial nucleus of the ventrobasal complex
VV	ventromedial portion of ventral principal part of the medial geniculate body
wm	white matter
ZI	zona incerta
z. i. caud.	caudal part of zona incerta
III EW	Edinger-Westphal nucleus of oculomotor complex
V mes	mesencephalic root of the trigeminal nerve
6m	medial frontal eye field
7p	area 7, posterior division
19	area 19
20a	area 20a
20b	area 20b
21a	area 21a
21b	area 21b
35/36	perirhinal cortex

ORIENTATION OF SECTION: A, anterior; D, dorsal; L, lateral; M, medial; P, posterior; V, ventral
Cortical layers: I, II, III, IV, V, VI

List of Species

Artiodactyla
  Miniature pig (*Sus scrofa*)
Birds
  Budgerigar (*Melopsittacus undulatus*)
  Canary (*Serenus canariensis*)
  Chicken (*Gallus domesticus*)
  Duck (*Anas domestica*)
    (*A. platyrhyncos*)
  Guinea fowl (*Numida meleagris*)
  Gull (*Larus ridibundus*)
  Jay (*Garrulus glandarius*)
  Mynah (*Gracula religiosa intermedia*)
  Pigeon (*Columba livia*)
  Starling (*Sturnus vulgaris*)
  Turtle dove (*Streptopelia decaocto*)
Carnivores
  Cat (*Felis cattus*)
  Dog (*Canis lupus*)
  Ferret (*Mustelis putoris*)
Cetacea
  Dolphin (*Stenella coeruleoalba*)
    (*T. gilli*)
  Dolphin (*Tursiops truncatus*)
    (*Steno attenuata*)
    (*S. bredanensis*)
  Porpoise (*Phocoena phocoena*)
Chiroptera
  Pallid bat (*Antrozous pallidus*)
  Little brown bat (*Myotis lucifugus*)
  Mustached bat (*Pteronotus parnellii*)
    (*P. rubiginosus*)
  Big brown bat (*Eptesicus fuscus*)
Fish
  Carp (*Cyprinus carpio*)
  Catfish (*Ictalurus punctatus*)
  Shark (*Charcharhinus melanopterus*)
    (*Mustelis henlei*)
    (*Negaprion acutidens*)
    (*Triakis semifasciata*)
Insectivores
  Hedgehog (*Hemiechinus aurita*)
  Tree shrew (*Tupaia glis*)
Lagomorph

Rabbit (*Lepus cuniluculus*)
Marsupialia
  Virginia opossum (*Didelphys virginiana*)
  Northern native cat (*Dasyurus hallucatus*)
  Phalanger (*Trichosurus vulpecula*)
Primates
  Cebus monkey (*Cebus apella*)
  Gibbon (*Hylobates lar*)
  Human (*Homo sapiens*)
  Loris (*Nycticebus coucang*)
  Marmoset (*Callithrix jacchus*)
  Owl monkey (*Aotus trivirgatus*)
  Rhesus monkey (*Macaca fascicularis*)
    (*M. fuscata*)
    (*M. mulatta*)
    (*M. nemestrina*)
Reptiles
  Iguana (*Iguana iguana*)
  Turtle (*Emys orbicularis*)
    (*Testudo horsfieldi*)
Rodents
  Guinea-pig (*Cavia porcellus*)
  Mole rat (*Spalax ehrenbergi*)
  Mongolian gerbil (*Meriones unguiculatus*)
  Rat (*Rattus norvegicus*)
  Squirrel (*Sciurus carolinensis*)

References

Abeles M, Goldstein MH Jr (1970) Functional architecture in cat primary auditory cortex: columnar organization and organization according to depth. J Neurophysiol 33:172–187.

Adrián HO, Lifschitz WM, Tavitas RJ, Galli FP (1966) Activity of neural units in medial geniculate body of cat and rabbit. J Neurophysiol 29:1046–1060.

Aitkin LM (1973) Medial geniculate body of the cat: responses to tonal stimuli of neurons in medial division. J Neurophysiol 36:275–283.

Aitkin LM (1986) The Auditory Midbrain. Structure and Function in the Central Auditory Pathways. Clifton, NJ: Humana Press.

Aitkin LM, Prain, SM (1974) Medial geniculate body: unit responses in the awake cat. J Neurophysiol 37:512–521.

Aitkin LM, Webster, WR (1972) Medial geniculate body of the cat: organization and responses to tonal stimuli of neurons in ventral division. J Neurophysiol 35:365–380.

Aitkin LM, Calford MB, Kenyon CE, Webster WR (1981) Some facets of the organization of the principal division of the cat medial geniculate body. In: Syka J, Aitkin LM (eds) Neuronal Mechanisms of Hearing. London: Plenum Publishing Company, pp. 163–181.

Aitkin LM, Dickhaus H, Schult W, Zimmerman M (1978) External nucleus of the inferior colliculus: auditory and spinal somatosensory afferents and their interactions. J Neurophysiol 41:837–847.

Aitkin LM, Irvine DRF, Nelson JE, Merzenich MM, Clarey JC (1986a) Frequency representation in the auditory midbrain and forebrain of a marsupial, the northern native cat (*Dasyurus hallucatus*). Brain Behav Evol 29:17–28.

Aitkin LM, Merzenich MM, Irvine DRF, Clarey JC, Nelson JE (1986b) Frequency representation in auditory cortex of the common marmoset (*Callithrix jacchus jacchus*). J Comp Neurol 252:175–185.

Aitkin LM, Kudo M, Irvine DRF (1988) Connections of the primary auditory cortex in the common marmoset, *Callithrix jacchus jacchus*. J Comp Neurol 269:235–248.

Albert ML, Bear D (1974) Time to understand. A case study of word deafness with reference to the role of time in auditory comprehension. Brain 97:373–384.

Allard TT, Clark SA, Jenkins WM, Merzenich MM (1985) Syndactyly results in the emergence of double-digit receptive fields in somatosensory cortex in adult owl monkeys. Proc Soc Neurosci 11:965.

Allman J, Miezin F, McGuiness E (1985) Stimulus specific responses from beyond the classical receptive field: neurophysiological mechanisms for local-global comparisons in visual neurons. Annu Rev Neurosci 8:407–430.

Allon N, Yeshurun Y, Wollberg Z (1981) Responses of single cells in the medial geniculate body of awake squirrel monkeys. Expl Brain Res 41:222–232.

Altman J, Bayer SA (1979a) Development of the diencephalon in the rat. IV. Quantitative study of the time of origin of neurons and the internuclear chronological gradients in the thalamus. J Comp Neurol 188:455–472.

Altman J, Bayer SA (1979b) Development of the diencephalon in the rat. V. Thymidine-radiographic observations on internuclear and intranuclear gradients in the thalamus. J Comp Neurol 188:473–500.

Altman J, Bayer SA (1981) Time of origin of neurons of the rat inferior colliculus and the relations between cytogenesis and tonotopic order in the auditory pathway. Expl Brain Res 42:411–423.

Altman J, Bayer SA (1988a) Development of the rat thalamus: I. Mosaic organization of the thalamic neuroepithelium. J Comp Neurol 275:346–377.

Altman J, Bayer SA (1988b) Development of the rat thalamus: II. Time and site of origin and settling pattern of neurons derived from the anterior lobule of the thalamic neuroepithelium. J Comp Neurol 275:378–405.

Altman J, Bayer SA (1989a) Development of the rat thalamus: IV. The intermediate lobule of the thalamic neuroepithelium, and the time and site of origin and settling pattern of neurons of the ventral nuclear complex. J Comp Neurol 284:534–566.

Altman J, Bayer SA (1989b) Development of the rat thalamus: V. The posterior lobule of the thalamic neuroepithelium and the time and site of origin and settling patterns of neurons of the medial geniculate body. J Comp Neurol 284:567–580.

Altman JA, Syka J, Shmigidina GN (1970) Neuronal activity in the medial geniculate body of the cat during monaural and binaural stimulation. Expl Brain Res 10:81–93.

Andersen RA, Roth GL, Aitkin LM, Merzenich MM (1980a) The efferent projections of the central nucleus of the inferior colliculus in the cat. J Comp Neurol 194:649–662.

Andersen RA, Knight PL, Merzenich MM (1980b) The thalamocortical and corticothalamic connections of AI, AII, and the anterior auditory field (AAF) in the cat: evidence for two largely segregated systems of connections. J Comp Neurol 194:663–701.

Andrew J, Watkins ES (1969) A Stereotaxic Atlas of the Human Thalamus and Adjacent Structures. A Variability Study. Baltimore, MD: Williams & Wilkins Company.

Andrews RJ, Knight RT, Kirby RP (1990) Evoked potential mapping of auditory and somatic sensory cortices in the miniature swine. Neurosci Lett 114:27–31.

Arnault P, Roger M (1990) Ventral temporal cortex in the rat: connections of secondary auditory areas Te2 and Te3. J Comp Neurol 302:110–123.

Asanuma A, Wong D, Suga N (1983) Frequency and amplitude representations in anterior primary auditory cortex of the mustached bat. J Neurophysiol 50:1182–1196.

Ashe JH, McKenna TM, Weinberger NM (1989) Cholinergic modulation of frequency receptive fields in auditory cortex: II. Frequency-specific effects of anticholinesterases provide evidence for a modulatory action of endogenous ACh. Synapse 4:44–54.

Avedaño C, Llamas A (1984) Thalamic and nonthalamic projections to association areas of the cat's cerebral cortex. In: Reinoso-Suárez F, Ajmone-Marsan C (eds) Cortical Integration. New York: Raven Press, pp. 195–211.

Back SA, Gorenstein C (1990) Fluorescent histochemical localization of neutral endopeptidase-24.11 (enkephalinase) in the rat brain stem. J Comp Neurol 296:130–158.

Bailey P, von Bonin G (1951) The Isocortex of Man. Urbana, IL: University of Illinois Press.

Barbas H (1986) Pattern in the laminar origin of corticocortical connections. J Comp Neurol 252:415–422.

Barth DS, Di D (1990) Three-dimensional analysis of auditory-evoked potentials in rat neocortex. J Neurophysiol 64:1527–1536.

Batzri-Izraeli R, Kelly JB, Glendenning KK, Masterton RB, Wollberg Z (1990) Auditory cortex of the long-eared hedgehog (*Hemiechinus auritus*) I. Boundaries and frequency representation. Brain Behav Evol 36:237–248.

Baughman RW, Gilbert CD (1981) Aspartate and glutamate as possible neurotransmitters in the visual cortex. J Neurosci 1:427–438.

Beaton R, Miller JM (1975) Single cell activity in the auditory cortex of the unanesthetized monkey: correlation with stimulus controlled behavior. Brain Res 100:543–562.

Beaulieu C, Colonnier M (1989) Effects of the richness of the environment on six different cortical areas of the cat cerebral cortex. Brain Res 495:382–386.

Belekhova MG, Zharskaja VD, Khachunts AS, Gaidaenko GV, Tumanova NL (1985) Connections of the mesencephalic, thalamic and telencephalic auditory centers in turtles. Some structural bases for audiosomatic interrelations. J für Hirnforsch 26:127–152.

Békésy von G (1960) Experiments in Hearing. Trans, Wever EG (ed). New York: Robert E. Krieger Publishing Company (1980 reprint of 1960 edition).

Benson DA, Hienz RD (1978) Single-unit activity in the auditory cortex of monkeys selectively attending left vs. right ear stimuli. Brain Res 159:307–320.

Bentivoglio M, Molinari M, Minciacchi D, Macchi G (1983) Organization of the cortical projections of the posterior complex and intralaminar nuclei of the thalamus as studied by means of retrograde tracers. In: Macchi G, Rustioni A, Spreafico R (eds) Somatosensory Integration in the Thalamus. Amsterdam: Elsevier Publishing Company, pp. 337–363.

Berkley KJ (1980) Spatial relationships between the terminations of somatic sensory and motor pathways in the rostral brainstem of cats and monkeys I. Ascending somatic sensory inputs to lateral diencephalon. J Comp Neurol 193:283–317.

Berkowitz A, Suga N (1989) Neural mechanisms of ranging are different in two species of bats. Hear Res 41:255–264.

Bigalke-Kunz B, Rübsamen R, Döorscheidt GJ (1987) Tonotopic organization and functional characterization of the auditory thalamus in a songbird, the European starling. J Comp Physiol A 161:255–265.

Bignall KE (1969) Bilateral temporofrontal projections in the squirrel monkey: origin, distribution and pathways. Brain Res 13:319–327.

Bignall KE (1970) Auditory input to frontal polysensory cortex of the squirrel monkey: possible pathways. Brain Res 19:77–86.

Bignall KE, Singer P (1967) Auditory, somatic and visual input to association and motor cortex of the squirrel monkey. Exp Neurol 18:300–312.

Blum PS, Day MJ, Carpenter MB, Gilman S (1979) Thalamic components of the ascending vestibular system. Exp Neurol 54:587–603.

Bolz J, Gilbert CD (1989) The role of horizontal connections in generating long receptive fields in the cat visual cortex. Eur J Neurosci 1:263–268.

Bolz J, Gilbert CD, Wiesel TN (1989) Pharmacological analysis of cortical circuitry. Trends Neurosci 12:292–296.

Bonke BA, Bonke D, Scheich H (1979) Connectivity of the auditory forebrain nuclei in the guinea fowl (*Numida meleagris*). Cell Tissue Res 200:101–121.

Bonke D, Scheich H, Langner G (1979) Responsiveness of units in the auditory neostriatum of the guinea fowl (*Numida meleagris*) to species-specific calls and synthetic stimuli. I. Tonotopy and functional zones of field L. J Comp Physiol A 132:243–255.

Bonke D, Bonke BA, Langner G, Scheich H (1981) Some aspects of functional organization of the auditory neostriatum in the guinea fowl. In: Syka J, Aitkin L (eds) Neuronal Mechanisms of Hearing. New York: Plenum Press, pp. 323–327.

Borbély AA (1970) Changes in click-evoked responses as a function of depth in auditory cortex of the rat. Brain Res 21:217–247.

Bourk TR, Mielcarz JP, Norris BE (1981) Tonotopic organization of the anteroventral cochlear nucleus of the cat. Hear Res 4:215–241.

Bowman EM, Olson CR (1988a) Visual and auditory association areas of the cat's posterior ectosylvian gyrus: thalamic afferents. J Comp Neurol 272:15–29.

Bowman EM, Olson CR (1988b) Visual and auditory association areas of the cat's posterior ectosylvian gyrus: cortical afferents. J Comp Neurol 272:30–42.

Braak H (1978a) On magnopyramidal temporal fields in the human brain — probable morphological counterparts of Wernicke's sensory speech region. Anat Embryol 152:141–169.

Braak H (1978b) On the pigmentarchitectonics of the human telencephalic cortex. In: Brazier MAB, Petsche H (eds) Architectonics of the Cerebral Cortex. New York: Raven Press, pp. 137–157.

Braak H, Bachmann A (1985) The percentage of projection neurons and inter-neurons in the human lateral geniculate nucleus. Human Neurobiol 4:91–95.

Braak H, Braak E (1984) Neuronal types in the neocortex-dependent lateral territory of the human thalamus. A Golgi-pigment study. Anat Embryol 169:61–72.

Bradford HF (1986) Chemical Neurobiology. An Introduction to Neurochemistry. New York: WH Freeman and Company.

Brandner S, Redies H (1990) The projection of the medial geniculate body to field AI: organization in the isofrequency dimension. J Neurosci 10:50–61.

Brauth SE (1990) Investigation of central auditory nuclei in the budgerigar with cytochrome oxidase. Brain Res 508:142–146.

Brauth SE, McHale CM (1988) Auditory pathways in the budgerigar. II. Intra-telencephalic pathways. Brain Behav Evol 32:193–207.

Brauth SE, McHale CM, Brasher CA, Dooling RJ (1987) Auditory pathways in the budgerigar. I. Thalamo-telencephalic projections. Brain Behav Evol 30:174–199.

Brawer JR, Morest DK (1975) Relations between auditory nerve endings and cell types in the cat's anteroventral cochlear nucleus seen with the Golgi method and Nomarski optics. J Comp Neurol 160:491–506.

Brawer JR, Morest DK, Kane EC (1974) The neuronal architecture of the cochlear nucleus of the cat. J Comp Neurol 155:251–300.

Brodal P (1972) The corticopontine projection in the cat. The projection from the auditory cortex. Arch Ital Biol 110:119–144.

Brodmann K (1909) Vergleichende Lokalisationslehre der Grosshirnrinde in ihren Prinzipien dargestellt auf Grund des Zellenbaues. Leipzig: J.A. Barth.

Bronchti G, Heil P, Scheich H, Wollberg Z (1989) Auditory pathway and auditory activation of primary visual targets in the blind mole rat (Spalax ehrenbergi): I. 2-deoxyglucose study of subcortical centers. J Comp Neurol 284:253–274.

Brugge JF (1975) Mechanisms of coding information in the auditory system. Acta Symbol 6:35–63.

Brugge JF (1982) Auditory cortical areas in primates. In: Woolsey CN (ed) Cortical Sensory Organization, Vol 3, Multiple Auditory Areas. Clifton, NJ: Humana Press, pp. 59–70.

Brugge JF, Dubrovsky NA, Aitkin LM, Anderson DJ (1971) Sensitivity of single neurons in the auditory cortex of cat to binaural tone stimulation; effects of varying interaural time and intensity. J Neurophysiol 32:1005–1024.

Brugge JF, Orman SS, Coleman JR, Chan JKC, Phillips DP (1985) Binaural interactions in cortical area AI of cats reared with unilateral atresia of the external ear canal. Brain Res 20:275–287.

Brysch W, Brysch I, Creutzfeldt OD, Schlingensiepen R, Schlingensiepen K-H (1990) The topology of the thalamo-cortical projections in the marmoset monkey (Callithrix jacchus). Expl Brain Res 81:1–17.

Buchwald JS, Halas ES, Schramm S (1966) Changes in cortical and subcortical unit activity during behavioral conditioning. Physiol Behav 1:11–22.

Bullock TH, Corwin JT (1979) Acoustic evoked activity in the brain in sharks. J Comp Physiol A 129:223–234.

Burger PM, Mehl E, Cameron PL, Maycox PR, Baumert M, Lottspeich F, De Camilli P, Jahn R (1989) Synaptic vesicles immunoisolated from rat cerebral cortex contain high levels of glutamate. Neuron 3:715–720.

Burton H, Jones EG (1976) The posterior thalamic region and its cortical projection in new world and old world monkeys. J Comp Neurol 168:249–302.

Burton H, Mitchell G, Brent D (1982) Second somatic sensory area in the cerebral cortex of cats: somatotopic organization and cytoarchitecture. J Comp Neurol 210:109–135.

Caird D, Scheich H, Klinke R (1991) Functional organization of auditory cortical fields in the Mongolian gerbil (*Meriones unguiculatus*): binaural 2-deoxyglucose patterns. J Comp Physiol A 168:13–26.

Calford MB (1983) The parcellation of the medial geniculate body of the cat defined by the auditory response properties of single units. J Neurosci 3:2350–2364.

Calford MB, Aitkin LM (1983) Ascending projections to the medial geniculate body of the cat: evidence for multiple, parallel auditory pathways through the thalamus. J Neurosci 3:2365–2380.

Calford MB, Webster WR (1981) Auditory representation within principal division of cat medial geniculate body: an electrophysiological study. J Neurophysiol 45:1013–1028.

Campain R, Minckler J (1976) A note on the gross configuration of the human auditory cortex. Brain Lang 3:318–323.

Campbell AW (1905) Histological Studies on the Localisation of Cerebral Function. Cambridge: Cambridge University Press.

Campbell CBG, Boord RL (1974) Central auditory pathways of nonmammalian vertebrates. In: Keidel WD, Neff WD (eds) Handbook of Sensory Physiology, Vol V, Part 1, Auditory System. Anatomy, Physiology (Ear). Berlin: Springer-Verlag, pp. 337–362.

Campbell G, Frost DO (1988) Synaptic organization of anomalous retinal projections to the somatosensory and auditory thalamus: target-controlled morphogenesis of axon terminals and synaptic glomeruli. J Comp Neurol 272:383–408.

Campbell MJ, Lewis DA, Foote SL, Morrison JH (1987) Distribution of choline acetyltransferase-, serotonin-, dopamine-β-hydroxylase-, tyrosine hydroxylase-immunoreactive fibers in monkey primary auditory cortex. J Comp Neurol 261:209–220.

Cant NB (1982) Identification of cell types in the anteroventral cochlear nucleus that project to the inferior colliculus. Neurosci Lett 32:241–246.

Cant NB (1992) Cochlear nuclei—cell types and connectivity. In: Popper AN, Fay RR, Webster DB (eds) Springer Handbook of Auditory Research, Vol. 1, The Mammalian Auditory Pathway: Neuroanatomy. Springer-Verlag, New York and Berlin.

Casseday JH, Diamond IT, Harting JK (1976) Auditory pathways to the cortex in *Tupaia glis*. J Comp Neurol 166:303–340.

Casseday JH, Kobler JB, Isbey SF, Covey E (1989) Central acoustic tract in an echolocating bat: an extralemniscal auditory pathway to the thalamus. J Comp Neurol 287:247–259.

Celesia GG (1976) Organization of auditory cortical areas in man. Brain 99:403–414.

Celesia, GG, Puletti F (1969) Auditory cortical areas of man. Neurol 19:211–220.

Celesia GG, Broughton RJ, Rasmussen T, Branch C (1968) Auditory evoked responses from the exposed human cortex. Electroencephal and Clin Neurophysiol 24:458–466.

Chow KL, Leiman A (eds) (1970) The structural and functional organization of the neocortex. Neurosci Res Prog Summ 5:153–220.

Cipolloni PB, Pandya DN (1985) Topography and trajectory of commissural fibers of the superior temporal region in the rhesus monkey. Expl Brain Res 57:381–389.

Cipolloni PB, Pandya DN (1989) Connectional analysis of the ipsilateral and contralateral afferent neurons of the superior temporal region in the rhesus monkey. J Comp Neurol 281:567–585.

Cipolloni PB, Keller A (1989) Thalamocortical synapses with identified neurons in monkey primary auditory cortex: a combined Golgi/EM and GABA/peptide immunocytochemistry study. Brain Res 492:347–355.

Cipolloni PB, Peters A (1979) The bilaminar and banded distribution of the callosal terminals in the posterior neocortex of the rat. Brain Res 176:33–47.

Cipolloni PB, Peters A (1983) The termination of callosal fibres in the auditory cortex of the rat. A combined Golgi-electron microscope and degeneration study. J Neurocytol 12:713–726.

Clarey JC, Irvine DRF (1986a) Auditory response properties of neurons in the anterior ectosylvian sulcus. Brain Res 386:12–19.

Clarey JC, Irvine DRF (1986b) Auditory response properties of neurons in the claustrum and putamen of the cat. Expl Brain Res 61:432–437.

Clarke S, Innocenti GM (1986) Organization of immature interhemispheric connections. J Comp Neurol 251:1–22.

Clarke S, Innocenti GM (1990) Auditory neurons with transitory axons to visual areas form short permanent connections. Eur J Neurosci 2:227–242.

Clements M, Kelly JB (1978) Directional responses by kittens to an auditory stimulus. Develop Psychobiol 11:505–511.

Clerici WJ, Coleman JR (1990) Anatomy of the rat medial geniculate body: I. Cytoarchitecture, myeloarchitecture and neocortical connectivity. J Comp Neurol 297:14–31.

Clerici WJ, McDonald AJ, Thompson R, Coleman JR (1990) Anatomy of the rat medial geniculate body: II. Dendritic morphology. J Comp Neurol 297:32–54.

Clugnet M-C, LeDoux JE (1990) Synaptic plasticity in fear conditioning circuits: induction of LTP in the lateral nucleus of the amygdala by stimulation of the medial geniculate body. J Neurosci 10:2818–2824.

Clugnet M-C, LeDoux JE, Morrison SF (1990) Unit responses evoked in the amygdala and striatum by electrical stimulation of the medial geniculate body. J Neurosci 10:1055–1061.

Code RA, Winer JA (1985) Commissural neurons in layer III of cat primary auditory cortex (AI): pyramidal and non-pyramidal cell input. J Comp Neurol 242:485–510.

Code RA, Winer JA (1986) Columnar organization and reciprocity of commissural connections in cat primary auditory cortex (AI). Hear Res 23:205–222.

Colavita FB, Szeligo FV, Zimmer SD (1974) Temporal pattern discrimination in cats with insular-temporal lesions. Brain Res 79:153–156.

Colombo M, D'Amato MR, Rodman HR, Gross CG (1990) Auditory association cortex lesions impair auditory short-term memory. Science 247:336–338.

Colwell S (1975) Thalamocortical-corticothalamic reciprocity: a combined anterograde-retrograde tracer technique. Brain Res 92:443–449.

Conel JL (1959) The Postnatal Development of the Human Cerebral Cortex, Vol VI, The Cortex of the Twenty-Four-Month Infant. Cambridge: Harvard University Press.

Conley M, Kupersmith AC, Diamond IT (1991) Organization of the auditory thalamic sector of the thalamic reticular nucleus in Galago: comparison of the reciprocal connexions between the reticular nucleus and the ventral and magnocellular divisions of the medial geniculate complex. Eur J Neurosci 3:1089–1103.

Cooper ERA (1948) The development of the human auditory pathway from the cochlear ganglion to the medial geniculate body. Acta Anat 57:99–122.

Cooper JR, Bloom FE, Roth RH (1982) The Biochemical Basis of Neuropharmacology, fourth ed. New York: Oxford University Press.

Cornwell P, Ravizza R, Payne B (1984) Extrinsic visual and auditory cortical connections in the 4-day-old kitten. J Comp Neurol 229:97–120.

Cotter JR, Laemle LK (1990) Cholecystokinin (CCK)-like immunoreactivity in the brain of the little brown bat (*Myotis Lucifugus*) (sic). J für Hirnforsch 31:87–97.

Covenas R, Romo R, Cheramy A, Cesselin F, Conrath M (1986) Immunocytochemical study of enkephalin-like cell bodies in the thalamus of the cat. Brain Res 377:355–361.

Crabtree JW, Killackey HP (1989) The topographic organization and axis of projection within the visual sector of the rabbit's thalamic reticular nucleus. Eur J Neurosci 1:94–109.

Creutzfeldt O, Hellweg F-C, Schreiner Chr (1980) Thalamocortical transformation of responses to complex auditory stimuli. Expl Brain Res 39:87–104.

Crunelli V, Leresche N (1991) A role for GABA$_B$ receptors in excitation and inhibition of thalamocortical cells. Trends Neurosci 14:16–21.

Cusick CG, MacAvoy MG, Kaas JH (1985) Interhemispheric connections of cortical sensory areas in tree shrews. J Comp Neurol 235:111–128.

Dahlström A, Fuxe K (1964) Evidence for the existence of monoamine containing neurons in the central nervous system. I. Demonstration of monoamines in the cell bodies of brain stem neurons. Acta Physiol Scand 62 Supplement 232:1–55.

D'Amato RJ, Blue ME, Largent BL, Lynch DR, Ledbetter DJ, Molliver ME, Snyder SH (1987) Ontogeny of the serotonergic projection to rat neocortex: transient expression of a dense innervation to primary sensory areas. Proc Natl Acad Sci (USA) 84:4322–4326.

Davis TL, Sterling P (1979) Microcircuitry of cat visual cortex: classification of neurons in layer IV of area 17, and identification of the patterns of lateral geniculate input. J Comp Neurol 188:599–628.

Deacon TW, Eichenbaum H, Rosenberg P, Eckmann KW (1983) Afferent connections of the perirhinal cortex in the rat. J Comp Neurol 220:168–190.

De Carlos JA, Lopez-Mascaraque L, Ramón y Cajal-Agüeras S, Valverde F (1987) Chandelier cells in the auditory cortex of monkey and man: a Golgi study. Expl Brain Res 66:295–302.

de Courten C, Garey LJ (1982) Morphology of the neurons in the human lateral geniculate nucleus and their normal development. A Golgi study. Expl Brain Res 47:159–171.

DeFelipe J, Conley M, Jones EG (1986) Long-range focal collaterlization of axons arising from corticocortical cells in monkey sensory-motor cortex. J Neurosci 6:3749–3766.

Dehay C, Kennedy H, Bullier J (1988) Characterization of transient cortical projections from auditory, somatosensory, and motor cortices to visual areas 17, 18, and 19 in the kitten. J Comp Neurol 272:68–89.

Demeter S, Rosene DL, Van Hoesen GW (1990) Fields of origin and pathways of interhemispheric commissures in the temporal lobe of macaques. J Comp Neurol 302:29–53.

Deschênes M, Hu B (1990) Electrophysiology and pharmacology of corticothalamic input to lateral thalamic nuclei: an intracellular study in the cat. Eur J Neurosci 2:140–152.

Dewson JH III, Pribram KH, Lynch JC (1969) Effects of ablations of temporal cortex upon speech sound discrimination in the monkey. Expl Neurol 24:579–591.

Dewulf A (1971) Anatomy of the Normal Human Thalamus. Topometry and Standardized Nomenclature. Amsterdam: Elsevier Publishing Company.

DeYoe EA, Van Essen DC (1988) Concurrent processing streams in monkey visual cortex. Trends Neurosci 11:219–226.

Diamond DM, Weinberger NM (1984) Physiological plasticity of single neurons in auditory cortex of the cat during acquisition of the pupillary conditioned response: II. Secondary field (AII). Behav Neurosci 98:189–210.

Diamond DM, Weinberger NM (1986) Classical conditioning rapidly induces specific changes in frequency receptive fields of single neurons in secondary and ventral ectosylvian auditory cortical fields. Brain Res 372:357–360.

Diamond DM, Weinberger NM (1989) Role of context in the expression of learning-induced plasticity of single neurons in auditory cortex. Behav Neurosci 103:471–494.

Diamond IT (1983) Parallel pathways in the auditory, visual and somatic systems. In: Macchi G, Rustioni A, Spreafico R (eds) Somatosensory Integration in the Thalamus. Amsterdam: Elsevier Publishing Company, pp. 251–272.

Diamond IT, Utley JD (1963) Thalamic retrograde degeneration study of sensory cortex in opossum. J Comp Neurol 120:129–160.

Diamond IT, Jones EG, Powell TPS (1968) Interhemispheric fiber connections of the auditory cortex of the cat. Brain Res 11:177–193.

Diamond IT, Jones EG, Powell TPS (1969) The projection of the auditory cortex upon the diencephalon and brain stem of the cat. Brain Res 15:305–340.

Dickson JW, Gerstein GL (1974) Interactions between neurons in auditory cortex of the cat. J Neurophysiol 37:1239–1261.

Disterhoft JF, Olds J (1972) Differential development of conditioned unit changes in thalamus and cortex of rat. J Neurophysiol 35:665–679.

Disterhoft JF, Stuart DK (1976) Trial sequence of changed unit activity in auditory system of alert rat during conditioned response acquisition and extinction. J Neurophysiol 39:266–281.

Domenici L, Waldvogel HJ, Matute C, Streit P (1988) Distribution of GABA-like immunoreactivity in the pigeon brain. Neurosci 25:931–950.

Downman CBB, Woolsey CN, Lende RA (1960) Auditory areas I, II and Ep: cochlear representation, afferent paths and interconnections. Bull Johns Hopkins Hosp 106:127–142.

Dreher B, Cottee LJ (1975) Visual receptive-field properties of cells in area 18 of cat's cerebral cortex before and after acute lesions in area 17. J Neurophysiol 38:735–750.

Dreher B, Leventhal AG, Hale PT (1980) Geniculate input to cat visual cortex: a comparison of area 19 with areas 17 and 18. J Neurophysiol 44:804–826.

Dykes RW, Herron P, Lin C-S (1986) Ventroposterior thalamic regions projecting to cytoarchitectonic areas 3a and 3b in the cat. J Neurophysiol 56:1521–1541.

Dykes RW, Rasmussen DD, Hoeltzell PB (1980) Organization of primary somatosensory cortex in the cat. J Neurophysiol 43:1527–1546.

Ebner FF (1969) A comparison of primitive forebrain organization in metatherian and eutherian mammals. Ann N Y Acad Sci 167:241–257.

Eccles JC (1966) Cerebral synaptic mechanisms. In: Eccles JC (ed) Brain and Conscious Experience. New York: Springer-Verlag, pp. 24–58.

Echteler SM (1984) Connections of the auditory midbrain in a teleost fish, *Cyprinus carpio*. J Comp Neurol 230:536–551.

Echteler SM (1985) Organization of central auditory pathways in a teleost fish, *Cyprinus carpio*. J Comp Physiol A 156:267–280.

Eckenstein FP, Baughman RW, Quinn J (1988) An anatomical study of cholinergic innervation in rat cerebral cortex. Neurosci 25:457–474.

Edeline J-M (1990) Frequency-specific plasticity of single unit discharges in the rat medial geniculate body. Brain Res 529:109–119.

Edeline J-M, Dutrieux G, Neuenschwander-El Massioui N (1988) Multiunit changes in hippocampus and medial geniculate body in free-behaving rats during acquisition and retention of a conditioned response to tone. Behav Neural Biol 50:61–79.

Edeline J-M, Neuenschwander-El Massioui N, Dutrieux G (1990) Discriminative long-term retention of rapidly induced multiunit changes in the hippocampus, medial geniculate and auditory cortex. Behav Brain Res 39:145–155.

Eidelberg D, Galaburda AM (1982) Symmetry and asymmetry in the human posterior thalamus. I. Cytoarchitectonic analysis in normal persons. Arch Neurol 39:325–332.

Elberling C, Bak C, Kofoed B, Lebech J, Saermark K (1982) Auditory magnetic fields. Source location and 'tonotopical organization' in the right hemisphere. Scand Audiol 11:61–65.

Evans EF, Whitfield IC (1964) Classification of unit responses in the auditory cortex of unanaesthetized and unrestrained cat. J Physiol (London) 171:476–493.

Evans EF, Ross HF, Whitfield IC (1965) The spatial distribution of unit characteristic frequency in the primary auditory cortex of the cat. J Physiol (London) 179:238–247.

Fabri M, Burton H (1991) Topography of connections between primary somatosensory cortex and posterior complex in rat: a multiple fluorescent tracer study. Brain Res 538:351–357.

Fallon JH, Leslie FM (1986) Distribution of dynorphin and enkephalin peptides in the rat brain. J Comp Neurol 249:293–336.

Feng JZ, Brugge JF (1983) Postnatal development of auditory callosal connections in the kitten. J Comp Neurol 214:416–426.

Ferrer I, Perera M (1988) Structure and nerve cell organization in the cerebral cortex of the dolphin *Stenella coeruleoalba* a Golgi study. With special attention to the primary auditory area. Anat Embryol 178:161–173.

Ferster D, LeVay S (1978) The axonal arborizations of lateral geniculate neurons in the striate cortex of the cat. J Comp Neurol 182:923–944.

Fisken RA, Garey LJ, Powell TPS (1975) The intrinsic, association and commissural connections of area 17 of the visual cortex. Phil Trans R Soc London, Ser B 272:487–536.

Fitzpatrick D, Lund JS, Schmechel DE, Towles AC (1987) Distribution of GABAergic neurons and axon terminals in the macaque striate cortex. J Comp Neurol 264:73–91.

Fitzpatrick D, Diamond IT, Raczkowski D (1989) Cholinergic and monoaminergic innervation of the cat's thalamus: comparison of the lateral geniculate nucleus with other principal sensory nuclei. J Comp Neurol 288:647–675.

FitzPatrick KA, Imig TJ (1980) Auditory cortico-cortical connections in the owl monkey. J Comp Neurol 192:589–610.

Fonnum F (1984) Glutamate: a neurotransmitter in mammalian brain. J Neurochem 42:1–11.

Fonnum F, Storm-Mathisen J, Divac I (1981) Biochemical evidence for glutamate as neurotransmitter in corticostriatal and corticothalamic fibres in rat brain. Neuroscience 6:863–873.

Foote SL, Freedman R, Oliver AP (1975) Effects of putative transmitters on neuronal activity in monkey auditory cortex. Brain Res 86:229–242.

Forbes A, Sherrington CS (1914) Acoustic reflexes in the decerebrate cat. Am J Physiol 35:367–376.

Forbes BF, Moskowitz N (1977) Cortico-cortical connections of the superior temporal gyrus in the squirrel monkey. Brain Res 136:547–552.

Foster RE, Hall WC (1978) The organization of central auditory pathways in a reptile, *Iguana iguana*. J Comp Neurol 178:783–832.

Friedman DP, Murray EA, O'Neill JB, Mishkin M (1986) Cortical connections of the somatosensory fields of the lateral sulcus of macaques: evidence for a corticolimbic pathway for touch. J Comp Neurol 252:323–347.

Frost DO (1981) Orderly anomalous retinal projections to the medial geniculate, ventrobasal, and lateral posterior nuclei of the hamster. J Comp Neurol 203:227–256.

Frost DO (1986) Development of anomalous retinal projections to nonvisual thalamic nuclei in Syrian hamsters: a quantitative study. J Comp Neurol 252:95–105.

Frost DO, Metin C (1985) Induction of functional retinal projections to the somatosensory system. Nature 317:162–164.

Frostig RD, Gottlieb Y, Vaadia E, Abeles M (1983) The effects of stimuli on the activity and functional connectivity of local neuronal groups in the cat auditory cortex. Brain Res 272:211–221.

Fullerton BC (1978) Morphological studies of the inferior colliculus and the medial geniculate body in the rhesus monkey and the albino rat. Doctoral dissertation, Boston University, Boston, pp. 1–192.

Funkenstein HH, Winter P (1973) Responses to acoustic stimuli of units in the auditory cortex of awake squirrel monkeys. Expl Brain Res 18:464–488.

Fuxe K (1965) The distribution of monoamine terminals in the central nervous system. Acta Physiol Scand 64:37–78.

Fuzessery ZM, Feng AS (1983) Mating call selectivity in the thalamus and midbrain of the leopard frog (*Rana p. pipiens*): single and multiunit analysis. J Comp Physiol A 150:333–344.

Gabriel M, Miller JD, Saltwick SE (1976) Multiple-unit activity of the rabbit medial geniculate nucleus in conditioning, extinction, and reversal. Physiol Psychol 4:124–134.

Galaburda AM (1986) Role of the thalamus in auditory lateralization–histologic data. Rev Neurol 142:441–444.

Galaburda AM, Pandya DN (1983) The intrinsic architectonic and connectional organization of the superior temporal region of the rhesus monkey. J Comp Neurol 221:169–184.

Galaburda A, Sanides F (1980) Cytoarchitectonic organization of the human auditory cortex. J Comp Neurol 190:597–610.

Galambos R, Myers RE, Sheatz GC (1961) Extralemniscal activation of auditory cortex in cats. Am J Physiol 200:23–28.

Galazyuk AV, Volkov IO (1988) Tonotopic organization of dorsocaudal zone of AII cortical area in the cat. Neurophysiol 20:220–227 (original in Russian, abstract and figure legends in English).

Galli F, Lifschitz W, Adrian H (1971) Studies on the auditory cortex of rabbit. Expl Neurol 30:324–335.

Games KD, Winer JA (1988) Layer V in rat auditory cortex: projections to the inferior colliculus and contralateral cortex. Hear Res 34:1–26.

Gates GR, Aitkin LM (1982) Auditory cortex in the marsupial possum Trichosurus vulpecula. Hear Res 7:1–11.

Gerren RA, Weinberger NM (1983) Long term potentiation in the magnocellular medial geniculate nucleus of the anesthetized cat. Brain Res 265:138–142.

Geschwind N, Levitsky W (1968) Human brain: left-right asymmetries in temporal speech region. Science 161:186–187.

Gilbert CD, Wiesel TN (1983) Clustered intrinsic connections in cat visual cortex. J Neurosci 3:1116–1133.

Girard P, Bullier J (1989) Visual activity in area V2 during reversible inactivation of area 17 in the macaque monkey. J Neurophysiol 62:1287–1302.

Giuffrida R, Rustioni A (1989) Glutamate and aspartate immunoreactivity in corticospinal neurons of rats. J Comp Neurol 288:154–164.

Glassman RB, Forgus MW, Goodman JE, Glassman HN (1975) Somesthetic effects of damage to cats' ventrobasal complex, medial lemniscus or posterior group. Expl Neurol 48:460–492.

Gloor P (1990) Experiential phenomena of temporal lobe epilepsy. Facts and hypotheses. Brain 113:1673–1694.

Goldstein MH Jr, Hall JL, Butterfield BO (1968) Single-unit activity in the primary auditory cortex of unanesthetized cats. J Acoust Soc Am 43:444–455.

Gottlieb Y, Vaadia E, Abeles M (1989) Single unit activity in the auditory cortex of a monkey performing a short term memory task. Expl Brain Res 74:139–148.

Graham J (1977) An autoradiographic study of the efferent connections of the superior colliculus in the cat. J Comp Neurol 173:629–654.

Granda RH, Crossland WJ (1989) GABA-like immunoreactivity of neurons in the chicken diencephalon and mesencephalon. J Comp Neurol 287:455–469.

Graybiel AM (1972a) Some ascending connections of the pulvinar and nucleus lateralis posterior of the thalamus in the cat. Brain Res 44:99–125.

Graybiel AM (1972b) Some fiber pathways related to the posterior thalamic region in the cat. Brain Beh Evol 6:363–393.

Gross CG, Weiskrantz L (1962) Evidence for dissociation of impairment on auditory discrimination and delayed response following lateral frontal lesions in monkeys. Expl Neurol 5:453–476.

Gross NB, Lifschitz WS, Anderson DJ (1974) The tonotopic organization of the auditory thalamus of the squirrel monkey (Saimiri sciureus). Brain Res 65:323–332.

Haight JR, Neylon L (1978) An atlas of the dorsal thalamus of the marsupial brush-tailed possum, Trichosurus vulpecula. J Anat (London) 126:225–245.

Haight JR, Neylon L (1981) A description of the dorsal thalamus of the marsupial native cat, Dasyurus viverrinus (Dasyuridae). Brain Behav Evol 19:155–179.

Hall JC, Feng AS (1987) Evidence for parallel processing in the frog's auditory thalamus. J Comp Neurol 258:407–419.

Hall JL II, Goldstein MH Jr (1968) Representation of binaural stimuli by single units in primary auditory cortex of unanesthetized cats. J Acoust Soc Am 43:456–461.

Hari R, Hämäläinen M, Kaukoranta E, Mäkelä J, Joutsiniemi SL, Tiihonen J (1989a) Selective listening modifies activity of the human auditory cortex. Expl Brain Res 74:463–470.

Hari R, Joutsiniemi SL, Hämäläinen M, Vilkman V (1989b) Neuromagnetic responses of human auditory cortex to interruptions in a steady rhythm. Neurosci Lett 99:164–168.

Hashikawa T (1983) The inferior colliculopontine neurons of the cat in relation to other collicular descending neurons. J Comp Neurol 219:241–249.

Hashikawa T, Kawamura K (1983) Retrograde labeling of ascending and descending neurons in the inferior colliculus. A fluorescent double labeling study in the cat. Expl Brain Res 49:457–461.

Hashimoto T (1980) Information processing of speech sounds in the medial geniculate and the inferior colliculus. Proc Jpn Acad Sci 56:294–299.

Heffner HE (1987) Ferrier and the study of auditory cortex. Arch Neurol 44:218–221.

Heffner HE, Heffner RS (1986a) Hearing loss in Japanese macaques following bilateral auditory cortex lesions. J Neurophysiol 55:256–271.

Heffner HE, Heffner RS (1986b) Effect of unilateral and bilateral auditory cortex lesions on the discrimination of vocalizations by Japanese macaques. J Neurophysiol 56:683–701.

Heffner HE, Heffner RS (1989) Effect of restricted cortical lesions on absolute threshold and aphasia-like behavior in Japanese macaques. Behav Neurosci 103:156–169.

Heffner HE, Masterton B (1975) Contribution of auditory cortex to sound localization in the monkey (Macaca mulatta). J Neurophysiol 38:1340–1358.

Heffner RS, Heffner HE (1984) Hearing loss in dogs after lesions of the brachium of the inferior colliculus and medial geniculate. J Comp Neurol 230:207–217.

Heil P, Scheich H (1991a) Functional organization of the avian auditory cortex analogue. I. Tonotopic organization of isointensity bandwidth. Brain Res 539:110–120.

Heil P, Scheich H (1991b) Functional organization of the avian auditory cortex analogue. II. Topographic distribution of latency. Brain Res 539:121–125.

Hellweg FC, Koch R, Vollrath M (1977) Representation of the cochlea in the neocortex of guinea pigs. Expl Brain Res 29:467–474.

Hendry SHC, Jones EG (1991) GABA neuronal subpopulations in cat primary auditory cortex – co-localization with calcium binding proteins. Brain Res 543:45–55.

Hendry SHC, Jones EG, DeFelipe J, Schmechel D, Brandon C, Emson PC (1984) Neuropeptide-containing neurons of the cerebral cortex are also GABAergic. Proc Natl Acad Sci (USA) 81:6526–6530.

Hendry SHC, Schwark HD, Jones EG, Yan J (1987) Numbers and proportions of GABA-immunoreactive neurons in different areas of monkey cerebral cortex. J Neurosci 7:1503–1519.

Henkel CK (1983) Evidence of sub-collicular auditory projections to the medial geniculate nucleus in the cat: an autoradiographic and horseradish peroxidase study. Brain Res 259:21–30.

Herbert H, Aschoff A, Ostwald J (1991) Topography of projections from the auditory cortex to the inferior colliculus in the rat. J Comp Neurol 304:103–122.

Hikosaka O, Sakamoto M, Usui S (1989) Functional properties of monkey caudate neurons II. Visual and auditory responses. J Neurophysiol 61:799–813.

Hirai T, Jones EG (1989a) A new parcellation of the human thalamus on the basis of histochemical staining. Brain Res Rev 14:1–34.

Hirai T, Jones EG (1989b) Distribution of tachykinin- and enkephalin-immunoreactive fibers in the human thalamus. Brain Res Rev 14:35–52.

Hocherman S, Gilat E (1981) Dependence of auditory cortex evoked unit activity on interstimulus interval in the cat. J Neurophysiol 45:987–997.

Hocherman S, Benson DA, Goldstein MH Jr, Heffner HE, Hienz RD (1976) Evoked unit activity in auditory cortex of monkeys performing a selective attention task. Brain Res 117:51–68.

Holstege G, Collewijn H (1982) The efferent projections of the nucleus of the optic tract and the superior colliculus in the rabbit. J Comp Neurol 209:139–175.

Horikawa J, Ito S, Hosokawa Y, Homma T, Murata K (1988) Tonotopic representation in the rat auditory cortex. Proc Jpn Acad Sci B 64:260–263.

Horner K, de Ribaupierre Y, de Ribaupierre F (1983) Neural correlates of cubic difference tones in the medial geniculate body of the cat. Hear Res 11:343–357.

Hornung J-P, Törk I, De Tribolet N (1989) Morphology of tyrosine hydroxylase-immunoreactive neurons in the human cerebral cortex. Expl Brain Res 76:12–20.

Hörster W, Ettlinger G (1987) Unilateral removal of the posterior insula or of area SII: inconsistent effects on tactile, visual and auditory performance in the monkey. Behav Brain Res 26:1–17.

Hose B, Langner G, Scheich H (1987) Topographic representation of periodicities in the forebrain of the mynah bird: one map for pitch and rhythm? Brain Res 422:367–373.

Hough HB, Wolff HG (1939) The relative vascularity of subcortical ganglia of cat's brain; the putamen, globus pallidus, substantia nigra, red nucleus, and geniculate bodies. J Comp Neurol 71:427–436.

Houser CR, Vaughn JE, Barber RP, Roberts E (1980) GABA neurons are the major cell type of the nucleus reticularis thalami. Brain Res 200:345–354.

Hu H, Jayarman A (1986) The projection pattern of the suprageniculate nucleus to the caudate nucleus in cats. Brain Res 368:201–203.

Huang C-M, Liu G (1985) Electrophysiological mapping of the auditory areas in the cerebellum of the cat. Brain Res 335:121–129.

Hubel DH, Wiesel TN (1959) Receptive fields of single neurones in the cat's striate cortex. J Physiol (London) 148:574–591.

Hubel DH, Wiesel TN (1962) Receptive fields, binocular interaction and functional architecture in the cat's visual cortex. J Physiol (London) 160:106–154.

Hubel DH, Wiesel TN (1965) Receptive fields and functional architecture in two non-striate visual areas (18 and 19) of the cat. J Neurophysiol 28:229–289.

Hubel DH, Wiesel TN (1970) The period of susceptibility to the physiological effects of unilateral eye closure in kittens. J Physiol (London) 206:419–436.

Hubel DH, Wiesel TN, LeVay S (1977) Plasticity of ocular dominance columns in monkey striate cortex. Phil Trans R Soc London, Ser B 278:377–409.

Huchton DM, Larue DT, Sun JY-M, Winer JA (1991) The organization of GA-BAergic neurons in the cat medial geniculate body: a quantitative immunocytochemical study of post-embedded material. Proc Soc Neurosci 17:300.

Huffman RF, Henson OW Jr (1990) The descending auditory pathway and acousticomotor systems: connections with the inferior colliculus. Brain Res Rev 15:295–323.

Humphrey AL, Sur M, Uhlrich DJ, Sherman SM (1985) Termination patterns of individual X- and Y-cell axons in the visual cortex of the cat: projections to area 18, to the 17/18 border region, and to both areas 17 and 18. J Comp Neurol 233:190–212.

Hurd LB II, Eldred WD (1989) Localization of GABA- and GAD-like immunoreactivity in the turtle retina. Vis Neurosci 3:9–20.

Hutson KA (1988) Connections of the auditory midbrain: efferent projections of the dorsal nucleus of the lateral lemniscus, the nucleus sagulum, and the origins of the GABAergic commissure of Probst. Doctoral dissertation, The Florida State University College of Letters and Sciences, Tallahassee, pp. 1–165.

Imig TJ, Adrián HO (1977) Binaural columns in the primary auditory field (A1) of cat auditory cortex. Brain Res 138:241–257.

Imig TJ, Brugge JF (1978) Sources and terminations of callosal axons related to binaural and frequency maps in primary auditory cortex of the cat. J Comp Neurol 182:637–660.

Imig TJ, Morel A (1983) Organization of the thalamocortical auditory system in the cat. Ann Rev Neurosci 6:95–120.

Imig TJ, Morel A (1984) Topographic and cytoarchitectonic organization of thalamic neurons related to their targets in low-, middle-, and high-frequency representations in cat auditory cortex. J Comp Neurol 227:511–539.

Imig TJ, Morel A (1985a) Tonotopic organization in lateral part of posterior group of thalamic nuclei in the cat. J Neurophysiol 53:836–851.

Imig TJ, Morel A (1985b) Tonotopic organization in ventral nucleus of medial geniculate body in the cat. J Neurophysiol 53:309–340.

Imig TJ, Reale RA (1980) Patterns of cortico-cortical connections related to tonotopic maps in cat auditory cortex. J Comp Neurol 192:293–332.

Imig TJ, Reale RA (1981) Ipsilateral corticocortical projections related to binaural columns in cat primary auditory cortex. J Comp Neurol 203:1–14.

Imig TJ, Ruggero MA, Kitzes LM, Javel E, Brugge JF (1977) Organization of auditory cortex in the owl monkey (*Aotus trivirgatus*). J Comp Neurol 171:111–128.

Imig TJ, Morel A, Kauer CD (1982) Covariation of distribution of callosal cell bodies and callosal axon terminals in layer III of cat primary auditory cortex. Brain Res 251:157–159.

Imig TJ, Irons WA, Samson FR (1990) Single-unit selectivity to azimuthal direction and sound pressure level of noise bursts in cat high-frequency primary auditory cortex. J Neurophysiol 63:1448–1466.

Imig TJ, Clarey JC, Barone P (1992) Physiology of the thalamus and cortex. In: Popper AN, Fay RR, Webster DB (eds) Springer Series in Auditory Research, Vol 2, The Physiology of the Mammalian Auditory Central Nervous System. Springer-Verlag, New York and Berlin (in press).

Innocenti GM, Clarke S (1984) Bilateral transitory projections to visual areas from auditory cortex in kittens. Develop Brain Res 14:143–148.

Irvine DRF (1986) The Auditory Brainstem. A Review of the Structure and Function of Auditory Brainstem Processing Mechanisms. In: Ottoson D (ed) Progress in Sensory Physiology 7:1–279.

Irvine DRF, Huebner H (1979) Acoustic response characteristics of neurons in nonspecific areas of cat cerebral cortex. J Neurophysiol 42:107–122.

Ivarsson C, de Ribaupierre Y, de Ribaupierre F (1988) Influence of auditory localization cues on neuronal activity in the auditory thalamus of the cat. J Neurophysiol 59:586–606.

Jacobowitz DM, Winsky L (1991) Immunocytochemical localization of calretinin in the forebrain of the rat. J Comp Neurol 304:198–218.

Jansen KRL, Faull RLM, Dragunow M (1989) Excitatory amino acid receptors in the human cerebral cortex: a quantitative autoradiographic study comparing the distributions of [³H]TCP, [³H]glycine, L-[³H]glutamate, [³H]AMPA and [³H]kainic acid binding sites. Neuroscience 32:587–607.

Jarrell TW, Gentile CG, Romanski LM, McCabe PM, Schneiderman N (1987) Involvement of cortical and thalamic auditory regions in retention of differential bradycardiac conditioning to acoustic conditioned stimuli in rabbits. Brain Res 412:285–294.

Jen PH-S, Sun X, Lin PJJ (1989) Frequency and space representation in the primary auditory cortex of the frequency modulating bat Eptesicus fuscus. J Comp Physiol A 165:1–14.

Jenkins WM, Masterton RB (1982) Sound localization: effects of unilateral lesions on central auditory system. J Neurophysiol 47:987–1016.

Jenkins WM, Merzenich MM (1984) Role of cat primary auditory cortex for sound-localization behavior. J Neurophysiol 52:819–847.

Johansson O, Hökfelt T, Elde RP (1984) Immunohistochemical distribution of somatostatin-like immunoreactivity in the central nervous system of the adult rat. Neuroscience 13:265–339.

Jones EG (1985) The Thalamus. New York: Plenum Press.

Jones EG, Burton H (1974) Cytoarchitecture and somatic sensory connectivity of thalamic nuclei other than the ventrobasal complex in the cat. J Comp Neurol 154:395–432.

Jones EG, Burton H (1976) Areal differences in the laminar distribution of thalamic afferents in cortical fields of the insular, parietal, and temporal regions of primates. J Comp Neurol 168:197–248.

Jones EG, Hendry SHC (1989) Differential calcium binding protein immunoreactivity distinguishes classes of relay neurons in monkey thalamic nuclei. Eur J Neurosci 1:222–246.

Jones EG, Powell TPS (1969) An electron microscopic study of the mode of termination of cortico-thalamic fibres within the sensory relay nuclei of the thalamus. Proc R Soc London, Ser B 172:173–185.

Jones EG, Powell TPS (1970) An anatomical study of converging sensory pathways within the cerebral cortex of the monkey. Brain 93:793–820.

Jones EG, Powell TPS (1971) An analysis of the posterior group of thalamic nuclei on the basis of its afferent connections. J Comp Neurol 143:185–216.

Jones EG, Powell TPS (1973) Anatomical organization of the somatosensory cortex. In: Iggo A (ed) Handbook of Sensory Physiology, Vol 2, Somatosensory System. Berlin: Springer-Verlag, pp. 579–620.

Jones EG, Rockel AJ (1971) The synaptic organization in the medial geniculate body of afferent fibres ascending from the inferior colliculus. Zeit für Zellforsch und Mikroskop Anat 113:44–66.

Jordan H (1973) The structure of the medial geniculate nucleus (MGN): a cyto- and myeloarchitectonic study in the squirrel monkey. J Comp Neurol 148:469–480.

Jouandet ML, Garey LJ, Lipp J-P (1984) Distribution of cells of origin of the corpus callosum and anterior commissure in the marmoset monkey. Anat Embryol 169:45–59.

Kaas JH, Sur M, Nelson RJ, Merzenich MM (1981) The postcentral somatosensory cortex. Multiple representations of the body in primates. In: Woolsey CN (ed) Cortical Sensory Organization, Vol 1, Multiple Somatic Areas. Clifton, NJ: Humana Press, pp. 29–45.

Kamiya H, Itoh K, Yasui Y, Ino T, Mizuno N (1988) Somatosensory and auditory relay nucleus in the rostral part of the ventrolateral medulla: a morphological study in the cat. J Comp Neurol 273:421–435.

Karten HJ (1967) The organization of the ascending auditory pathway in the pigeon (Columba livia) I. Diencephalic projections of the inferior colliculus (nucleus mesencephalicus lateralis, pars dorsalis). Brain Res 6:409–427.

Karten HJ (1968) The ascending auditory pathway in the pigeon (Columba livia) II. Telencephalic projections of the nucleus ovoidalis thalami. Brain Res 11:134–153.

Kaukoranta E, Sams M, Hari R, Hämäläinen M, Näätänen R (1989) Reactions of human auditory cortex to a change in tone duration. Hear Res 41:15–22.

Kelley DB, Nottebohm F (1979) Projections of a telencephalic auditory nucleus-field L-in the canary. J Comp Neurol 183:455–469.

Kelly JB, Judge PW (1985) Effects of medial geniculate lesions on sound localization by the rat. J Neurophysiol 53:361–372.

Kelly JB, Kavanagh GL (1986) Effects of auditory cortical lesions on pure-tone sound localization by the albino rat. Behav Neurosci 100:569–575.

Kelly JP, Wong D (1981) Laminar connections of the cat's auditory cortex. Brain Res 212:1–15.

Kelly JB, Larue DT, Sally SL, Cheff SJ, Kleinberg JL, Winer JA (1993) Origins of medial geniculate body projections to physiologically defined zones of rat primary auditory cortex (in preparation).

Kesarev VS, Malofeyeva LI, Trykova OV (1977) Ecological specificity of cetacean neocortex. J für Hirnforsch 18:447–460.

Kiang NY-S, with the assistance of Watanabe T, Thomas EC, Clark LF (1965) Discharge Patterns of Single Fibers in the Cat's Auditory Nerve. MIT Research Monograph No. 35. Cambridge: MIT Press.

King AJ, Palmer AR (1985) Integration of visual and auditory information in bimodal neurones in the guinea-pig superior colliculus. Expl Brain Res 60:492–500.

Kisvárday ZF, Adams CBT, Smith AD (1986) Synaptic connections of axo-somatic (chandelier) cells in human epileptic temporal cortex. Neuroscience 19:1179–1186.

Kisvárday ZF, Gulyas A, Beroukas D, North JB, Chubb IW, Somogyi P (1990) Synapses, axonal and dendritic patterns of GABA-immunoreactive neurons in human cerebral cortex. Brain 113:793–812.

Kitzes LM, Wrege KS, Cassady JM (1980) Patterns of responses of cortical cells to binaural stimulation. J Comp Neurol 192:455–472.

Knight PL (1977) Representation of the cochlea within the anterior auditory field (AAF) of the cat. Brain Res 130:447–467.

Knight RT, Brailowsky S (1990) Auditory evoked potentials from the primary auditory cortex of the cat: topographic and pharmacological studies. Electroenceph Clin Neurophysiol 77:225–232.

Knight RT, Scabini D, Woods DL (1989a) Prefrontal gating of auditory transmission in humans. Brain Res 504:338–342.

Knight RT, Scabini D, Woods DL, Clayworth CC (1989b) Contributions of temporal-parietal junction to the human auditory P3. Brain Res 502:109–116.

Kobler JB, Isbey SF, Casseday JH (1987) Auditory pathways to the frontal cortex of the mustache bat, *Pteronotus parnellii*. Science 236:824–826.

König N, Marty R (1974) On the functions and structure of deep layers of immature auditory cortex. J Physiol (Paris) 68:145–155.

König N, Pujol R, Marty R (1972) A laminar study of evoked potentials and unit responses in the auditory cortex of the postnatal cat. Brain Res 36:469–473.

König N, Roch G, Marty R (1975) The onset of synaptogenesis in rat temporal cortex. Anat Embryol 148:73–87.

Krieg WJS (1946) connections of the cerebral cortex I. The albino rat. A. Topography of the cortical areas. J Comp Neurol 84:221–275.

Kruger L (1959) The thalamus of the dolphin (*Tursiops truncatus*) and comparison with other mammals. J Comp Neurol 111:133–194.

Kudo M, Niimi K (1980) Ascending projections of the inferior colliculus in the cat: an autoradiographic study. J Comp Neurol 191:545–556.

Kudo M, Aitkin LM, Nelson JE (1989) Auditory forebrain organization of an Australian marsupial, the northern native cat (*Dasyurus hallucatus*). J Comp Neurol 279:28–42.

Kudo M, Glendenning KK, Frost SB, Masterton RB (1986) Origin of mammalian thalamocortical projections. I. Telencephalic projections of the medial geniculate body in the opossum (*Didelphis virginiana*). J Comp Neurol 245:176–197.

Kuffler S (1953) Discharge patterns and functional organization of mammalian retina. J Neurophysiol 16:37–68.

Kujirai K, Suga N (1983) Tonotopic representation and space map in the nonprimary auditory cortex of the mustached bat. Auris Nasus Larynx 10:9–24.

Kuriki S, Murase M (1989) Neuromagnetic study of the auditory responses in right and left hemispheres of the human brain evoked by pure tones and speech sounds. Expl Brain Res 77:127–134.

Kurokawa T, Yoshida K, Yamamoto T, Oka H (1990) Frontal cortical projections from the suprageniculate nucleus in the rat, as demonstrated by the PHA-L method. Neurosci Lett 120:259–262.

Lackner JR (1973) Visual rearrangement affects auditory localization. Neuropsychologia 11:29–32.

Lackner JR, Shenker B (1985) Proprioceptive influences on auditory and visual spatial localization. J Neurosci 5:579–583.

Landry P, Deschênes M (1981) Intracortical arborizations and receptive fields of identified ventrobasal thalamocortical afferents to the primary somatic sensory cortex in the cat. J Comp Neurol 199:345–372.

Larue DT, Huchton DM, Sun JY-M, Winer JA (1992) The GABAergic organization of the cat medial geniculate body: a quantitative study with post-embedding immunocytochemistry (in preparation).

Lauter JL, Herscovitch P, Formby C, Raichle ME (1985) Tonotopic organization in human auditory cortex revealed by positron emission tomography. Hear Res 20:199–205.

LeDoux JE, Sakaguchi A, Reis DJ (1984) Subcortical efferent projections of the medial geniculate nucleus mediate emotional responses conditioned to acoustic stimuli. J Neurosci 4:683–698.

LeDoux JE, Ruggiero DA, Reis DJ (1985) Projections to the subcortical forebrain from anatomically defined regions of the medial geniculate body in the rat. J Comp Neurol 242:182–213.

LeDoux JE, Iwata J, Pearl D, Reis DJ (1986a) Disruption of auditory but not visual learning by destruction of intrinsic neurons in the rat medial geniculate body. Brain Res 371:395–399.

LeDoux JE, Sakaguchi A, Iwata J, Reis DJ (1986b) Interruption of projections from the medial geniculate body to an archi-neostriatal field disrupts the classical conditioning of emotional responses to acoustic stimuli. Neuroscience 17:615–627.

LeDoux JE, Ruggiero DA, Forest R, Stornetta R, Reis DJ (1987) Topographic organization of convergent projections to the thalamus from the inferior colliculus and spinal cord in the rat. J Comp Neurol 264:123–146.

LeDoux JE, Cicchetti P, Xagoraris A, Romanski LM (1990a) The lateral amygdaloid nucleus: sensory interface of the amygdala in fear conditioning. J Neurosci 10:1062–1069.

LeDoux JE, Farb C, Ruggiero DA (1990b) Topographic organization of neurons in the acoustic thalamus that project to the amygdala. J Neurosci 10:1043–1054.

Leppelsack H-J (1974) Funktionelle Eigenschaften der Hörbahn im Feld L des Neostriatum caudale des Staren. J Comp Physiol A 88:271–320.

Leppelsack H-J (1978) Unit responses to species-specific sounds in the auditory forebrain center of birds. Fed Proc 37:2336–2341.

Leppelsack H-J, Vogt M (1976) Responses of auditory neurons in the forebrain of a songbird to stimulation with species-specific sounds. J Comp Physiol A 107:263–274.

LeVay S, Gilbert CD (1976) Laminar patterns of geniculocortical projection in the cat. Brain Res 113:1–19.

Levey AI, Hallanger AE, Wainer BH (1987) Choline acetyltransferase immunoreactivity in the rat thalamus. J Comp Neurol 257:317–332.

Liegeois-Chauvel C, Morin C, Musolino A, Bancaud J, Chauvel P (1989) Evidence of a contribution of the auditory cortex to audiospinal facilitation in man. Brain 112:375–391.

Liegeois-Chauvel C, Musolino A, Chauvel P (1991) Localization of the primary auditory cortex in man. Brain 114:139–153.

Llinás R, Sugimori M (1980) Electrophysiological properties of in vitro Purkinje cell somata in mammalian cerebellar slices. J Physiol (London) 305:171–195.

Luethke LE, Krubitzer LA, Kaas JH (1985) Connections of auditory cortex in squirrels. Proc Soc Neurosci 11:33.

Luethke LE, Krubitzer LA, Kaas JH (1988) Cortical connections of electrophysiologically and architectonically defined subdivisions of auditory cortex in squirrels. J Comp Neurol 268:181–203.

Luethke LE, Krubitzer LA, Kaas JH (1989) Connections of primary auditory cortex in a new world monkey, Saguinus. J Comp Neurol 285:487–513.

Lund JS, Henry GH, MacQueen CL, Harvey AR (1979) Anatomical organization of the primary visual cortex (area 17) of the cat. A comparison with area 17 of the monkey. J Comp Neurol 184:599–618.

Mäkelä JP (1988) Contra- and ipsilateral auditory stimuli produce different activation patterns at the human auditory cortex. A neuromagnetic study. Eur J Physiol 412:12–16.

Mäkelä JP, Hari R, Linnankivi A (1987) Different analysis of frequency and amplitude modulations of a continuous tone in the human auditory cortex. A neuromagnetic study. Hear Res 27:257–264.

Mäkelä JP, Hari R, Leinonen L (1988) Magnetic responses in the human auditory cortex to noise/square wave transitions. Electroenceph Clin Neurophysiol 69:423–430.

Mäkelä JP, Karmos G, Molnár M, Csépe V, Winkler I (1990) Steady-state responses from the cat auditory cortex. Hear Res 45:41–50.

Malach R, Ebert R, Van Sluyters RC (1984) Recovery from effects of brief monocular deprivation in the kitten. J Neurophysiol 51:538–551.

Manley JA, Mueller-Preuss P (1978a) Response variability in the mammalian auditory cortex: an objection to feature detection? Fed Proc 37:2355–2359.

Manley JA, Müller-Preuss P (1978b) Response variability of auditory cortex cells in the squirrel monkey to constant acoustic stimuli. Expl Brain Res 32:171–180.

Markowitsch HJ, Emmans D, Irle D, Streicher M, Preilowski B (1985) Cortical and subcortical afferents of the primate's temporal pole: a study of rhesus monkeys, squirrel monkeys, and marmosets. J Comp Neurol 242:425–458.

Martin KAC (1988) From single cells to simple circuits in the cerebral cortex. Quart J Expl Physiol 73:637–702.

Masterton RB, Hodos W, Jerison H (1976) (eds) Evolution, Brain, and Behavior: Persistent Problems. Hillsdale, NJ: Lawrence Erlbaum Associates.

Matsubara JA, Phillips DP (1988) Intracortical connections and their physiological correlates in the primary auditory cortex (AI) of the cat. J Comp Neurol 268:38–48.

McCormick DA (1989) GABA as an inhibitory transmitter in human cerebral cortex. J Neurophysiol 62:1018–1027.

McCormick DA, Feeser HR (1990) Functional implications of burst firing and single spike activity in lateral geniculate relay neurons. Neuroscience 39:103–113.

McCormick DA, Pape H-C (1990) Noradrenergic and serotonergic modulation of a hyperpolarization-activated cation current in thalamic relay neurones. J Physiol (London) 431:319–342.

McCormick DA, Prince DA (1986) Acetylcholine produces burst firing in thalamic reticular neurones by activating a potassium conductance. Nature 319:402–405.

McCormick DA, Prince DA (1987) Actions of acetylcholine in the guinea-pig and cat medial and lateral geniculate nuclei, in vitro. J Physiol (London) 431:319–342.

McGuire BA, Hornung J-P, Gilbert CD, Wiesel TN (1984) Patterns of synaptic input to layer 4 of cat striate cortex. J Neurosci 4:3021–3033.

McKenna TM, Ashe JH, Hui GK, Weinberger NM (1988) Muscarinic agonists modulate spontaneous and evoked unit discharge in auditory cortex of cat. Synapse 2:54–68.

McKenna TM, Ashe JH, Weinberger NM (1989) Cholinergic modulation of frequency receptive fields in auditory cortex: I. Frequency-specific effects of muscarinic agonists. Synapse 4:30–43.

McMullen NT, Glaser EM (1982a) Morphology and laminar distribution of non-pyramidal neurons in the auditory cortex of the rabbit. J Comp Neurol 208:85–106.

McMullen NT, Glaser EM (1982b) Tonotopic organization of rabbit auditory cortex. Expl Neurol 75:208–220.

McMullen NT, Glaser EM (1988) Auditory cortical responses to neonatal deafening: pyramidal neuron spine loss without changes in growth or orientation. Expl Brain Res 72:195–200.

McMullen NT, Glaser EM, Tagamets M (1984) Morphology of spine-free non-pyramidal neurons in rabbit auditory cortex. J Comp Neurol 222:383–396.

McMullen NT, Goldberger B, Suter CM, Glaser EM (1988) Neonatal deafening alters nonpyramidal dendrite orientation in auditory cortex: a computer microscope study. J Comp Neurol 267:91–106.

Mehler WR, Feferman ME, Nauta WJH (1960) Ascending axon degeneration following anterolateral cordotomy. An experimental study in the monkey. Brain 83:718–750.

Meredith MA, Clemo HR (1989) Auditory cortical projection from the anterior ectosylvian sulcus (field AES) to the superior colliculus in the cat: an anatomical and electrophysiological study. J Comp Neurol 289:687–707.

Merzenich MM, Brugge JF (1973) Representation of the cochlear partition on the superior temporal plane of the macaque monkey. Brain Res 50:275–296.

Merzenich MM, Knight PL, Roth GL (1975) Representation of cochlea within primary auditory cortex in the cat. J Neurophysiol 38:231–249.

Merzenich MM, Kaas JH, Roth GL (1976) Auditory cortex in the grey squirrel: tonotopic organization and architectonic fields. J Comp Neurol 166:387–402.

Merzenich MM, Kaas JH, Wall J, Nelson RJ, Sur M, Felleman D (1983) Topographic reorganization of somatosensory cortical areas 3b and 1 in adult monkeys following restricted deafferentation. Neuroscience 8:33–55.

Merzenich MM, Jenkins WM, Middlebrooks JC (1984) Observations and hypotheses on special organizational features of the central auditory nervous system. In: Edelman GM, Gall WE, Cowan WM (eds) Dynamic Aspects of Neocortical Function. New York: John Wiley & Sons, pp. 397–424.

Mesulam M-M, Pandya DN (1973) The projection of the medial geniculate complex within the sylvian fissure of the rhesus monkey. Brain Res 60:315–333.

Metherate R, Weinberger NM (1989) Acetylcholine produces stimulus-specific receptive field alterations in cat auditory cortex. Brain Res 480:372–377.

Metherate R, Ashe JH, Weinberger NM (1990) Acetylcholine modifies neuronal acoustic rate-level functions in guinea pig auditory cortex by an action at muscarinic receptors. Synapse 6:364–368.

Meyer G, Albus K (1981) Spiny stellates as cells of origin of association fibres from area 17 to area 18, in the cat's neocortex. Brain Res 210:335–341.

Meyer G, Castañeyra-Perdomo A, Ferres-Torres R (1984) A type of apparently axonless granule cell in the cat auditory cortex. Anat Embryol 170:319–320.

Meyer G, González-Hernández TH, Ferres-Torres R (1989) The spiny stellate neurons in layer IV of the human auditory cortex. A Golgi study. Neuroscience 33:489–498.

Middlebrooks JC, Knudsen EI (1984) A neural code for auditory space in the cat's superior colliculus. J Neurosci 4:2621–2634.

Middlebrooks JC, Knudsen EI (1987) Changes in external ear position modify the spatial tuning of auditory units in the cat's superior colliculus. J Neurophysiol 57:672–687.

Middlebrooks JC, Pettigrew JD (1981) Functional classes of neurons in primary auditory cortex of the cat distinguished by sensitivity to sound location. J Neurosci 1:107–120.

Middelbrooks JC, Zook JM (1983) Intrinsic organization of the cat's medial geniculate body identified by projections to binaural response-specific bands in the primary auditory cortex. J Neurosci 3:203–225.

Middlebrooks JC, Dykes RW, Merzenich MM (1980) Binaural response-specific bands in primary auditory cortex (AI) of the cat: topographic organization orthogonal to isofrequency contours. Brain Res 181:31–48.

Miller JM, Beaton RD, O'Connor T, Pfingst BE (1974) Response pattern complexity of auditory cells in the cortex of unanesthetized monkeys. Brain Res 69:101–113.

Mitani A, Itoh K, Nomura S, Kudo M, Kaneko T, Mizuno N (1984) Thalamocortical projections to layer I of the primary auditory cortex in the cat: a horseradish peroxidase study. Brain Res 310:347–350.

Mitani A, Shimokouchi M, Itoh K, Nomura S, Kudo M, Mizuno N (1985) Morphology and laminar organization of electrophysiologically identified neurons in primary auditory cortex in the cat. J Comp Neurol 235:430–447.

Molinari M, Hendry SHC, Jones EG (1987) Distribution of certain neuropeptides in the primate thalamus. Brain Res 426:270–289.

Montero VM (1983) Ultrastructural identification of axon terminals from the thalamic reticular nucleus in the medial geniculate body in the rat: an EM autoradiographic study. Expl Brain Res 51:338–342.

Moore JK (1980) The primate cochlear nuclei: loss of lamination as a phylogenetic process. J Comp Neurol 193:609–629.

Moore JK, Karapas F, Moore RY (1977) Projections of the inferior colliculus in insectivores and primates. Brain Behav Evol 14:301–327.

Moore RY, Goldberg JM (1966) Projections of the inferior colliculus in monkey. Expl Neurol 14:429–438.

Morán MA, Mufson EJ, Mesulam M-M (1987) Neural inputs into the temporopolar cortex of the rhesus monkey. J Comp Neurol 256:88–103.

Morel A (1980) Codage des sons dans le corps génouille median du chat: évaluation de l'organisation tonotopique di ses différents noyaux. Thèse de l'Université de Lausanne, Faculté des Sciences, Janis Druck + Verlag Zurich, pp. 1–154.

Morel A, Imig TJ (1987) Thalamic projections to fields A, AI, P, and VP in the cat auditory cortex. J Comp Neurol 265:119–144.

Morest DK (1964) The neuronal architecture of the medial geniculate body of the cat. J Anat (London) 98:611–630.

Morest DK (1965a) The laminar structure of the medial geniculate body of the cat. J Anat (London) 99:143–160.

Morest DK (1965b) The lateral tegmental system of the midbrain and the medial geniculate body: study with Golgi and Nauta methods in the cat. J Anat (London) 99:611–634.

Morest DK (1971) Dendrodendritic synapses of cells that have axons: the fine structure of the Golgi type II cell in the medial geniculate body of the cat. Zeit für Anat Entwicklungsgeschichte 133:216–246.

Morest DK (1974) LCN's in the medial geniculate body of the cat. Neurosci Res Prog Bull 13:367–377.

Morest DK (1975) Synaptic relations of Golgi type II cells in the medial geniculate body of the cat. J Comp Neurol 162:157–194.

Morest DK, Winer JA (1986) The comparative anatomy of neurons: homologous neurons in the medial geniculate body of the opossum and the cat. Adv Anat Embryol Cell Biol 97:1–96.

Morgane PJ, Jacobs MS, Galaburda A (1985) Conservative features of neocortical evolution in dolphin brain. Brain Behav Evol 26:176–184.

Moryś J, Słoniewska P, Narkiewicz O, Pilgrim C (1987) Origin and terminals of the pretectothalamoclaustral connections in the cat. Folia Morphol Warszawa 46:129–140.

Mountcastle VB (1957) Modality and topographic properties of single neurons of cat's somatic sensory cortex. J Neurophysiol 20:408–434.

Mountcastle VB, Henneman E (1949) The representation of tactile sensibility in the thalamus of the monkey. J Comp Neurol 97:409–440.

Mudry KM, Constantine-Paton M, Capranica RR (1977) Auditory sensitivity of the diencephalon of the leopard frog *Rana p. pipiens*. J Comp Physiol A 114:1–13.

Mugnaini E, Oertel WH (1985) An atlas of the distribution of GABAergic neurons and terminals in the rat CNS as revealed by GAD immunohistochemistry. In: Björklund A and Hökfelt T (eds) Handbook of Chemical Neuroanatomy, Vol 4, GABA and Neuropeptides in the CNS, Part I. Amsterdam: Elsevier Science Publishers B.V., pp. 436–608.

Mugnaini E, Osen KK, Dahl A-L, Friedrich VL Jr, Korte G (1980) Fine structure of granule cells and related interneurons (termed Golgi cells) in the cochlear nuclear complex of cat, rat and mouse. J Neurocytol 9:537–570.

Müller CM (1987) Differential effects of acetylcholine in the chicken auditory neostriatum and hyperstriatum ventral — studies in vivo and in vitro. J Comp Physiol A 161:857–866.

Müller CM (1988) Distribution of GABAergic perikarya and terminals in the centers of the higher auditory pathway of the chicken. Cell Tissue Res 252:99–106.

Müller CM, Leppelsack H-J (1985) Feature extraction and tonotopic organization in the avian auditory forebrain. Expl Brain Res 59:587–599.

Müller CM, Scheich H (1987a) GABAergic inhibition increases the neuronal selectivity to natural sounds in the avian auditory forebrain. Brain Res 414:376–380.

Müller CM, Scheich H (1987b) GABA mediates interaural inhibition in the avian auditory forebrain. Naturwiss 74:602–604.

Müller SC, Scheich H (1985) Functional organization of avian auditory field L. A comparative 2DG study. J Comp Physiol A 156:1–12.

Müller-Preuss P (1986) On the mechanism of call coding through auditory neurons in the squirrel monkey. Eur Arch Psych Neurol 236:50–55.

Müller-Preuss P, Jurgens U (1976) Projections from the 'cingular' vocalization area in the squirrel monkey. Brain Res 103:29–43.

Müller-Preuss P, Ploog D (1981) Inhibition of auditory cortical neurons during phonation. Brain Res 215:61–76.

Mulligan KA, Törk I (1988) Serotoninergic innervation of the cat cerebral cortex. J Comp Neurol 270:86–110.

Mysliveček J (1983) Development of the auditory evoked response in the auditory cortex in mammals. In: Romand R (ed) Development of Auditory and Vestibular Systems. New York: Academic Press, pp. 167–209.

Neal JW, Pearson RCA, Powell TPS (1986) The relationship between the auditory cortex and the claustrum in the cat. Brain Res 366:145–151.

Neff WD, Diamond IT, Casseday JH (1975) Behavioral studies of auditory discrimination: central nervous system. In: Keidel WD, Neff WD (eds) Handbook of Sensory Physiology, Vol V, Part 2, Auditory System. Anatomy, Physiology (Ear). Berlin: Springer-Verlag, pp. 307–400.

Neville HJ, Schmidt A, Kutas M (1983) Altered visual-evoked potential in congenitally deaf adults. Brain Res 266:127–132.

Newman JD (1970) Midbrain regions relevant to auditory communication in songbirds. Brain Res 22:259–261.

Newman JD, Lindsley DF (1976) Single unit analysis of auditory processing in squirrel monkey frontal cortex. Expl Brain Res 25:169–181.

Newman JD, Symmes D (1974) Arousal effects on unit responsiveness to vocalizations in squirrel monkey auditory cortex. Brain Res 78:125–138.

Newman JD, Wollberg Z (1973) Responses of single neurons in the auditory cortex of squirrel monkeys to variants of a single call type. Expl Neurol 40:821–824.

Newman DB, Hilleary SK, Ginsberg CY (1989) Nuclear terminations of corticoreticular fiber systems in rats. Brain Behav Evol 34:223–264.

Neylon L, Haight JR (1983) Neocortical projections of the suprageniculate and posterior thalamic nuclei in the marsupial brush-tailed possum, *Trichosurus vulpecula* (Phalangeridae), with a comparative commentary on the organization of the posterior thalamus in marsupial and placental mammals. J Comp Neurol 217:357–375.

Nieuwenhuys R, Veening JG, Van Domburg P (1988/1989) Core and paracores: some new chemoarchitectural entities in the mammalian neuraxis. Acta Morphol Neerland Scand 26:131–163.

Niimi K, Naito F (1974) Cortical projections of the medial geniculate body in the cat. Expl Brain Res 19:326–342.

Niimi K, Matsuoka H (1979) Thalamocortical organization of the auditory system in the cat studied by retrograde axonal transport of horseradish peroxidase. Adv Anat Embryol Cell Biol 57:1–56.

Niimi K, Matsuoka H, Yamazaki Y, Matsumoto H (1981) Thalamic afferents to the visual cortex in the cat studied by retrograde axonal transport of horseradish peroxidase. Brain Behav Evol 18:114–139.

Niimi K, Ono K, Kusunose M (1984) Projections of the medial geniculate nucleus to layer 1 of the auditory cortex in the cat traced with horseradish peroxidase. Neurosci Lett 45:223–228.

Norita M, Katoh Y (1987) The GABAergic neurons and axon terminals in the lateralis medialis-suprageniculate nuclear complex of the cat: GABA-immunocytochemical and WGA-HRP studies by light and electron microscopy. J Comp Neurol 253:54–67.

Norita M, Katoh Y (1988) Synaptic organization of the lateralis medialis-suprageniculate nuclear (LM-Sg) complex in the cat. Prog Brain Res 75:109–118.

Northcutt RG (1981) Audition and the central nervous system of fishes. In: Tavolga WN, Popper AN, Fay RR (eds) Hearing and Sound Communication in Fishes. New York: Springer-Verlag, pp. 331–353.

Ogren MP, Hendrickson AE (1979) The structural organization of the inferior and lateral subdivisions of the *Macaca* monkey pulvinar. J Comp Neurol 188:147–178.

Ohara PT, Lieberman AR, Hunt SP, Wu J-Y (1983) Neural elements containing glutamic acid decarboxylase (GAD) in the dorsal lateral geniculate nucleus of the rat: immunohistochemical studies by light and electron microscopy. Neuroscience 8:189–211.

Ohara PT, Chazal H, Ralston HJ III (1989) Ultrastructural analysis of GABA-immunoreactive elements in the monkey thalamic ventrobasal complex. J Comp Neurol 283:541–558.

Oleson TD, Ashe JH, Weinberger NM (1975) Modification of auditory and somatosensory system activity during pupillary conditioning in the paralyzed cat. J Neurophysiol 38:1114–1139.

Oliver DL (1982) A Golgi study of the medial geniculate body of the tree shrew (*Tupaia glis*). J Comp Neurol 209:1–16.

Oliver DL (1984) Neuron types in the central nucleus of the inferior colliculus that project to the medial geniculate body. Neuroscience 11:409–424.

Oliver DL, Hall WC (1978a) The medial geniculate body of the tree shrew, *Tupaia glis*. I. Cytoarchitecture and midbrain connections. J Comp Neurol 182:423–458.

Oliver DL, Hall WC (1978b) The medial geniculate body of the three shrew, *Tupaia glis*. II. Connections with the neocortex. J Comp Neurol 182:459–494.

Oliver DL, Morest DK (1984) The central nucleus of the inferior colliculus in the cat. J Comp Neurol 222:237–264.

Oliver DL, Merzenich MM, Roth GL, Hall WC, Kaas JH (1976) Tonotopic organization and connections of primary auditory cortex in the tree shrew. Anat Rec 184:491.

Oliver DL, Winer JA, Beckius GH, Saint Marie RL (1992) Morphology of GABAergic neurons in the cat inferior colliculus. (To be submitted).

Olsen JF (1986) Processing of biosonar information by the medial geniculate body of the mustached bat. Doctoral dissertation, Washington University, St. Louis, pp. 1–325.

Olson CR, Graybiel AM (1987) Ectosylvian visual area of the cat: location, retinotopic organization, and connections. J Comp Neurol 261:277–294.

O'Neill WE, Suga N (1979) Target range-sensitive neurons in the auditory cortex of the mustache bat. Science 203:69–73.

Ong WY, Garey LJ (1990) Neuronal architecture of the human temporal cortex. Anat Embryol 181:351–364.

Orban GA, Hoffmann K-P, Duysens J (1985) Velocity sensitivity in the cat visual system. I. Responses of LGN cells to moving bar stimuli: a comparison with cortical areas 17 and 18. J Neurophysiol 54:1026–1049.

Oyanagi K, Ohama E, Ikuta F (1989) The auditory system in methyl mercury intoxication: a neuropathological investigation of 14 autopsy cases in Niigata, Japan. Acta Neuropathol 77:561–568.

Pallas SL, Roe AW, Sur M (1990) Visual projections induced into the auditory pathway of ferrets. I. Novel inputs to primary auditory cortex (AI) from the LP/pulvinar complex and the topography of the MGN-AI projection. J Comp Neurol 298:50–68.

Palmer AR, King AJ (1985) A monaural space map in the guinea pig superior colliculus. Hear Res 17:267–280.

Pandya DN, Sanides F (1973) Architectonic parcellation of the temporal operculum in rhesus monkey and its projection pattern. Zeit für Anat Entwicklungsgeschichte 139:127–161.

Pandya DN, Hallett M, Mukherjee SK (1969) Intra- and interhemispheric connections of the neocortical auditory system in the rhesus monkey. Brain Res 14:49–65.

Pantev C, Hoke M, Lehnertz K, Lütkenhöner B (1989a) Neuromagnetic evidence of an amplitopic organization of the human auditory cortex. Electroenceph Clin Neurophysiol 72:225–231.

Pantev C, Hoke M, Lehnertz K, Lütkenhöner B, Anogianakis G, Wittkowski W (1989b) Tonotopic organization of the human auditory cortex revealed by transient auditory evoked magnetic fields. Electroenceph Clin Neurophysiol 69:160–170.

Pantev C, Hoke M, Lütkenhöner B, Lehnertz K (1989c) Tonotopic organization of the auditory cortex: pitch versus frequency representation. Science 246:486–488.

Papez JW (1929) Central acoustic tract in cat and man. Anat Rec 42:60.

Patterson HA (1976) An anterograde degeneration and retrograde axonal transport study of the cortical projections of the rat medial geniculate body. Doctoral dissertation, Boston University Graduate School, Boston, pp. 1–171.

Paula-Barbosa MM, Feyo PB, Sousa-Pinto A (1975) The association connexions of the suprasylvian fringe (SF) and other areas of the cat auditory cortex. Expl Brain Res 23:535–554.

Payne BR, Pearson HE, Cornwell P (1988) Neocortical connections in fetal cats. Neurosci Res 5:513–543.

Pelleg-Toiba R, Wollberg Z (1989) Tuning properties of auditory cortex cells in the awake squirrel monkey. Expl Brain Res 74:353–364.

Penfield W, Perot P (1963) The brain's record of auditory and visual experience. Brain 86:595–697.

Penny GR, Conley M, Schmechel DE, Diamond IT (1984) The distribution of glutamic acid decarboxylase immunoreactivity in the diencephalon of the opossum and the rabbit. J Comp Neurol 228:38–56.

Perrott DR, Saberi K, Brown K, Strybel TZ (1990) Auditory psychomotor coordination and visual search performance. Percep Psychophys 48:214–226.

Peters A, Harriman KM (1988) Enigmatic bipolar cell of rat visual cortex. J Comp Neurol 267:409–432.

Peters A, Regidor J (1981) A reassessment of the forms of nonpyramidal neurons in area 17 of cat visual cortex. J Comp Neurol 203:685–716.

Peterson BA, Winer JA (1988) Projections of the cat medial geniculate body to the primary auditory cortex. Proc Soc Neurosci 14:492.

Peterson BA, Winer JA (1989) GABA-immunoreactive neurons and puncta in layers IV and V of cat primary auditory cortex (AI). Proc Soc Neurosci 15:1110.

Peterson BA, Winer JA (1993) Laminar and areal patterns of input to auditory cortical fields from the cat medial geniculate body (in preparation).

Peterson BA, Prieto JJ, Winer JA (1990) GABA-immunoreactive axon terminals in cat primary auditory cortex (AI). Proc Soc Neurosci 16:796.

Petrides M, Pandya DN (1988) Association fiber pathways to the frontal cortex from the superior temporal region in the rhesus monkey. J Comp Neurol 273:52–66.

Petrusz, P, Merchenthaler I, Maderdrut JL (1985) Distribution of enkephalin-containing neurons in the central nervous system. In: Björklund A, Hökfelt T (eds) Handbook of Chemical Neuroanatomy, Vol 4, GABA and Neuropeptides in the CNS, Part I. Amsterdam: Elsevier Science Publishers B.V., pp. 273–334.

Pfingst BE, O'Connor TA (1981) Characteristics of neurons in auditory cortex of monkeys performing a simple auditory task. J Neurophysiol 45:16–34.

Pfingst BE, O'Connor TA, Miller JM (1977) Response plasticity of neurons in auditory cortex of the rhesus monkey. Expl Brain Res 29:393–404.

Phillips DP, Farmer ME (1990) Acquired word deafness, and the temporal grain of sound representation in the primary auditory cortex. Behav Brain Res 40: 85–94.

Phillips DP, Hall SE (1990) Response timing constraints on the cortical representation of sound time structure. J Acoust Soc Am 88:1403–1411.

Phillips DP, Irvine DRF (1981) Responses of single neurons in physiologically defined area AI of cat cerebral cortex: sensitivity to interaural intensity differences. Hear Res 4:299–307.

Phillips DP, Irvine DRF (1982) Properties of single neurons in the anterior auditory field (AAF) of the cat cerebral cortex. Brain Res 248:237–244.

Phillips DP, Kelly JB (1988) Coding of tone-pulse amplitude by single neurons in auditory cortex of albino rats (Rattus norvegicus). Hear Res 37:267–279.

Phillips DP, Orman SS (1984) Responses of single neurons in posterior field of cat auditory cortex to tonal stimulation. J Neurophysiol 51:147–163.

Poggio GF, Mountcastle VB (1960) A study of the functional contributions of the lemniscal and spinothalamic systems to somatic sensibility. Central nervous mechanisms in pain. Bull Johns Hopkins Hosp 106:266–316.

Popowits JM, Larue DT, Winer JA (1988) Glutamate is a major transmitter in the rat medial geniculate body. Proc Soc Neurosci 14:490.

Powell TPS, Erulkar SD (1962) Transneuronal cell degeneration in the auditory relay nuclei of the cat. J Anat (London) 96:249–268.

Preuss A, Müller-Preuss P (1990) Processing of amplitude modulated sounds in the medial geniculate body of squirrel monkeys. Expl Brain Res 79:207–211.

Prieto JJ, Peterson BA, Winer JA (1990) The GABAergic neurons in cat primary auditory cortex. Proc Soc Neurosci 16:796.

Prieto JJ, Peterson BA, Winer JA (1992a) Morphology and spatial distribution of GABAergic neurons in cat primary auditory cortex (AI). (in preparation).

Prieto JJ, Peterson BA, Winer JA (1992b) Laminar distribution and neuronal targets of GABAergic axon terminals in cat primary auditory cortex (AI). (in preparation).

Prieto JJ, Winer JA (1991) Neurons of layer VI in cat primary auditory cortex (AI): study with Golgi method and GABA and GAD immunocytochemistry. Proc Soc Neurosci 17:301.

Pritz MB (1974a) Ascending connections of a midbrain auditory area in a crocodile, Caiman crocodilus. J Comp Neurol 153:179–198.

Pritz MB (1974b) Ascending connections of a thalamic auditory area in a crocodile, Caiman crocodilus. J Comp Neurol 153:199–214.

Pritz MB, Stritzel ME (1988) Thalamic nuclei that project to the reptilian telencephalon lack GABA and GAD immunoreactive neurons and puncta. Brain Res 457:154–159.

Pujol R (1972) Development of tone-burst responses along the auditory pathway in the cat. Acta Otolaryngol 74:383–391.

Raczkowski D, Diamond IT, Winer J (1976) Organization of thalamo-cortical auditory system in the cat studied with horseradish peroxidase. Brain Res 101:345–354.

Rajan R, Aitkin LM, Irvine DRF, McKay J (1990a) Azimuthal sensitivity of neurons in primary auditory cortex of cats. I. Types of sensitivity and the effects of variations in stimulus parameters. J Neurophysiol 64:872–887.

Rajan R, Aitkin LM, Irvine DRF (1990b) Azimuthal sensitivity of neurons in primary auditory cortex of cats. II. Organization along frequency-band strips. J Neurophysiol 64:888–902.

Ramón y Cajal S (1900) Estudios sobre la corteza humana III: Corteza acústica. Revista Trimestral Micrográfica, Madrid 5:129–183. Reprinted in DeFelipe J and Jones EG (1988) (eds and trans) Cajal on the Cerebral Cortex. An Annotated Translation of the Compete Writings. In: Corsi P, Jones EG, Shepherd GM (eds) History of Neuroscience, Vol 1, pp. 251–288.

Ramón y Cajal S (1911) Histologie du Système Nerveux de l'Homme et des Vértébres, trans Azoulay L. Maloine, Paris (reprinted in 1972 by Consejo Superior de Investigaciones Çientificas).

Reale RA, Brugge JF (1990) Auditory cortical neurons are sensitive to static and continuously changing interaural phase cues. J Neurophysiol 64:1247–1260.

Reale RA, Imig TJ (1980) Tonotopic organization in auditory cortex of the cat. J Comp Neurol 182:265–291.

Reale RA, Imig TJ (1983) Auditory cortical field projections to the basal ganglia of the cat. Neuroscience 8:67–86.

Reale RA, Kettner RE (1986) Topography of binaural organization in primary auditory cortex of the cat: effects of changing interaural intensity. J Neurophysiol 56:663–682.

Reale RA, Brugge JF, Feng JZ (1983) Geometry and orientation of neuronal processes in cat primary auditory cortex (A I) related to characteristic-frequency maps. Proc Nat Acad Sci (USA) 80:5449–5453.

Reale RA, Brugge JF, Chan JCK (1987) Maps of auditory cortex in cats reared after unilateral cochlear ablation in the neonatal period. Develop Brain Res 34:281–290.

Rebillard G, Rebillard M, Carlier E, Pujol R (1976) Histo-physiological relationships in the deaf white cat auditory system. Acta Otolaryngol 82:48–56.

Rebillard G, Carlier E, Rebillard M, Pujol R (1977) Enhancement of visual responses on the primary auditory cortex of the cat after an early destruction of cochlear receptors. Brain Res 129:162–164.

Rebillard G, Rebillard M, Pujol R (1980) Factors affecting the recording of visual-evoked potentials from the deaf cat primary auditory cortex. Brain Res 188:252–254.

Redies H, Sieben U, Creutzfeldt OD (1989a) Functional subdivisions in the auditory cortex of the guinea pig. J Comp Neurol 282:473–488.

Redies H, Brandner S, Creutzfeldt OD (1989b) Anatomy of the auditory thalamocortical system of the guinea pig. J Comp Neurol 282:489–511.

Reinoso-Suárez F (1984) Connectional patterns in parietotemporoccipital association cortex of the feline cerebral cortex. In: Reinoso-Suárez F and Ajmone-Marsan C (eds) Cortical Integration. New York: Raven Press, pp. 255–278.

Revishchin AV, Garey LJ (1990) The thalamic projection to the sensory neocortex of the porpoise, Phocoena phocoena. J Anat (London) 169:85–102.

Rhode WS (1992) Physiology of the cochlear nuclei. In: Popper AN, Fay RR, Webster DB (eds) Springer Series in Auditory Research, Vol 2, The Physiology of the Mammalian Auditory Central Nervous System. New York: Springer-Verlag (in press).

Rhode WS, Oertel D, Smith PH (1983) Physiological properties of cells labeled intracellulary with horseradish peroxidase in cat ventral cochlear nucleus. J Comp Neurol 213:448–463.

Rinvik E, Ottersen OP, Storm-Mathisen J (1987) Gamma-aminobutyrate-like immunoreactivity in the thalamus of the cat. Neuroscience 21:781–805.

Rioch D McK (1929) Studies on the diencephalon of carnivora. Part I. The nuclear configuration of the thalamus, epithalamus, and hypothalamus of the dog and cat. J Comp Neurol 49:1–119.

RoBards MJ, Watkins DW III, Masterton RB (1976) An anatomical study of some somesthetic afferents to the intercollicular terminal zone of the midbrain of the opossum. J Comp Neurol 170:499–524.

Robertson D, Irvine DRF (1989) Plasticity of frequency organization in auditory cortex of guinea pigs with partial unilateral deafness. J Comp Neurol 282:456–471.

Robertson RT, Gragnola TG, Yu J (1990) Patterns of transiently expressed acetylcholinesterase activity in cerebral cortex and dorsal thalamus of rats with cytotoxin-induced microencephaly. Int J Develop Neurosci 8:223–232.

Robin DA, Tranel D, Damasio H (1990) Auditory perception of temporal and spectral events in patients with focal left and right cerebral lesions. Brain Lang 39:539–555.

Robson JA (1983) The morphology of corticofugal axons to the dorsal lateral geniculate nucleus in the cat. J Comp Neurol 216:89–103.

Robson JA (1984) Reconstructions of corticogeniculate axons in the cat. J Comp Neurol 225:193–200.

Rockel AJ, Hiorns RW, Powell TPS (1980) The basic uniformity in structure of the neocortex. Brain 103:221–244.

Roda JM, Reinoso-Suárez F (1983) Topographic organization of thalamic projections to the cortex of the anterior ectosylvian sulcus in the cat. Expl Brain Res 49:131–139.

Rodrigues-Dagaeff C, Simm G, de Ribaupierre Y, Villa A, de Ribaupierre F, Rouiller EM (1989) Functional organization of the ventral division of the medial geniculate body of the cat: evidence for a rostro-caudal gradient of response properties and cortical projections. Hear Res 39:103–125.

Roe AW, Pallas SL, Hahm J-O, Sur M (1990) A map of visual space induced in primary auditory cortex. Science 250:818–820.

Roger M, Arnault P (1989) Anatomical study of the connections of the primary auditory area in the rat. J Comp Neurol 287:339–356.

Roland PE, Skinhøj E, Lassen NA (1981) Focal activations of human cerebral cortex during auditory discrimination. J Neurophysiol 45:1139–1151.

Romani GL, Williamson SJ, Kaufman L (1982) Tonotopic organization of the human auditory cortex. Science 216:1339–1340.

Room P, Groenenwegen HJ (1986) Connections of the parahippocampal region. I. Cortical afferents. J Comp Neurol 251:415–450.

Rose JE, Woolsey CN (1949) The relations of thalamic connections, cellular structure, and evocable electrical activity in the auditory region of the cat. J Comp Neurol 91:441–466.

Rose JE, Woolsey CN (1958) Cortical projections and functional organization of thalamic auditory system of cat. In: Harlow HF, Woolsey CN (eds) Biological and Biochemical Bases of Behavior. Madison: University of Wisconsin Press, pp. 127–150.

Roucoux-Hanus M, Boisacq-Schepens N (1977) Ascending vestibular projections: further results at cortical and thalamic levels in the cat. Expl Brain Res 29:283–292.

Rouiller EM, de Ribaupierre F (1985) Origins of afferents to physiologically defined regions of the medial geniculate body of the cat: ventral and dorsal divisions. Hear Res 19:97–114.

Rouiller EM, de Ribaupierre F (1990) Arborization of corticothalamic axons in the auditory thalamus of the cat: a PHA-L tracing study. Neurosci Lett 108:29–35.

Rouiller EM, Colomb E, Capt M, de Ribaupierre F (1985) Projections of the reticular complex of the thalamus onto physiologically characterized regions of the medial geniculate body. Neurosci Lett 53:227–232.

Rouiller EM, Hornung JP, de Ribaupierre F (1989a) Extrathalamic ascending projections to physiologically identified fields of the cat auditory cortex. Hear Res 40:233–246.

Rouiller EM, Rodrigues-Dagaeff C, Simm G, de Ribaupierre Y, Villa A, de Ribaupierre F (1989b) Functional organization of the medial division of the medial geniculate body of the cat: tonotopic organization, spatial distribution of response properties and cortical connections. Hear Res 39:127–146.

Rubel EW (1978) Ontogeny of structure and function in the vertebrate auditory system. In: Jacobson M (ed) Handbook of Sensory Physiology, Vol IX, Development of Sensory Systems. New York: Springer-Verlag, pp. 135–237.

Rudell AP, Eberle LP (1985) Acoustic facilitation of the Hoffman reflex. Expl Neurol 89:592–602.

Russchen FT (1982) Amygdalopetal projections in the cat. II. Subcortical afferent connections. A study with retrograde tracing techniques. J Comp Neurol 207:157–176.

Rustioni A, Cuénod M (1982) Selective retrograde transport of D-aspartate in spinal interneurons and cortical neurons of rats. Brain Res 236:143–155.

Ryan AF, Woolf NK, Sharp FR (1982) Tonotopic organization in the central auditory pathway of the Mongolian gerbil: a 2-deoxyglucose study. J Comp Neurol 207:369–380.

Ryan AF, Miller JM, Pfingst BE, Martin GK (1984) Effects of reaction time performance on single-unit activity in the central auditory pathway of the rhesus macaque. J Neurosci 4:298–308.

Ryugo DK, Fekete DM (1982) Morphology of primary axosomatic endings in the anteroventral cochlear nucleus of the cat: a study of the endbulbs of Held. J Comp Neurol 210:239–257.

Ryugo DK, Killackey HP (1974) Differential telencephalic projections of the medial and ventral divisions of the medial geniculate body of the rat. Brain Res 82:173–177.

Ryugo DK, Weinberger NM (1976) Corticofugal modulation of the medial geniculate body. Expl Neurol 51:377–391.

Ryugo DK, Weinberger NM (1978) Differential plasticity of morphologically distinct populations in the medial geniculate body of the cat during classical conditioning. Behav Biol 22:275–301.

Saini KD, Leppelsack H-J (1977) Neuronal arrangement of the auditory field L of the neostriatum of the starling. Cell Tissue Res 176:309–316.

Saini KD, Leppelsack H-J (1981) Cell types of the auditory caudomedial neostriatum of the starling (Sturnus vulgaris). J Comp Neurol 1981:209–229.

Sakai K, Salvert D, Kitahama K, Kimura H, Maeda T, Jouvet M (1983) Projections ascendantes et descendantes des neurones de l'hypothalamus posterior immunoreactifs à la serotonine après administration 5-hydroxytrytophane chez le chat. Comptes Rendus Academie des Sciences (Paris) Série III 296:1013–1018.

Sakurai Y (1990) Cells in rat auditory system have sensory-delay correlates during the performance of an auditory working memory task. Behav Neurosci 104:856–868.

Sally SL, Kelly JB (1988) Organization of auditory cortex in the albino rat: sound frequency. J Neurophysiol 59:1627–1638.

Scheel M (1988) Topographic organization of the auditory thalamocortical system in the albino rat. Anat Embryol 179:181–190.

Scheibel ME, Scheibel AB (1966) Patterns of organization in specific and non-specific thalamic fields. In: Purpura DP, Yahr MD (eds) The Thalamus. New York: Columbia University Press, pp. 13–46.

Scheibner T, Törk I (1987) Ventromedial mesencephalic tegmental (VMT) projections to ten functionally different cortical areas in the cat: topography and quantitative analysis. J Comp Neurol 259:247–265.

Scheich H (1983) Two columnar systems in the auditory neostriatum of the chick: evidence from 2-deoxyglucose. Expl Brain Res 51:99–105.

Scheich H, Bonke BA (1981) Tone-versus FM-induced patterns of excitation and suppression in the 14-C-2-deoxyglucose labeled auditory cortex of the guinea fowl. Expl Brain Res 44:445–449.

Scheich H, Langner G, Bonke D (1979) Responsiveness of units in the auditory neostriatum of the guinea fowl (Numida meleagris) to species-specific calls and synthetic stimuli. II. Discrimination of Iambus-like calls. J Comp Physiol A 132:257–276.

Schiffmann S, Campistron G, Tugendhaft P, Brotchi J, Flament-Durand J, Geffard M, Vanderhaeghen J-J (1988) Immunocytochemical detection of GABAergic nerve cells in the human temporal cortex using a direct γ-aminobutyric acid antiserum. Brain Res 442:270–278.

Schreiner CE, Cynader MS (1984) Basic functional organization of second auditory cortical field (AII) of the cat. J Neurophysiol 51:1284–1305.

Schreiner CE, Mendelson JR (1990) Functional topography of cat primary auditory cortex: distribution of integrated excitation. J Neurophysiol 64:1442–1459.

Schreiner CE, Urbas JV (1986) Representation of amplitude modulation in the auditory cortex of the cat. I. The anterior auditory field (AAF). Hear Res 21:227–241.

Schreiner CE, Urbas JV (1988) Representation of amplitude modulation in the auditory cortex of the cat. II. Comparison between cortical fields. Hear Res 32:49–64.

Schweitzer L, Cant NB (1984) Development of the cochlear innervation of the dorsal cochlear nucleus of the hamster. J Comp Neurol 225:228–243.

Segraves MA, Rosenquist AC (1982a) The distribution of the cells of origin of callosal projections in cat visual cortex. J Neurosci 2:1079–1089.

Segraves MA, Rosenquist AC (1982b) The afferent and efferent callosal connections of retinotopically defined areas in cat cortex. J Neurosci 2:1090–1107.

Seldon HL (1981) Structure of human auditory cortex. I. Cytoarchitectonics and dendritic distribution. Brain Res 229:277–294.

Seltzer B, Pandya DN (1978) Afferent cortical connections and architectonics of the superior temporal sulcus and surrounding cortex in the rhesus monkey. Brain Res 149:1–24.

Seltzer B, Pandya DN (1989a) Frontal lobe connections of the superior temporal sulcus in the rhesus monkey. J Comp Neurol 281:97–113.

Seltzer B, Pandya DN (1989b) Intrinsic connections and architectonics of the superior temporal sulcus in the rhesus monkey. J Comp Neurol 290:451–471.

Shamma SA, Symmes D (1985) Patterns of inhibition in auditory cortical cells in awake squirrel monkeys. Hear Res 19:1–13.

Shneiderman A, Oliver DL (1989) EM autoradiographic study of the projections of the dorsal nucleus of the lateral lemniscus — a possible source of inhibitory inputs to the inferior colliculus. J Comp Neurol 286:28–47.

Shosaku A, Kayama Y, Sumitomo I, Sugitani M, Iwama K (1989) Analysis of recurrent inhibitory circuit in rat thalamus: neurophysiology of the thalamic reticular nucleus. Prog Neurobiol 32:77–102.

Shute CCD, Lewis PR (1967) Ascending cholinergic limbic system — neocortical olfactory and subcortical projections. Brain 90:497–521.

Sillito AM, Kemp JA (1983) Cholinergic modulation of the functional organization of the cat visual cortex. Brain Res 289:143–155.

Sindberg RM, Thompson RF (1962) Auditory response fields in ventral temporal and insular cortex of cat. J Neurophysiol 25:21–28.

Smith DE, Moskowitz N (1979) Ultrastructure of layer IV of the primary auditory cortex of the squirrel monkey. Neuroscience 4:349–359.

Smith Y, Séguéla P, Parent A (1987) Distribution of GABA-immunoreactive neurons in the thalamus of the squirrel monkey (*Saimiri sciureus*). Neuroscience 22:579–591.

Sousa-Pinto A (1973a) Cortical projections of the medial geniculate body in the cat. Adv Anat Embryol Cell Biol 48:1–42.

Sousa-Pinto A (1973b) The structure of the first auditory cortex (AI) in the cat. I. Light microscopic observations on its structure. Arch Ital Biol 111:112–137.

Sousa-Pinto A, Paula-Barbosa MM, Matos MDC (1975) A Golgi and electron microscopical study of nerve cells in layer I of the cat auditory cortex. Brain Res 95:443–458.

Sovijärvi ARA (1975) Detection of natural complex sounds by cells in the primary auditory cortex of the cat. Acta Physiol Scand 93:318–335.

Sovijärvi ARA, Hyvärinen J (1974) Auditory cortical neurons in the cat sensitive to the direction of sound source movement. Brain Res 73:455–471.

Špaček J, Lieberman AR (1974) Ultrastructural and three-dimensional organization of synaptic glomeruli in rat somatosensory thalamus. J Anat (London) 117:486–516.

Starr A, Don M (1972) Responses of squirrel monkey medial geniculate body to binaural clicks. J Neurophysiol 35:501–517.

Steinschneider M, Arezzo JC, Vaughan HG Jr (1990) Tonotopic features of speech-evoked activity in primate auditory cortex. Brain Res 519:158–168.

Stępień I, Stępień L, Łubińska E (1990) Function of dog's auditory cortex in tests involving auditory location cues and directional instrumental responses. Acta Neurobiol Exp 50:1–12.

Stępniewska I, Rajkowska G (1989) The sensory projections to the frontal association cortex in the dog. Acta Neurobiol Exp 49:299–310.

Steriade M, Llinás R (1988) The functional states of the thalamus and the associated neuronal interplay. Physiol Rev 68:649–742.

Stiebler I (1987) A distinct ultrasound-processing area in the auditory cortex of the mouse. Naturwiss 74:96–97.

Stone J (1983) Parallel Processing in the Visual System. The Classification of Retinal Ganglion Cells and Its Impact on the Neurobiology of Vision. In:

Blakemore C (ed) Perspectives on Vision Research. New York and London: Plenum Press.

Striedter GF (1990a) The diencephalon of the channel catfish, *Ictalurus punctatus* I. Nuclear organization. Brain Behav Evol 36:329–354.

Striedter GF (1990b) The diencephalon of the channel catfish, *Ictalurus punctatus* II. Retinal, tectal, cerebellar and telencephalic connections. Brain Behav Evol 36:355–377.

Suga N (1978) Specialization of the auditory system for reception and processing of species-specific sounds. Fed Proc 37:2342–2354.

Suga N (1984a) Neural mechanisms of complex-sound processing for echolocation. Trends Neurosci 7:20–27.

Suga N (1984b) The extent to which biosonar information is represented in the bat auditory cortex. In: Edelman GM, Gall WE, Cowan WM (eds) Dynamic Aspects of Neocortical Function. New York: John Wiley & Sons, pp. 315–373.

Suga N (1988) What does single-unit analysis in the auditory cortex tell us about information processing in the auditory system? In: Rakic P, Singer W (eds) Neurobiology of Neocortex. Chichester: John Wiley & Sons Limited, pp. 331–350.

Suga N, Horikawa J (1986) Multiple time axes for representation of echo delays in the auditory cortex of the mustached bat. J Neurophysiol 55:776–805.

Suga N, O'Neill WE, Kujirai K, Manabe T (1983) Specificity of combination sensitive neurons for processing of complex biosonar signals in auditory cortex of the mustached bat. J Neurophysiol 49:1573–1626.

Suga N, Hiwa H, Taniguchi I, Margoliash D (1987) The personalized auditory cortex of the mustached bat: adaptation for echolocation. J Neurophysiol 58:643–654.

Supple WF Jr, Kapp BS (1989) Response characteristics of neurons in the medial component of the medial geniculate nucleus during Pavlovian differential fear conditioning in rabbits. Behav Neurosci 103:1276–1286.

Sur M, Garraghty PE, Roe AW (1988) Experimentally induced visual projections into auditory thalamus and cortex. Science 242:1437–1441.

Sutter ML, Schreiner CE (1991) Physiology and topography of neurons with multipeaked tuning curves in cat primary auditory cortex. J Neurophysiol 65: 1207–1226.

Swarbrick L, Whitfield IC (1972) Auditory cortical units selectively responsive to stimulus "shape." J Physiol (London) 224:68–69.

Symmes D, Anderson KV (1967) Reticular modulation of higher auditory centers in monkey. Expl Neurol 18:161–176.

Symmes D, Alexander GE, Newman JD (1980) Neural processing of vocalizations and artificial stimuli in the medial geniculate body of the squirrel monkey. Hear Res 3:133–146.

Tago H, McGeer PL, McGeer EG, Akiyama H, Hersh LB (1989) Distribution of choline acetyltransferase positive structures in the rat brainstem. Brain Res 495:271–297.

Takeuchi A (1987) The transmitter role of glutamate in the nervous system. Jpn J Physiol 37:559–572.

Tamai Y, Miyashita E (1989) Subcortical connections of an 'oculomotor' region in the ventral bank of the anterior ectosylvian sulcus in the cat. Neurosci Res 7:249–256.

Tarlov EV, Moore RY (1966) The tecto-thalamic connections in the brain of the rabbit. J Comp Neurol 126:403–422.

Teas DC, Kiang NY-s (1964) Evoked responses from the auditory cortex. Expl Neurol 10:91–119.

Tebēcis AK (1967) Are 5-hydroxytryptamine and noradrenaline inhibitory transmitters in the medial geniculate nucleus? Brain Res 6:780–782.

Tebēcis AK (1970a) Effects of monoamines and amino acids on medial geniculate neurones of the cat. Neuropharmacol 9:381–390.

Tebēcis AK (1970b) Properties of cholinoceptive neurones in the medial geniculate nucleus. Brit J Pharmacol 38:117–137.

Teich AH, McCabe PD, Gentile CC, Schneiderman LS, Winters RW, Lisowsky DR, Schneiderman N (1989) Auditory cortex lesions prevent the extinction of Pavlovian differential heart rate conditioning to tonal stimuli in rabbits. Brain Res 480:210–218.

Thompson RF, Sindberg RM (1960) Auditory response fields in association and motor cortex of cat. J Neurophysiol 23:87–105.

Tiihonen J, Hari R, Kaukoranta E, Kajola M (1989) Interaural interaction in the human auditory cortex. Audiology 28:37–48.

Tsuzuki K, Suga N (1988) Combination-sensitive neurons in the ventroanterior area of the auditory cortex of the mustached bat. J Neurophysiol 60:1908–1923.

Tunturi AR (1971) Classification of neurons in the ectosylvian auditory cortex of the dog. J Comp Neurol 142:153–166.

Tusa RJ, Palmer LA, Rosenquist AC (1981) Multiple cortical visual areas. Visual field topography in the cat. In: Woolsey CN (ed) Cortical Sensory Organization, Vol 2, Multiple Visual Areas. Clifton, NJ: Humana Press, pp. 1–31.

Uhl GR, Goodman RR, Kuhar MJ, Childers SR, Snyder SH (1979) Immunohistochemical mapping of enkephalin containing cell bodies, fibers and nerve terminals in the brain stem of the rat. Brain Res 166:75–94.

Ulinski PS (1983) Dorsal Ventricular Ridge. A Treatise on Forebrain Organization in Birds and Reptiles. In: Northcutt RG (ed) Wiley Series in Neurobiology. New York: John Wiley & Sons.

Ungerstedt U (1971) Stereotaxic mapping of monamine pathways in rat brain. Acta Physiol Scand Suppl 367:1–48.

Usami S, Hozawa J, Tazawa M, Igarashi M, Thompson GC, Wu J-Y, Wenthold RJ (1989) Immunocytochemical study of the GABA system in chicken vestibular endorgans and the vestibular ganglion. Brain Res 503:214–218.

Van Buren JM, Borke RC (1972) Variations and Connections of the Human Thalamus. New York: Springer-Verlag.

Vaughan DW (1983) Thalamic and callosal connections of the rat auditory cortex. Brain Res 260:181–189.

van Noort J (1969) The Structure and Connections of the Inferior Colliculus. An Investigation of the Lower Auditory System. The Netherlands: Van Gorcum & Company, NV.

Vincent SR, Hökfelt T, Skirboll LR, Wu J-Y (1983) Hypothalamic γ-aminobutyric acid neurons project to the neocortex. Science 220:1309–1311.

von Economo C (1929) The Cytoarchitectonics of the Human Cerebral Cortex, trans Parker S. London: Humphrey Milford and Oxford University Press.

Walker AE (1938) The Primate Thalamus. Chicago and London: University of Chicago Press.

Wallach H (1940) The role of head movements and vestibular and visual cues in sound localization. J Expl Psychol 27:339–368.

Warr WB (1982) Parallel ascending pathways from the cochlear nucleus: neuroanatomical evidence of functional specialization. In: Neff WD (ed) Contributions to Sensory Physiology 7, New York: Academic Press, pp. 1–38.

Wegener JG (1973) The sound localizing behavior of normal and brain damaged monkeys. J Aud Res 13:191–219.

Wegener JG (1976) Auditory and visual discrimination following lesions of the anterior supratemporal plane in monkeys. Neuropsychol 14:161–173.

Weinberger NM (1982) Sensory plasticity and learning: the magnocellular medial geniculate nucleus of the auditory system. In: Woody CD (ed) Conditioning: Representation of Involved Neural Functions. New York and London: Plenum Press, pp. 697–710.

Weinberger NM, Diamond DM (1987) Physiological plasticity in auditory cortex: rapid induction by learning. Prog Neurobiol 29:1–55.

Weinberger NM, Hopkins W, Diamond DM (1984) Physiological plasticity of single neurons in auditory cortex of the cat during acquisition of the pupillary conditioned response: I. Primary field (AI). Behav Neurosci 98:171–188.

Weinberger NM, Imig TJ, Lippe WR (1972) Modification of unit discharges in the medial geniculate nucleus by click-shock pairing. Expl Neurol 36:46–58.

Weiskrantz L, Mishkin M (1958) Effects of temporal and frontal cortical lesions on auditory discrimination in monkeys. Brain 81:406–414.

Wenstrup JJ, Winer JA (1987) Projections to the medial geniculate body from physiologically defined frequency representations of the mustached bat's inferior colliculus. Proc Soc Neurosci 13:324.

Wepsic JG, Sutin J (1964) Posterior thalamic and septal influence upon pallidal and amygdaloid slow-wave and unitary activity. Expl Neurol 10:67–80.

Whitfield IC (1982) Coding in the auditory cortex. In: Neff WD (ed) Contributions to Sensory Physiology 6, New York: Academic Press, pp. 159–178.

Whitfield IC, Purser D (1972) Microelectrode study of the medial geniculate body in unanaesthetized free-moving cats. Brain Behav Evol 6:311–322.

Whitley JM, Henkel CK (1984) Topographical organization of the inferior colliculus projection and other connections of the ventral nucleus of the lateral lemniscus in the cat. J Comp Neurol 229:257–270.

Wickesberg RE, Oertel D (1988) Tonotopic projection from the dorsal to the anteroventral cochlear nucleus of mice. J Comp Neurol 268:389–399.

Winer JA (1984a) Anatomy of layer IV in cat primary auditory cortex (AI). J Comp Neurol 224:535–567.

Winer JA (1984b) The pyramidal cells in layer III of cat primary auditory cortex (AI). J Comp Neurol 229:476–496.

Winer JA (1984c) The non-pyramidal neurons in layer III of cat primary auditory cortex (AI). J Comp Neurol 229:512–530.

Winer JA (1984d) The human medial geniculate body. Hear Res 15:225–247.

Winer JA (1984e) Identification and structure of neurons in the medial geniculate body projecting to primary auditory cortex (AI) in the cat. Neuroscience 13: 395–413.

Winer JA (1985a) Structure of layer II in cat primary auditory cortex (AI). J Comp Neurol 238:10–37.

Winer JA (1985b) The medial geniculate body of the cat. Adv Anat Embryol Cell Biol 86:1–98.

Winer JA (1986) Neurons accumulating [³H]gamma-aminobutyric acid (GABA) in supragranular layers of cat primary auditory cortex (AI). Neuroscience 19:771–793.

Winer JA (1991) Anatomy of the medial geniculate body. In: Altschuler RA, Bobbin RP, Clopton BM, Hoffman DW (eds) Neurobiology of Hearing, Vol 2, The Central Auditory System. New York: Raven Press, pp. 293–333.

Winer JA (1992) Neuronal organization of layer V in cat primary auditory cortex (AI). (in preparation).

Winer JA, Larue DT (1987) Patterns of reciprocity in auditory thalamocortical and corticothalamic connections: study with horseradish peroxidase and autoradiographic methods in the rat medial geniculate body. J Comp Neurol 257:282–315.

Winer JA, Larue DT (1988) Anatomy of glutamic acid decarboxylase (GAD) immunoreactive neurons and axons in the rat medial geniculate body. J Comp Neurol 278:47–68.

Winer JA, Larue DT (1989) Populations of GABAergic neurons and axons in layer I of rat auditory cortex. Neuroscience 33:499–515.

Winer JA, Larue DT (1992) Evolution of GABAergic organization in the mammalian medial geniculate complex. (in preparation).

Winer JA, Morest DK (1983a) The medial division of the medial geniculate body of the cat: implications for thalamic organization. J Neurosci 3:2629–2651.

Winer JA, Morest DK (1983b) The neuronal architecture of the dorsal division of the medial geniculate body of the cat. A study with the rapid Golgi method. J Comp Neurol 221:1–30.

Winer JA, Morest DK (1984) Axons of the dorsal division of the medial geniculate body of the cat: a study with the rapid Golgi method. J Comp Neurol 224:344–370.

Winer JA, Peterson BA (1988) Origins of auditory corticothalamic projections onto the cat medial geniculate body. Proc Soc Neurosci 14:492.

Winer JA, Wenstrup JJ (1992a) The neurons of the medial geniculate body in the mustached bat (Pteronotus parnellii). (submitted).

Winer JA, Wenstrup JJ (1992b) Cytoarchitecture of the medial geniculate body of the mustached bat (Pteronotus parnellii). (submitted).

Winer JA, Morest DK, Diamond IT (1988) A cytoarchitectonic atlas of the medial geniculate body of the opossum, Didelphys virginiana, with a comment on the posterior intralaminar nuclei of the thalamus. J Comp Neurol 274:422–448.

Winer JA, Diamond IT, Raczkowski D (1977) Subdivisions of the auditory cortex of the cat: the retrograde transport of horseradish peroxidase to the medial geniculate body and posterior thalamic nuclei. J Comp Neurol 176:387–418.

Winer JA, Wenstrup JJ, Larue DT (1992) Patterns of GABAergic immunoreactivity define subdivisions of the mustached bat's medial geniculate body. J Comp Neurol 319:172–190.

Winguth SD, Winer JA (1986) Corticocortical connections of cat primary auditory cortex (AI): laminar organization and identification of supragranular neurons projecting to area AII. J Comp Neurol 248:36–56.

Wise LZ, Irvine DRF (1983) Auditory response properties of neurons in deep layers of cat superior colliculus. J Neurophysiol 49:674–685.

Witter MP, Groenenwegen HJ (1986) Connections of the parahippocampal cortex in the cat. III. Cortical and thalamic efferents. J Comp Neurol 252:1–31.

Witter MP, Groenenwegen HJ, Lopes da Silva FH, Lohman AHM (1989) Functional organization of the extrinsic and intrinsic circuitry of the parahippocampal region. Prog Neurobiol 33:161–254.

Wong D, Kelly JP (1981) Differentially projecting cells in individual layers of the auditory cortex: a double-labeling study. Brain Res 230:362–366.

Wong D, Shannon SL (1988) Functional zones in the auditory cortex of the echolocating bat, *Myotis lucifugus*. Brain Res 453:349–352.

Wong WC (1967) The tangential organization of dendrites and axons in three auditory areas of the cat's cerebral cortex. J Anat (London) 101:419–433.

Woody CD, Gruen E, Melamed O, Chizhevsky V (1991) Patterns of unit activity in the rostral thalamus of cats related to short-latency discrimination between different auditory stimuli. J Neurosci 11:48–58.

Woolsey CN, Walzl EM (1942) Topical projection of nerve fibers from local regions of the cochlea to the cerebral cortex of the cat. Bull Johns Hopkins Hosp 71:315–344.

Wu MF, Mallick BN, Siegel JM (1989) Lateral geniculate spikes, muscle atonia and startle response elicited by auditory stimuli as a function of stimulus parameters and arousal state. Brain Res 499:7–17.

Wuarin JP, Kim YI, Cepeda C, Tasker JG, Walsh JP, Peacock WJ, Buchwald NA, Dudek FE (1990) Synaptic transmission in human neocortex removed for treatment of intractable epilepsy in children. Ann Neurol 28:503–511.

Yeshurun Y, Wollberg Z, Dyn N, Allon N (1985) Identification of MGB cells by Volterra kernels. I. Prediction of responses to species specific vocalizations. Biol Cybernet 51:383–390.

Yeterian EH, Pandya DN (1989) Thalamic connections of the cortex of the superior temporal sulcus in the rhesus monkey. J Comp Neurol 282:80–97.

Yingcharoen K, Rinvik E, Storm-Mathisen J, Ottersen OP (1989) GABA, glycine, glutamate, aspartate and taurine in the perihypoglossal nuclei: an immunocytochemical investigation with particular reference to the issue of amino acid colocalization. Expl Brain Res 78:345–357.

Zeki S, Shipp S (1988) The functional logic of cortical connections. Nature 335:311–317.

Zilles K (1985) The Cortex of the Rat. A Stereotaxic Atlas. Berlin: Springer-Verlag.

Zook JM, Winer JA, Pollak GD, Bodenhamer RD (1985) Topology of the central nucleus of the mustache bat's inferior colliculus: correlation of single unit properties and neuronal architecture. J Comp Neurol 231:530–546.

7
Organization of Olivocochlear Efferent Systems in Mammals

W. BRUCE WARR

1. Introduction

The decade of the eighties witnessed important advances in our under-standing of virtually every aspect of hearing, including the basic inner-vation pattern of the organ of Corti. With the aid of such powerful tech-niques as retrograde and anterograde transport of tracers, anatomical reconstruction from serial images, the combination of single unit elec-trophysiology with intracellular fiber staining and immunocytochemical localization of neuroactive substances, these research endeavors revealed a quite unexpected degree of specificity in the detailed anatomy of various types of auditory neurons, both afferent and efferent. Taken in their en-tirety, these new findings have required major modifications in our views of the underlying neuronal mechanisms involved in the process of hearing.

In addition, major advances in our understanding of the physiology of the cochlear efferents have also occurred in the last few years. Although the function of the cochlear efferents is beyond the scope of this chapter, these physiological studies indicate that efferent fibers contacting outer hair cells suppress cochlear sensitivity through a mechanism involving contractile responses of outer hair cells (Kim 1986; Brownell 1990; Dallos et al. 1991). Detailed discussions of the functional role of cochlear effer-ents will be found in Guinan (1988), Liberman (1988, 1991) and Rajan (1990). In addition, an intriguing synthesis of the way in which olivo-cochlear and middle ear reflex loops may combine to affect peripheral acoustic transmission has been presented by Kiang et al. (1987).

The purpose of this chapter is to summarize and evaluate the literature on the efferent innervation of the mammalian auditory receptor. This review covers the neuroanatomy of the cochlear efferent neurons, includ-ing their (1) distribution and numbers; (2) cellular morphology and af-ferent inputs; (3) projections; and (4) immunocytochemistry. Consider-ation will also be given to species differences, including humans, and to intraspecies variability, where possible. Using the foregoing list of topics as a basic format, this chapter consists of descriptions of, and contrasts between, what have come to be known as the lateral and medial olivo-cochlear systems, LOC and MOC, respectively. These comprise the two basic groups of neurons which provide the efferent innervation of the mammalian organ of Corti. Evidence suggesting the need for recognizing further subgroups of efferent neurons will also be presented. For the sake of clarity and ease of access, the subject matter will be grouped under descriptive headings.

This chapter does not cover all of the descending auditory pathways of the brain. For this the reader is referred to a recent review by Huffman and Henson (1990) and to a forthcoming chapter (Spangler and Warr 1991). Nor does the present chapter cover the efferent innervation of the lateral line organs or auditory hair cell receptor organs of nonmammalian vertebrates, but the interested reader is again referred to recent reviews (Fritzsch et al. 1989; Meredith 1988; Will and Fritzsch 1988; Roberts and Meredith 1989). Finally, this chapter does not deal with the development of the cochlear efferents, but this topic is the subject of a forthcoming review of auditory development in general (Walsh and Romand 1992).

2. Definition of the Lateral and Medial Efferent Systems

Considerable evidence has accumulated in support of the existence of two separate systems of olivocochlear neurons, LOC and MOC, which can be distinguished both according to their projections and, as their names imply, according to the location of their respective neuronal cell bodies in the lateral and medial regions of the superior olivary complex (Warr 1975; Warr and Guinan 1979).

Chief among the reasons for distinguishing between the LOC and MOC systems is that tract-tracing studies in the cat had demonstrated that LOC neurons project mainly to the ipsilateral cochlea and terminate in the inner hair cell region, whereas MOC neurons project mainly to the con-tralateral cochlea and terminate beneath the outer hair cells (Warr and Guinan 1979; Warr 1980; Guinan et al. 1983). This observation accords well with the fact that only two basic kinds of efferent fibers can be recognized in the organ of Corti with single-fiber labeling techniques: those that terminate in the inner hair cell region, and those that terminate in the outer hair cell region (Liberman and Brown 1986; Brown 1987; Wilson et al. 1991). In addition, electron microscopic analysis of the organ of Corti shows that the efferent innervation consists morphologically of two different fiber types: a population of thin fibers synapsing mainly with the afferent nerve fibers in the region underneath the inner hair cells (Smith 1961; Smith and Rasmussen 1963; Spoendlin 1966; Iurato et al. 1978; Liberman 1980a), and a population of relatively thick fibers syn-apsing at the base of the outer hair cells (Kimura and Wersall 1962; Wright and Preston 1973; Nakai and Igarashi 1974; Iurato et al. 1978; Ginzberg and Morest 1984).

3. Distribution of OC Neurons in the Brainstem

3.1 Origin and Course of the Olivocochlear Bundle

The origin and course of the olivocochlear bundle (OCB) in the cat is shown schematically in Figure 7.1. Both the thin, unmyelinated axons of LOC neurons and the thicker, myelinated axons of MOC neurons travel dorsomedially (Osen and Roth 1969), often passing through the neuropil of adjacent nuclei and sometimes meandering quite close to the midline (Thompson and Thompson 1986; Warren and Liberman 1989). All eventually gather into a compact bundle beneath the facial genu in close proximity to the floor of the fourth ventricle where fibers destined for the cochlea on the opposite side turn medially and cross the midline. Those destined for the same side turn laterally and join with fibers from the opposite side. In its further course, the OCB breaks up into a series of fascicles (not shown in Fig. 7.1), which weave through the facial root before again condensing into a compact bundle in the root of the vestibular nerve (Rasmussen 1946; White and Warr 1983; Thompson and Thompson 1986). Some OCB fibers send collaterals to the cochlear nucleus which terminate mainly in granule cell regions (Brown et al. 1988a, 1991; Godfrey et al. 1990). For detailed descriptions of the various routes taken by these collaterals in the rat, the reader should consult White and Warr (1983), Osen et al. (1984), and Godfrey et al. (1987).

The OCB exits from the brain as a ventral component of the inferior division of the vestibular nerve. Just distal to the saccular ganglion, the OCB leaves the saccular nerve in Oort's vestibulocochlear anastomosis (VCA) and pursues a mainly apical course in the intraganglionic spiral bundle, but species differences exist in the detailed course of the OCB beyond the VCA (Rasmussen 1953; Gacek 1961; Arnesen and Osen 1978, 1984; Brown 1987).

Evidence that the axons of MOC neurons are myelinated comes from the direct examination of labeled fibers (Fig. 7.8) and tracing of these fibers to their cells of origin after physiological recording and intracellular marking with HRP (Liberman and Brown 1986). Light and electron microscopic evidence that LOC neurons have unmyelinated axons was provided by Ryan et al. (1987) following selective retrograde labeling of LOC neurons with D-[^{3}H]aspartic acid. Furthermore, as shown in Figure 7.2, the fact that the average number of myelinated and unmyelinated fibers in the VCA in the cat closely matches the average number of MOC and LOC neurons, respectively, labeled by retrograde transport from the cochlea, is important evidence that the two populations in fact correspond.

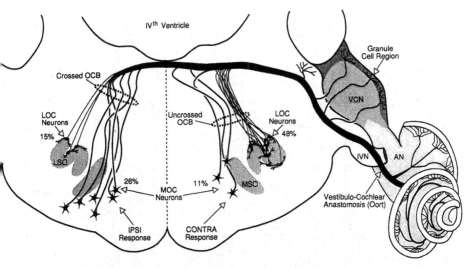

FIGURE 7.1. Origins and course of the olivocochlear bundle in the cat, based in part on retrograde labeling with HRP injections into the cochlea which determined the percent of crossed and uncrossed LOC and MOC neurons (Warren and Liberman 1989) and in part on combined single unit recordings and HRP-staining of OCB fibers in the VCA to determine the location of cell bodies of origin driven by ipsilateral (IPSI) or contralateral (CONTRA) sound (Liberman and Brown 1986). Thin lines connecting LOC neurons represent thin unmyelinated fibers; thick lines connecting MOC neurons represent myelinated fibers. As is true of all species studied to date, LOC neurons form the single largest group of OC neurons (48% + 15% = 63%); these project predominantly ipsilaterally (48:15 = 3.2:1) whereas MOC neurons project predominantly to the contralateral side (26:11 = 2.4:1). (Diagram generously provided by M.C. Liberman.)

Higher totals obtained for myelinated fibers in the VCA than for labeled MOC cell bodies could in part be accounted for by the existence of a small percentage of MOC neurons which project to both cochleas (see Section 3.4).

3.2 Maps of OC Neurons in Different Species

Using the retrograde transport method, OC neurons have been labeled, mapped and counted in eleven species, including the bats *Rhinolophus rouxi* (a species similar to the horseshoe bat, *Rhinolophus ferrumequinum*), *Phylostomus discolor, Tadarida* sp., *Rhinopoma* sp.) (Aschoff and Ostwald 1987), Pteronotus parnelii, the Mustache Bat (Bishop and Henson 1987), the cat (Warr 1975; Warr et al. 1982, 1986; Adams 1983; Warren and Liberman 1989), the gerbils *Meriones unguiculatus, Meriones tristrami,* and *Pachyuromys duprasi* (Aschoff et al. 1988), the guinea pig

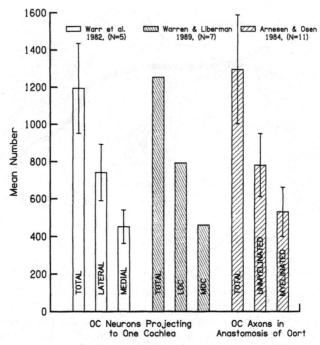

FIGURE 7.2. Mean numbers of retrogradely labeled OC neurons in the superior olivary complex (left and center) and of OC axons in the vestibulocochlear anatomosis of Oort (right) on the cat. Range bars represent one standard deviation. In data from Warr et al. 1982, counts from one aberrant case (Case #51) were excluded. For the same reason, case #2534 was excluded from Arnesen and Osen (1984). Data from Warren and Liberman (1989) was obtained by means of a digitizer tablet from their Figure 10C. (Modified from Warr et al. 1986, Figure 2 and printed here with permission of the publisher.)

(Robertson 1985; Aschoff and Ostwald 1987, 1988; Robertson et al. 1987a; Tokunaga 1988), the mouse (Campbell and Henson 1988), the rat (White and Warr 1983; Aschoff and Ostwald 1988; Robertson et al. 1989; Vetter and Mugnaini 1990) and the squirrel monkey, *Saimiri sciureus* (Thompson and Thompson 1986).

The range of species studied can be extended to the chinchilla if maps obtained with acetylcholinesterase histochemistry are included (Osen et al. 1984) and to Old World monkeys, the chimpanzee and humans, if published abstracts lacking maps are included (Strominger et al. 1981; Adams 1986; Parnes et al. 1986). In addition, counts of myelinated and unmyelinated fibers in the VCA have been made in human (Arnesen 1984).

The distribution and number of OC neurons in species for which maps are available have been summarized in Table 7.1 and will be discussed below.

3.3 The General Pattern of Distribution and Some Exceptions

In all species studied to date, OC neurons appear to be readily classifiable as LOC or MOC based on their location within the superior olivary complex. Specifically, LOC neurons are characterized by having a more or less intimate spatial relationship with the lateral superior olivary nucleus (LSO) or its afferents from the ipsilateral ventral cochlear nucleus, whereas the location of MOC neurons, while varying greatly in different species, is always in the medial, ventral or rostral periolivary zones as identified by Osen and her colleagues (Osen et al. 1984). Ipsilaterally, LOC neurons appear to be distributed along the full extent of the characteristic frequencies (CF) represented in the LSO, but those on the contralateral side are usually limited to regions of the nucleus representing lower frequencies (Warr 1975; Robertson et al. 1987a; Ryan et al. 1987). Comparable differences in distribution for crossed versus uncrossed MOC neurons have not been reported, probably because the tonotopic organization of the medial periolivary region is not known in most species.

The data in Table 7.1 support three conclusions: (1) there are always more LOCs than MOCs; (2) LOCs always project predominantly to the

TABLE 7.1. Distribution of olivocochlear neurons in various species.

Species	Ipsilateral SOC		Contralateral SOC		Approximate total number
	LOC[a]	MOC[a]	MOC[a]	LOC[a]	
Cat	48	11	26	15	1,366 (VCA)
Gerbil (*Meriones unguic.*)	75	8.5	11.5	5	995 (DY)
Guinea pig	51	17	31	1	2,346 (FB)
Monkey (squirrel)	55	10	15	20	880 (FB and DY)
Mouse	65	8.5	26	0.5	475 (HRP)
Rat	54.3	21.3	24.4	0	1,055 (FB and DY)
Bat (mustached)	75	7.5	17.5	0	1,585 (HRP)
Bat (horseshoe)	100	0	0	0	917 (FB)

[a] Expressed as percent of total population.

Data are arranged in columns from left to right according to their relative positions in the brainstem. Counts obtained from retrograde transport of fluorescent dyes were included where available. Total numbers of OC neurons in different species shown in the table are not necessarily comparable since they are based on the retrograde transport of either HRP (cat, mouse, and mustached bat) or the more sensitive fluorescent dyes (gerbil, guinea pig, squirrel monkey, rat, and horseshoe bat). Data derived from the following sources: cat, (Arnesen and Osen 1984; Warren and Liberman 1989); gerbil (Aschoff et al. 1988); guinea pig (Aschoff and Ostwald 1987); squirrel monkey (Thompson and Thompson 1986); mouse (Campbell and Henson 1988); rat (Robertson et al. 1989); mustached bat (Bishop and Henson 1987); horseshoe bat (Aschoff and Ostwald 1987). Abbreviations: DY, diamidino yellow; FB, fast blue; HRP, horseradish peroxidase; LOC, lateral olivocochlear neurons; MOC, medial olivococlear neurons.

same side; and (3) MOCs always project predominantly to the opposite side, although the proportion of crossed to uncrossed neurons is usually more balanced than it is for LOC neurons, with the exception of squirrel monkey and mouse. That LOC neurons outnumber MOC neurons appears to apply to the OCB in humans as well, because the VCA contains more unmyelinated than myelinated fibers, the mean ratio being 3:1 (Arnesen 1984). This ratio is higher than in the cat, where it is approximately 1.7:1 (Arnesen and Osen 1984; Warren and Liberman 1989).

Some investigators have questioned the use of the term "medial" to characterize cells which may often be located not medially, but rostrally and laterally in the superior olivary complex. Indeed, some MOC neurons in the rostral periolivary region may be situated as far lateral as LOC neurons (Adams 1983; Aschoff and Ostwald 1987; Aschoff et al. 1988). But these periolivary neurons are innervated predominantly from the contralateral cochlear nucleus, and this fact is a decisive criterion for MOC neurons. Nevertheless, certain workers have preferred to distinguish "small" neurons in the LSO from "large" extra-LSO OC neurons, but this terminology is not strictly accurate either, since in most species the cell sizes of the two populations overlap (see 4.2 below).

Another semantic issue concerns the rostral periolivary zone which some investigators identify as the caudal part of the ventral nucleus of the lateral lemniscus (VNLL). However, MOC neurons of the ventral nucleus of the trapezoid body (VNTB) form an apparently continuous column of loosely arranged, large multipolar cells which extends to about the level of the densely populated columnar (or "strip") region of the VNLL proper (White and Warr 1983; Guinan et al. 1983), and so little would appear to be gained by making this VNTB/VNLL distinction. It should be noted, however, that in guinea pig, this rostral area contains about 20% of the MOC neuronal population (Robertson et al. 1987a; Stopp 1990) and 50% of those MOCs which project simultaneously to both cochleas (Robertson et al. 1987b). While this suggests some degree of functional specialization for this region, evidence from the cat suggests that such specialization may merely reflect a tonotopic representation of low CFs (Guinan et al. 1984; Liberman and Brown 1986).

3.3.1   Is the Variability of OC Populations Exceptional?

The absolute numbers of OC neurons counted and their proportional distributions in the lateral and medial regions for any given species are characterized by large standard deviations (Fig. 7.2) and the occasional appearance of frankly aberrant cases. This is in marked contrast to the motoneuronal populations contributing to the rat sciatic nerve, the standard deviations of which are less than 5% of the mean (Swett et al. 1986) or to cat stapedial neuronal populations which, under optimal experimental conditions, may exhibit a range of less than 2% of the mean (Joseph et al. 1985).

The following examples bear witness to the variability of OC neuronal populations. Fiber counts of the cat's VCA in 12 specimens yielded totals ranging from 860–1,810, a mean of 1356 and a standard deviation of 300, or 22% of the mean (Arnesen and Osen 1984). In an HRP study in the cat, one animal was observed to have substantially more MOC than LOC neurons, which is a reversal of the norm (Warr et al. 1982). Retrograde labeling studies in guinea pigs show great variability in total counts of OC neurons (Robertson et al. 1987a; Tokunaga 1988), and individual rats have sometimes been found to have more uncrossed than crossed MOCs, yet another reversal of the norm (Aschoff and Ostwald 1988; Vetter and Mugnaini 1990). Such within-group variability has received comment from several investigators who have generally interpreted it as methodological in origin, as some of it is; witness the consistently larger counts obtained with fluorescent dyes than with HRP (Robertson et al. 1987a). However, there is evidence that much of this intraspecies variability is real, especially in the case of fiber counts of the VCA where the efficiency of retrograde transport or histochemical variables are not at issue.

In light of the foregoing considerations, any interspecific comparison of the number of OC neurons should ideally be performed on data obtained with methods of comparable sensitivity. Therefore, each of the estimated totals of OC neurons in Table 7.1 is qualified parenthetically with an indication of the method by which the number was obtained.

It should be noted that, in spite of the shortcomings of the available data, Bishop and Henson (1987) show that disparities between species in counts of LOC and MOC are markedly reduced when each is divided by the appropriate basilar membrane length.

3.3.2   Are OC Neuron Maps Species Specific?

The details of where OC neurons are actually located within the superior olivary complex is seemingly unique to each species. There does, however, appear to be some predictability in the location of LOC neurons. In rodents, more than 90% of all LOC neurons are found within the LSO proper, whereas in primates and bats (Strominger et al. 1981; Adams 1986; Thompson and Thompson 1986; Aschoff and Ostwald 1987; Bishop and Henson 1987), they are located in a nucleus dorsal to the LSO. In the cat, LOC neurons lie in the hilus and on the margins of the LSO, but extend dendrites into the LSO neuropil (see Fig. 7.1).

The LSO is the only nucleus of the superior olivary complex which is known to receive a complete ipsilateral frequency representation from the ventral cochlear nucleus (Guinan 1972; Tsuchitani 1977). LOC neurons, likewise, appear to be distributed throughout this frequency representation and to project tonotopically to the entire length of the cochlea. Perhaps these functional relationships account for the intimate association of LOC neurons with this nucleus.

As for the distributions of MOCs across species, the one feature all MOC neurons seem to have in common is that they inhabit regions which are traversed by trapezoid body fibers ascending from the cochlear nuclei of both, but principally the opposite, sides. In addition, these regions are traversed by descending fibers from the inferior colliculus and by fibers belonging to olivary neurons destined for various targets local and distant. Based upon their location and typically long dendrites, MOC neurons should be among the best informed auditory neurons in the brainstem. However, no satisfactory explanation can be offered at present to account for the varied configurations of MOC neuronal distributions across species, perhaps because information on the basic connectional anatomy of the superior olivary complex in most species is still fragmentary.

3.3.2.1 Variation on the Basic Plan

Interspecific variation in OC neuron maps is of two types: (1) the natural deletion of one class of OC neuron, as observed only in the bat, Rhinolophus, which lacks MOC neurons entirely (Aschoff and Ostwald 1987) and lacks a corresponding efferent innervation of its outer hair cells (Bruns and Schmieszek 1980); and (2) the natural deletion of LOC neurons with crossed projections, as reported in rat (White and Warr 1983; Aschoff and Ostwald 1988; Vetter and Mugnaini 1990), and bat (Aschoff and Ostwald 1987; Bishop and Henson 1987).

The presence or absence of LOC neurons with crossed projections, which constitute 15% of the total OC population in the cat, is an aspect of considerable variation across the species listed in Table 7.1. The few neurons in this group are generally located in the lateral, low-frequency portion of the LSO in the cat as well as in other species.

There is a report (Thompson and Thompson 1986) that one small subgroup of squirrel monkey LOC neurons, distinguished by the large size of its neuronal cell bodies and their location lateral to the LSO (hence the name LSO-L), has strongly crossed projections (73%). However, because these cells comprise only 8% of the entire OC population, the LOC population as a whole remains predominantly uncrossed, in spite of this anomalous subgroup (see Table 7.1). A possibly related datum is that severing the crossed OCB in this same species resulted in the degeneration of 70% of the fibers in the inner spiral bundle beneath the inner hair cells in the basal turn (Nakai and Igarashi 1974), an effect far more profound than reported in any other animal (Kimura and Wersall 1962; Smith and Rasmussen 1963; Spoendlin 1966; Wright and Preston 1973; Iurato et al. 1978). Whether this finding is connected to the crossed projections of LSO-L will obviously need to be determined by more specific techniques. As expected, most of the efferent terminals on the outer hair cells in the basal turn also degenerated.

3.3.3   Are Ambivalently-Projecting MOC Neurons Binaural?

A second anomaly found in the squirrel monkey consists of a slight divergence from the general rule that MOC neurons have predominantly crossed projections. Specifically, a group of MOC neurons situated dorsal to the medial superior olivary nucleus (MSO) and having polar dendrites extending in the lateral and medial directions, designated MSO-d, was found to have bilaterally symmetrical projections, that is, equal numbers of cells with crossed and uncrossed projections (Thompson and Thompson 1986). One explanation for this ambivalence of projections may relate to the probable afferent input to these neurons, which, based upon their dendritic morphology and location dorsal to the medial superior olivary nucleus (MSO), are likely to be binaural and tuned to low frequencies, as is true of MSO neurons themselves (Chapter 4, this volume). The ambivalent sidedness of projections from the MSO-d cell group may be a reflection of the binaural inputs to this cell group.

3.4 MOC Neurons with Bilateral Projections

Robertson (Robertson and Gummer 1985; Robertson et al. 1987b) and his associates have suggested, based upon a similarity in numbers, that physiologically identified, binaurally-excitable MOC neurons in the guinea pig may correspond to a MOC subgroup identified histologically as having axons which bifurcate and send branches to both cochleas, certainly the extreme case of ambivalence of the kind discussed above. Interestingly, MSO-d neurons in the squirrel monkey were noted to be especially likely to have bilateral cochlear projections, as shown by double retrograde labeling, as explained below (Thompson and Thompson 1986).

Neurons with projections to both cochleas can be demonstrated by the double retrograde labeling technique which involves the injection of different tracers into the two cochleas. Double-labeled MOC neurons have been found in all species in which this technique has been applied, including cat (Adams 1982), rat (Aschoff and Ostwald 1988; Robertson et al. 1989), guinea pig (Robertson et al. 1987b), and squirrel monkey (Thompson and Thompson 1986). Although detailed maps of double labeled OC neurons are generally lacking, all appear to be MOCs and most are located in the ventral nucleus of the trapezoid body or the caudal part of the ventral nucleus of the lateral lemniscus. Such neurons appear to comprise about 5% or less of the medial OC neurons in a given species, except in the squirrel monkey where they are estimated to comprise at least 10%. Although not particularly numerous, these peculiar cells may point to an unsuspected importance of binaural interactions in the functioning of the MOC system, particularly for frequencies below 2 kHz, to which most binaurally driven MOC units in the cat are tuned (Liberman and Brown 1986).

3.5 A Connectionistic Interpretation of OC Neuron Maps

As originally suggested by Rasmussen, OC neurons appear to project preferentially to the cochlea from which they receive their main source of stimulation (Rasmussen 1960). This has now been directly confirmed for MOC neurons by means of single-unit recordings and retrograde labeling of the parent cell bodies, as indicated in Figure 7.1. The results show that the majority of units (approximately 2.5:1), do in fact respond to tonal stimuli delivered to the ear ipsilateral to the recording site in the VCA (Liberman and Brown 1986) and comprise the MOC portion of the crossed OCB.

It is also clear from Figure 7.1, however, that some considerable proportion of MOC neurons (see Table 7.1), project to the cochlea from which they cannot be unilaterally driven, and that these fibers originate from MOC neurons ipsilateral to the recording site in the VCA (Liberman and Brown 1986). These latter MOC neurons travel in the uncrossed OCB in company with the much larger contingent of uncrossed LOC axons and form the final neuronal link in a pathway interconnecting the two cochleas, as Fex originally observed and which has since been repeatedly confirmed (Fex 1965; Buno 1978; Cody and Johnson 1982; Robertson and Gummer 1985; Warren and Liberman 1989).

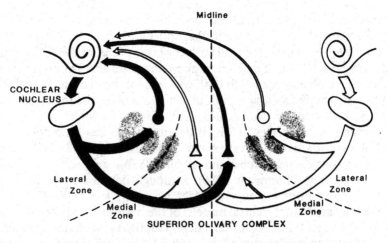

FIGURE 7.3. Organization of ascending inputs to olivocochlear neurons terminating on the left side. Solid arrows and symbols represent major pathways which are driven by acoustic stimulation of the left cochlea. Open arrows and symbols represent the smaller intercochlear pathways driven by stimulation of the right cochlea. A convergence of inputs from the two ears is indicated in the Medial Zone, accounting for the binaural responses of medial OC neurons. (From Spangler and Warr 1991, reprinted with permission)

A few MOC neurons, not indicated in Figure 7.1, have also been shown to respond equally well to tones in either ear (Robertson and Gummer 1985; Liberman and Brown 1986).

The ascending auditory connections which are presumed to underlie these phenomena are diagrammed in Figure 7.3. Lateral OC neurons lie in the projection field of the ipsilateral ventral cochlear nucleus, and medial OC neurons lie within the projection field of the contralateral ventral cochlear nucleus (Guinan et al. 1972; Warr 1982; Thompson and Thompson 1991). Paradoxically, therefore, the pathway connecting MOC neurons to their destination in the opposite cochlea, crosses the midline twice, first in the trapezoid body and once again with the fibers in the crossed OCB. A better name for this fiber contingent might be "double-crossed OCB"!

The small arrows branching towards the medial zone imply that, in fact, binaural convergence occurs there (Thompson and Thompson 1991). Although the MOC neurons driven by tones to either ear comprise a fairly small group, as discussed in Section 3.3.3 above, the responses of virtually all MOC neurons exhibit binaural facilitation to either tones or noise, depending on the magnitude of unit CF (Liberman 1988). The connectional basis of these response properties is not yet understood.

The responses of lateral OC neurons are still totally unknown, but based upon the fact that they are contacted by projections of the ipsilateral ventral cochlear nucleus and not by contralateral projections (Thompson and Thompson 1991), firing would presumably be driven by ipsilateral sounds, as are the principal cells of the LSO (Guinan et al. 1972). Thus, the predominant projections of LOC neurons are to the cochlea from which they receive auditory input. Like the MOC system, there is usually, but not always, a smaller contingent which projects contrariwise (see Table 7.1), and thus forms an LOC intercochlear pathway.

4. Morphology of OC Neurons

4.1 Cell Bodies and Dendrites of LOC Neurons

In all species studied, LOC neurons are smaller than MOC neurons (Warr 1975; Adams 1983; White and Warr 1983; Bishop and Henson 1987; Thompson and Thompson 1991) and in rodents, they are among the smallest neurons in the LSO itself (Helfert and Schwartz 1987; Tokunaga 1988). Their shape is most often described as ovoid or fusiform and they have an eccentric nucleus indented by infoldings of the nuclear membrane (Spangler et al. 1986; White 1986; Helfert et al. 1988). Because of their small size, scanty cytoplasm and the paucity of Nissl substance (Adams 1983; Helfert et al. 1988) they are inconspicuous in Nissl-stained sections, and their existence went undetected until the acetylcholinesterase (AChE)

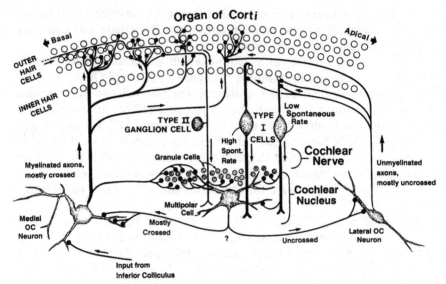

FIGURE 7.4. Schematic diagram summarizing known and hypothesized connections of olivocochlear neurons. Reflex circuits of a representative LOC neuron (right) and a representative MOC neuron (left) are depicted. A cochlear nuclear multipolar cell, with dendrites extending into the granule cell region, is shown hypothetically linking the periphery to the two kinds of OC neurons and receiving feedback in the form of a collateral from the MOC neuron. A projection by this multipolar neuron to OC neurons has yet to be demonstrated. (See text and Figure 7.11 for further details.)

technique was applied (Osen and Roth 1969). In some species, such as the bat *Pteronotus parnelii*, all of the LOC neurons appear to be similar (Bishop and Henson 1987), whereas in others there are a few larger neurons, as shown in Figure 7.5, not so intimately associated with the LSO (Adams 1983; White and Warr 1983; Thompson and Thompson 1986; Vetter and Mugnaini 1990). These resemble MOC neurons in their dendritic architecture and immunocytochemical reactivity to antibodies to choline acetyltransferase, but not antibodies to calcitonin gene-related peptide (see below).

The dendrites of LOC neurons are slender, cylindrical, relatively nontapering, and extend considerable distances within the LSO or its environs. As shown in Figure 7.5A and 7.4B, the dendrites of small LOC neurons in the rat span the width of the LSO parallel to the isofrequency planes of this nucleus (Tsuchitani 1977; Friauf and Ostwald 1988; Sanes and Rubel 1988). Because these dendrites are oriented anterior posteriorly in sagittal sections (Adams 1983; White and Warr 1983; Thompson and Thompson 1991; Vetter and Mugnaini submitted), these LOC neurons appear to have discoid dendritic fields, as do the principal cells of the LSO (Scheibel and Scheibel 1974; Sanes et al. 1990).

Figure 7.5A also shows that there is a gradient of decreasing LOC cell density from the high (upper left) to low frequency (lower right) within the rat LSO, as similarly observed in the gerbil (Ryan et al. 1987). In the cat, on the other hand, there appears to be a gradient of LOC density opposite to that found in rat and gerbil, although only the LOC dendrites penetrating the LSO are available as an index of probable characteristic frequency in this species (Warr 1975). In any case, the gradient in the cat accords well with the basal to apical increase in fiber counts of the inner spiral bundle (Spoendlin 1970) and of the efferent innervation density in the inner hair cell region in the same species (Liberman et al. 1990), as shown in Figure 7.7.

4.1.1 Inputs to LOC Neurons

Afferent innervation to LOC neurons has been found to arise in the ipsilateral posteroventral cochlear nucleus, as shown by light-microscopic tract-tracing studies in the guinea pig (Thompson and Thompson 1988, 1991), horseshoe bat (Vater and Feng 1990), and the cat (Warr 1982). Axon-tracing experiments in the guinea pig show that these inputs are associated exclusively with the LOC dendrites (Thompson and Thompson 1991) and the same evidence indicates that LOC neurons receive no projections from the contralateral posteroventral cochlear nucleus. In comparison to MOC neurons in the same cases, the innervation of LOCs is decidedly sparse. In addition, the possibility exists that other parts of the ventral cochlear nucleus or the ventral nucleus of the trapezoid body may also project to LOC neurons because they both send fibers to the LSO (Cant and Casseday 1986; Warr and Spangler 1989).

Consistent with the foregoing observations, electron microscopy of LOC neurons shows that the cell soma is virtually lacking in synaptic contacts and that the dendrites too are only sparsely innervated. The morphology of synaptic terminals contacting LOC neurons consists of a single type of bouton which contains small, round-oval (pleomorphic) synaptic vesicles and some larger vesicles with a dense core (Spangler et al. 1986; White 1986; Helfert et al. 1988). This morphological type of bouton is virtually identical to one of the three types of boutons found on the principal neurons of the LSO of the rat and cat (White 1983; Cant 1984) suggesting that LSO principal cells and LOC neurons may share a common input, probably originating from multipolar cells in the ventral cochlear nucleus (Fig. 7.4). The spherical and globular cells (Osen 1969; Cant and Casseday 1986) of the anteroventral cochlear nucleus are probably not a source of inputs to LOC neurons because the large round vesicles which characterize their terminals have not yet been found in contact with them (Cant 1984; Spangler et al. 1986; Helfert et al. 1988).

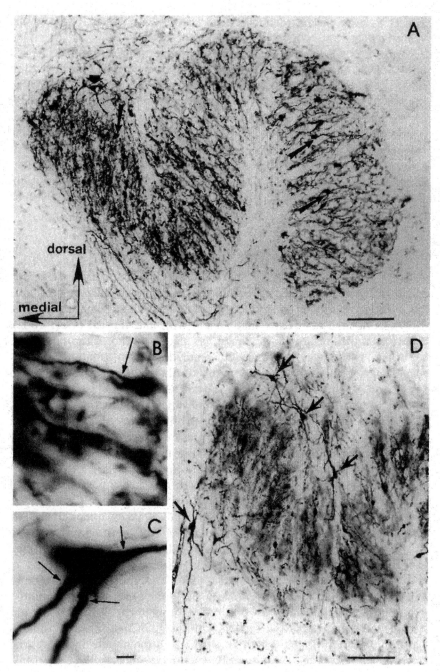

FIGURE 7.5. Caption on facing page.

4.2 Cell Bodies and Dendrites of MOC Neurons

MOC neurons are generally large to medium-sized stellate or triangular cells (Fig. 7.5C) which have large, irregular clumps of Nissl substance in their ample cytoplasm. They are usually among the largest cells found in the periolivary region (Adams 1983; Tokunaga 1988). As a population, MOC neurons are heterogeneous in size and shape (Adams 1983) so that LOC and MOC size distributions may overlap either considerably, as in squirrel monkey (Thompson and Thompson 1986) slightly, as in cat and guinea pig (Warr 1980; Robertson 1985; Tokunaga 1988) or virtually not at all, as in rat (White and Warr 1983; Aschoff and Ostwald 1987).

The dendrites of medial OC neurons are irregularly radiated, branched, and gradually tapered (Fig. 7.4 and 7.4C) as they spread out across the full span of trapezoid body fibers in which they are embedded (Osen and Roth 1969; Adams 1983; Osen et al. 1984; Vetter and Mugnaini submitted).

4.2.1 Inputs to MOC Neurons

Inputs to MOC neurons have been traced from the posteroventral cochlear nucleus bilaterally, but mainly (78%) from the nucleus of the opposite side (Thompson and Thompson 1991) and from the central nucleus of the ipsilateral inferior colliculus (Faye-Lund 1986; Vetter and Saldana 1990). The anteroventral cochlear nucleus has also been cited as a probable source of inputs to MOC neurons, based on fiber degeneration experiments (Warr 1982). Most terminals appear to contact MOC dendrites where electron microscopic study shows the majority of synaptic endings are located (Spangler et al. 1986; White 1986; Helfert et al. 1988). Input descending from the inferior colliculus also terminates on dendrites, but only on those extending ventrally towards the surface of the brainstem. Liberman (1988) has shown that MOC neurons are particularly sensitive to noise presented to the cochlea opposite to that from which they exhibit

---

FIGURE 7.5 Light micrographs of cellular morphology of LOC and MOC neurons in the rat, as revealed by retrograde transport of a cholera toxin conjugated with HRP (Vetter and Mugnaini, submitted). A. LOC neurons (long arrows) within the LSO ipsilateral to the injection form a dense dendritic plexus oriented across the width of the nucleus, parallel to the insofrequency planes of this nucleus. Larger LOC neurons (short arrow) are found in the margins of the medial lamina of the LSO. Frontal section. Bar = 100 mμm. B. LOC cell body extending a long, slender, nontapering dendrite. Frontal section. Scale same as in C. C. MOC cell body extending several thick, tapering dendrites (arrows) Bar = 5 mμm. D. Micrograph of the medial limb of the LSO showing a few large LOC neurons, termed shell neurons by Vetter and Mugnaini (submitted), scattered near its margins. Dendrites of some shell neurons penetrate the LSO neuropil. Frontal section. Bar = 100 mμm. (Plate generously assembled and supplied by D.E. Vetter.)

excitation and sharp tuning to tonal stimuli. The neuroanatomical basis for these responses to noise is unknown, but would appear to require the convergence and summation of inputs.

Consistent with the variety of their inputs, as many as four distinct types of synaptic endings contact MOC neurons, the most prevalent of which closely resembles the type found on LOC neurons, at least in the gerbil (Helfert et al. 1988), but similar studies in cat and rat are less clear on this (Spangler et al. 1985; White 1986). Nevertheless, Figure 7.4 suggests a possible projection, indicated with a question mark, from the same multipolar cell in the ventral cochlear nucleus to both LOC and MOC neurons. In addition, White (1986) found a type of bouton on MOC neurons in the rat which bears a strong resemblance to the calyces of Held in the medial nucleus of the trapezoid body of this species. Since globular cells of the ventral cochlear nucleus are the source of the calyces of Held, this same cell type is implicated as a source of inputs to MOC neurons in the rat.

5. Projections of OC Neurons

As pointed out above, a fundamental distinction between LOC and MOC neurons relates to the differences in their projections to the inner versus outer hair cell regions, respectively. Although substantial evidence has accumulated which confirms these basic differences in projections, some questions still remain about how absolute the differences are.

5.1 The Projections of LOC Neurons

The projections of LOC neurons are predominantly ipsilateral and travel in the OCB as unmyelinated fibers. Once in the cochlea, these thin (<1 mμ) fibers enter the organ of Corti and form the bulk of the inner spiral bundle and all of the tunnel spiral bundle (Guinan et al. 1983; Spangler and Warr 1987; Liberman et al. 1990). In the cat, LOC neurons are preferentially distributed in low frequency regions of the LSO, both ip-

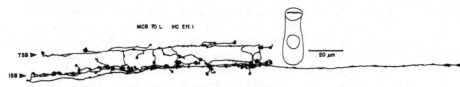

FIGURE 7.6. Camera lucida drawing of the terminal region of an HRP-labeled unidirectional LOC efferent fiber in the inner spiral bundle of the guinea pig. An inner hair cell is outlined for reference. Basal is to the left. (Reproduced from Brown, 1987, with the author's permission.)

silaterally and contralaterally (Warr 1975). This may account for the fact that the fiber population of the cat's inner spiral bundle increases nearly fivefold from base to apex (Spoendlin 1970). There is also an overall basal to apical increase in the efferent innervation density of radial afferent fibers, as shown in Figure 7.7 (Liberman et al. 1990).

5.1.1 Types of Terminal Fibers

Two types of inner hair cell efferent fibers, unidirectional and bidirectional have recently been identified in the basal turn of the guinea pig cochlea (Brown 1987). The great majority of fibers were classified as unidirectional (84%) because they spiralled either apically or basally less than 1 mm and had discrete (approximately 150 mμ), highly branched and varicose, terminal arbors, portions of which often contributed to the tunnel spiral bundle (Fig. 7.6). Bidirectional fibers, in contrast, bifurcated on entering the organ of Corti, sending long (>1 mm) spiral branches both apically and basally which form terminals all along their course. These fibers were more likely than unidirectional fibers to travel substantial distances in the tunnel spiral bundle. Brown (1987) notes in addition that bidirectional fibers often occurred in pairs or triplets, each of which was a terminal branch of a common parent axon. Both types of efferent fibers are presumed to originate from LOC neurons, but no information is yet available on any morphological differences between their respective cells of origin.

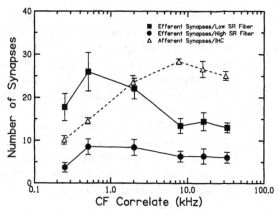

FIGURE 7.7. Comparison of the afferent and efferent innervation densities, apex to base, in the inner hair cell region of the cat, as determined by reconstructions from serial electron micrographs. The efferent innervation densities of the two categories of spontaneous rate (SR) radial afferent fibers are plotted separately. (Replotted from Figures 3A and 8A in Liberman et al. 1990).

5.1.2   Specificity of Postsynaptic Targets of LOC Fibers

The principal postsynaptic targets of this projection are the radial fibers, i.e., the peripheral processes of type I spiral ganglion cells (Spoendlin 1970; Kiang et al. 1982), two types of which have been recognized on the basis of their morphology and corresponding auditory sensitivity and spontaneous firing rate (SR) (Liberman 1980b, 1982; Liberman and Oliver 1984; see Chapter 2, this volume). Thick, high-SR, low-threshold fibers terminate preferentially on the pillar side of the inner hair cell, whereas thin, high-threshold, low-SR fibers terminate preferentially on the modiolar side, as shown schematically in Figure 7.4. Radial fibers synapsing on the low SR, modiolar side of the hair cell receive substantially greater number of efferent synapses than do those contacting the pillar side, most strikingly in the apical half of the cochlea, as shown graphically in Figure 7.7 (Liberman 1980a; Liberman et al. 1990). In addition, the efferent synapses on low-SR radial afferents often completely surround the postsynaptic element and are situated closer to the receptoneural terminal swelling than is the case for synapses on high-SR fibers (Liberman 1980a). LOC efferent synapses in the cat consist of electrondense puncta located within vesicle filled varicosities at points of intimate contact with radial fibers (Spoendlin 1984; Liberman et al. 1990). Electron microscopic profiles of these unusually *en passant* contacts often contain both clear and dense-cored vesicles ranging in size from 20–50 and 70–120 nanometers, respectively (Pujol and Lenoir 1986).

5.1.3   Lesser Cochlear Targets of LOC Innervation

Lesser postsynaptic targets of LOC axons are inner hair cells and MOC axons, the latter of which receive contacts as they traverse the tunnel of Corti in close proximity to the tunnel spiral bundle (Iurato et al. 1978; Liberman 1980a). As for the former, reconstructions from serial sections in the cat indicate that although all inner hair cells are contacted by at least a few efferent terminals, synaptic specializations are rare except in the most apical location examined (0.25 kHz) (Liberman et al. 1990). As for the latter, Brown (1987) has observed that MOC efferent fibers pass directly through the tunnel spiral bundle on their way to the outer hair cells and holds open the possibility that if most or all of the terminals and swellings found on tunnel spiral fibers are actually presynaptic, some form of axo-axonic interaction between the two kinds of efferents in the tunnel of Corti could occur (see Fig. 7.4). This would, presumably, not apply in the rat which lacks a tunnel spiral bundle.

5.1.4   Topographic Organization of LOC Projections

The existence of an orderly topographic projection from LOC neurons to discrete locations along the length of the organ of Corti has been demonstrated by anterograde transport in the cat (Guinan et al. 1984) and retrograde axonal transport in the guinea pig (Stopp 1983; Robertson et al. 1987c). Tracer injections into the lateral, low frequency, part of the LSO (see Section 4.1) produced a peak in terminal labeling in the expected region of the cochlear apex ipsilaterally, whereas progressively more medial injections produced peaks in appropriate regions more basally. Crossed LOC projections were preferentially to the apex of the cochlea, which is consistent with the location of contralaterally projecting LOC neurons in the low frequency portion of the LSO (Guinan et al. 1984). In the guinea pig, injections of diamidino yellow into single turns of the guinea pig cochlea produced isolated bands of retrogradely labeled LOC neurons in the LSO. Occasional labeled LOC neurons were found to be displaced from the major band of labeling. These may represent neurons with bidirectional fibers described by Brown in this species (Brown 1987).

5.1.5   LOC Collaterals to the Cochlear Nucleus?

Finally, the existence of collaterals from LOCs to the cochlear nucleus has been somewhat inconsistently reported, suggesting that their number is probably small and variable across species. Indeed, Godfrey et al. (1987) found, at most, only modest depletions of choline acetyltransferase in various microsamples of the rat cochlear nucleus after complete section of the OCB. Nevertheless, collateral innervation to what appears to be the medial cochlear nerve root region of the ventral cochlear nucleus, an important site of termination for low CF MOC collaterals in rodents (Brown et al. 1991), was found by the technique of retrograde/anterograde labeling of LOCs by D-[³H]aspartic acid in gerbil (Ryan et al. 1987, 1990). In addition, using the immunocytochemical localization of calcitonin gene-related peptide, a specific marker for LOC cell bodies, Lu et al. (1987) demonstrated calcitonin gene-related peptide-like immunoreactive axon terminals mainly in the dorsal cochlear nucleus of the cat, but in rat they were less restricted. The existence of collaterals from LOC axons to the cochlear nucleus has been denied by all investigators working in guinea pigs (Brown et al. 1988a; Winter et al. 1989). The cochlear nuclear collaterals of MOC neurons are discussed in Section 5.2.5 below.

5.1.6   Effects of LOC De-Efferentation

Beyond the direct effects of sectioning the OCB, such as the degeneration of all efferent fibers and terminals in the organ of Corti (Spoendlin 1966), a variety of secondary effects also occur, both acute and chronic. Six days after complete section of the efferent bundle or selective destruction of LOC neurons in the LSO, severe, transient swelling occurs in the radial fibers beneath the inner hair cells (Bodian and Gucer 1980; Spangler and Warr 1987). Based on the proliferation of filaments and miscellaneous organelles they observed, Bodian and Gucer interpreted this remarkable phenomenon as a unique kind of transient denervation atrophy.

Three or more weeks after cutting the entire OCB, its effects on the response characteristics of large populations of single units in the cochlear nerve were assessed (Liberman 1990). The findings showed that while response thresholds, tuning curves and rate-level functions were normal, there was a significant decrease in average rates of spontaneous activity in the de-efferented animals. In addition, Liberman noted that during unit recording sessions, there was an unusual tendency for the unit thresholds in these animals to shift progressively upwards, starting with the low frequencies and gradually spreading basally, until further recording was impossible.

5.2   Projections of MOC Neurons

5.2.1   Course and Branching Patterns of MOC Efferents

The projections of MOC neurons are predominantly contralateral and travel in the OCB and the VCA as the myelinated fiber component (Liberman and Brown 1986; Warr et al. 1986; Brown 1987). Except for the collaterals they send to the nucleus Y and to cochlear nucleus (see Section 5.2.5 below) these fibers do not branch until they are distal to the VCA. Major branching occurs in the intraganglionic spiral bundle (Fig. 7.8) from which daughter branches make their way through the osseous spiral lamina, either radically (cat and guinea pig) or spirally (human). After entering the organ of Corti, the fibers reach the outer hair cells as upper tunnel radial fibers (Spoendlin 1966; Ishii et al. 1967).

There are species differences in the exact course the fibers take to the outer hair cells. In the cat few, if any, MOC fibers join the inner spiral bundle (Liberman 1980a; Liberman and Brown 1986), whereas in guinea pig and rat many do (Robertson 1984; Robertson and Gummer 1985; Brown 1987; Warr, unpublished observations). Brown (1989) found that MOC fibers in the guinea pig do not supply terminal branches in their course through the inner spiral bundle and travel well below the level where most synaptic contacts occur with radial fibers. However, White et al. (1986) found vesicle-filled varicosities along MOC fibers in the inner spiral bundles of the guinea pig which met some, but not all of the criteria of presynaptic specializations.

Similar possibilities have been raised by the observations of Ryan and Schwartz (1986) in the gerbil, where only MOC neurons are retrogradely labeled from the cochlea with the GABA analogue, tritiated nipecotic acid, and presynaptic elements in the inner spiral bundle are labeled in electron microscopic autoradiographs.

As shown in Figure 7.8 from a study in the cat, these fibers may supply terminals to 8 or 10 discrete clusters of outer hair cells, spanning a distance of as much as 2.8 mm of the organ of Corti, a distance estimated

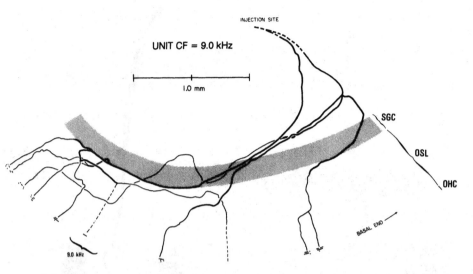

FIGURE 7.8. Camera lucida tracing of the peripheral portion of an HRP-labeled MOC fiber in the osseous spiral lamina and organ of Corti of the cat. The CF of the efferent fiber was 9.0 kHz. Thick lines denote myelinated portions; thinner lines denote unmyelinated portions. The locations of outer hair cells innervated by this fiber are indicated by dots at the bottom margin. The bracket on the lower left indicates the region of the organ of Corti generating afferent responses with CFs of 9 kHz. Abbreviations: SGC, spiral ganglion cells; OHC, outer hair cell; OSL, osseous spiral lamina. (Reproduced from Liberman and Brown 1987, with permission.)

to represent as much as one octave (Liberman and Brown 1986). In guinea pigs the pattern is the same, although the terminal fields are less than 1 mm in length, at least in the basal turn from which most data were obtained (Brown 1987, 1989). In contrast, mouse upper tunnel radial fibers usually form only one terminal that contacts one hair cell (Wilson et al. 1991), and in the mustache bat, each outer hair cell is contacted by only one efferent terminal (Bishop and Henson 1987a). Figure 7.9 illustrates the tendency for individual upper tunnel radial fibers in the cat to innervate clusters of outer hair cells and shows that in the apical turn (right panel), not all terminals are confined to the bases of the outer hair cells, as recently documented quantitatively (Liberman et al. 1990).

The fine structure of these synaptic terminals shows them to be located preferentially on the modiolar side of the outer hair cell base, and to exhibit all the usual features of a chemical synapse as well as postsynaptic specialization unique to outer hair cells, the subsynaptic cistern (Spoendlin 1969; Bodian 1983; Kimura 1984; Pujol and Lenoir 1986).

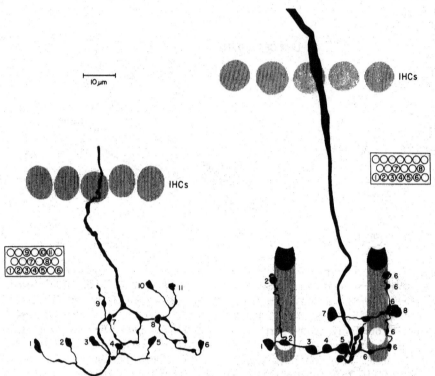

FIGURE 7.9. Camera lucida tracings of terminal branches from each of two efferent fibers. The fiber on the right was from the apical half of the cochlea (roughly the 1.5 kHz region). The one on the left was from the basal half (roughly the 22 kHz region of the cochlea). Each sketch illustrates only one of many branches from each neuron. The boxed inset on each side schematizes the distribution of endings to the three rows of outer hair cells in each case. (Reproduced from Liberman and Brown 1987, with permission.)

5.2.2 Distribution of MOC Innervation from Base to Apex

There are remarkably few reports which have attempted to quantify the generally accepted fact that the density of efferent innervation of the outer hair cell region varies according to basal-to-apical location, hair cell row, and is subject to individual variability at both the basal and apical extremities of its range (Churchill and Schuknecht 1959; Ishii et al. 1967; Ginzberg and Morest 1984; Guinan et al. 1984; Fex and Altschuler 1986). To address this problem in the cat, Liberman and his co-workers recently used several counting and measurement strategies at both the light and electron microscopic levels which resulted in a series of highly consistent curves quantifying the efferent innervation of the outer hair cells in three animals (Liberman et al. 1990). As shown in Figure 7.10, the number of efferent endings per hair cell attains a broad maximum at approximately the 10 kHz region of the cochlea and displays a clear decreasing gradient according to hair cell row. These findings are very similar to the pattern derived by radiolabeling the MOC neurons unilaterally and pooling silver grain counts from autoradiographs of the two cochleas (Guinan et al. 1984; Liberman et al. 1990). Interestingly, as also shown in Figure 7.10, the corresponding afferent innervation of the outer hair cells displays a somewhat complementary pattern.

5.2.3 Synaptic Contacts with Spiral Afferent Fibers

A minor postsynaptic target of MOC fibers is the outer spiral fibers. As shown schematically in Figure 7.4, these fibers are the peripheral processes of type II spiral ganglion cells. They cross the floor of the tunnel of Corti and spiral basally beneath the outer hair cells, to which they provide their afferent innervation (Simmons and Liberman 1988). Electron microscopic studies have identified efferent synapses in contact with the spiral fibers in both of the locations in which they travel as a discrete population (Dunn and Morest 1975; Eybalin et al. 1988). However, a quantitative light microscopic survey using synaptophysin antibodies to label all vesiculated efferent terminals, found that immunoreactive varicosities in contact with the spiral afferent fibers were rare in the basal half of the cochlea but numbered as high as 120/mm at the 0.5 kHz region of the cat's cochlea (Liberman et al. 1990). This same reference has been cited previously in this chapter as showing that a number of aberrations

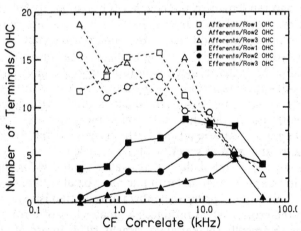

FIGURE 7.10. Comparison of the afferent and efferent innervation densities, apex to base, contacting the three rows of outer hair cells (OHC) in the cat, as determined by reconstructions of semi electron micrographs at eight locations along the basilar membrane. Only efferent terminals containing synaptic vesicles were counted. (Replotted from Figure 3B and 12B in Liberman et al. 1990).

and oddities of efferent innervation are a virtually exclusive feature of the apex of the cochlea, including supranuclear terminals on both inner and outer hair cells.

5.2.4 Tonotopic Organization of MOC Projections

Two lines of evidence indicate that the MOC efferent innervation is tonotopically organized: mapping terminals belonging to single MOC fibers of known characteristic frequency (CF) and axonal transport of tracers either from MOC neurons to the cochlea, or vice versa.

The single fiber mapping technique has provided direct evidence that MOC fibers terminate at cochlear locations corresponding to their CF. However, only about a dozen fibers have been completely labeled and successfully traced to all of their postsynaptic outer hair cells (Robertson 1984; Robertson and Gummer 1985; Liberman and Brown 1986; Brown 1989). One apparently completely labeled MOC fiber from the cat is presented in Figure 7.8. This unit was tuned to 9.0 kHz and its terminals contact outer hair cells spanning the appropriate cochlear place. Most fibers, like the one in Figure 7.8, tend to have more of their terminal arbor basal to, than apical to, the characteristic place of their respective CFs. This bias in the terminal field will be referred to as "basalward displacement" and is pointed out because of its possible significance in the functioning of the MOC system in relation to a theorized role of outer hair cells as a "cochlear amplifier" (Neely and Kim 1983; Kim 1986).

The second line of evidence for tonotopy of projections was obtained by tracing MOC fibers from their origins in the superior olive to their terminals along the length of the cochlea (Guinan et al. 1984). In light of information gathered since then that MOC neurons are sharply tuned and have limited terminations along the length of the cochlea, it can be deduced that the topographic map relating site of injection to site of termination in the cat is in fact a tonotopic map. Guinan et al. (1984) related the dorsomedial periolivary region of the cat with basal projections and the ventral nucleus of the trapezoid body with apical projections. It is worth noting that this MOC tonotopic map coexists with maps of quite different spatial orientation in the nearby medial nucleus of the trapezoid body and medial superior olivary nucleus.

In addition, it was found that there was a tendency for the crossed projections from a given locus to terminate more basally than the uncrossed projections. This would appear to confirm a well-established observation that cutting the crossed OCB produces a greater loss of terminals in the base than more apically (Nakai and Igarashi 1974; Iurato et al. 1978).

Finally, retrograde transport of tracers injected into single turns of the guinea pig cochlea support certain of the tonotopic features observed in the cat (Robertson et al. 1987c). For example, basal injections labeled many cells in the dorsomedial periolivary nucleus and apical injections vastly reduced the number of neurons labeled contralaterally, mainly by reducing the labeling in the dorsomedial periolivary nucleus. However, a clear tonotopic map of the sort clearly discernable in the LSO in these experiments was not obtained. Subsequent to these experiments, a medio-lateral, high-to-low tonotopic organization of the ventral nucleus of the trapezoid body in the rat was reported (Warr and Spangler 1989; Saldana 1990). Even if a similar organization were present in the guinea pig, it might be difficult to recognize using the retrograde transport method.

5.2.5 MOC Collaterals to the Cochlear Nucleus

Collateral branches to the cochlear nucleus from OCB fibers have been described in most species studied, including the cat (Brown et al. 1988a), the rat (White and Warr 1983; Osen et al. 1984; Godfrey et al. 1987), the guinea pig (Winter et al. 1989), the gerbil (Ryan et al. 1987; Brown et al. 1988a), the mouse (Brown et al. 1988a), but not apparently in humans which lack a cochlear nucleus granule cell layer (Moore and Osen 1979). In all species, the collaterals are most readily demonstrated to be branches of the thick, myelinated fibers of the MOC contingent of the OCB (but see Section 5.1.5). In addition, Ryan et al. (1990) report a collateral pro-jection to nucleus Y in gerbil.

The distribution of MOC collaterals is somewhat variable across spe-cies, but always involves arborizations in portions of the granule cell domain, a cellular lamina of varying depth and development comprised of granule cells, that surrounds virtually the entire cochlear nuclear com-plex in most species (Mugnaini et al. 1980; Osen et al. 1984). It is relevant to point out that the central processes of type II ganglion cells also ar-borize, in part, among the granule cells, suggesting a convergence of in-formation relating to outer hair cell function in this region, as dia-grammed schematically in Figure 7.4 (Brown et al. 1988b; Ryan et al. 1990).

Recent work in the mouse shows that collaterals arborize in two granule cell regions, a ventromedial one near the cochlear nerve root and a dor-solateral one, the superficial granule cell lamina, and that a tonotopic arrangement exists (Benson and Brown 1990; Brown et al. 1991).

The postsynaptic target of these MOC collaterals has been shown to be not the granule cells themselves, but large dendrites of multipolar cells in the ventral cochlear nucleus. Since the morphology of the synapses was determined to be of the excitatory type, Benson and Brown (1990) proposed an intriguing hypothesis, illustrated in Figure 7.11, that MOC collaterals to the multipolar cells may function to compensate for the well-known suppressive action of these fibers on the sensitivity of cochlear nerve fibers. Thus, the amount of negative feedback reaching the cochlea could be balanced by an equivalent amount of positive feedback provided by MOC collaterals to multipolar neurons, the outputs of which would

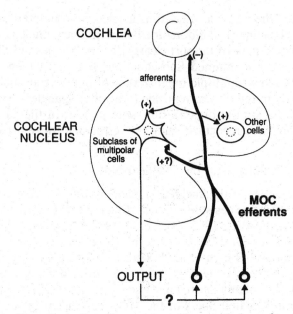

FIGURE 7.11. Hypothesis that allows output from a subclass of cochlear nucleus multipolar cells to be independent of the effect of olivocochlear feedback at the cochlea. Primary afferents are known to excite cochlear-nucleus neurons (+). MOC efferents are known to provide feedback to the cochlea (−) that decreases the activity of the afferents, thus indirectly decreasing activity in most cochlear-nucleus neurons. MOC branches to a subclass of multipolar neurons are postulated to be excitatory (+?) and provide enough excitation to keep OUTPUT independent of the effect of MOC feedback at the cochlea. The OUTPUT of those multipolar cells could then truly reflect the intensity of sound at the cochlea. (Reproduced from Benson and Brown 1990, with permission.)

reflect the true sound level. Figure 7.11 also suggests that this class of multipolar cell may regulate the overall level of MOC activity by direct projections to MOC neurons. However direct evidence for this circuit is lacking.

6. Immunocytochemistry of OC Neurons

During the last few years, attempts to localize neurotransmitter-related substances in OC fibers using immunocytochemical (ICC) techniques have been a major focus of several leading auditory research groups (Fex and Altschuler 1986; Eybalin and Altschuler 1990; Vetter et al. 1991). Simultaneously, parallel studies using tract-tracing (Warr 1975; Warr and Guinan 1979; Guinan et al. 1983) and single-fiber marking techniques (Robertson 1984; Liberman and Brown 1986; Brown 1987) were revealing that, at least according to conventional morphological criteria, there appeared to be only two kinds of efferent fibers. Thus, the concept of two separate efferent systems gained acceptance and has proved useful in interpreting the patterns of immunoreactivity (IR) observed in the organ of Corti (Fex and Altschuler 1986; Eybalin and Altschuler 1990).

There is good evidence that most cochlear efferent fibers exhibit IR for choline acetyltransferase (ChAT), the enzyme that catalyzes synthesis of acetylcholine and is a definitive marker for cholinergic neurons (Altschuler et al. 1985; Eybalin and Pujol 1987; Vetter et al. 1991).

A perhaps more startling outcome of the application of ICC to the efferent innervation, however, is the discovery of the simultaneous containment or colocalization of several neuroactive substances in individual LOC neurons, but not usually in MOC neurons. Colocalized with ChAT IR in LOC cell bodies are IR to: enkephalin (Altschuler et al. 1984); dynorphin (Abou-Madi et al. 1987) and calcitonin gene-related peptide (Vetter et al. 1991). The reader is referred to reviews of this topic for further details (Fex and Altschuler 1986; Eybalin and Altschuler 1990).

The question of whether the two efferent systems can be further subdivided on the basis of their IR to various other neurotransmitter-related antibodies has lately been answered in the affirmative. That is, the ICC technique has identified subgroups of LOC and MOC fibers which differ from the majority in that they exhibit IR to glutamic acid decarboxylase (GAD) and γ-aminobutyric acid (GABA) (Fex and Altschuler 1984; Fex et al. 1986; Thompson et al. 1986; Eybalin et al. 1988; Usami et al. 1988; Whitlon and Sobkowicz 1989; Vetter et al. 1991).

In most species, these fibers are reportedly few in number and distribute to the inner and outer hair cell regions in the upper turns of the cochlea. The rat appears to be exceptional in that about one half of the inner spiral bundle, from base to apex, is described as showing GAD or GABA IR (Vetter et al.1991). Thus, ICC has been the main reason for considering further subdivision of the LOC and MOC systems.

7. A Schema of Cochlear Innervation

This review of the anatomy of the efferent innervation of the organ of Corti would be incomplete without attempting to integrate the structural pattern of its neuronal elements with those belonging to the afferent system, which over the last decade have also been described in unprecedented detail (Kiang et al. 1982; Spoendlin 1984; Simmons and Liberman 1988; Chapter 2, this volume). It is of obvious functional interest to assemble what is known about the spatial relationships prevailing among neural elements which have the same CF. To do this, however, it is necessary to combine data from the cat and guinea pig, and so the result is intended only as schema that will be corrected and refined as new findings are obtained. The fact that two of the four types of neurons innervating the organ of Corti have unmyelinated axons and have never been characterized electrophysiologically, namely the type II afferent and LOC efferent, is no deterrent because the association of these cell types with other neurons of known CF, type I ganglion cells and LSO principal cells, respectively, provides circumstantial evidence of their tuning.

The hypothetical isofrequency unit of cochlear innervation is presented in Figure 7.12. At the center of the figure is shown the inner hair cell of a given CF in exclusive presynaptic relation to a population of radial fibers belonging to type I ganglion cells, as originally demonstrated by Spoendlin (1971). The innervation ratio of inner hair cell to radial afferent fibers peaks near the middle of the cat's cochlea (Fig. 7.7). The type II ganglion cell comprises less than 10% of the cells in the spiral ganglion and provides the afferent innervation of the outer cells (Spoendlin 1971; Kiang et al. 1982). The peripheral processes of these cells form the outer spiral fibers that travel basally before sending branches to a number of outer hair cells that are nearly always all confined to a single given row (Simmons and Liberman 1988). For clarity, only one type II ganglion cell is shown, but outer hair cells are presynaptic to many different type II ganglion cells, the innervation ratio depending on outer hair cell row and basal to apical location as shown in Figure 7.10.

In Figure 7.12, the efferent innervation is also shown in highly simplified form, although many lateral and medial OC neurons are presynaptic to each of their respective targets. Specifically, LOC neurons provide from 0 to approximately 30 efferent terminals on each radial afferent fiber (see Fig. 7.7), but for the sake of clarity, only one efferent terminal arbor in the inner spiral bundle is represented, and no attempt was made to show a bidirectional LOC efferent fiber, which in guinea pig, comprise about 15% of the total LOC population (Brown 1987). Finally, the complete terminal arbor of an MOC efferent is shown. Note that although MOC efferents are sharply tuned, their terminal arbors are distributed to several isolated clusters of outer hair cells spread out over some considerable length of the organ of Corti. Presumably, other MOC neurons with similar CFs would fill in the gaps and supplement the innervation of the outer hair cells, each of which may receive up to eight or so efferent contacts (Fig. 7.10). A frankly speculative feature of the schematic diagram is the exaggerated basalward displacement shown for the distribution of the MOC terminals on the outer hair cells. Although MOC fibers often exhibit such basalward displacement relative to their CF (Fig. 7.8), they apparently do not always do so (Liberman and Brown 1986). Moreover, the technical difficulties involved in obtaining complete tracings of individual MOC terminal arbors and simultaneous marking of afferent fibers of the same CF, means that there are few data from which to generalize. The diagram also minimizes the apical spread of the MOC terminal arbor, not only for clarity but also to stress the overlapping

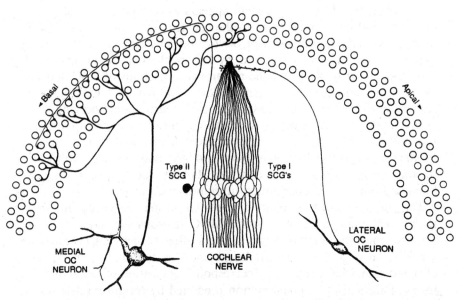

FIGURE 7.12. A hypothetical isofrequency unit of afferent and efferent innervation from the middle of the cochlea. The unit is comprised of two types of afferent and two types of efferent fibers tuned to the same frequency. Note the basal displacement and partial spatial overlap of MOC efferent and type-II afferent innervation patterns on the outer hair cells relative to the corresponding inner hair cell and lateral efferent neuron. (See text for further details.)

distributions of MOC efferents and type II afferents on outer hair cells basal to the cochlear place of their corresponding type I ganglion cells. Such a displacement has been an important feature of computer models that simulate the theorized role of outer hair cells as amplifiers of basilar membrane motion (Neely and Kim 1983; Kim 1986).

8. Summary

Mammalian OC neurons appear to conform to a stereotypical pattern in distribution, morphology, and projections to specific targets in the organ of Corti, although deletions of certain features or even cell types may occur. Regardless of species, LOC neurons are always smaller but more numerous than MOC neurons. The projections of LOC neurons is always strongly to radial afferent fibers in the ipsilateral cochlea, whereas those of MOC neurons are mostly to the outer hair cells of the contralateral cochlea, but the projections are more evenly distributed between the two sides.

The anatomy of the LOC system appears to be precisely tonotopic in its inputs and in its projections to the cochlea. The association of LOC neurons with the lateral superior olivary nucleus may relate to the full frequency representation received by this nucleus from the ipsilateral ventral cochlear nucleus and the reciprocal projections of LOC neurons to the entire length of the organ of Corti. The fact that LOC neurons contain a variety of neurotransmitters and neuromodulators and have unmyelinated axons suggest a slow and chemically somewhat complex, modulatory role directed at radial afferent fibers, particularly the low spontaneous group, which they most strongly innervate. An apparently separate group of LOC neurons express GABAergic as opposed to cholinergic properties, as determined immunocytochemically.

The anatomy and physiology of the MOC system is more completely understood than the LOC system. MOC neurons are physiologically sharply tuned and may innervate clusters of outer hair cells spanning as much as an octave of cochlear length, but the terminal arbor is often centered somewhat basal to the corresponding location of radial afferent fibers of similar CF, a phenomenon predicted by certain models which hypothesize an active role of outer hair cells in cochlear mechanics. The efferent innervation of outer hair cells is greatest in the mid-high frequency region of the cochlea and declines both apically and basally. The position of MOC neurons and their large dendritic arbors make them potentially capable of receiving inputs from a variety of ascending, descending and local auditory sources representing activity in both ears, but the organization of these inputs is not yet well understood. Like LOC neurons, most MOC neurons project to the cochlea from which they receive their afferent innervation; however, a substantial number do not and comprise an intercochlear effector pathway by which sound in one ear can influence the responses in the other ear. A small contingent of GABAergic MOC fibers terminate on outer hair cells mainly in the apical region of the cochlea.

For the present, the concept of LOC and MOC systems provides a satisfactory framework for integrating what is known about the structure and function of mammalian cochlear efferent neurons, although immunocytochemical techniques suggest that in addition to cholinergic OC neurons, GABAergic subsystems exist.

Acknowledgments. It is a pleasure to thank Bernd Fritzsch for his critical reading of the manuscript, Walt Jesteadt and Theresa Dethlefs for their assistance in replotting data, and T.E. Benson, M.C. Brown, M.C. Liberman, E. Mugnaini, and D.E. Vetter for generously supplying illustrations. Finally, the editorial assistance of Charlotte Lieser and the bibliographic software support of Jo E. Peters is greatly appreciated.

References

Abou-Madi L, Pontarotti P, Tramu G, Cupo A, Eybalin M (1987) Coexistence of putative neuroactive substances in lateral olivocochlear neurons of rat and guinea pig. Hear Res 30:135–146.

Adams JC (1982) Collaterals of labyrinthine efferent axons. Soc Neurosci Abs 8:149.

Adams JC (1983) Cytology of periolivary cells and the organization of their projections in cat. J Comp Neurol 215:275–289.

Adams JC (1986) Cells of origin of cochlear efferents in human. Assoc Res Otolaryngol Abs 9:5.

Altschuler RA, Fex J, Parakkal MH, Eckenstein F (1984) Colocalization of enkephalin-like and choline acetyltransferase-like immunoreactivities in olivocochlear neurons of the guinea pig. J Histochem Cytochem 32:839–843.

Altschuler RA, Kachar B, Rubio JA, Parakkal MH, Fex J (1985) Immunocytochemical localization of choline acetyltransferase-like immunoreactivity in the guinea pig cochlea. Brain Res 338:1–11.

Arnesen AR (1984) Fiber population of the vestibulocochlear anastomosis in humans. Acta Otolaryngol (Stockh) 98:501–518.

Arnesen AR, Osen KK (1978) The cochlear nerve in the cat: Topography, cochleotopy, and fiber spectrum. J Comp Neurol 178:661–673.

Arnesen AR, Osen KK (1984) Fiber population of the vestibulocochlear anastomosis in the cat. Acta Otolaryngol (Stockh) 98:225–269.

Aschoff A, Ostwald J (1987) Different origins of cochlear efferents in some bat species, rats, and guinea pigs. J Comp Neurol 264:56–72.

Aschoff A, Ostwald J (1988) Distribution of cochlear efferents and olivo-collicular neurons in the brainstem of rat and guinea pig: A double labeling study with fluorescent tracers. Exp Brain Res 71:241–251.

Aschoff A, Muller M, Ott H (1988) Origin of cochlear efferents in some gerbil species. Exp Brain Res 71:252–262.

Benson TE, Brown MC (1990) Synapses formed by olivocochlear axon branches in the mouse cochlear nucleus. J Comp Neurol 295:52–70.

Bishop AL, Henson OW Jr (1987) The efferent cochlear projections of the superior olivary complex in the mustached bat. Hear Res 31:175–182.

Bishop AL, Henson OW Jr (1987a) The olivocochlear system in Doppler-shift compensating bats. In: Nachtigal PE (ed) Animal Sonar Systems II, Vol. II. New York: Plenum Press.

Bodian D (1983) Electron microscopic atlas of the simian cochlea. Hear Res 9:201–246.

Bodian D, Gucer C (1980) Denervation study of synapses of organ of Corti of old world monkeys. J Comp Neurol 192:785–796.

Brown MC (1987) Morphology of labeled efferent fibers in the guinea pig cochlea. J Comp Neurol 260:605–618.

Brown MC (1989) Morphology and response properties of single olivocochlear fibers in the guinea pig. Hear Res 40:93–109.

Brown MC, Pierce S, Berglund AM (1991) Cochlear-nucleus branches of thick (medial) olivocochlear fibers in the mouse: A cochleotopic projection. J Comp Neurol 303:300–315.

Brown MC, Liberman MC, Benson TE, Ryugo DK (1988a) Brainstem branches of olivocochlear axons in cats and rodents. J Comp Neurol 278:591–603.

Brown MC, Berglund AM, Kiang NYS, Ryugo DK (1988b) Central trajectories of type II spiral ganglion neurons. J Comp Neurol 278:581–590.

Brownell WE (1990) Outer hair cell electromotility and otoacoustic emissions. Ear Hear 11:82–92.

Bruns V, Schmieszek E (1980) Cochlear innervation in the greater horseshoe bat: Demonstration of an acoustic fovea. Hear Res 3:27–43.

Buno W (1978) Auditory-nerve fiber activity influenced by contralateral ear sound stimulation. Exp Neurol 59:62–74.

Campbell JP, Henson MM (1988) Olivocochlear neurons in the brainstem of the mouse. Hear Res 35:271–274.

Cant NB (1984) The fine structure of the lateral superior olivary nucleus of the cat. J Comp Neurol 227:63–77.

Cant NB, Casseday JH (1986) Projections from the anteroventral cochlear nucleus to the lateral and medial superior olivary nuclei. J Comp Neurol 247:457–477.

Churchill JA, Schuknecht HF (1959) The relationship of the acetylcholinesterase in the cochlea to the olivocochlear bundle. Henry Ford Hosp Med Bull 7:202–205.

Cody AR, Johnstone BM (1982) Acoustically evoked activity of single efferent neurons in the guinea pig cochlea. J Acoust Soc Am 72:280–282.

Dallos P, Evans BN, Hallworth R (1991) Nature of the motor element in electrokinetic shape changes of cochlear outer hair cells. Nature 350:155–157.

Dunn RA, Morest DK (1975) Receptor synapses without synaptic ribbons in the cochlea of the cat. Proc Natl Acad Sci USA 72:3599–3603.

Eybalin M, Altschuler RA (1990) Immunoelectron microscopic localization of neurotransmitters in the cochlea. J Elect Micros Tech 15:209–224.

Eybalin M, Pujol R (1987) Choline acetyltransferase (ChAT) immunoelectron microscopy distinguishes at least three types of efferent synapses in the organ of Corti. Exp Brain Res 65:261–270.

Eybalin M, Parnaud C, Geffard M, Pujol R (1988) Immunoelectron microscopy identifies several types of GABA-containing efferent synapses in the guinea pig organ of Corti. Neurosci 24:29–38.

Faye-Lund H (1986) Projection from the inferior colliculus to the superior olivary complex in the albino rat. Anat Embryol 175:35–52.

Fex J (1965) Auditory activity in uncrossed centrifugal cochlear fibers in cat. Acta Physiol Scand 64:43–57.

Fex J, Altschuler RA (1984) Glutamic acid decarboxylase immunoreactivity of cochlear neurons in the organ of Corti of guinea pig and rat. Hear Res 15:123–131.

Fex J, Altschuler RA (1986) Neurotransmitter related immunocytochemistry of the organ of Corti. Hear Res 22:249–263.

Fex J, Altschuler RA, Kachar B, Wenthold RJ, Zempel JM (1986) GABA visualized by immunocytochemistry in the guinea pig cochlea in axons and endings of efferent neurons. Brain Res 366:106–117.

Friauf E, Ostwald J (1988) Divergent projections of physiologically characterized rat ventral cochlear nucleus neurons as shown by intra-axonal injection of horseradish peroxidase. Exp Brain Res 73:263–284.

Fritzsch B, Dubuc R, Ohta H, Grillner S (1989) Efferents to the labyrinth of the river lamprey (Lampetra fluviatilis) as revealed with retrograde tracing techniques. Neurosci Lett 96:241–246.

Gacek RR (1961) The efferent cochlear bundle in man. Arch Otolaryngol 74:690–694.

Ginzberg RD, Morest DK (1984) Fine structure of cochlear innervation in the cat. Hear Res 14:109–127.

Godfrey DA, Park-Hellendall JL, Dunn JD, Ross CD (1987) Effect of olivocochlear bundle transection of choline acetyltransferase activity in the rat cochlear nucleus. Hear Res 28:237–251.

Godfrey DA, Beranek KL, Carlson L, Parli JA, Dunn JD, Ross CD (1990) Contribution of centrifugal innervation to choline acetyltransferase activity in the cat cochlear nucleus. Hear Res 49:259–280.

Guinan JJ Jr (1988) Physiology of the olivocochlear efferents. In: Syka J, Masterton RB (eds) Auditory Pathways: Structure and Function, New York: Plenum Press, pp. 253–267.

Guinan JJ Jr, Norris BE, Guinan SS (1972) Single auditory units in the superior olivary complex. II: Location of unit categories and tonotopic organization. Int J Neurosci 4:147–166.

Guinan JJ Jr, Warr WB, Norris BE (1983) Differential olivocochlear projections from lateral versus medial zones of the superior olivary complex. J Comp Neurol 221:358–373.

Guinan JJ Jr, Warr WB, Norris BE (1984) Topographic organization of the olivocochlear projections from the lateral and medial zones of the superior olivary complex. J Comp Neurol 226:21–27.

Helfert RH, Schwartz IR (1987) Morphological features of five neuronal classes in the gerbil lateral superior olive. Am J Anat 179:55–69.

Helfert RH, Schwartz IR, Ryan AF (1988) Ultrastructural characterization of gerbil olivocochlear neurons based on differential uptake of 3H-d-aspartic acid and a wheatgerm agglutinin-horseradish peroxidase conjugate of the cochlea. J Neurosci 8:3111–3123.

Huffman RF, Henson OW Jr (1990) The descending auditory pathway and acousticomotor systems: connections with the inferior colliculus. Brain Res Rev 15:295–323.

Ishii T, Murakami Y, Balogh KJ (1967) Acetylcholinesterase activity in the efferent nerve fibers of the human inner ear. Ann Otol Rhinol Laryngol 76:69–82.

Iurato S, Smith CA, Eldredge DH, Henderson D, Carr C, Ueno Y, Cameron S, Richter R (1978) Distribution of the crossed olivocochlear bundle in the chinchilla's cochlea. J Comp Neurol 182:57–76.

Joseph MP, Guinan JJ Jr, Fullerton BC, Norris BE, Kiang NYS (1985) Number and distribution of stapedius motoneurons in cats. J Comp Neurol 232:43–54.

Kiang NYS, Guinan JJ Jr, Liberman MC, Brown MC, Eddington DK (1987) Feedback control mechanisms of the auditory periphery: Implications for cochlear implants. In: Banfai P (ed) Cochlear Implant: Current Situation. International Cochlear Implant Symposium.

Kiang NYS, Rho JM, Northrup CC, Liberman MC, Ryugo DK (1982) Hair-cell innervation by spiral ganglion cells in adult cat. Science 217:175–177.

Kim DO (1986) Active and nonlinear cochlear biomechanics and the role of outer-hair-cell subsystem in the mammalian auditory system. Hear Res 22:105–114.

Kimura RS (1984) Sensory and Accessory Epithelia of the Cochlea. In: Friedman I, Ballantyne J (eds) Ultrastructural Atlas of the Inner Ear, 1st Ed, Vol 1. London: Butterworth.

Kimura RS, Wersall J (1962) Termination of the olivocochlear bundle in relation to the outer hair cells of the organ of Corti in guinea pigs. Acta Otolaryngol (Stockh) 55:11–32.

Liberman MC (1980a) Efferent synapses in the inner hair cell area of the cat cochlea: An electron microscopic study of serial sections. Hear Res 3:189–204.

Liberman MC (1980b) Morphological differences among radial afferent fibers in the cat cochlea: an electron-microscopic study of serial sections. Hear Res 3:45–63.

Liberman MC (1982) Single-neuron labeling in the cat auditory nerve. Science 216:1239–1241.

Liberman MC (1988) Response properties of cochlear efferent neurons: Monaural vs. binaural stimulation and the effects of noise. J Neurophysiol 60:1779–1798.

Liberman MC (1990) Effects of chronic cochlear de-efferentation on auditory-nerve response. Hear Res 49:209–224.

Liberman MC (1991) The olivocochlear efferent bundle and the susceptibility of the inner ear to acoustic injury. J Neurophysiol 65:123–132.

Liberman MC, Brown MC (1986) Physiology and anatomy of single olivocochlear neurons in the cat. Hear Res 24:17–36.

Liberman MC, Oliver ME (1984) Morphometry of intracellularly labeled neurons of the auditory nerve: correlations with functional properties. J Comp Neurol 223:163–176.

Liberman MC, Dodds LW, Pierce S (1990) Afferent and efferent innervation of the cat cochlea: Quantitative analysis with light and electron microscopy. J Comp Neurol 301:443–460.

Lu SM, Schweitzer L, Cant NB, Dawbarn D (1987) Immunoreactivity to calcitonin gene-related peptide in the superior olivary complex and cochlea of cat and rat. Hear Res 31:137–146.

Meredith GE (1988) Comparative view of the central organization of afferent and efferent circuitry for the inner ear. Acta Biol Hung 39:229–249.

Moore JK, Osen KK (1979) The cochlear nuclei in man. Am J Anat 154:393–418.

Mugnaini E, Warr WB, Osen KK (1980) Distribution and light microscopic features of granule cells in the cochlear nuclei of cat, rat and mouse. J Comp Neurol 191:581–606.

Nakai Y, Igarashi M (1974) Distribution of the crossed olivocochlear bundle terminals in the squirrel monkey cochlea. Acta Otolaryngol (Stockh) 77:393–404.

Neely ST, Kim DO (1983) An active cochlear model showing sharp tuning and high sensitivity. Hear Res 9:123–130.

Osen KK (1969) Cytoarchitecture of the cochlear nuclei in the cat. J Comp Neurol 136:453–484.

Osen KK, Roth K (1969) Histochemical localization of cholinesterases in the cochlear nuclei of the cat, with notes on the origin of acetylcholinesterase-positive afferents and the superior olive. Brain Res 16:165–185.

Osen KK, Mugnaini E, Dahl A-L, Christiansen AH (1984) Histochemical localization of acetylcholinesterase in the cochlear and superior olivary nuclei. A reappraisal with emphasis on the cochlear granule cells system. Arch Ital Biol 122:169–212.

Parnes SM, Strominger NL, Silver SM, Strominger RN (1986) Studies of the primate olivocochlear bundle. Assoc Res Otolaryngol Abs 9:37.

Pujol R, Lenoir M (1986) The four types of synapses in the organ of Corti. In: Altschuler RA, Bobbin RP, Hoffman DW (eds) Neurobiology of Hearing: The Cochlea. New York: Raven Press.

Rajan R (1990) Functions of the efferent pathways to the mammalian cochlea. In: Rowe M, Aitkin L (eds) Neurology and Neurobiology, Vol 56: Information Processing in Mammalian Auditory and Tactile Systems. New York: Alan R Liss, Inc.

Rasmussen GL (1946) The olivary peduncle and other fiber projections of the superior olivary complex. J Comp Neurol 84:141–219.

Rasmussen GL (1953) Further observations on the efferent cochlear bundle. J Comp Neurol 99:61–74.

Rasmussen GL (1960) Efferent fibers of the cochlear nerve and cochlear nucleus. In: Windle WF, Rasmussen GL (eds) Neural Mechanisms of Auditory and Vestibular Systems. Springfield, IL: Carles C. Thomas.

Roberts BL, Meredith GE (1989) The efferent system. In: Coombs S, Gorner P, Munz H (eds) The Mechanosensory Lateral Line. New York: Springer-Verlag.

Robertson D (1984) Horseradish peroxidase injection of physiologically characterized afferent and efferent neurones in the guinea pig spiral ganglion. Hear Res 15:113–121.

Robertson D (1985) Brainstem location of efferent neurones projecting to the guinea pig cochlea. Hear Res 20:79–84.

Robertson D, Gummer M (1985) Physiological and morphological characterization of efferent neurons in the guinea pig cochlea. Hear Res 20:63–77.

Robertson D, Cole KS, Harvey AR (1987a) Brainstem organization of efferent projections to the guinea pig cochlea studied using the fluorescent tracers fast blue and diamidino yellow. Exp Brain Res 66:449–457.

Robertson D, Cole KS, Corbett K (1987b) Quantitative estimate of binaurally projecting medial olivocochlear neurons in the guinea pig brainstem. Hear Res 27:177–181.

Robertson D, Anderson C-J, Cole KS (1987c) Segregation of efferent projections to different turns of the guinea pig cochlea. Hear Res 25:69–76.

Robertson D, Harvey AR, Cole KS (1989) Postnatal development of the efferent innervation of the rat cochlea. Develop Brain Res 47:197–207.

Ryan AF, Schwartz IR (1986) Nipecotic acid: Preferential accumulation in the cochlea by GABA uptake systems and selective retrograde transport to brainstem. Brain Res 399:399–403.

Ryan AF, Keithley EM, Wang Z-X, Schwartz IR (1990) Collaterals from lateral and medial olivocochlear efferent neurons innervate different regions of the cochlear nucleus and adjacent brainstem. J Comp Neurol 300:572–582.

Ryan AF, Schwartz IR, Helfert RH, Keithley E, Wang Z-X (1987) Selective retrograde labeling of lateral olivocochlear neurons in the brainstem based on preferential uptake of 3H-D-aspartic acid in the cochlea. J Comp Neurol 255:606–616.

Saldana E (1990) The rat colliculo-olivary projection is tonotopic. Soc Neurosci Abs 16:716.

Sanes DH, Rubel EW (1988) The ontogeny of inhibition and excitation in the gerbil lateral superior olive. J Neurosci 8:682–700.

Sanes DH, Goldstein NA, Ostad M, Hillman DE (1990) Dendritic morphology of central auditory neurons correlates with their tonotopic position. J Comp Neurol 294:443–454.

Scheibel ME, Scheibel AB (1974) Neuropil organization in the superior olive of the cat. Exp Neurol 43:339–348.

Simmons DD, Liberman MC (1988) Afferent innervation of outer hair cells in adult cats: I. Light microscopic analysis of fibers labeled with horseradish peroxidase. J Comp Neurol 270:132–144.

Smith CA (1961) Innervation pattern of the cochlea. The internal hair cell. Ann Otol Rhinol Laryngol 70:504–527.

Smith CA, Rasmussen GL (1963) Recent observation of the olivocochlear bundle. Ann Otol Rhinol Laryngol 72:489–505.

Spangler KM, Warr WB (1991) The descending auditory system. In: Altschuler R, Hoffman DW, Bobbin RP, Clopton BM (eds) The Neurobiology of Hearing, 1st Ed, Vol II. New York: Raven Press.

Spangler KM, White JS, Warr WB (1985) The light and electron microscopic features of olivocochlear neurons in the cat. Anat Rec 211:182.

Spangler KM, White JS, Warr WB (1986) Electron microscopic features of axon terminals on olivocochlear neurons in the cat. Assoc Res Otolaryngol Abs 9:37–38.

Spangler KS, Warr WB (1987) Transneuronal changes in cochlear radial afferent fibers following destruction of lateral olivocochlear neurons. Soc Neurosci Abs 13:1258.

Spoendlin H (1966) The organization of the cochlear receptor. Adv Oto-Rhino-Laryngol 13:1–114.

Spoendlin H (1970) Structural basis of peripheral frequency analysis. In: Plomp R, Smoorenburg FG (eds) Frequency Analysis and Periodicity Detection in Hearing. Leiden: Sijthoff.

Spoendlin H (1971) Degeneration behavior of the cochlear nerve. Arch klin exp Ohr- Nas-u KehlkHeilk 200:275–291.

Spoendlin H (1984) Primary neurons and synapses. In: Friedman I, Ballantyne J (eds) Ultrastructural Atlas of the Inner Ear. London: Butterworths.

Stopp PE (1983) The distribution of the olivo-cochlear bundle and its possible role in frequency/intensity coding. In: Klinke R, Hartmann R (eds) Hearing—Physiological Bases and Psychophysics. Berlin: Springer-Verlag.

Stopp PE (1990) The problem of obtaining reproducible quantitative data on the olivocochlear pathway as exemplified in the guinea pig. Eur Arch Otorhinolaryngol 247:29–32.

Strominger NL, Silver SM, Truscott TC, Goldstein JC (1981) The cells of origin of the olivocochlear bundle in new and old world monkeys. Anat Rec 199:246.

Swett JE, Wikholm RP, Blanks RHI, Swett AL, Conley LC (1986) Motoneurons of the rat sciatic nerve. Exp Neurol 93:227–252.

Thompson AM, Thompson GC (1988) Neural connections identified with PHA-L anterograde and HRP retrograde tract-tracing techniques. J Neurosci Methods 25:13–17.

Thompson AM, Thompson GC (1991) Posteroventral cochlear nucleus projections to olivocochlear neurons. J Comp Neurol 303:267–285.

Thompson GC, Thompson AM (1986) Olivocochlear neurons in the squirrel monkey brainstem. J Comp Neurol 254:246–258.

Thompson GC, Cortez AM, Igarashi M (1986) GABA-like immunoreactivity in the squirrel monkey organ of Corti. Brain Res 372:72–79.

Tokunaga A (1988) Superior olivary and lateral lemniscal neurons projecting to the cochlea in the guinea pig. Neurosci Res 6:20–30.

Tsuchitani C (1977) Functional organization of lateral cell groups of cat superior olivary complex. J Neurophysiol 40:296-318.

Usami S, Igarashi M, Thompson GC (1988) Light- and electron-microscopic study of gamma-aminobutyric-acid-like immunoreactivity in the guinea pig organ of Corti. ORL 50:162-169.

Vater M, Feng AS (1990) Functional organization of ascending and descending connections of the cochlear nucleus of horseshoe bats. J Comp Neurol 292:373-395.

Vetter DE, Mugnaini E (1990) An evaluation of retrograde tracing methods for the identification of chemically distinct cochlear efferent neurons. Arch Ital Biol 128:331-353.

Vetter DE, Mugnaini E Distribution and dendritic features of three groups of rat olivocochlear neurons: A study with two retrograde cholera toxin tracers. Anat Embryol (submitted).

Vetter DE, Saldana E (1990) Descending input from the central nucleus of the inferior colliculus to the medial olivocochlear system in rat: a combined Pha-L and CT-HRP study. Soc Neurosci Abs 16:716.

Vetter DE, Adams JC, Mugnaini E (1991) Chemically distinct rat olivocochlear neurons. Synapse 7:21-43.

Walsh EJ, Romand R (1992) Functional development of the cochlea and the cochlear nerve. In: Romand R (ed) Development of Auditory and Vestibular Systems, 2nd Ed. Amsterdam: Elsevier.

Warr WB (1975) Olivocochlear and vestibular efferent neurons of the feline brain stem: their location, morphology and number determined by retrograde axonal transport and acetylcholinesterase histochemistry. J Comp Neurol 161:159-182.

Warr WB (1980) Efferent components of the auditory system. Ann Otol Rhinol Laryngol 89 (Suppl 74):114-120.

Warr WB (1982) Parallel ascending pathways from the cochlear nucleus: Neuroanatomical evidence of functional specialization. Contrib Sens Physiol 7:1-38.

Warr WB, Guinan JJ Jr (1979) Efferent innervation of the organ of Corti: two separate systems. Brain Res 173:152-155.

Warr WB, Spangler KM (1989) A novel projections of the ventral nucleus of the trapezoid body in the rat. Soc Neurosci Abs 15:745.

Warr WB, Guinan JJ Jr, White JS (1986) Organization of the efferent fibers: The lateral and medial olivocochlear systems. In: Altschuler RA, Hoffman DW, Bobbin RP (eds) Neurobiology of Hearing: The Cochlea, 1st Ed. New York: Raven Press.

Warr WB, White JS, Nyffeler MJ (1982) Olivocochlear neurons: Quantitative comparison of the lateral and medial efferent systems in adult and newborn cats. Soc Neurosci Abs 8:346.

Warren EH III, Liberman MC (1989) Effects of contralateral sound of auditory-nerve responses. I. Contributions of cochlear efferents. Hear Res 37:89-104.

White JS (1983) Fine structure of the lateral superior olivary nucleus. Soc Neurosci Abs 9:765.

White JS (1986) Differences in the ultrastructure of labyrinthine efferent neurons in the albino rat. Assoc Res Otolaryngol Abs 9:34-35.

White JS, Warr WB (1983) The dual origins of the olivocochlear bundle in the albino rat. J Comp Neurol 219:203-214.

White JS, Robertson D, Warr WB (1986) Electron-microscopic observations on an HRP-filled, physiologically characterized medial olivocochlear axon in the guineapig cochlea. Soc Neurosci Abs 12:1264.

Whitlon DS, Sobkowicz HM (1989) GABA-like immunoreactivity in the cochlea of the developing mouse. J Neurocytol 18:505–518.

Will U, Fritzsch B (1988) The eighth nerve of amphibians: Peripheral and central distribution. In: Fritzsch B, Ryan MJ, Wilczynski W, Hetherington TE, Walkowiak W (eds) The evolution of the amphibian auditory system. New York: John Wiley & Sons.

Wilson JL, Henson MM, Henson OW Jr (1991) Course and distribution of efferent fibers in the cochlea of the mouse. Hear Res (in press).

Winter IM, Robertson D, Cole KS (1989) Descending projections from auditory brainstem nuclei to the cochlea and cochlear nucleus of the guinea pig. J Comp Neurol 280:143–157.

Wright CG, Preston RE (1973) Degeneration and distribution of efferent nerve fibers in the guinea pig organ of Corti. A light and scanning electron microscopic study. Brain Res 58:37–59.

Subject Index

Explanation of index:
Major neuroanatomical structures are indexed individually, however, often under abbreviations (e.g., superior olive for superior olivary complex). Cell groups within nuclear complexes are indexed individually if they are covered extensively in separate sections of chapters. However, they are also indexed under the name of their major group (e.g., Lateral superior olivary nucleus will be indexed under Superior olive.

Species are indexed under their common name and only for major nuclear groups and pathways. Primates other than humans are grouped together as non-human primates and all species of bats are grouped together as bat.

Author Index

For multiauthored papers only the first three authors are listed.